Biotechnology & Genetic Engineering Reviews

Volume 16

Biotechnology & Genetic Engineering Reviews

Volume 16

Editor:
STEPHEN E. HARDING

Professor of Physical Biochemistry, University of Nottingham

Associate Editor:
MICHAEL P. TOMBS

Special Professor, University of Nottingham

Intercept

Andover

British Library Cataloguing in Publication Data
Biotechnology & genetic engineering reviews.—
    Vol. 16
        1. Biotechnology—Periodicals

A CIP catalogue record for this book is available from the British Library
ISBN 1–898298–58–0
ISSN 0264–8725

Published in April 1999 by Intercept Limited,
PO Box 716, Andover, Hants. SP10 1YG, England

Typeset in Times by
Ann Buchan (Typesetters), Shepperton, Middlesex.
Printed by Athenaeum Press Ltd, Newcastle-upon-Tyne.

Contents

Contributors

PATRICIA AYOUBI, *Department of Microbiology and Molecular Genetics, Oklahoma State University, Stillwater, OK 74078, USA*

CHRISTIAN BARNETT, *Department of Life Sciences, The Nottingham Trent University, Clifton Lane, Clifton, Nottingham, NG11 8NS, UK*

STEPHEN BEVAN, *Blond McIndoe Centre, Queen Victoria Hospital, East Grinstead, West Sussex, RH19 3DZ, UK and Institute of Cancer Research, Haddow Laboratories, 15 Cotswold Road, Sutton, Surrey, SM2 5NG, UK*

H. FRANCES J. BLIGH, *Division of Plant Science, School of Biological Sciences, University of Nottingham, Nottingham, NG7 2RD, UK*

BERIT BORGERSEN, *Pronova Biomedical, Gaustadalleen 21, N-0371 Oslo, Norway*

ANTHONY R.M. COATES, *Department of Medical Microbiology, St George's Hospital Medical School, Cranmer Terrace, London, SW17 0RE, UK*

HELMUT CÖLFEN, *Max-Planck Institute of Colloids and Interfaces, Colloid Chemistry Department, Kantstrasse 55, D-14513, Teltow-Seehof, Federal Republic of Germany*

CRISTINA COLOMINAS, *Glycobiology Institute, Department of Biochemistry, University of Oxford, South Parks Road, Oxford, OX1 3QU, UK*

S.S. (BOB) DAVIS, *School of Pharmacy, University of Nottingham, University Park, Nottingham, NG7 2RD, UK and DanBiosyst (UK) Ltd., Albert Einstein Centre, Highfields Science Park, Nottingham*

MATTHEW P. DEACON, *NCMH unit, University of Nottingham, School of Biological Sciences, Sutton Bonington, Leics., LE12 5RD, UK and School of Pharmacy, University of Nottingham, University Park, Nottingham, NG7 2RD, UK*

MICHAEL DORNISH, *Pronova Biomedical, Gaustadalleen 21, N-0371 Oslo, Norway*

RAYMOND A. DWEK, *Glycobiology Institute, Department of Biochemistry, University of Oxford, South Parks Road, Oxford, OX1 3QU, UK*

ANNA EBRINGEROVÁ, *Institute of Chemistry, Slovak Academy of Science, 842 38 Bratislava, Slovak Republic*

NEIL ERRINGTON, *University of Nottingham, School of Biological Sciences, Sutton Bonington, Leics., LE12 5RD, UK*

ANA FERNANDES, *Centre for Drug Delivery Research, School of Pharmacy, University of London, 29–39 Brunswick Square, London, WC1N 1AX, UK*

IMMO FIEBRIG, *NCMH unit, University of Nottingham, School of Biological Sciences, Sutton Bonington, Leics., LE12 5RD, UK and School of Pharmacy, University of Nottingham, University Park, Nottingham, NG7 2RD, UK*

FELIX FRANKS, *BioUpdate Foundation, 7 Wootton Way, Cambridge, CB3 9LX, UK*

GREGORY GREGORIADIS, *Centre for Drug Delivery Research, School of Pharmacy, University of London, 29–39 Brunswick Square, London, WC1N 1AX, UK*

ALEXEY E. GRISHCHENKO, *Department and Institute of Physics of St Petersburg University, Ulianovskaya str. 1, Petergof, 198904, St Petersburg, Russia*

ANNIKEN HAGEN, *Pronova Biomedical, Gaustadalleen 21, N-0371 Oslo, Norway*

STEPHEN E. HARDING, *NCMH unit, University of Nottingham, School of Biological Sciences, Sutton Bonington, Leics., LE12 5RD, UK*

EDMUND HART, *Glycobiology Institute, Department of Biochemistry, University of Oxford, South Parks Road, Oxford, OX1 3QU, UK*

BRIAN HENDERSON, *Cellular Microbiology Research Group, Division of Surgical Sciences, Eastman Dental Institute, University College London, 256 Gray's Inn Road, London, WC1X 8LD, UK*

ZDENA HROMÁDKOVÁ, *Institute of Chemistry, Slovak Academy of Science, 842 38 Bratislava, Slovak Republic*

CLEANTHES ISRAILIDES, *Institute of Technology of Agricultural Products, National Agricultural Research Foundation, 1, S. Venizelou St., Lycovrissi 141 23, Athens, Greece*

ALEJANDRO G. MARANGONI, *Department of Food Science, University of Guelph, Guelph, Ontario, N1G 2W1, Canada*

ROBIN MARTIN, *Blond McIndoe Centre, Queen Victoria Hospital, East Grinstead, West Sussex, RH19 3DZ, UK*

PAOLO MASCAGNI, *Italfarmaco Research Centre, Via Lavoratori, 54, Cinisello Balsamo 20092, Milan, Italy*

TAJ S. MATTU, *Glycobiology Institute, Department of Biochemistry, University of Oxford, South Parks Road, Oxford, OX1 3QU, UK*

BRENDA McCORMACK, *Centre for Drug Delivery Research, School of Pharmacy, University of London, 29–39 Brunswick Square, London, WC1N 1AX, UK*

IAN A. McKAY, *Centre For Cutaneous Research, St Bartholomew's and The Royal London School of Medicine & Dentistry, 2 Newark Street, London, E1 2AT, UK*

ANAND MEHTA, *Glycobiology Institute, Department of Biochemistry, University of Oxford, South Parks Road, Oxford, OX1 3QU, UK*

MALINI MITAL, *Centre for Drug Delivery Research, School of Pharmacy, University of London, 29–39 Brunswick Square, London, WC1N 1AX, UK*

ANDREW J. MORT, *Department of Biochemistry and Molecular Biology, Oklahoma State University, Stillwater, OK 74078, USA*

GHISLAIN OPDENAKKER, *Glycobiology Institute, Department of Biochemistry, University of Oxford, South Parks Road, Oxford, OX1 3QU, UK and Rega Institute for Medical Research, University of Leuven, Minderbroedersstraat 10, B-3000 Leuven, Belgium*

GEORGES M. PAVLOV, *Department and Institute of Physics of St Petersburg University, Ulianovskaya str. 1, Petergof, 198904, St Petersburg, Russia*

ROLF A. PRADE, *Department of Microbiology and Molecular Genetics, Oklahoma State University, Stillwater, OK 74078, USA*

PAULINE M. RUDD, *Glycobiology Institute, Department of Biochemistry, University of Oxford, South Parks Road, Oxford, OX1 3QU, UK*

BERNARD SCANLON, *Department of Life Sciences, The Nottingham Trent University, Clifton Lane, Clifton, Nottingham, NG11 8NS, UK*

HOWARD SIMONS, *University of Nottingham, School of Biological Sciences, Sutton Bonington, Leics., LE12 5RD, UK*

ØYVIND SKAUGRUD, *Pronova Biomedical, Gaustadalleen 21, N-0371 Oslo, Norway*

ALAN SMITH, *Department of Life Sciences, The Nottingham Trent University, Clifton Lane, Clifton, Nottingham, NG11 8NS, UK*

STEVEN M. SMITH, *Institute of Cell and Molecular Biology, University of Edinburgh, Mayfield Road, Edinburgh, EH9 3JH, UK*

IAN W. SUTHERLAND, *Institute of Cell and Molecular Biology, Edinburgh University, Mayfield Road, Edinburgh, EH9 3JH, UK*

TAKESHI TAKAHA, *Biochemical Research Laboratories, Ezaki Glico Co Ltd., Utajima, Nishiyodogawaku, Osaka 555, Japan*

GREGORY A. TUCKER, *University of Nottingham, School of Biological Sciences, Sutton Bonington, Leics., LE12 5RD, UK*

WENDY M. WILLIS, *Department of Food Science, University of Guelph, Guelph, Ontario, N1G 2W1, Canada*

DONGFENG ZHAN, *Department of Biochemistry and Molecular Biology, Oklahoma State University, Stillwater, OK 74078, USA*

XIAOQIN ZHANG, *Centre for Drug Delivery Research, School of Pharmacy, University of London, 29–39 Brunswick Square, London, WC1N 1AX, UK*

NICOLE ZITZMANN, *Glycobiology Institute, Department of Biochemistry, University of Oxford, South Parks Road, Oxford, OX1 3QU, UK*

Editor's Foreword

Professor Michael Tombs has now retired as Editor of *Biotechnology and Genetic Engineering Reviews*. During his period as Editor he continued the work of the founding editor Professor Russell in establishing *BGER* as a leader in the field with a successful blend of industrial and biomedically based invited articles written by distinguished operators. I do not propose to change this successful formula, with the exception that after consultation with the Editorial Board all articles from Volume 16 onwards are now sent to an independent reviewer prior to acceptance for publication.

Although invited reviews form the cornerstone of *BGER* we will continue to welcome contributions from other workers in the field for consideration for publication: those considering submitting a review should contact either myself (email: Steve.Harding@nottingham.ac.uk; Fax: 0044–115–951–6142) or any member of the Editorial Board.

Stephen E. Harding
Nottingham, October 1998

1
Glycoproteins: Rapid Sequencing Technology for N-linked and GPI Anchor Glycans

PAULINE M. RUDD[1]*, TAJ S. MATTU[1], NICOLE ZITZMANN[1], ANAND MEHTA[1], CRISTINA COLOMINAS[1], EDMUND HART[1], GHISLAIN OPDENAKKER[1,2] AND RAYMOND A. DWEK[1]

Glycobiology Institute, Department of Biochemistry, University of Oxford, South Parks Road, Oxford, OX1 3QU, UK and [2]Rega Institute for Medical Research, University of Leuven, Minderbroedersstraat 10, B-3000 Leuven, Belgium

Introduction

Recent advances in oligosaccharide sequencing technology have made routine glycan analysis of glycoproteins and the glycan moieties of glycosylphosphatidylinositol (GPI) anchors a real possibility for many laboratories. The strategies described here have been developed specifically to address the need for a rapid and robust method of glycan analysis which can be applied routinely to microgram levels of glycoproteins with the minimum requirement for specialised equipment or expertise. This strategy also allows the possibility of isolating individual sugars of particular interest for more detailed analysis. We discuss the N-glycosylation of CD48, IgG and IgA1, the Fab and Fc fragments of IgG and also the small (S) and middle (M) glycoproteins of the hepatitis B virus coat protein. We describe the analysis of the major O-glycans of neutrophil gelatinase B and also present a novel method for analysing the glycans

Abbreviations: 2AB: 2 aminobenzamide, AHM: anhydromannitol, CHO-K1: Chinese hamster ovary type K1, ES-MS: electrospray mass spectrometry, GPC: gel permeation chromatography, GPI: glycosylphosphatidylinositol, gu: glucose unit, HBV: hepatitis B virus, HPAEC: high performance anion exchange chromatography, HPLC: high performance liquid chromatography, MALDI-TOF MS: matrix assisted laser desorption ionisation time-of-flight mass spectrometry, NP: normal phase, PVDF: polyvinylidene difluoride, OGS: Oxford GlycoSciences, S, M and L: small, middle and large glycoproteins of HBV respectively, TIMP-1: tissue inhibitor metalloproteinase-1, VSG: variant surface glycoprotein.
Abbreviations used for enzymes: ABS: *Arthrobacter ureafaciens* sialidase; BTG: bovine testes β-galactosidase; BEF: bovine epididymis α-fucosidase; SPH: *Streptococcus pneumoniae* β-N-acetylhexosaminidase.
Abbreviations used for describing oligosaccharide structures: A(1–4) indicates the number of antennae linked to the trimannosyl core; G(0–4) indicates the number of terminal galactose residues in the structure; F: fucose; B: bisecting N-acetylglucosamine (GlcNAc); GalNAc: N-acetylgalactosamine; S: sialic acid; G, Gal: galactose; M, Man: mannose; H: hexose; N: N-acetylhexosamine.

*To whom correspondence may be addressed.

attached to the GPI anchor of a variant surface glycoprotein of *Trypanosoma brucei* directly from a Western blot. The underlying aim of these analytical strategies is to obtain oligosaccharide sequencing data which, in combination with oligosaccharide and protein structural data, can be visualised in a molecular model. In this way, oligosaccharides can be viewed in the context of the proteins to which they are attached, and some insight can be gained into the roles which sugars might play in the structure and function of the glycoproteins.

Glycoproteins generally exist as populations of glycosylated variants (glycoforms) of a single homogeneous polypeptide. Although the same glycosylation machinery is available to all proteins which enter the secretory pathway in a given cell, most glycoproteins emerge with characteristic glycosylation patterns and heterogeneous populations of glycans at each glycosylation site (Review: Rudd and Dwek, 1997). The composition of the glycoform populations and the role that heterogeneity plays in the function of glycoproteins are important questions for protein chemists and glycobiologists alike. It is only when glycoproteins are viewed in their entirety that the full significance of glycosylation for the function of the molecule can be appreciated. However, while peptide sequencing is routinely available to protein chemists, the robust, rapid and automated technology for oligosaccharide sequencing which we describe here has been developed only recently.

In this review we describe some of the recent advances in technology which enable N- and O-linked glycans and GPI anchor glycans to be analysed rapidly at the sub-picomole level. These include in-gel enzymatic release of N-linked sugars, release of GPI anchor glycans from Western blots, fluorescent labelling of glycan pools for high sensitivity and using a single HPLC run to make preliminary structural assignments of both neutral and charged sugars. We also show how enzyme arrays are used to analyse simultaneously the total sugars released from glycoproteins.

In the first instance the primary sequence of the protein determines whether or not it will be modified by the addition of N- or O-linked oligosaccharides or a GPI anchor. N-linked sugars are added to the amide side-chain of some Asn residues which form part of the triplet AsnXaaSer/Thr, while O-linked sugars can be added to the hydroxyl side-chain of some Ser, Thr or hydroxyproline residues. To receive a GPI anchor the protein must contain a signal sequence which commonly consists of 12–20 hydrophobic residues at the C-terminus of the primary translation product preceded by a polar region of amino acids (Ferguson, 1991; McConville and Ferguson, 1993). This signal sequence is cleaved and replaced by a pre-assembled GPI precursor. In the fully formed anchor the C-terminal cleavage amino acid (restricted to Cys, Asp, Asn, Gly, Arg or Ser) is linked via ethanolamine phosphate to a glycan with a conserved backbone sequence (Manα1→2Manα1→6Manα1→4GlcNH$_2$). Here we discuss some applications of the new analytical strategies for glycan synthesis (Rudd *et al.*, 1997) and show how they have been used to characterise the N-glycans of CD48, the S and M surface antigen glycoproteins of hepatitis B virus, and to compare the glycosylation of serum IgA1 with that of IgG and of IgG Fab with that of IgG Fc. We also discuss the analysis of the major O-glycans of neutrophil gelatinase B, and of the GPI anchor glycans of a variable surface glycoprotein (VSG) of *Trypanosoma brucei*. Importantly, we show how an analysis of the oligosaccharides combined with protein structural data can lead to insights into the structural roles and biological functions of the sugars within the context of the structure and function of the protein to which they are attached.

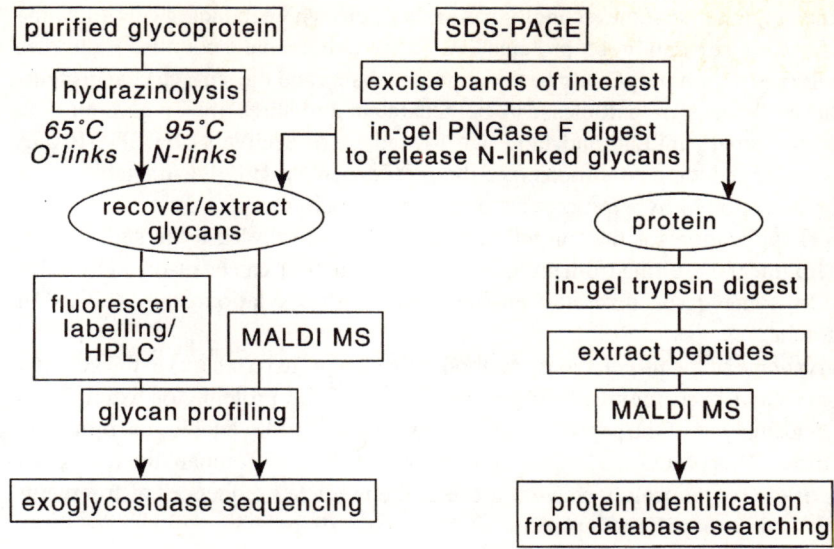

Figure 1.   Strategies for glycan analysis and protein identification.

Strategies for analysing N- and O-linked oligosaccharides

Analysing the sugars attached to glycoproteins involves (i) releasing the sugars from the protein (ii) labelling the total glycan pool (iii) analysing the components of the pool by predictive HPLC or MALDI-TOF MS to obtain preliminary structural information (iv) confirming the preliminary assignments by simultaneous sequencing of the glycan pool with exoglycosidase arrays. A straightforward approach which is generally applicable to all glycoproteins is shown in *Figure 1* and discussed below.

RELEASE OF N- AND O-LINKED GLYCANS FROM GLYCOPROTEINS

(i) Enzymatic release and analysis of N-linked oligosaccharides from protein bands on SDS-PAGE gels.   An 'in-gel' release method has been developed to release N-glycans from Coomassie blue stained protein bands on SDS-PAGE gels by peptide-N-glycosidase F (PNGase F) (Küster *et al.*, 1997). The released glycan pool can be analysed by MALDI-TOF MS and after fluorescent labelling, by HPLC. The individual sugars in the pool can be analysed simultaneously using enzyme arrays (Guile *et al.*, 1996) monitored either by MALDI-TOF MS or HPLC.

The development of this method has opened the way to analysing glycans from biologically important molecules which are difficult to purify or where amounts are limited. In many cases releasing sugars directly from SDS-PAGE gels eliminates the need for extensive protein purification. For example, when gelatinase B from some cell types is purified on gelatin-sepharose it is contaminated with its natural inhibitor, tissue inhibitor metalloproteinase-1 (TIMP-1) (Opdenakker *et al.*, 1991). Although this enzyme-inhibitor complex is stable, SDS-PAGE can be used to resolve the two proteins. N-linked oligosaccharides can then be released directly from the gel for analysis (Rudd *et al.*, 1997).

Another application of the technology is in situations where reducing gels allow the straightforward separation of proteins into their component subunits. Examples include the resolution of IgG into heavy and light chains and the surface coat proteins of hepatitis B virus or particles can be separated into three major glycoprotein components, small, middle and large (S, M and L, respectively) by SDS-PAGE analysis under reducing conditions (see the sections on 'In-gel glycan analysis . . .' and 'The detection of hyperglucosylated glycan . . .' below).

A third application for the 'in-gel' release method is to release the sugars from peptide fragments resulting from protease digests when these can be resolved by SDS-PAGE. In some cases this may enable a rapid glycosylation site analysis of glycoproteins.

An advantage of the in-gel release method is that the protein remains in the gel after the sugars have been removed. In-gel proteolysis of the protein, for example by trypsin, and analysis of the peptide fragments by nanospray MS, enables the protein to be identified. This is achieved by using a protein data base to compare the molecular weights of the tryptic fragments with those of the predicted sequences of fragments from tryptic digests of known proteins (Küster *et al.*, 1996).

(ii) Chemical release of N- and O-glycans from glycoproteins using anhydrous hydrazine.   Hydrazinolysis is a general method for chemically and non-selectively releasing N- and O-glycans from glycoproteins (Patel and Parekh, 1994). The analysis of the N-glycans of IgA1 and IgG released by automated hydrazinolysis (GlycoPrep 1000 Oxford GlycoSciences (OGS)) is discussed in the section on 'IgA1 N-linked sugars are more processed . . .' below.

Hydrazine release of O-glycans is best achieved by manual hydrazinolysis since the conditions can be optimised to minimise the 'peeling' of oligosaccharides associated with the degradation of the terminal monosaccharide (N-acetyl galactosamine in the case of O-glycans) when it is substituted in the 3-position. The release and analysis of the O-glycans from human neutrophil gelatinase B is discussed in the section on 'Analysis of the O-glycans . . .' below. Base catalysed reductive β-elimination is also commonly used to release O-glycans. Although this minimises 'peeling' the reduction of the aldehyde group of the reducing monosaccharide to an alditol precludes subsequent derivitisation which involves a reductive amination step. The inability to label the oligosaccharides with fluorescent tags, such as 2-aminobenzamide (2AB) (see the following section) limits the sensitivity of detection and in many cases prevents the analyses of biologically relevant samples.

LABELLING THE GLYCAN POOL

Oligosaccharides released with hydrazine or PNGase F exist as an equilibrium between the cyclic (hydroxyl at C1) and opened ring form (aldehyde at C1) of the reducing GlcNAc residue (*Figure 2*). The opened ring form can be derivatised at C1 by fluorophores such as 2-aminobenzamide (2AB) via a reductive amination reaction which initially forms a Schiff's base with the sugar. The reaction is driven to completion by the reduction of the Schiff's base to an amine functionality, therefore favouring the forward direction of the tautomerisms.

Although mass spectromety can be applied to unlabelled sugars, in most other

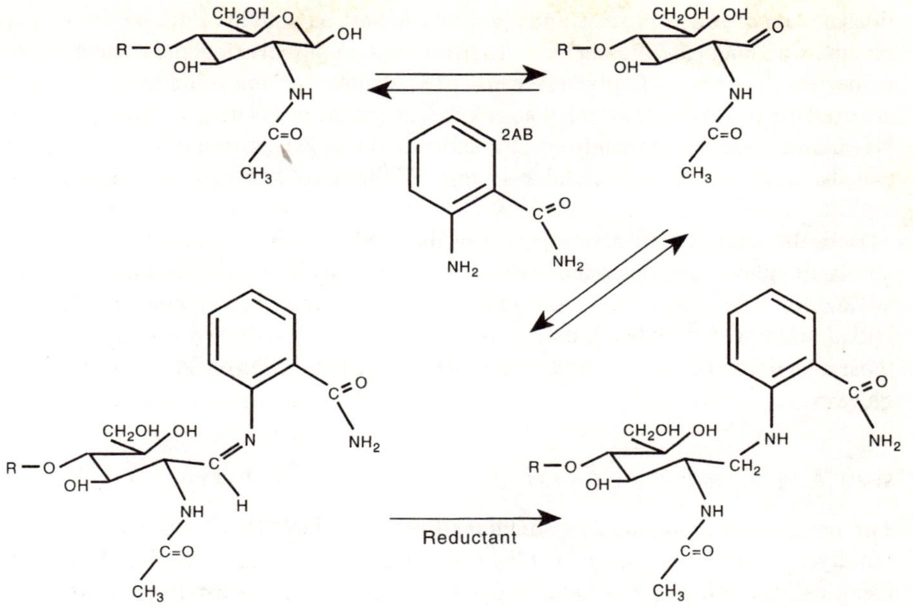

Figure 2. Mechanism for the reductive amination of an oligosaccharide with the fluorophore 2-aminobenzamide. The figure indicates that the reaction can only take place when the terminal GlcNAc residue in the chitobiose core is in the 'ring opened form'. R indicates the remainder of the oligosaccharide.

techniques the reducing termini of the sugars need to be tagged to allow sensitive detection. A range of fluorescent molecules is available which allow detection of sugars in the femtomole range. A suitable label must have a high molar labelling efficiency, and must also label the oligosaccharide components of a glycan pool non-selectively so that they are detected in their correct molar proportions. 2-amino-benzamide (2AB) fulfils these requirements and is compatible with a range of separation techniques including weak anion exchange and reverse phase HPLC, high performance anion exchange chromatography, MALDI-TOF MS and electrospray mass spectrometry (ES-MS) as well as BioGel P4 gel permeation chromatography (P4 GPC) (Bigge *et al.*, 1995) and normal phase (NP) HPLC.

RESOLUTION OF GLYCAN POOLS AND STRUCTURAL ASSIGNMENTS USING NP HPLC

NP HPLC is performed using a column with polar functional groups which interact with the hydroxyl groups on the sugars. Glycans are resolved on the basis of hydrophilicity. Sugars applied in high concentrations of organic mobile phase adsorb to the column surface and are eluted with an aqueous gradient. In general, larger oligosaccharides are more hydrophilic than small ones, and require higher concentrations of aqueous solvent to elute them.

A sensitive and reproducible NP HPLC technology has been developed (Guile *et al.*, 1996) using the GlycoSep N column (OGS). This system is capable of resolving sub-picomole quantities of mixtures of fluorescently labelled neutral and acidic N- and O-glycans simultaneously and in their correct molar proportions. Elution positions are expressed as glucose units (gu) by comparison with the elution positions of glucose oligomers (dextran ladder). The contribution of individual monosaccharides to the overall gu value of a given glycan can be calculated and these incremental values are used to predict the structure of an unknown sugar from its gu value. The GlycoSep N column is able to resolve arm specific substitutions of galactose and linkage position can also have an effect on retention giving a further level of specificity (Guile *et al.*, 1996).

Both the increased resolving power of the predictive HPLC technology and the ability to analyse sialylated and neutral sugars in one run represent a considerable advance over the classical approach to glycan analysis which depends on P4 GPC. The HPLC strategy is now being extended to the analysis of O-linked glycans, such as those attached to human neutrophil gelatinase B (see the section on 'Analysis of the O-glucans . . .' below).

SIMULTANEOUS SEQUENCING OF OLIGOSACCHARIDES USING ENZYME ARRAYS

The preliminary assignment of structures from the initial HPLC run are confirmed rapidly by sequencing all of the oligosaccharides in a glycan pool simultaneously using enzyme arrays. Enzymatic analysis of oligosaccharides using highly specific exoglycosidases is a powerful means of determining the sequence and structure of glycan chains. However, until recently, it was necessary to isolate single sugars from the glycan pool for digestion with exoglycosidases, either sequentially or in arrays. The high resolving power of the NP HPLC system allowed a new approach to be developed (Guile *et al.*, 1996). This involves the simultaneous analysis of the total glycan pool by digesting aliquots with a set of enzyme arrays. After overnight incubation the products of each digestion are analysed by NP HPLC or by MALDI-TOF MS. On the HPLC system, structures are assigned to each peak from a knowledge of the specificity of the enzymes and the incremental values of individual monosaccharide residues. To illustrate this technique, the rapid profiling and simultaneous analysis of the major N-glycans attached to rat CD48 expressed in CHO cells is discussed in the section on 'Simultaneous analysis of the N-glycan pool . . .' below.

Applications of this technology

In the examples below we demonstrate how rapid oligosaccharide sequencing of fluorescently labelled sugars and the use of an oligosaccharide data base (which gives the sizes of the sugars) can be combined with protein structural data to give insights into roles for glycosylation in the function of individual proteins.

IgA1 N-LINKED SUGARS ARE MORE PROCESSED THAN THOSE OF IgG. THIS IMPLIES THAT THE IgA SUGARS CANNOT BE CONSTRAINED BETWEEN THE CH2 DOMAINS AS IS THE CASE FOR IgG

The annotated glycosylation profiles in *Figure 3a* indicate that, in contrast to IgG, most of the N-linked sugars attached to IgA1 are galactosylated and sialylated (Mattu *et al.*, 1998). This suggests that there are significant differences in the accessibility of

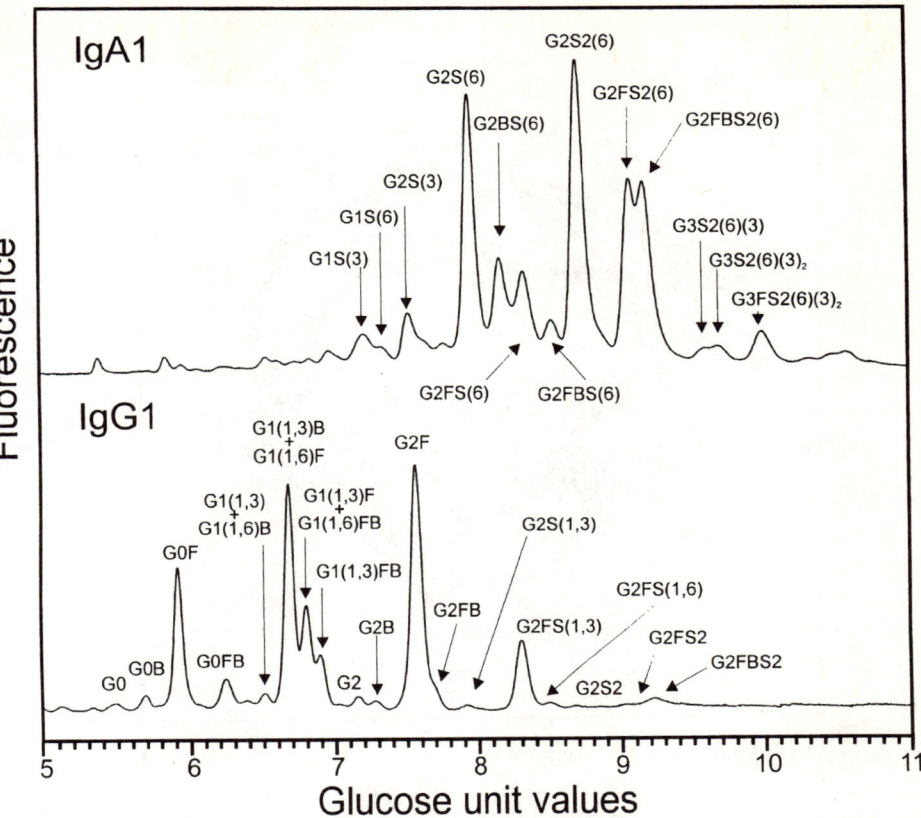

Figure 3a. The glycan analysis of the N-linked sugars from IgA1 and serum IgG. The hydrazine released 2AB labelled glycans from the IgA1 subclass purified from human serum (top) and total serum IgG (containing all subclasses) were resolved by normal phase HPLC. In contrast to IgG, the majority of IgA1 glycans are sialylated and fully galactosylated. The data have been plotted using a new computer program known as the Lineariser (E. Hart – unpublished data). This program automatically determines the gu values by comparison with a dextran standard, and can also transform a chromatogram plot to a linear gu scale. The use of an axis linear in gu values rather than elution times brings the advantage that a given gu increment, such as is produced by the action of a particular enzyme, is of constant length over the full range of the chromatogram. This allows for a more direct comparison of glycosylation profiles. G(0–3) indicates the number of terminal galactose residues in the structure; F: fucose; B: bisecting N-acetyl glucosamine; S: sialic acid.

the sugars to the glycosylating enzymes in the two immunoglobulins. The crystal structure of IgG Fc (Deisenhofer, 1981) indicates that the conserved sugars at Asn297SerThr are contained in the interstitial space between the CH2 domains (*Figure 3b*). The structure shows that the sugars are also involved in non-covalent interactions with the protein surface (Padlan, 1991) which further limits their accessibility to the glycosyltransferases. In order to compare the location of the IgG Fc oligosaccharides with those in IgA1 Fc, for which no protein structure is available, a molecular model was constructed using the primary sequence and disulphide bond pattern of IgA1, the crystal structure of IgG1 Fc and the glycan sequencing data (Mattu *et al.*, 1998). The model of IgA1 Fc shows that, as a consequence of the disulphide bonding arrangement, the amide side chains of Asn263LeuThr are pointing away from the protein and therefore the IgA1 Fc N-linked glycans cannot be

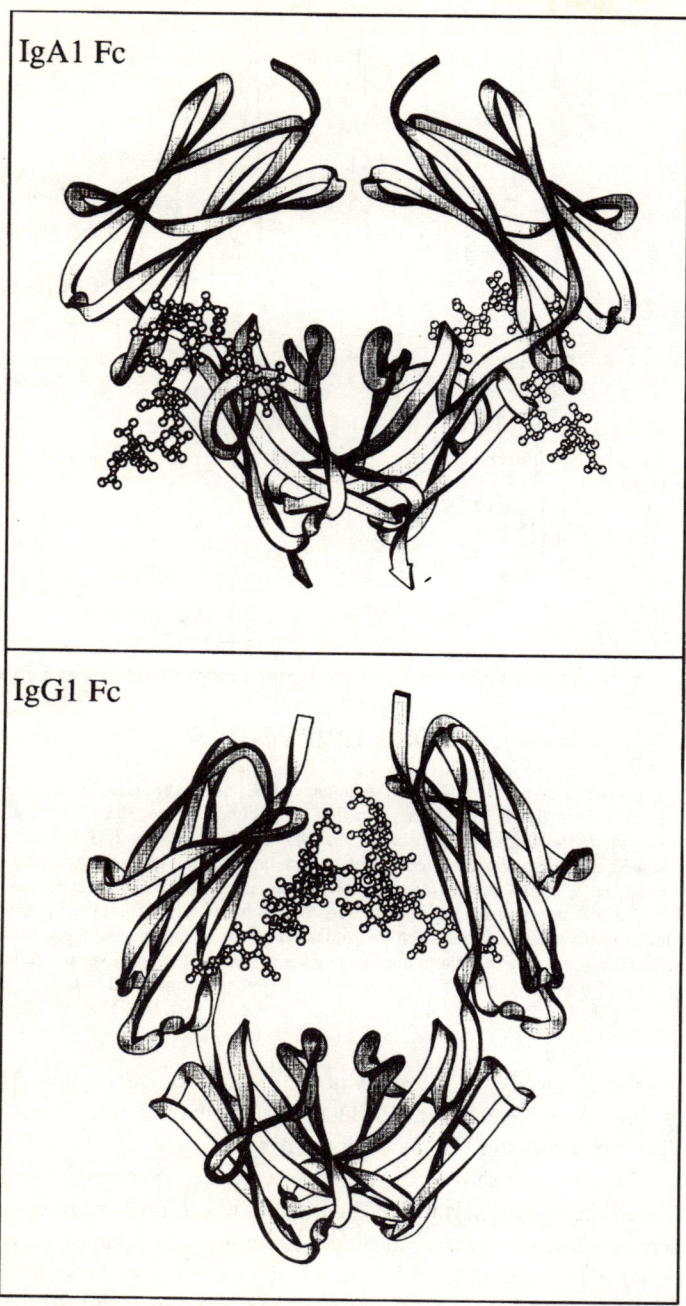

Figure 3b. Comparison of the molecular models of IgA1Fc and IgG1Fc. In contrast to the IgGFc glycans, which are contained in the interstitial space between the CH2 domains (Deisenhofer, 1981), the N-glycans attached to IgA1 are exposed on the outside of the CH2 domains (Mattu *et al.*, 1998). This is the result of the different location of the glycosylation sites in the CH2 domains and the altered quaternary structure of IgA1 Fc. The IgG molecule modelled here is of the subclass type 1.

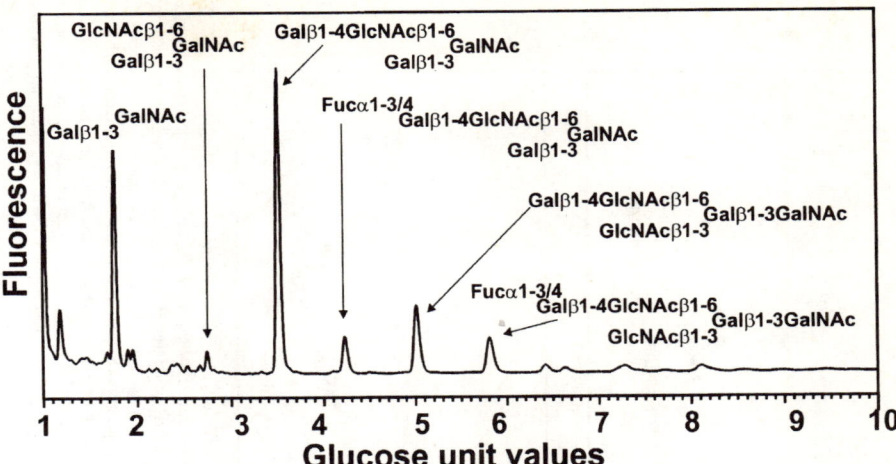

Figure 4. The normal phase HPLC profile of the desialylated O-glycan pool released from human neutrophil gelatinase B.

contained in the interstitial space between the CH2 domains (*Figure 3b*). They are therefore more accessible to the glycosylating enzymes, in particular the galactosyl and sialyl transferases, than those on IgG Fc. The analysis of the IgA sugars (*Figure 3a*) confirms this and molecular modelling (*Figure 3b*) indicates that the IgA1 Fc glycans are too large to fit into the predicted space between the two CH2 domains.

The differences in the structure of the Fc regions in IgA1 and IgG have implications for the functions of the sugars. In IgG Fc the protein-oligosaccharide interactions play a role in maintaining the relative geometry of the domains (Rudd *et al.*, 1991). This is consistent with the finding that, as a result of perturbation of a region proximal to the receptor binding sites (Burton and Woof, 1992), both non-glycosylated and degalactosylated IgG bind less efficiently to the Fcγ receptors (Nose and Wigzell, 1983; Leatherbarrow *et al.*, 1985; Tsuchiya *et al.*, 1989). In contrast, for IgA1, no reduction in binding to its Fc receptor was detected in CHO-K1 mutants in which the N-linked sugars sites in the CH2 domains had been deleted (Mattu *et al.*, 1998). This suggests that in IgA1 the sugars in the CH2 domains do not play a role in maintaining domain structure. However, just as the multiple O-glycans in the hinge have been shown to protect the hinge region of IgA1 against non-specific proteolysis (Mestecky and Kilian, 1985), so the N-glycans are expected to shield large areas of the Fc region from proteases.

ANALYSIS OF THE O-GLYCANS FROM HUMAN NEUTROPHIL GELATINASE B SUGGESTS THAT THEY MAY PRODUCE AN EXTENDED AND RIGID REGION OF THE PEPTIDE

Gelatinase B is a multidomain metalloproteinase (MMP 9) which cleaves extracellular matrix substrates such as denatured collagens (gelatins) after these have been clipped by collagenases, stromelysin or other metalloproteinases. Gelatinase B contains seven protein domains, three potential N-glycosylation sites, a Pro/Ser/Thr rich region and a number of isolated serine and threonine residues. The

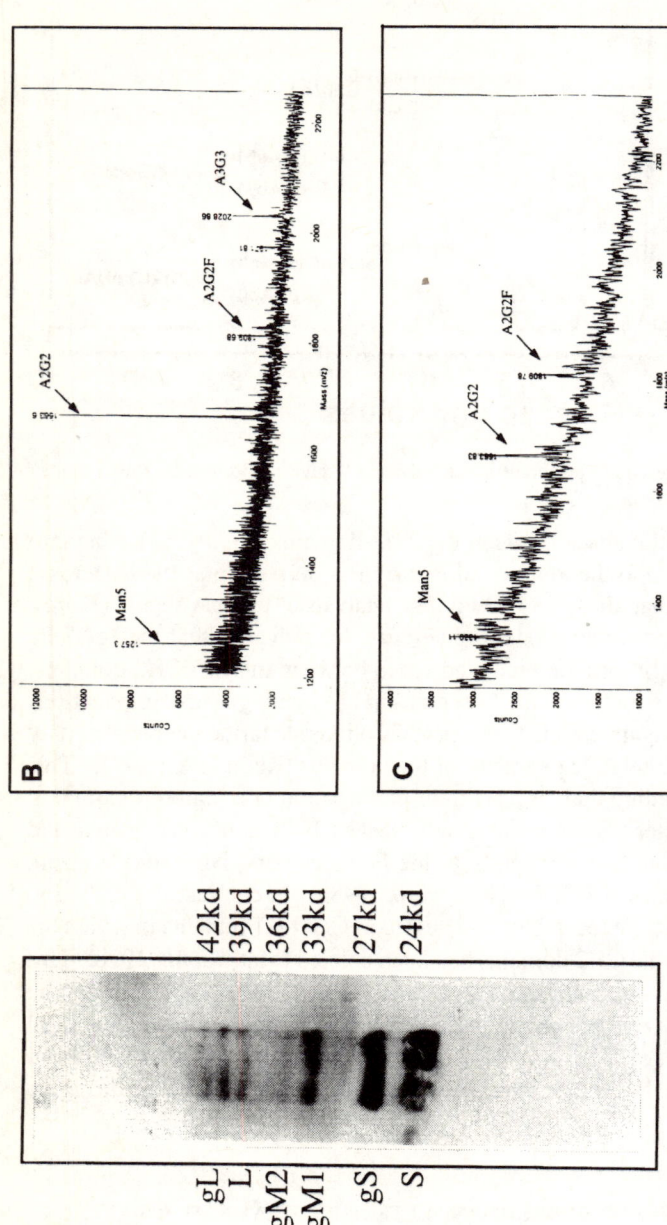

Figure 5A–C.   Analysis of the glycosylation of the individual HBV envelope proteins by in-gel glycan analysis. A) The HBV envelope glycoproteins as resolved by a 12.5% SDS-PAGE gel and stained with Coomassie brilliant blue (see text). The proteins as indicated are from top to bottom: gL, glycosylated HBV L protein; L, unglycosylated HBV L protein; gM2, doubly glycosylated HBV M protein; gM1, singly glycosylated HBV M protein; gS, glycosylated HBV S protein; S, unglycosylated HBV S protein. B) The glycans associated with band gS (the 27kDa S protein) as determined by the in-gel release method. Arrows indicate the glycans: Man5, high mannose type glycan with five mannose residues; A2G2, biantennary complex type glycan; A2G2F, fucosylated biantennary complex type glycan; A3G3, triantennary complex type glycan. C) The glycans associated with gM1 (the 33 kDa M protein) as determined by the in-gel release method. Arrows indicate the glycans: Man5, high mannose type glycan with five mannose residues; A2G2, biantennary complex type glycan; A2G2F, fucosylated biantennary complex type glycan. Note that the signal intensity is much reduced compared with 5B but that this reduction correlates with the band intensity seen in *Figure 5A*. Also, while the gS band consists of predominantly the biantennary complex type glycan, the gM1 band contains much a greater amount of the fucosylated biantennary complex type glycan.

O-glycans, which are expected to be located mainly in the Ser/Thr/Pro rich domain, were released by manual hydrazinolysis, profiled (*Figure 4*) and analysed using enzyme arrays (Rudd, P.M., Mattu, T.S., Opdenakker, G. and Dwek, R.A. Manuscript in preparation). *Figure 4* shows that the glycans in the major peaks contained core types 1, 2 and 6. The attachment of these glycans to the Ser/Thr/Pro rich domain may produce an extended and rigid region of the peptide. Electron microscopic studies indicate that the extension contributed per residue in an O-glycosylated peptide varies from 0.2–0.25nm in CD43 (Cyster *et al.*, 1991) and mucins (Shogren *et al.*, 1989; Jentoft *et al.*, 1990).

IN-GEL GLYCAN ANALYSIS OF THE HEPATITIS B VIRUS (HBV) SUB-VIRAL PARTICLES LOCALISES FUCOSYLATED GLYCANS TO THE HBV M PROTEIN

HBV, which is a major etiological agent of liver disease and hepatocellular carcinoma (Sherker *et al.*, 1990), encodes for three envelope glycoproteins proteins: large (L), middle (M) and small (S). These are derived from a single open reading frame through the utilization of alternative translational start sites (Heerman *et al.*, 1992). All three glycoproteins contain a common site of N-glycan attachment at Asn146 of the S domain with the M protein containing an additional site at Asn4 of the pre-S2 domain. In addition to being the major component of the viral envelope, these glycoproteins are secreted in the form of smaller non-infectious sub-viral particles. These particles, which lack DNA, are secreted in vast excess compared to the viral particle.

The glycosylation of the individual HBV glycoproteins can be studied by the in-gel release method described (Küster *et al.*, 1997). Briefly, the envelope glycoproteins associated with subviral particles from Hep G2.2.2.15 cells were separated on an SDS polyacrylamide gel and the glycans removed *in situ* with PNGase F. The glycans were subsequently desalted and analyzed by mass spectrometry. *Figure 5* shows the major glycan structures found on the HBV S (gp27) and M (gp33) proteins respectively from Hep G2.2.15 cells (*Figure 5A*). The major glycan structures found on the S protein are biantennary, fucosylated biantennary, and triantennary glycan structures (*Figure 5B*). Although the glycan structures found on the M protein are similar (*Figure 5C*), the proportions are different. The M protein appears to contain much more of the fucosylated biantennary glycan. As these are the same glycans found on sub-viral particles from Hep G2.2.15 cells (Mehta *et al.*, 1997), this provides evidence that the fucosylated biantennary glycan is found predominantly on the HBV M protein. *Figure 5* also shows the sensitivity of this technique. The mass spectrum seen in *Figure 5C* was determined from less than 1μg of protein. Therefore this technique can be used to examine the glycosylation of individual constituents of large oligomeric structures and from small quantities of starting material.

THE DETECTION OF HYPERGLUCOSYLATED GLYCAN IN THE SERUM OF WOOD-CHUCKS TREATED WITH THE GLUCOSIDASE INHIBITOR N-NONYL-DNJ ACTS AS A SURROGATE MARKER FOR DRUG EFFICACY

The high sensitivity of glycan analysis using fluorescently labelled sugars and NP HPLC (femtomole range) is demonstrated in this study in which hyperglucosylated

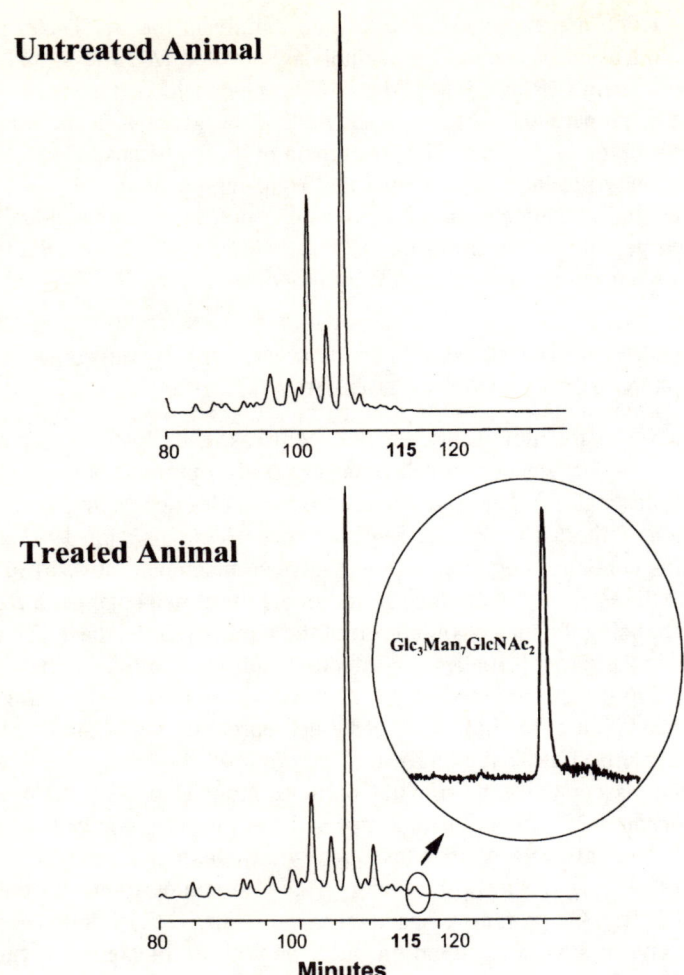

Figure 5D.   Serum glycan profiles from untreated woodchuck (top) and treated woodchuck 4 weeks post treatment. NP HPLC profile of serum glycans from untreated animal and treated animal 4 weeks after treatment with the α-glucosidase inhibitor N-nonyl-deoxynorjirimycin. The X-axis shows the retention time of the glycans on the HPLC column. Note that while after 115 minutes, there are no peaks in the untreated animal, the treated animal has a small peak at approximately 116.5 minutes. This peak, which has been shown by oligosaccharide sequencing to be the Glc$_3$Man$_7$GlcNAc$_2$ structure is indicated. Glucosidase inhibition does not lead a 100% production of the Glc$_3$Man$_7$GlcNAc$_2$ structure due to the presence of a shunt pathway (Moore *et al.*, 1990).

glycans were detected in serum from woodchucks that had been treated with the glucosidase inhibitor N-nonyl-deoxynojirimycin (Block *et al.*, 1998). The total glycans associated with the glycoproteins from 15 μl of woodchuck serum were released with hydrazine using a Glycoprep 1000 machine (OGS). Released glycans were fluorescently labelled at their reducing end with 2AB using the Signal Labelling Kit (OGS) and subsequently analyzed by NP HPLC. The glycan profiles from the serum of the untreated and treated animal are compared in *Figure 5D*. The result of glucosidase inhibition is that treated animals contain small amounts of the

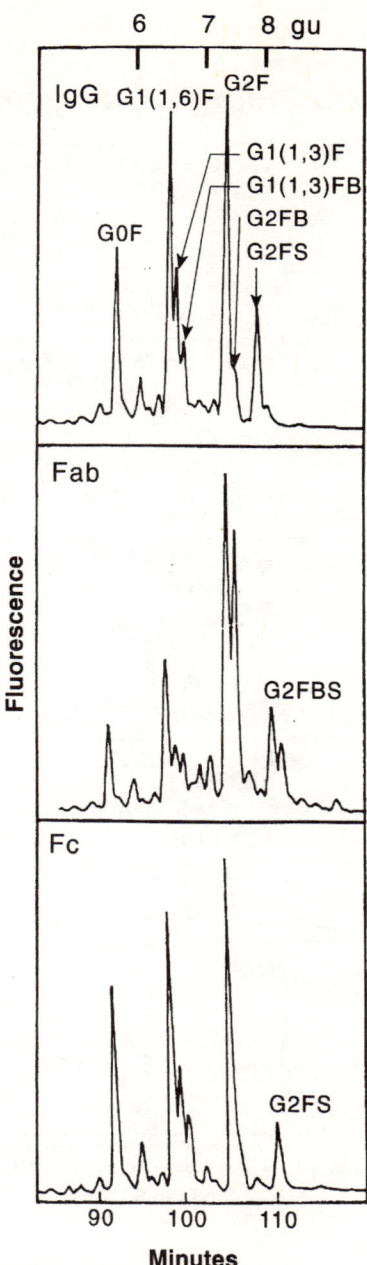

Figure 6. Analysis of oligosaccharides released from IgG, Fab and Fc. Panels a-c show the NP HPLC profiles of the 2AB labelled oligosaccharides released from normal serum IgG, Fab and Fc, respectively.

$Glc_3Man_7GlcNAc_2$ structure which represents less than 2% of the total serum glycan pool. The levels of $Glc_3Man_7GlcNAc_2$ correlated with the reduction in viremia in individual animals treated with the inhibitor (Block *et al.*, 1998). Thus, in this model, the high sensitivity of this technique allows the detection of $Glc_3Man_7GlcNAc_2$ structures to act as a surrogate marker for drug efficacy.

Sequencing CD48 glycans

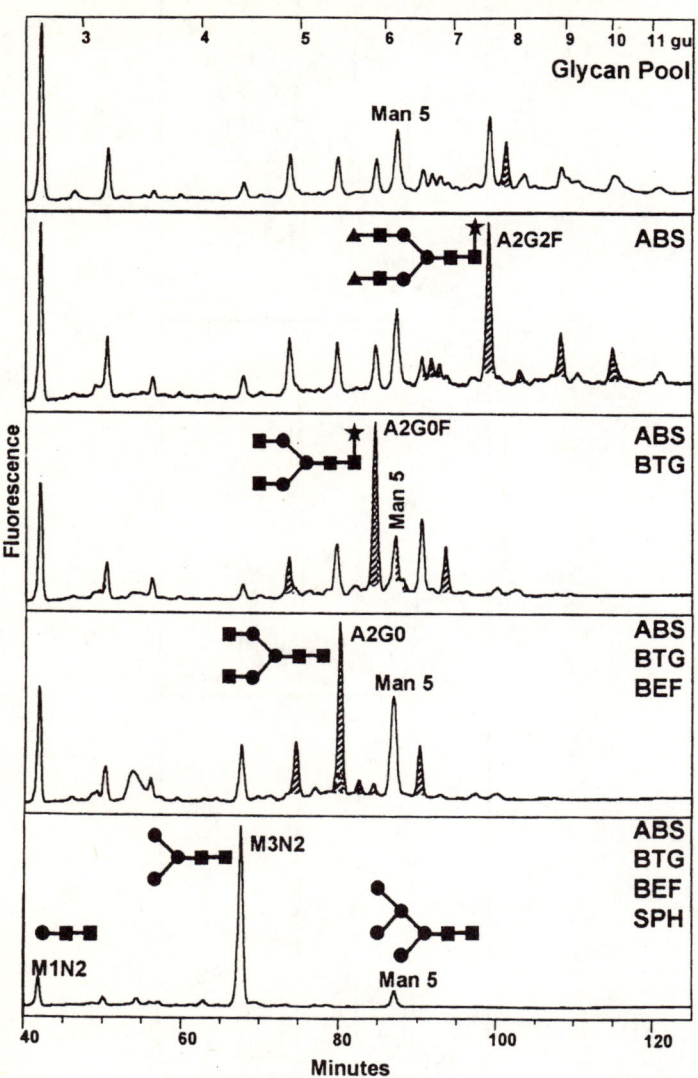

Figure 7. Simultaneous analysis of N-glycans released from rat CD48 (expressed in CHO cells) using enzyme arrays. The figure shows the HPLC analysis of the total glycan pool and the products resulting from the digestion of four aliquots of the total CD48 glycan pool with a series of enzyme arrays. The particular enzyme array which produced each profile is shown on the appropriate panel. The shaded areas define the peaks which contain glycans which were subsequently digested by the additional enzyme present in the next array. The gu value of each peak was calculated by comparison with the dextran hydrolysate ladder shown at the top of the figure. Structures were assigned from the gu values, previously determined incremental values for monosaccharide residues (Guile *et al.*, 1996) and the known specificity of the exoglycosidase enzymes. The structures of the most abundant glycan populations are shown. ABS:*Arthrobacter ureafaciens* sialidase; BTG: bovine testes β-galactosidase; BEF: bovine epididymus α-fucosidase; SPH: *Streptococcus pneumoniae* β-N-acetylhexosaminidase.

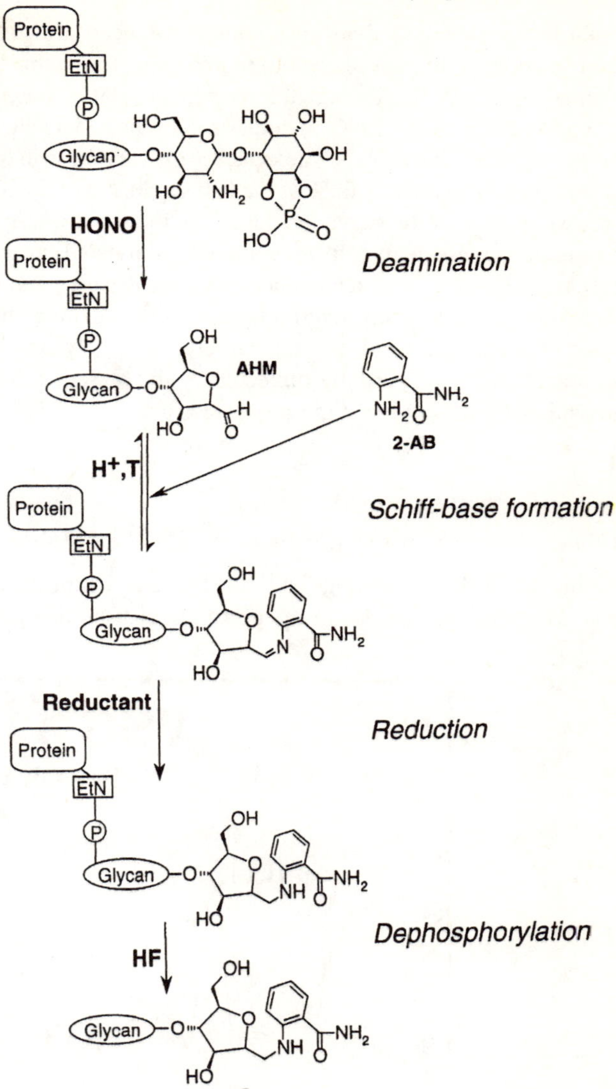

2AB labelled
GPI neutral glycan

Figure 8A. General scheme for preparing 2AB labelled GPI neutral glycans 'on the blot'. The GPI-anchored protein is run on an SDS-PAGE gel and blotted onto a PVDF membrane. Nitrous acid (HONO) deamination converts the GlN residue to 2,5-anhydromannose (AHM) which is reacted with 2AB by reductive amination. These steps are performed while the protein is still attached to the PVDF membrane. The 2AB labelled glycan is subsequently released by dephosphorylation with cold aqueous hydrofluoric acid (HF).

COMPARISON OF THE N-GLYCOSYLATION PROFILES OF IgG FAB AND FC INDICATE THAT THE FAB GLYCANS ARE MORE ACCESSIBLE TO SOME GLYCOSYLATING ENZYMES

A comparison of the high resolution HPLC profiles of oligosaccharides released from glycoproteins gives a rapid insight into alterations of glycosylation with disease and in some cases into pathogenesis. In particular, decreases in the galactosylation of serum IgG in rheumatoid arthritis, which have been shown to correlate with disease severity and activity, have been monitored using this technique (Routier *et al.*, 1998; Wormald *et al.*, 1997). The alterations in glycosylation are mainly confined to the Fc region (Youings *et al.,* 1996). Comparison of the sugars attached to normal IgG Fab and Fc, indicates that there is 'site' specific glycosylation in IgG (*Figure 6*). In normal serum IgG approximately 60% of the Fab sugars are of the digalactosylated complex type, while 68% of the sugars on the Fc are mono- or non-galactosylated. In the Fab there is an increase in glycoforms containing sialylated sugars and in those containing bisecting GlcNAc. These data indicate that the Fab glycans are more accessible to the GlcNAcV, galactosyl and sialyl transferases which attach bisecting GlcNAc, galactose and silica acid residues, respectively, to the glycan chain. In contrast, the Fc oligosaccharides are partially protected by the protein structure at the stages of the biosynthetic pathway when these enzymes act.

SIMULTANEOUS ANALYSIS OF THE N-GLYCAN POOL FROM RAT CD48 EXPRESSED IN CHINESE HAMSTER OVARY (CHO) CELLS REVEALS EXTENSIVE HETEROGENEITY

Soluble rat CD48, containing 5 N-linked glycosylation sites, was expressed in CHO cells. The oligosaccharides in the glycan pool were analysed simultaneously using

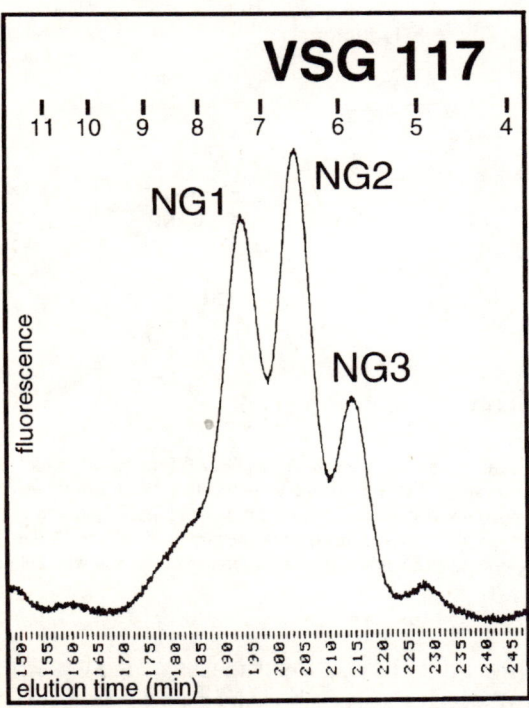

Figure 8B. Bio-Gel P4 gel filtration chromatography and identification of 2AB labelled GPI neutral glycans from purified *T. brucei* VSG 117, prepared 'on the blot'. An electroblot band of 5 µg (100 pmol) VSG 117 was treated as described in the legend to *Figure 8A* and the resulting 2AB labelled GPI neutral glycans were analysed by Bio-Gel P4 gel filtration chromatography. The structures represented by peaks NG1, NG2 and NG3 are shown in *Figure 8C*. The small numbers at the top indicate the glucose units (gu).

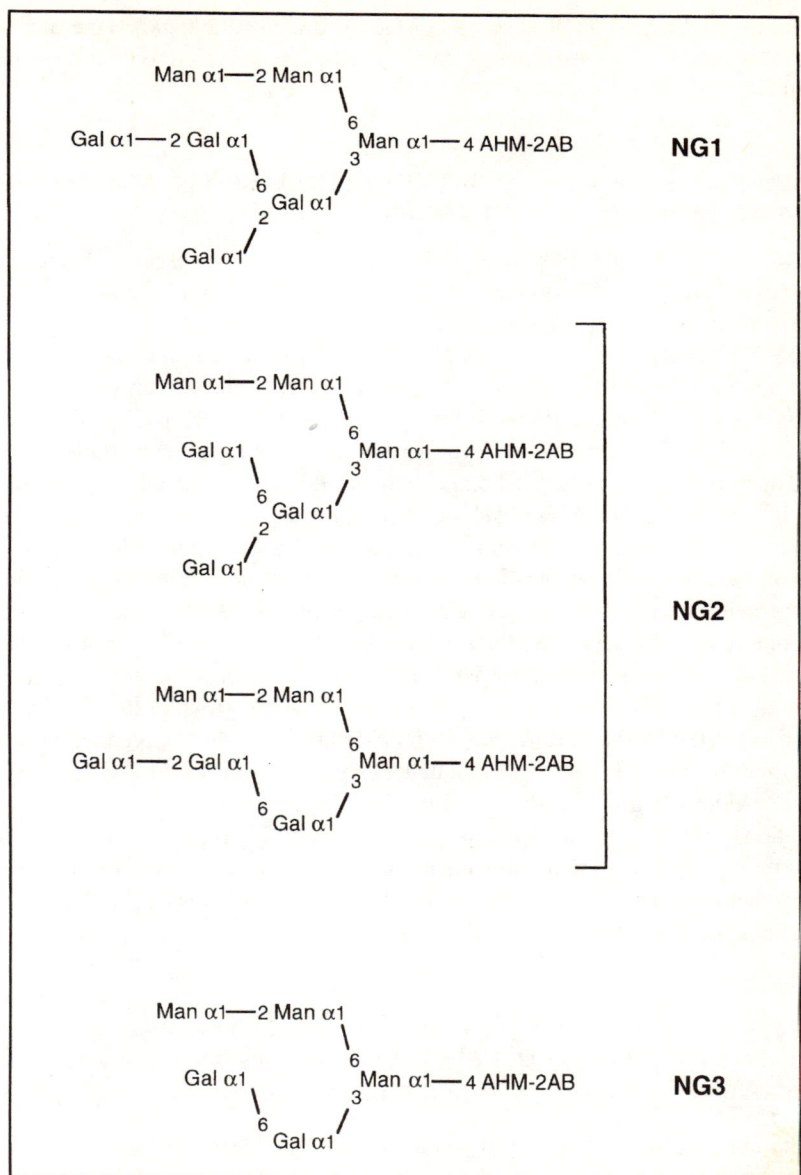

Figure 8C. Structures of the VSG 117 neutral glycans (NGs). The structures shown were determined by Ferguson (1988).

enzyme arrays (*Figure 7*) (Rudd *et al.*, submitted). In molecular modelling studies the range of complex, hybrid and oligomannose structures were found to shield large areas of the two IgSF domains of which CD48 is composed (Rudd *et al.*, submitted). CD48 is the major ligand for CD2 and this cell adhesion pair mediates the precise alignment of the cell surfaces of cytolytic T-lymphocytes carrying the TCR complex with those of target cells carrying loaded HLA class 1 molecules (Dustin *et al.*, 1996). The glycosylation analysis suggests that the sugars protect CD48 against proteases and play a role in the packing of CD48 on the cell surface. In addition the sugars at the sites close to the membrane may serve to orientate the face of the protein which contains the binding sites.

A NOVEL STRATEGY FOR ANALYSING GPI ANCHOR GLYCANS FROM MICROGRAM QUANTITIES OF PROTEIN ON ELECTROBLOTS

A novel strategy for the analysis of 2AB labelled GPI anchor glycans directly from electroblots, known as the 'on the blot' method (Zitzmannn and Ferguson, 1999) has recently been developed (*Figure 8A*).

ES MS methods to characterise the phosphatidylinositol components (Treumann *et al.*, 1997) and chromatographic methods to characterise the glycan components of GPI anchors (Treumann *et al.*, 1997; Ferguson, 1992; Schneider and Ferguson, 1995) have been described previously. However, the methods for glycan characterisation relied on introducing a tritium radiolabel into the GPI glycan with all of the associated complications of handling NaB3H4 reductant. This disadvantage led to the development of a new method for labelling GPI anchor glycans using the fluorophore 2AB that can be used both on purified glycoproteins and on microgram quantities of glycoproteins immobilised on polyvinylidene difluoride (PVDF) membranes after electroblotting (Zitzmannn, N. PhD Thesis, 1997; Zitzmannn and Ferguson, 1999).

The 2AB labelling procedure depends upon the presence of non-N-acetylated glucosamine (GlcN) in all GPI structures. The free amino group of the GlcN residue can be exploited by the nitrous acid deamination reaction that converts the GlcN to 2,5-anhydromannose (AHM) and simultaneously releases the inositol residue (*Figure 8A*). The aldehyde group of the AHM residue is then coupled to 2AB by reductive amination to give a stable covalent linkage between the GPI glycan and the fluorophore. The protein and ethanolamine substituents are subsequently removed by aqueous HF dephosphorylation to yield 2AB labelled GPI neutral glycans that can be characterised by liquid chromatography.

THE ANALYSIS OF THE GPI ANCHOR GLYCANS OF THE VARIANT SURFACE GLYCO-PROTEIN (VSG) FROM *T. BRUCEI*: THE α-GALACTOSE SIDEBRANCH HAS BEEN IMPLICATED IN THE PACKING OF VSG ON THE CELL SURFACE

Results obtained using 5 μg (100 picomoles) of VSG from *Trypanosoma brucei* (Ferguson *et al.*, 1988) following the 'on the blot' labelling method described in the previous section are shown in *Figure 8B*. The structures represented by peaks NG1, NG2 and NG3 are shown in *Figure 8C*. Their identity was confirmed by electrospray mass spectrometry analysis (*Figure 8D*).

GPI anchors afford a stable means of attachment of proteins to the membrane and can be viewed as an alternative mechanism to a single-pass hydrophobic transmembrane peptide domain. Although functions for the different glycan sidechains of GPI anchors are not generally known, the α-galactose sidebranch of the VSG GPI anchor has been implicated in the dense packing of the surface coat which acts as a barrier protecting the parasite from complement-mediated lysis in the host blood stream (Homans *et al.*, 1989).

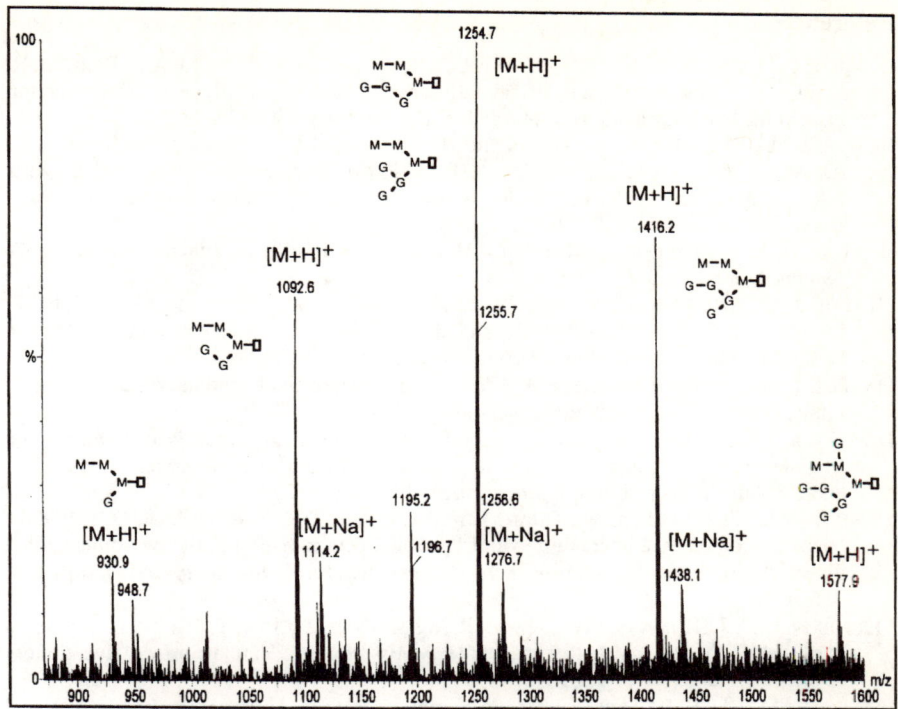

Figure 8D. ES MS of 2AB labelled VSG 117 neutral glycans (NGs). Positive ion mode electrospray mass spectrum of 2AB labelled VSG 117 NGs. 30 pmol were injected in methanol/acetic acid/water (1:1:1). The NG structures giving rise to major peaks are indicated. The empty square represents 2AB labelled anhydromannose.

Conclusion

There are three main post-translational modifications of proteins which involve glycosylation. The technology described here has been developed to enable all of these three types, N- and O-linked sugars and GPI anchor glycans, to be analysed rapidly and routinely. The minimum need is for an HPLC system equipped with an NP column and a fluorescence detector, a series of glycan standards and a set of endo- and exo-glycosidase enzymes are required. The ability to sequence N- and O-linked sugars directly from gels and GPI anchor glycans from blots has eliminated the need for exhaustive protein purification, making feasible the glycosylation analysis of biological samples available only in low yields.

The aim of developing such a strategy is to enable glycoproteins to be viewed in their entirety so that more insight can be gained into the complementary roles which sugars, proteins and anchors play in the structure and function of the glycoprotein.

References

BIGGE, J. C., PATEL, T. P., BRUCE, J. A., GOULDING, P. N., CHARLES, S. M. AND PAREKH, R. B. (1995). Nonselective and efficient fluorescent labeling of glycans using 2-amino benzamide and anthranilic acid. *Analytical Biochemistry* **230**, 229–238.

BLOCK, T., LU, X., MEHTA, A., BLUMBERG, B.S., TENNANT, B., EBLING, M., KORBA, B., LANSKY, D.M., JACOB, G.S. AND DWEK, R.A. (1998). Treatment of chronic hepadnavirus infection in a woodchuck animal model with an inhibitor of protein folding and trafficking. *Nature Medicine* **4**(5), 610–614.

BURTON, D. R. AND WOOF, J. M. (1992). Human antibody effector function. *Advances in Immunology* **51**, 1–84.

CYSTER, J. C., SHOTTON, D. M. AND WILLIAMS, A. F. (1991). The dimensions of the T-lymphocyte glycoprotein leukosialin and identification of linear protein epitopes that can be modified by glycosylation. *EMBO Journal* **10**, 893–902.

DAVIS, S.J. AND VAN DER MERWE, P.A. (1996). The structure and ligand interactions of CD2: implications for T-cell function. *Immunology Today* **17**, 177–187.

DEISENHOFER, J. (1981). Crystallographic refinement and atomic models of a human Fc fragment and its complex with fragment B of protein A from Staphylococcus aureus at 2.9- and 2.8-Angstroms resolution. *Biochemistry* **20**, 2361–2370.

DUSTIN, M.L., FERGUSON, L.M., CHAN, P.Y., SPRINGER, T.A., AND GOLAN, D.E. (1996). Visualization of CD2 interaction with LFA-3 and determination of the two-dimensional dissociation constant for adhesion receptors in a contact area. *Journal of Cell Biology* **132**, 456–474.

FERGUSON, M.A.J., HOMANS, S.W., DWEK, R.A. AND RADEMACHER, T.W. (1988). Glycosyl-phosphatidylinositol moiety that anchors *Trypanosoma brucei* Variant Surface Glycoprotein to the membrane. *Science* **239**, 753–759.

FERGUSON, M.A.J. (1991). Lipid anchors on membrane proteins. *Current Opinion in Structural Biology* **1**, 522–529.

FERGUSON, M. A. J. (1992). The chemical and enzymatic analysis of GPI fine structure. In *Lipid Modification of Proteins. A Practical Approach*. Eds. N.M. Hooper and A.J. Turner, pp 191–230. Oxford: IRL Oxford University Press.

GUILE, G.G., RUDD, P.M., WING, D.R., PRIME, S.B. AND DWEK, R.A. (1996). A Rapid High Resolution Method for Separating Oligosaccharide Mixtures and Analysing Sugarprints. *Analytical Biochemistry* **240**, 210–226.

HOMANS, S.W., EDGE, C.J., FERGUSON, M A. J., DWEK, R.A. AND RADEMACHER, T. (1989). Solution structure of the glycosylphosphatidylinositol membrane anchor glycan of *Trypanosoma brucei* variant surface glycoprotein. *Biochemistry* **28**, 2881–2887.

JENTOFT, N. (1990). Why are proteins O-glycosylated? *Trends in Biochemical Science* **15**, 291–294.

KÜSTER, B., WHEELER, S.F., HUNTER, A.P., DWEK, R.A. AND HARVEY, D.J. (1997). Sequencing of N-linked oligosaccharides directly from protein-gels: in-gel deglycosylation followed by matrix-assisted laser desorption/ionization mass spectrometry and normal-phase high-performance liquid chromatography. *Analytical Biochemistry* **250**, 82–101.

LEATHERBARROW, R.J. AND DWEK, R.A. (1984). Binding of complement sub component C1q to mouse IgG1, IgG2a and IgG2b: A novel C1q binding assay. *Molecular Immunology* **21**, 321–327.

MATTU, T.S., PLEASS, R.J., WILLIS, A., KILLIAN, M., WORMALD, M.R., LELLOUCH, A-M., RUDD, P.M., WOOF, J. AND DWEK, R.A. (1998). The glycosylation and structure of human serum IgA1, Fab abd Fc regions and the role of N-glycosylation on FcαR interactions. *Journal of Biological Chemistry* **273**, 2260–2272.

MCCONVILLE, M.J. AND FERGUSON, M.A.J. (1993). The structure, biosynthesis and function of glycosylated phosphatidylinositols in the parasitic protozoa and higher eukaryotes. *Biochemical Journal* **294**, 305–324.

MEHTA, A., LU, X., BLOCK, T.M., BLUMBERG, B.S. AND DWEK, R.A. (1997). Hepatitis B virus envelope proteins vary drastically in their sensitivity to glycan processing. *Proceedings of the National Acadamy of Sciences* (USA) **94**, 1822–1827.

MESTECKY, J. AND KILIAN, M. (1985). Immunoglobulin A (IgA). *Methods in Enzymology* 116, 37–75.

MOORE, S.E., SPIRO, R.G. (1990). Demonstration that Golgi endo-α-D-mannosidase provides a glucosidase independent pathway for the formation of complex N-linked oligosaccharides of glycoproteins. *Journal of Biological Chemistry* 265, 13104–13112.

NOSE, M. AND WIGZELL, H. (1983). Biological significance of carbohydrate chains on mono-clonal antibodies. *Proceedings of the National Academy of Sciences, USA* 80, 6632–6636.

OPDENAKKER, G., MASURE, S., PROOST, P., BILLIAU, A. AND VAN DAMME, J. (1991). Natural human monocyte gelatinase and its inhibitor. *FEBS Letters* 284, 73–78.

PADLAN, E.A. (1994). Anatomy of the antibody molecule. *Molecular Immunology* 31, 169–217.

PATEL, T.P. AND PAREKH, R.B. (1994). Release of oligosaccharides from proteins by hydrazinolysis. *Methods in Enzymology* 230, 57–66.

ROUTIER, F.H., HOUNSELL, E.F., RUDD, P.M., TAKASHI, N., BOND, A., HAY, F.C., AXFORD, J.A. AND JEFFERIS, R. Quantitation of human IgG glycoforms isolated from rheumatoid sera: a critical evaluation of different methods. *Journal of Immunological Methods* (in press).

RUDD, P.M., LEATHERBARROW, R.L., RADEMACHER, T.W. AND DWEK, R.A. (1991). Diversification of the IgG Molecule by Oligosaccharides. *Journal of Molecular Immunology* 28, 1369–1378.

RUDD, P.M. AND DWEK, R.A. (1997). Glycosylation: Heterogeneity and the 3D structure of the protein. *Critical Reviews in Biochemistry and Molecular Biology* 32, 1–100.

RUDD, P.M., GUILE, G.R., KÜSTER, B., HARVEY, D.J., OPDENAKKER, G. AND DWEK, R.A. (1997). Oligosaccharide Sequencing Technology. *Nature* 388, 205–208.

RUDD, P.M., WORMALD, M.R., HARVEY, D.J., DEVASHAYEM, M., McALISTER, M.S.B., BARCLAY, A.N., BROWN, M.H., DAVIS, S. J. AND DWEK, R.A. N-linked oligosaccharide processing in the Ly-6, scavenger receptor and immunoglobulin superfamilies is constrained by local or tertiary structure rather than by domain topology. (Manuscript submitted).

SCHNEIDER, P. AND FERGUSON, M.A.J. (1995). Microscale analysis of glycosylphosphatidylinositol structures. *Methods in Enzymology* 250, 614–630.

SHERKER, A. AND MARION, P. (1991). Hepadnaviruses and hepatocellular carcinoma. *Annual Reviews of Microbiology* 45, 475–508.

SHOGREN, R., GERKEN, T. A. AND JENTOFT, N. (1989). Role of glycosylation on the conformation and chain dimensions of O-linked glycoproteins: light-scattering studies of ovine submaxillary mucin. *Biochemistry* 28, 5525–5536.

TREUMANN, A., GUETHER, M.L S., SCHNEIDER, P. AND FERGUSON, M.A.J. (1997). Analysis of the carbohydrate and lipid components of glycosylphosphatidylinositol structures. In *Methods in Molecular Biology: Glycoanalysis Protocols*. Ed. E.F. Hounsell. Totowa, USA: Humana Press (in press).

TSUCHIYA, N., ENDO, T., MATSUTA, K., YOSHINOYA, S., AIKAWA, T., KOSUGE, E., TAKEUCHI, F., MIYAMOTO, T. AND KOBATA, A. (1989). Effects of galactose depletion from oligosaccharide chains on immunological activities of human IgG. *Journal of Rheumatology* 16, 285–290.

WORMALD, M.W., RUDD, P.M., HARVEY, D.H., CHANG, S-C., SCRAGG, I.G. AND DWEK, R.A. (1997). Variations in oligosaccharide-protein interactions in immunoglobulin G determine the site specific glycosylation profiles and modulate the dynamic motion of the Fc oligosaccharides. *Biochemistry* 36, 1370–1380.

YOUINGS, A., CHANG, S-C., DWEK, R.A. AND SCRAGG, I.G. (1996). Site specific glycosylation changes on human immunoglobulin G in pregnancy and rheumatoid arthritis. *Biochemical Journal* 314, 621–630.

ZITZMANNN, N. (1997). Studies on the Processing of Glycosylphosphatidylinositol Membrane Anchors in Transfected *Trypanosoma brucei*. *PhD thesis*, University of Dundee.

ZITZMANNN, N. AND FERGUSON, M.A.J. (1999). Analysis of the carbohydrate components of glycosylphosphatidylinositol structures using fluorescent labelling. *Methods of Lipid Purification* (in press).

2
Biomedical and Pharmaceutical Applications of Alginate and Chitosan

ØYVIND SKAUGRUD, ANNIKEN HAGEN, BERIT BORGERSEN AND MICHAEL DORNISH*

Pronova Biomedical, Gaustadalleen 21, N-0371 Oslo, Norway

Introduction

Biopolymers from marine sources have been studied and utilized in commercial applications and product development for a number of years. Pharmaceutical and medical uses of biopolymers have gained interest. Alginates and chitosan can be utilized as hydrophilic drug carriers and as matrix materials. In addition, these polysaccharides have potential as rate-controlling excipients in drug release systems. One of the most interesting applications of alginate is in the development of alginate-immobilized cells as artificial organs. For example, the potential use of alginate-encapsulated pancreatic islet cells for the treatment of type I diabetes could be a useful treatment for a large number of patients. This application, however, puts strict requirements on the encapsulation system as well as on the purity and documentation of the alginate used. Chitosan has shown promise in the development of non-parenteral delivery systems for challenging drugs. For example, chitosan salts have been shown to increase the transport of drugs across the nasal epithelial surface.

These, and other applications will be pursued in the near future. Such applications will require highly purified polymers with documented safety profiles. In addition, polymers used in pharmaceutical applications will also have to be acceptable to regulatory authorities. The characterization of key parameters for each polymer will be necessary.

Description of alginate and chitosan

ALGINATE

Alginate, extracted from brown algae, is a linear polymer composed of two uronic acid monosaccharides: D-mannuronic (M) and L-guluronic (G) acid linked by $\beta(1\rightarrow4)$ and $\alpha(1\rightarrow4)$ glycosidic bonds. The two monomers are arranged in homopolymeric blocks,

---

*To whom correspondence may be addressed.

Biotechnology and Genetic Engineering Reviews – Vol. 16, April 1999
0264–8725/99/15/23–40 $20.00 + $0.00 © Intercept Ltd, P.O. Box 716, Andover, Hampshire SP10 1YG, UK

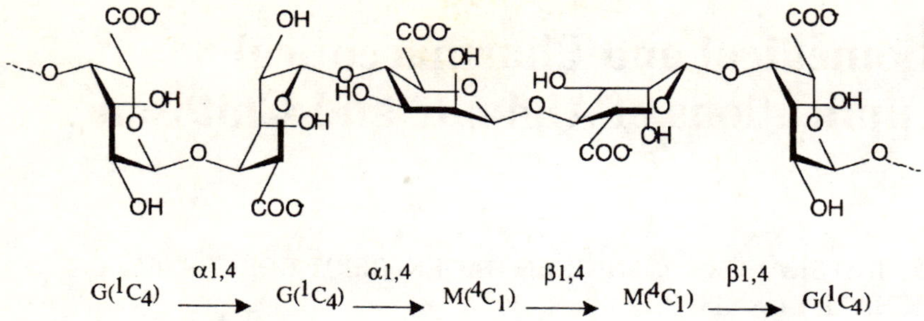

$$G(^1C_4) \xrightarrow{\alpha1,4} G(^1C_4) \xrightarrow{\alpha1,4} M(^4C_1) \xrightarrow{\beta1,4} M(^4C_1) \xrightarrow{\beta1,4} G(^1C_4)$$

G: Guluronate   M: Mannuronate

Figure 1.   Chemical structures of alginate.

M-blocks and G-blocks as well as sequences containing both monomers, MG-blocks (*Figure 1*).

Alginates isolated from different algae can vary both in monomer composition and block arrangement (*Table 1*), and these variations are also reflected in the properties of the alginate.

Whilst viscosity depends mainly on molecular size, the affinity for cations and the gel forming properties are mostly related to the block structure of repeating guluronic acid residues. When two guluronic acid residues are adjacent in the polymer, they form a binding site for polyvalent cations. The content of G-blocks is the main structural feature contributing to gel strength and stability of the gel. Reactivity with calcium, causing gel formation, is a direct function of the average length of the G blocks occurring in the polymer chain (*Figure 2*) (Smidsrød and Draget, 1996; Smidsrød and Skjåk-Bræk, 1990)

CHITOSAN

Chitosan is a high molecular weight cationic polysaccharide derived from crustacean shells by deacetylation of naturally occurring chitin. Chitosan is also a linear polymer which is composed of glucosamine and *N*-acetyl glucosamine units linked

Table 1.   Typical values for M (mannuronate) and G (guluronate) content in seaweed used for alginate production

Seaweed	M/G	%M	%G	%MM	%GG
Laminaria hyperborea (stem)	0.45	30	70	18	58
Laminaria hyperborea (leaf)	1.22	55	45	36	26
Laminaria digitata	1.22	55	45	39	29
Macrocystis pyrifera	1.50	60	40	40	20
Lessonia nigrescens	1.50	60	40	43	23
Ascophyllum nodosum	1.86	65	35	56	26
Laminaria japonica	1.86	65	35	48	18
Durvillea antarctica	2.45	71	29	58	16
Durvillea potarum	3.33	77	23	69	13

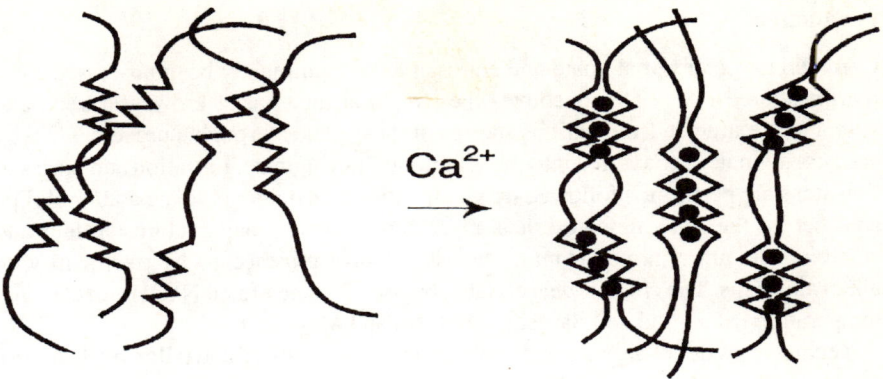

Figure 2. Crosslinking of alginate G-blocks with calcium.

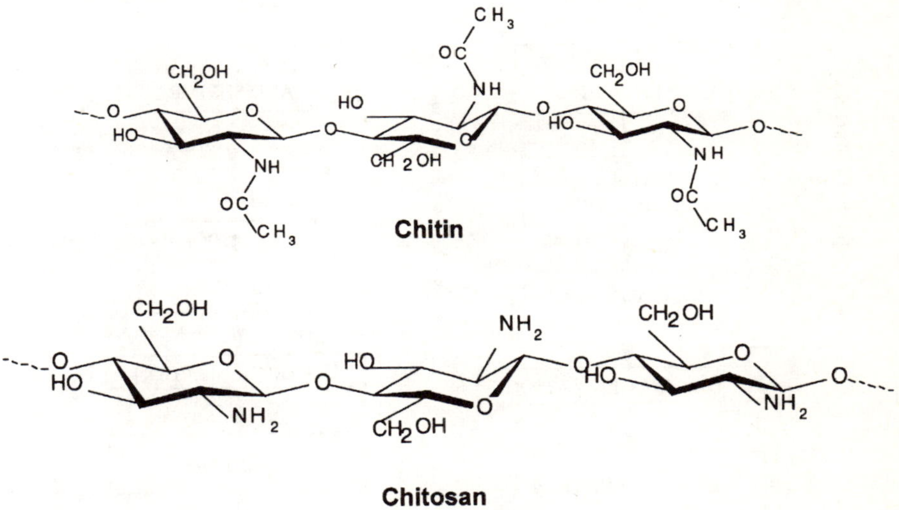

Chitin

Chitosan

Figure 3. Chemical structure of chitin and chitosan.

in a $\beta(1\rightarrow4)$ manner (*Figure 3*). The glucosamine and *N*-acetyl glucosamine can theoretically be arranged by a similar sequential structure as M and G in alginate. Most often, commercial products will have a random distribution of the remaining *N*-acetyl glucosamine units after deacetylation. The ratio between glucosamine and *N*-acetyl glucosamine is referred to as the degree of deacetylation. In solution, chitosan salts will carry a positive charge through protonation of the free amino group on glucosamine. Reactivity with negatively charged surfaces is a direct function of the positive charge density of chitosan. The cationic nature of chitosan gives this polymer a mucoadhesive property (Skaugrud, 1995; Allan *et al.*, 1984; Li *et al.*, 1992).

Production

Commercial grades of alginate and chitosan have traditionally been processed from marine sources. Alginate constitutes the structural material of brown seaweed and kelp and is extracted from various species of these plants through a process whereby calcium alginate is converted into the soluble sodium alginate. Filtration and purification steps are performed followed by precipitation as calcium alginate and/or alginic acid before the final alginate salt is made. For chitosan manufacture, protein and calcium salts are removed from the exoskeleton of crustaceans by treatment with alkali and acids. The chitin is deacetylated by use of concentrated NaOH and elevated temperature (Skaugrud and Sargent, 1990; No and Meyers, 1995).

Technical grades of alginate and chitosan are further purifed and tailor-made for use in pharmaceutical and biomedical applications by using various chemical treatment and filtration steps. Microfiltration is used to remove insoluble compounds while ultrafiltration removes low molecular weight compounds. For pharmaceutical use in particular, the endotoxin content is reduced to a level acceptable for use of these polymers in humans. Of additional importance is the specification of the range of molecular weight distribution and the monomer composition of these polymers. To meet regulatory requirements, the materials are manufactured in compliance with

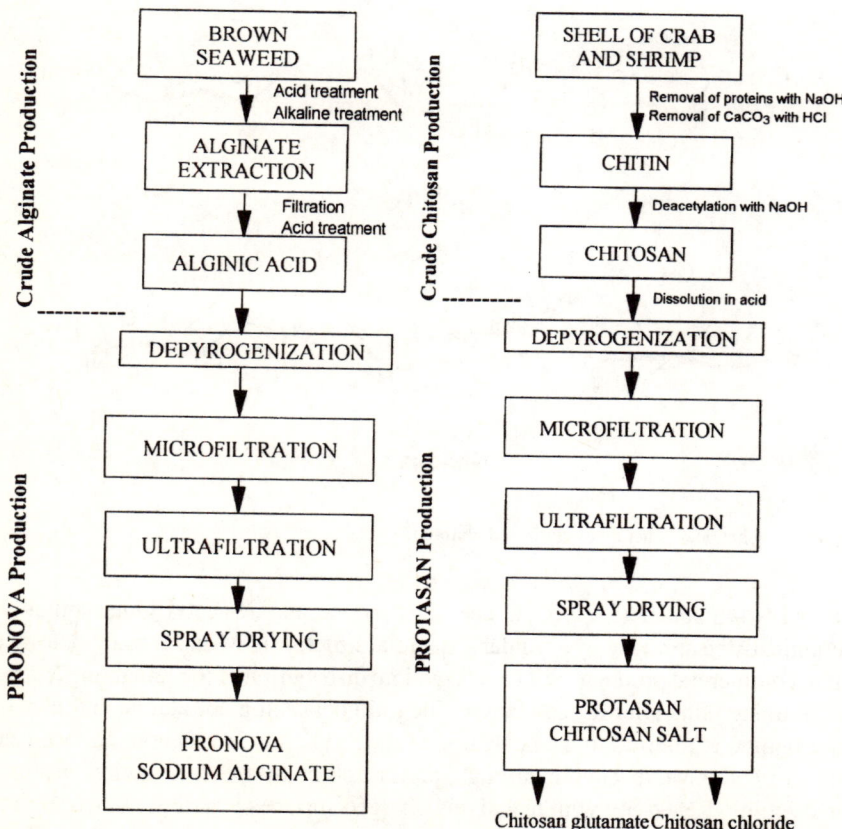

Figure 4.   Production scheme for PRONOVA™ sodium alginate and PROTASAN™ chitosan salts.

'Good Manufacturing Practice' (GMP) guidelines. Basic toxicological data and polymer properties are documented in 'Drug Master Files' (DMF).

While small quantities of polymers can be produced relatively easily at laboratory scale, large scale production must meet the supply requirements for commercial applications. The scale-up process is not always straight forward and the documentation required to satisfy the regulatory authorities can be considerable. Pronova Biomedical's experience with large scale production of alginate and chitosan together with experience in characterizing key parameters has made these biopolymers commercially available in standard and ultrapure grades. Currently sodium alginate is being produced under the PRONOVA™ trademark and the chloride and glutamate salts of chitosan are being produced under the PROTA-SAN™ trademark (*Figure 4*).

Characterization

COMPOSITION AND SEQUENTIAL STRUCTURE

The composition and sequential structure of alginate and chitosan can be determined by high resolution ¹H- and ¹³C-nuclear magnetic resonance spectroscopy (NMR). For alginate, techniques have been developed to determine the monad frequencies as well as diads and triads. Based on such measurements, parameters such as M/G ratio, G-content with consecutive G>1, and average length of blocks of consecutive G units can be calculated (*Figure 5*) (Grasdalen, 1983). For chitosan, the degree of deacetylation can be detected by ¹H- and ¹³C-NMR (Vårum *et al.*, 1991a,b).

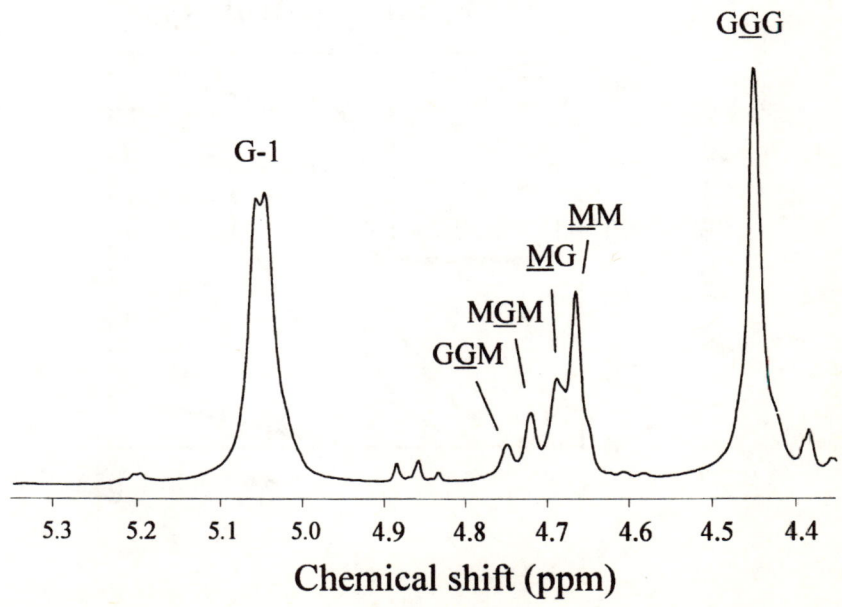

Figure 5.   Typical ¹H-NMR spectrogram of PRONOVA™ UP LVG alginate.

MOLECULAR WEIGHT AND POLYDISPERSITY

Commercial alginates, like polysaccharides in general, are polydisperse with respect to molecular weight (M). Therefore, the given M of an alginate always represents an average of all of the molecules in the population. The most common ways to express molecular weights are as the number average ($\bar{M}_n$) and the weight average ($\bar{M}_w$). The two averages are defined by the following equations:

$$\bar{M}_n = \frac{\sum_i N_i M_i}{\sum_i N_i} \quad \text{and} \quad \bar{M}_w = \frac{\sum_i w_i M_i}{\sum_i w_i} = \frac{\sum_i N_i M_i^2}{\sum_i N_i M_i}$$

where N_i = number of molecules having a specific molecular weight M_i
$\quad\quad w_i$ = weight of molecules having a specific molecular weight M_i

In a polydisperse molecule population the relation $\bar{M}_w > \bar{M}_n$ is always valid. The coefficient $\bar{M}_w / \bar{M}_n$ is referred to as the polydispersity index, and will typically be in the range 1.5–3.0 for commercial alginates. More methods exist for determination of molecular weights. The most common ones in use are calculations based on intrinsic viscosity and light scattering measurements.

SOLUBILITY

Sodium alginate is soluble in water, but will precipitate as alginic acid at low pH. The pK_a values of guluronic and mannuronic acid are 3.6 and 3.3, respectively. The solubility of chitosan depends, to a certain extent, on the molecular weight and degree of deacetylation (*Figure 6*). Dissolution of chitosan, however, requires an acidic

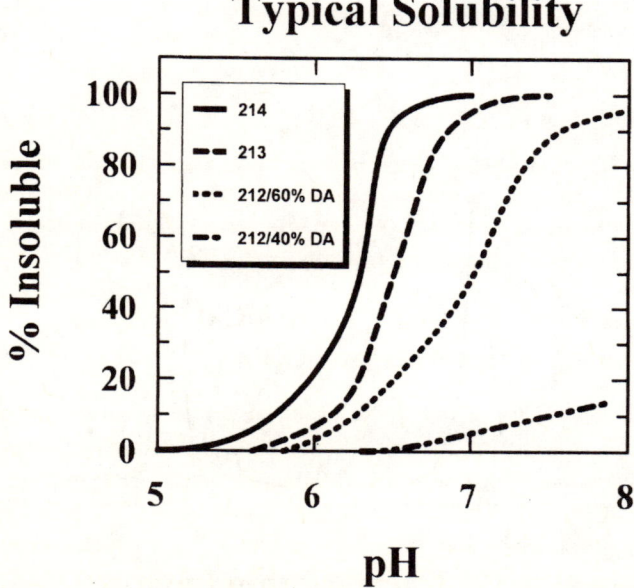

Figure 6.   Solubility of PROTASAN™ chitosan. The degree of deacetylation increases from 40% and 60% for type 212, through to approximately 85% for 213 to 95% for type 214.

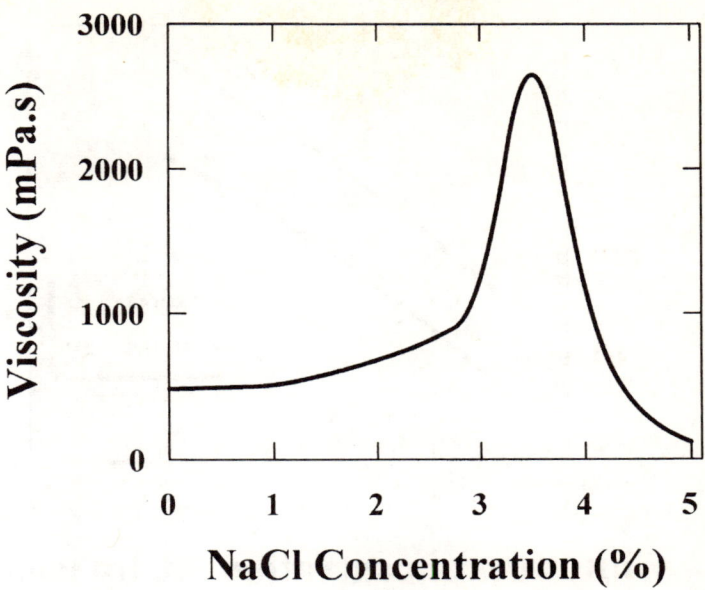

Figure 7.   Alginate viscosity as a function of ionic strength .

environment. PROTASAN™ grades contain a stoichiometric amount of an appropriate acid integrated with chitosan as a counterion and will easily dissolve in water.
The apparent pK_a value of the amino group of the glucosamine moiety is 6.5. In
aqueous media at acidic pH, the chitosan molecule will be highly positively
charged.

VISCOSITY

Viscosity is a function of the molecular weight of the biopolymer and its conformation
in solution. Under different conditions the flow characteristic (rheology) of the
alginate or chitosan solution will vary as, for example, on varying the ionic strength of
the polymer solution (*Figure 7*). *Figure 8* illustrates the dependency of viscosity
following increases in concentration for two chitosans.

POLYELECTROLYTIC PROPERTIES OF PROTASAN™

The apparent pK_a-value of the amino group of the glucosamine moiety is 6.5. In
aqueous media at acidic pH, the chitosan molecule will be highly positively charged.
The repelling effect of each positively charged deacetylated unit on neighboring
glucosamine units will result in an extended conformation of the polymer in solution.
The addition of salt will reduce this effect, resulting in a more random coil conformation of the molecule. At higher ionic strength a salting-out effect will occur, precipitating
the chitosan from the solution. This is shown as a reduction in solution viscosity in
Figure 9.

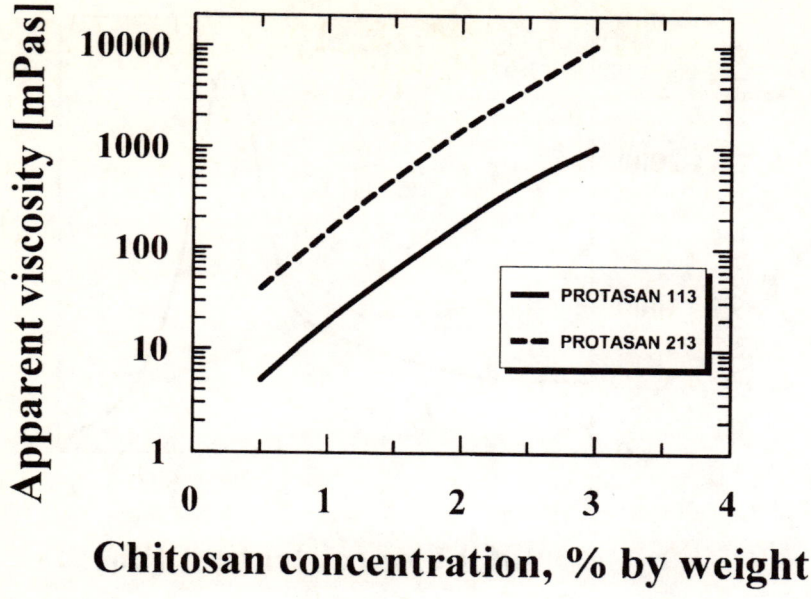

Figure 8. Chitosan viscosity as a function of concentration.

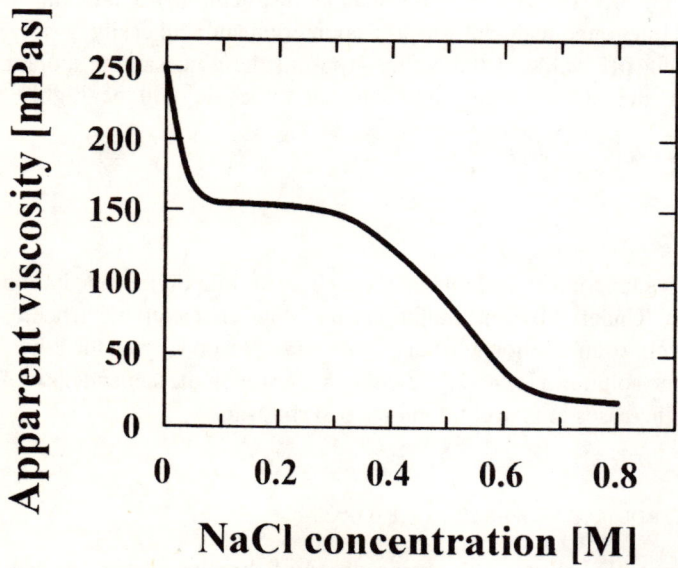

Figure 9. Viscosity of chitosan as a function of ionic strength.

Applications

A general overview of the characteristics of alginate and chitosan which can be used in biomedical and pharmaceutical applications is shown in *Table 2* (Skaugrud, 1995; Allan *et al.*, 1984; Li *et al.*, 1992).

Table 2.   Characteristics of alginate and chitosan useful for medical and pharmaceutical applications

Feature	Benefits	Alginate	Chitosan
Solubility	Control of dissolution in the gastrointestinal tract	+++	+++
Gel	Immobilization agent for drug delivery systems	+++	+
Electrolyte	Bioadhesive agent for drug delivery systems	–	+++
Swelling	Tablet disintegration. Matrix for drug delivery systems	+++	++
Film/fibre	Wound treatment. Encapsulation agent for drug delivery systems	+++	+++
Chelation	Cation binding	++	+
Viscosity	Suspensions	++	–
Biological	Immune Response	+++	+
Properties	Anti-microbial effect	–	+++

CELL IMMOBILIZATION WITH ALGINATE

The technique to immobilize cells, particularly pancreatic islet cells, in calcium alginate matrices was developed by Lim at the end of the 1970's (Lim and Sun, 1980). By coating the bead with polycations like poly-L-lysine, poly-L-ornithine, or chitosan, the strength of the surface coating as well as the capsule porosity can be controlled (*Figure 10*) (Skjåk-Bræk and Espevik, 1996).

NASAL DELIVERY OF CHALLENGING DRUGS

Polar drugs are not well absorbed across the nasal mucosa. These drugs include low molecular weight compounds as well as biotechnology products such as polypeptides

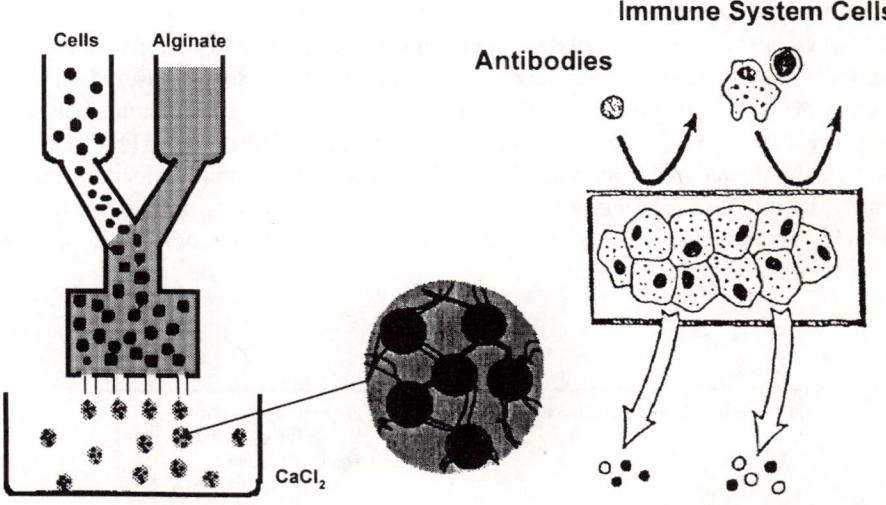

Figure 10.   Encapsulation of cells with alginate.

and proteins. Solutions and powder formulations based on chitosan salts of a molecular weight of greater than 100 kD and with a defined degree of deacetylation have been found to enhance the delivery of nasally administered polypeptides such as insulin, calcitonin, LHRH analogues, growth hormone as well as non-peptide polar compounds (Illum *et al.*, 1994; Aspden *et al.*, 1996, 1997).

Regulatory considerations

The underlying documentation of the GRAS ('Generally Recognized As Safe') status of alginate goes back prior to 1972 (21CFR 184.1724). In the US *Pharmacopoeia* alginate is still described as a poly mannuronate, even though the existence of guluronate and the block forming structure of alginate is now well-known. For chitosan no US monograph exists yet, however, a European monograph is under evaluation.

For pharmaceutical and biomedical applications of alginate and chitosan to be successful, regulatory issues will have to be addressed. There are three main areas in this respect which must be dealt with:

- characterization and functionality
- product reproducibility
- toxicology and long term safety

Safety is one of the biggest issues for the commercialization of alginate and chitosan for human medical applications. The development of new biomedical products involves several different aspects such as the property of the biopolymer, production quality and quantity, and clinical effects. The applications of alginate and chitosan to human studies will also involve regulatory approval by national and international authorities, such as the FDA in the United States. Regulatory issues that are important in the commercialization of alginate and chitosan for biomedical uses are (*Table 3*): Characterization and functionality, specifications of the product, and analysis using validated methods. Stability of the compound is of prime importance. Reproducibility of the manufacture of the compound is very important, and this is ensured under a series of GMP guidelines. Documentation of not only the manufacture but also specifications and safety of the product is described in a 'Drug Master File' (DMF), both in the US and in Europe. Drug Master Files form the backbone of documentation that is required for registering a product, either as an active drug or as an excipient. Finally, toxicology and safety covering basic studies and application-specific studies must be documented.

Table 3.   Regulatory Issues

Characterization and Functionality:	Product specification
	Validation of analytical methods
	Stability Studies
Manufacture:	Reproducibility (GMP)
	Documentation (DMF)
Toxicology and safety:	Basic studies
	Application-specific studies

As examples we now consider ultrapure alginate (PRONOVA™) and ultrapure chitosan (PROTASAN™). The reader is refererred to the two articles by Dornish *et al.* (1997a, b).

Pharmacokinetics of sodium alginate in mice.   In spite of various uses of alginate in the biomedical and pharmaceutical field for more than 50 years, no extensive toxicology documentation is yet available in the public domain. Pharmacokinetic studies of PRONOVA™ alginate have been carried out and the *Figures 11–14* and *Table 4* show the results following various bolus injections of a radiolabeled alginate (Hagen *et al.*, 1995 and unpublished results).

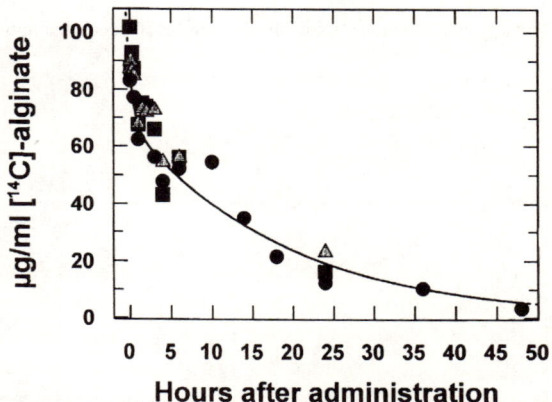

Figure 11.   The pharmacokinetics in mice following an intravenous bolus injection of 100 mg alginate. The profile appears to be biphasic, indicating a 2-compartment model. The initial half-life ($t_{1/2}$) is approximately 4 hours, while the secondary $t_{1/2}$ appears to be about 22 hours.

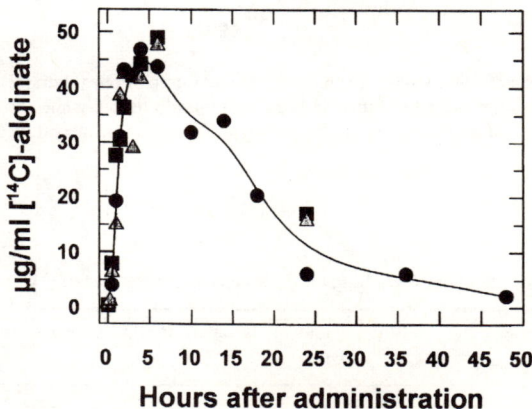

Figure 12.   The pharmacokinetics in mice following an intraperitoneal bolus injection of 100 mg alginate. The absorption reaches a maximum after 5–6 hours. Thereafter, the serum concentration declines with an apparent half-life of about 12.5 hours. The elimination following intraperitoneal administration may also occur in a biphasic fashion, similar to intravenous administration.

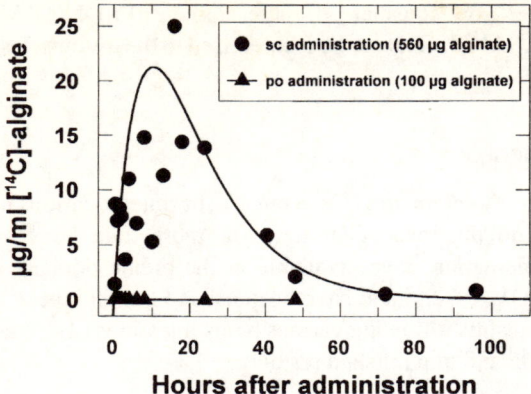

Figure 13.   The pharmacokinetics in mice following oral and subcutaneous administration of alginate.

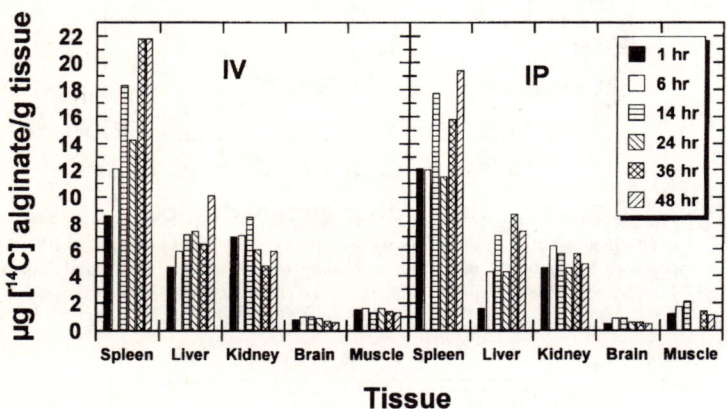

Figure 14.   Tissue distribution studies of intravenous and intraperitoneal administrated alginate. Only very small amounts of alginate were found in brain and muscle tissue, while the liver and the kidney contained moderate amounts. The spleen, not surprisingly, appeared to concentrate the alginate over a 48 hours period.

Table 4.   Pharmacokinetic parameters following various administrations of [^{14}C] alginate.

Pharmokinetic Parameter	Intravenous, (IP)	Intraperitoneal, (IV)	Subcutaneous, (SC)	Oral (per oral), (PO)
Area under the Curve (μg h ml^{-1})	1253	550	69	0
Volume of Distribution (l)	1.09	1.0	–	–
Clearance (l/h)	0.08	0.09	–	–
Bioavailability (%)	100	44	5	0

Calculations were performed by using the MK Model version 5.0 program (Biosoft, Cambridge, UK).

Table 5. Specifications of PROTASAN™ used in the studies

Specification	UP G 110	UP G 210	UP Cl 110
Batch number	705-583-04	610-583-07B	310-490-01
Apparent viscosity	2.8 mPa.s	91 mPa.s	12 mPa.s
Intrinsic viscosity	0.7 dl/g	6.9 dl/g	ND
Degree of deacetylation	83%	83%	87%
Acid content	41% glutamic	40% glutamic	19% chloride
Loss on drying	4.9%	4.5%	4.6%
Insolubles	<0.1%	<0.1%	<0.1%
Heavy metals	<27 ppm	<26 ppm	<30 ppm
Endotoxin content	195 EU/g	625 EU/g	250 EU/g
Microbiology	< 1 cfu/g	<1 cfu/g	<1 cfu/g

UP: Ultrapure grade. G: glutamate salt. Cl: Chloride salt.

SAFETY AND TOXICOLOGY OF PROTOSAN™

We review here some of the results reported in Dornish *et al.* (1997a,b). Specifications for the chitosan glutamate or chitosan chloride used are shown in *Table 5*. The 'UP' designation indicates an ultrapure quality. The G stands for glutamate while Cl indicates a chloride salt. The 110 is a low viscosity chitosan while 210 indicates a medium viscosity. The chitosan salts were made up in physiological saline solution (0.9% NaCl), or in cell culture medium. Safety and toxicological studies were performed in accordance with applicable guidelines.

Oral toxicity studies were performed in rats which received daily administration of chitosan glutamate solutions up to 600 mg/kg/day for 13 weeks. The results summarized in *Table 6* indicate no toxicological effects were seen in any group. No real change in body weight was found between treated and control animals. Blood chemistry was also comparative between the four groups.

Table 6. Oral toxicity of PROTASAN™ UP G 210. Daily oral administration (gavage) to rats for 13 weeks

Group Number	Group Designation	Dose level (mg/kg/day)	Dose volume (ml/kg/day)	Dose conc. (mg/ml)	Results
1	Control	0	10	0	No
2	Low dose	100	10	10	toxicological
3	Intermediate	300	10	30	effects seen
4	High dose	600	10	60	in any group

Gavage: Gastric incubation

Table 7. Intravenous toxicity of PROTASAN™ UP G 210 & UP G 110. One single intravenous administration to rats

Group Number	Group Designation	Dose level (mg/kg)	Dose conc. (mg/ml)	Results
1	Low viscosity (UP G 110)	25	5	No toxico-
2	Medium viscosity (UP G 210)	25	5	logical effects seen in any group

Table 8.   Intraperitoneal toxicity of PROTASAN™ UP G 210 & UP Cl 110. One single intraperitoneal administration to rats

Group Number	Group Designation	Dose level (mg/kg)	Dose conc. (mg/ml)	Results
1	Low dose	100	5	No toxicological
2	Intermediate dose	250	15	effects seen in
3	High dose	500	25	any group

An intravenous (IV) study was performed as a limit study investigating the effect of 25 mg/ml. Higher doses were toxic to animals. The cause of toxicity is most likely the aggregation of red blood cells by binding to chitosan resulting in blockage of capillaries. As seen in *Table 7*, no toxicological effects were seen at this dose level.

IP (intraperitoneal) studies showed that a bolus injection of up to 500 mg/kg chitosan glutamate or chitosan chloride does not result in any toxicological effects evaluated seven days after injection (*Table 8*). There was no change in body weight relative to control animals and no changes in physical or macroscopic appearance of the organs.

Two hypersensitization studies using PROTASAN™ UP G 210 were performed. A Magnusson and Kligman test (see Method B6 in Annex V of EEC Commission Directive 92/69 (1922) and as described in OECD Guidelines for Testing of Chemicals (1992) No. 406.) of the sensitizing potential in guinea pigs is a study in which the compound to be tested is intradermally injected to small areas on the back of the guinea pigs either with or without Freund's adjuvant. A topical application of a highly concentrated chitosan glutamate solution was given as the challenge at a separate site to the induction. Dichloronitrobenzene was used as the positive control. The results in *Table 9* show no hypersensitive reaction on application of the chitosan glutamate.

Anaphylactic shock is an antibody-medicated reaction leading to mortality. Guinea pigs were injected subcutaneously with chitosan glutamate as an induction. The challenge was given as an intravenous injection 21 days later. The positive control was ovalbumin which, on challenge, caused death in 8 of 10 animals. There was no mortality in the chitosan-treated groups and body weights were normal. There was some cyanosis directly after injection of chitosan salts, but this condition cleared by four hours after injection.

Table 9.   Hypersensitization studies with PROTASAN™ UP G 210

Study	Induction	Challenge	Results
Magnusson and Kligman	1 mg/ml UP G 210 Intradermal 2 × 0.1ml –Freund's –0.9%NaCl 50:50 NaCl/Freund's	60 mg/ml UP G 210 Topical 0.5ml 48 hr.	No mortality. Body wt. not affected Irritation during induction No delayed hypersensitization (Positive control: 10/10 with DCNB)
Anaphylactic Shock	10 mg/kg UP G 210 Subcutaneous Days 1 and 8	20 mg/kg IV Day 21	No mortality in treated groups. Body wt. not affected. Cyanosis in 5/10, normal after 4hr. (Positive control death in 8/10 using ovalbumin)

Table 10.   Nasal irritancy of PROTASAN™ UP G 210. Three daily intranasal administrations to the rat for one week

Group Number	Group Designation	Dose level (mg/rat/day)	Dose volume (μl/rat/day)	Dose conc. (mg/ml)	Results
1	Control	0	3 × 100μl	0	Congestion in all groups.
2	Low dose	1.5	3 × 100μl	5	Increase mucus, not dose related.
3	High dose	3	3 × 100μl	10	No toxic effect on ciliated cells

Table 11.   Other safety studies with PROTASAN™

Study	Dose	Schedule	Results
Ames test	Up to 5000 mg/plate UP G 210	±S9 activation TA98, TA100, TA1535, TA1537 & TA102 strains of *Salmonella*	No significant increases in the number of revertants observed. 5000μg/plate toxic to TA98 & TA1537
In vitro cell survival (Colony forming assay)	Up to 1 mg/ml	24 hr exposure 3T3 and V79 cells	Little effect on cell survival.

One of the prime applications of chitosan salts is in the field of drug delivery. The bioadhesive properties of chitosan can be used in nasal applications (Illum *et al.*, 1994). Safety of nasally administered chitosan glutamate was determined by treating rats with 0.5 or 1% chitosan glutamate solutions three times a day for seven consecutive days (*Table 10*). Sections of the nasal cavities were stained with hematoxylin and eosin. Histological examination of the nasal mucosa indicated some increase in the thickness of the mucus layer in chitosan-treated animals (*Figure 15*). Goblet cells increased the production of mucus, not unlike the reaction occurring in other nasal irritation reactions. This reaction could also be due to the treatment technique itself. One of the most important findings, however, was that ciliated cells appeared normal, there was no de-ciliation of these cells.

The Ames test is an evaluation of the mutagenic potential of a compound. In this study chitosan glutamate was incorporated into the dishes used for culturing various strains of *Salmonella*. The study also evaluated the potential of an S9 mitochondrial extract to metabolize the chitosan into a mutagenic compound. The results in *Table 11* show that chitosan glutamate induced no mutagenic effect up to a concentration of 5000 μg incorporated in each dish. The lack of mutagenicity was irrespective of the presence of an S9 mixture.

In vitro studies have shown that chitosan salts induce very little toxicity to cultured cells. In *Table 11*, the effect of a 24 hour incubation with chitosan salts on the mouse embryonal cell line 3T3 and the Chinese hamster lung cell line V79 were tested. Neither of these cell lines are considered cancerous or malignant.

Conclusions

The evaluation of ultrapure chitosan salts has shown these compounds to be well

Safety of PROTASAN™
Nasal irritancy study

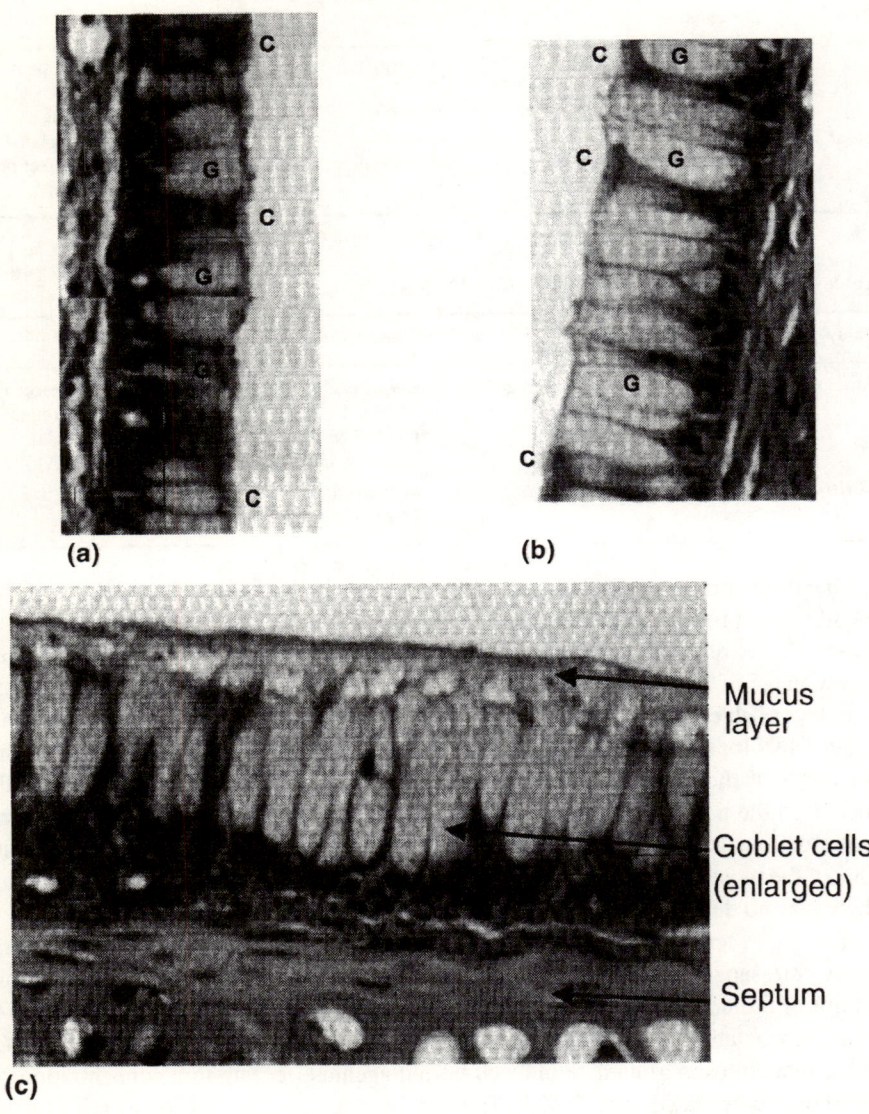

Figure 15. Micrographs showing the effect of chitosan glutamate on rat nasal epithelia. (a) control, (b) and (c) after 'high dose' of 1 mg/rat in right nostril, 3 times a day for a 7 day period. G: Goblet cells, C: Ciliated cells.

tolerated in safety and toxicology studies. These data are important for a further evaluation of the usefulness and applicability of chitosan in biomedical and pharmaceutical applications. In addition, safety studies of the types presented here are necessary for regulatory approval of the use of chitosan salts in humans.

Oral applications of chitosan have been previously reported by others and the safety of orally administered chitosan has been reviewed by Weiner (1992) and Hirano *et al.*

(1990). In work involving PROTASAN™ we have also shown that long-term (13 week) administration of chitosan glutamate had no deleterious effect in rats. Further toxicological evaluation in animals and man have shown that chitosan glutamate has no deleterious effects on the nasal mucosa nor on mucociliary transport (Aspden, 1997a,b). These findings are of importance for the development of chitosan in nasal drug delivery systems.

When chitosan was introduced as an industrial product in the early 1970's, the field of wound healing played a significant role in the initial commercial development of the biopolymer. For new pharmaceutical and biomedical applications of chitosan to be successful, studies like the ones presented here will be of importance. Moreover, regulatory issues, such as production process validation, quality control and product stability, will have to be addressed. In addition to characterization and functionality, the commercial manufacture of chitosan products for pharmaceutical use must also include product reproducibility and safety. PROTASAN™ salts are manufactured in accordance with GMP guidelines in order to ensure quality control and documentation for commercial grades of water-soluble chitosan salts.

References

ALLAN, G.G., ALTMAN, L.C., BENSINGER, R.E., GHOSH, D.K., HIRABAYASHI, Y., NEOGI, A.N. AND NEOGI, S. (1984). Biomedical applications of chitin and chitosan. In *Chitin, chitosan and related enzymes*. Ed. J.P. Zikakis, pp 119–133. New York: Academic Press.

ASPDEN, T.J., ILLUM, L. AND SKAUGRUD, Ø. (1997a). Chitosan as a nasal delivery system: evaluation of insulin absorption enhancement and effect on nasal membrane integrity using rat models. *European Journal of Pharmaceutical Science* 4, 23–31.

ASPDEN, T.J., MASON, J.D.T., JONES, N.S., LOWE, J., SKAUGRUD, Ø. AND ILLUM, L. (1997b). Chitosan as a nasal delivery system: The effect of chitosan solutions on *in vitro* and *in vivo* mucocilliary transport rates in human turbinates and volunteers. *Journal of Pharmaceutical Science* 86, 509–513.

DORNISH, M., HAGEN, A., HANSSON, E., PECHEUR, C., VERDIER, F. AND SKAUGRUD, Ø. (1997a). Safety of PROTASAN™: Ultrapure chitosan salts for biomedical and pharmaceutical use. In *Advances in Chitin Science*, vol. 2. Eds. A. Domard, G.A.F. Roberts and K.M. Vårum, pp 664–670. Lyon: Jacques Andre Publicher.

DORNISH, M., SKAUGRUD, Ø., ILLUM, L. AND DAVIS, S.S. (1997b). Nasal drug delivery with PROTASAN™. In *Advances in Chitin Science*, vol. 2. Eds. A. Domard, G.A.F. Roberts and K.M. Vårum, pp 694–697. Lyon: Jacques Andre Publicher.

GRASDALEN, H. (1983). High-field ^{1}H-n.m.r. spectroscopy of alginate: sequential structure and linkage conformations. *Carbohydrate Research* 118, 255–260.

HAGEN, A., SKJÅK-BRÆK, G. AND DORNISH, M. (1996). Pharmokinetics of sodium alginate in mice. *European Journal of Pharmaceutical Sciences* 4 (suppl.), S100.

HIRANO, S., SEINO, H., AKIYAMA, Y. AND NONAKA, I. (1990). Chitosan: A biocompatible material for oral and intravenous administrations. In *Progress in Biomedical Polymers*. Eds. C.G. Gebelein and R.L. Dunn, pp 283–290. New York: Plenum Press.

ILLUM, L., FARRAJ, N.F. AND DAVIS, S.S. (1994). Chitosan as a novel nasal delivery system for peptide drugs. *Pharmaceutics Research* 11, 1186–1189.

LI, Q., DUNN, E.T., GRANDMAISON, E.W. AND GOOSEN, M.F.A. (1992). Applications and properties of chitosan. *Journal of Bioactive Compatible Polymers* 7, 370–397.

LIM, F. AND SUN, A.M. (1980). Microencapsulated islets as a bioartificial endocrine pancreas. *Science* 210, 908–910.

NO, H.K. AND MEYERS, S.P. (1995). Preparation and characterization of chitin and chitosan – a review. *Journal of Aquatic Food Product Technology* 4, 27–52.

SKAUGRUD, Ø. AND SARGENT, G. (1990). Chitin and chitosan: Crustacean biopolymers with potential. In *Making Profits out of Seafood Wastes*. Ed. S. Keller, pp 61–69. Alaska: Alaska Sea Grant College Program AK-SG-.90–07.

SKAUGRUD, Ø. (1995). Drug delivery systems with alginate and chitosan. In *Excipients and Delivery Systems for Pharmaceutical Formulations*. Eds. D.R. Karsa and R.A. Stephenson, pp 96–107. Cambridge: Spec. Publ. Royal Soc. Chem. No 161.

SKJÅK-BRÆK, G. AND ESPEVIK, T. (1996). Application of alginate gels in biotechnology and biomedicine. *Carbohydrates in Europe* **14**, 19–25.

SMIDSRØD, O. AND DRAGET, K.I. (1996). Chemistry and physical properties of alginates. *Carbohydrates in Europe* **14**, 6–13.

SMIDSRØD, O. AND SKJÅK-BRÆK, G. (1990). Alginate as immobilization matrix for cells. *Trends in Biotechnology* **8**, 71–78.

VÅRUM, K.M., ANTHONSEN, M.W., GRASDALEN, H. AND SMIDSRØD, O. (1991). ^{13}C-N.m.r. studies of the acetylation sequences in partially *N*-deacetylated chitins (chitosans). *Carbohydrate Research* **217**, 19–27.

VÅRUM, K.M., ANTHONSEN, M.W., GRASDALEN, H. AND SMIDSRØD, O. (1991). Determination of the degree of *N*-acetylation and the distribution of *N*-acetyl groups in partially *N*-deacetylated chitins (chitosans) by high-field n.m.r. spectroscopy. *Carbohydrate Research* **211**, 17–23.

WEINER, M.L. (1992). An overview of the regulatory status and of the safety of chitin and chitosan as food and pharmaceutical ingredients. In *Advances in Chitin and Chitosan*. Eds. C.J. Brine, P.A. Sandford and J.P. Zikakis, pp 663–670. Essex: Elsevier Science Publishers Limited.

3
Biopolymer Mucoadhesives

STEPHEN E. HARDING[1]*, S.S. (BOB) DAVIS[2,3], MATTHEW P. DEACON [1,2] AND IMMO FIEBRIG[1,2]

[1]NCMH unit, University of Nottingham, School of Biological Sciences, Sutton Bonington, Leics., LE12 5RD, UK; [2]School of Pharmacy, University of Nottingham, University Park, Nottingham, NG7 2RD, UK and [3]DanBiosyst (UK) Ltd., Albert Einstein Centre, Highfields Science Park, Nottingham

Introduction

In this review we will consider how we can utilize the adhesive properties of certain types of biopolymer to increase the residence time of orally administered drugs as they pass through the stomach and small intestine. This utilization helps maximise the time window a drug spends near its site of optimum absorption. The review updates and builds on our earlier work (Fiebrig *et al.*, 1995a) which focussed more on the macroscopic aspects, and where an extensive table (Table 18.3) reviewed mucoadhesive performance; the reader is strongly recommended to cross-refer to that article. The current article will attempt to focus more on the molecular aspects of biopolymers interacting with mucus and its key macromolecular component – mucin. We will:

- Consider the principles of mucoadhesion and the strategy for the oral administration of drugs
- Consider absorption enhancement and the strategies used to delay gastrointestinal transit
- Look closely at the mucin substrate – on which there has been massive progress in our understanding of the molecular nature of this substance – and potential adhesive materials, particularly two types of polycationic polysaccharide and an unusual protein from the feet of mussels
- Consider the assay method: the 'macroscopic' mechanical/*in vivo* methods and the molecular mucin based methods
- Consider a '*Case*' study on the molecular hydrodynamics in some detail: namely mucoadhesive interactions with gastric mucin
- Conclude on the most likely candidate mucoadhesive(s) and how they could be constructed into delivery systems.

*To whom correspondence may be addressed.

Biotechnology and Genetic Engineering Reviews – Vol. 16, April 1999
0264–8725/99/15/41–86 $20.00 + $0.00 © Intercept Ltd, P.O. Box 716, Andover, Hampshire SP10 1YG, UK

Bio-adhesion and mucoadhesion

'*Adhesion*' is defined by the physicist as 'the molecular force of attraction in the area of contact between unlike bodies that acts to hold them together' (Webster, 1989) and '*bioadhesion*' is simply those adhesive phenomena where at least one of the adherents is biological (Kaelbe and Moacanin, 1977). The materials are attached to each other by interfacial forces for an 'extended' period of time (Gu, Robinson and Leung, 1988; Duchêne, Touchard and Peppas, 1988). Bioadhesive systems have been used for many years in the area of dentistry as denture adhesives (Wright, 1981; Hollingsbee and Timmins, 1990) with stoma based adhesives – such as karaya gum in Stomahesive® or synthetic polypectins (Winkler, 1986) – and also used for surgical applications such as the cyanoacrylates used as a 'surgical glue' (Wang, 1974, Harper and Ralston, 1983). Closely related to adhesion are film forming technologies such as the use of hydroxypropylmethyl cellulose for contact lens technology (see Silver *et al.*, 1994). Bacteria use polysaccharide adhesins to stick to surfaces, including the human gut. These adhesins are lectin-like, carbohydrate-binding molecules, expressed on the bacterial surface which bind specifically to sugar residues of mucins or other carbo-hydrates of the host cell surface (Beachey, 1980; Boedecker, 1984; Mergenhagen and Rosan, 1985; Hörstedt *et al.*, 1989). The term 'bioadhesion' has also been used to describe adhesive phenomena related to the ability of some non-biological macro-molecules and hydrocolloids to adhere to biological tissues for therapeutic purposes in medicine (Kaelbe and Moacanin, 1977; Peppas and Buri, 1985).

'*Mucoadhesion*' is a term used for a bioadhesive phenomenon where the biological substrate is a mucosal surface (Robinson, 1990), and concerns how advantage can be made for therapeutic purposes, especially for the delivery of drug formulations taken through the mouth (Fiebrig *et al.*, 1995a) and delivered to the eye (see Saettone *et al.*, 1989).

Oral administration of drugs

The local treatment of diseases can be unsatisfactory, because the drug may not stay at the site of action long enough for the desired effect (typically eye, mouth or vaginal cavity). Similarly systemic treatment via the oral route can be hampered because the drug may not stay at the optimal site of absorption long enough. Despite the advances in alternative delivery systems (e.g., nasal, pulmonary), oral administration of drugs is the most popular route of administration for both medical profession and patients alike. The administration of dosage forms via the mouth can easily be undertaken and is generally safe. By contrast, invasive methods (e.g. injection) usually require the assistance of trained personnel and the procedure always involves certain risks.

Oral drug administration begins with ingestion of the dosage form through the mouth (*Figure 1*) and from there it passes down the oesophagus and into the stomach. Little drug can be absorbed from the stomach, largely because of its relatively small surface area, particularly in the case of cationic drugs which will be mainly ionised in the acidic conditions of the stomach. Nevertheless, the stomach may represent a site for local treatment. The major site of drug absorption is the small intestine. Its large surface area (~100 m^2 in a healthy adult) makes it very efficient for the uptake of solutes (Bowman and Rand, 1980). Theoretically, drug absorption can occur along the

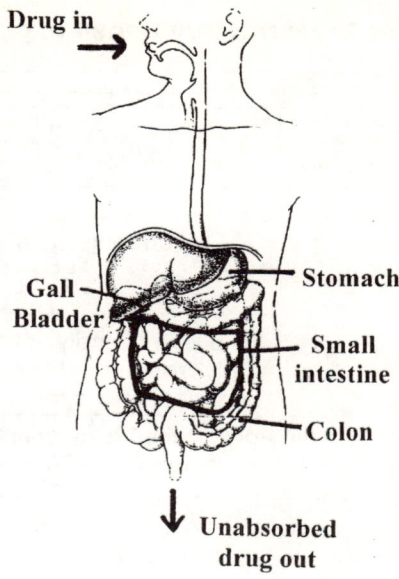

Drug in

Gall Bladder

Stomach

Small intestine

Colon

Unabsorbed drug out

Figure 1. Drug administration through the gastrointestinal tract.

entire length of the small intestine, however the majority of drugs are actually absorbed from the proximal small intestine (Booth, 1967). If the drug is poorly soluble, or is in the form of a controlled release dosage form, significant absorption of the drug may also occur in the large intestine (Davis, 1989). The limited surface area of the large intestine can be compensated by a long transit time. More recent studies with a once-a-day preparation (e.g. theophylline) have shown that therapeutic drug levels can be maintained for periods of up to 24 h (Gruber *et al.*, 1987) even though these systems are expected to have emptied from a fasted stomach, passed through the small intestine and have arrived at the ileocaecal junction after 4 h (Davis, 1985). Oral drug delivery ceases eventually with the faecal excretion of any unabsorbed drug.

Absorption enhancement in oral drug delivery systems

Too rapid a transit, or low residence time in the stomach and small intestine is just one of three major problems in oral delivery. Low appearance of the drug in the systemic circulation (bioavailability) can be due to *1.* rapid transit of the drug-containing delivery system past the ideal absorption site, *2.* rapid degradation of the drug in the gastrointestinal tract once it has been released (this can be serious for peptide drugs) and, *3.* low transmucosal permeability of the drug due to its size, ionisation, solubility or other characteristics of the drug molecule. Although this review concerns itself with addressing item *1.*, it turns out that the one of the best potential mucoadhesive materials (the polycationic polysaccharide *chitosan*) can also assist with solving the permeability problem, *3.*

A satisfactory resolution of problem *1.* will also be beneficial for polar drugs such as hydrochlorothiazide, which are only poorly absorbed from selected regions of the small intestine and whose bioavailability is believed to be dependent on the residence

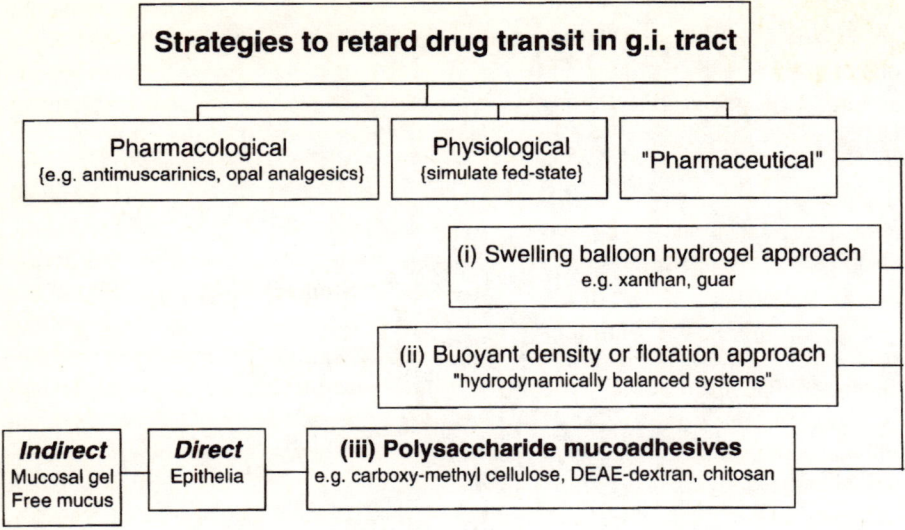

Figure 2.   Strategies for retarding gastrointestinal drug transit.

time of the dosage form at or *upstream* of its small intestinal absorption window (Beermann *et al.,* 1976; Lynch *et al.,* 1987). Resolution of problem *1.* will be particularly important in the case of controlled release drug delivery systems (DDS), designed to deliver drugs over extended periods of time (e.g. 12–24 h). Once in the colon, these DDS may be delivering a proportion of drug to a non-optimal site for absorption (Davis, 1985). An ideal oral sustained release dosage form should be comparable to an intravenous infusion, which continuously supplies the amount of drug needed to maintain constant plasma levels once a steady state is reached (Förster and Lippold, 1982)

Strategies for retarding drug transit through the gastrointestinal tract

The strategies for delaying drug transit through the gastrointestinal (g.i.) tract fall into one of three categories (*Figure 2*): *pharmacological, physiological* and *pharmaceutical*; the first two being less attractive than the last because of toxicity problems. (For another recent review of retentive systems see Hwang *et al.* (1998)). A pharmacological approach involves the co-administration or incorporation of a drug into the dosage form. This drug delays gastrointestinal emptying. Examples include antimuscarinics, e.g. propantheline, which are relaxants of the smooth muscle (Beermann and Groschinsky-Grind, 1978; Manninen *et al.,* 1973) or a drug that changes motility, e.g. opiate analgesics or derivatives such as loperamide (Minami and McCallum, 1984). However, the potential side effects that may arise from such treatments on a routine basis would not be acceptable for regulatory approval.

A physiological approach is the use of natural materials or fat derivatives such as triethanolamine myristate (Gröning and Heun, 1984, 1989), which stimulate the duodenal or jejunal receptors to slow gastric emptying. The use of large amounts of a 'volume filling' polymer such as polycarbophil (Harris *et al.,* 1990a,b) can also cause a 'slowing down' response in terms of gastrointestinal transit.

Pharmacological and physiological approaches thus set out to delay gastrointestinal transit by modification of the rate of gastric emptying using 'passage-delaying agents'. By contrast, the *pharmaceutical strategies* attempt to achieve the same objective by actually retaining the dosage form at or upstream of its absorption site for as long as possible. This is achieved by a particular physical or physicochemical characteristic. Mucoadhesion is one strategy (*Figure 2*).

(*i*) *Swelling balloon hydrogel.* If large enough, the formulation will not be expelled from the fasted stomach even when the pyloric sphincter is in its non-contracted state. The size of such systems has to increase after ingestion to an extent that gastric emptying is totally inhibited (Moës, 1993). The size-related retention of a dosage form in the stomach has been studied with various systems to include systems such as swelling balloon hydrogels (Park and Park, 1987) or unfolding stratified medicated polymer sheets (BE Patent No. 867, 692) or non-erodible or erodible tetrahedron shaped devices (Cargill *et al.,* 1988, 1989). These have never passed beyond the experimental stage and clinical data are unavailable. In any case these gastric retention devices may not be safe. The hazard of lodging in the oesophagus (Kikendall *et al.,* 1983; Al-Dujaili *et al.,* 1983; Wilson, 1990) or permanent retention in the stomach with cumulative effects (Brahams, 1984; Vere, 1984) could lead to life-threatening problems.

Another approach uses dosage forms of moderately high density, based on the premise that high density formulations remain in the stomach longer than conventional formulations, since they would be localised in the lower part of the antrum provided the density exceeds that of the normal stomach contents, i.e. > 1.4 g/ml (Bechgaard and Ladefoged, 1978). The effectiveness of this approach has not been confirmed on a broad basis and the evidence remains controversial (Moës, 1993).

(*ii*) *Buoyant density/flotation approach.* This approach uses buoyant dosage forms which float on the gastric contents as a result of their relatively low density. Floating dosage forms have been discussed extensively by Moës (1993): The first floating dosage forms (F forms) (Sheth and Tossounian, 1984), also called 'hydrodynamically balanced systems' (HBS), were able to maintain their low density while a polymer hydrated and built a gelled barrier at the outer surface. Hoffmann-LaRoche produced patents for floating drug delivery systems and *in vivo* studies on diazepam HBS capsules such as Valium® CR and Valrelease® and the L-dopa plus benserazide containing formulation Madopar® HBS (Prolopa® HBS). Moës (1993) has attempted to clarify the conflicting views on the gastric retention capabilities of floating systems resulting from a number of *in vivo* trials by different authors (Müller-Lissner and Blum, 1981; Davis *et al.,* 1986; Timmermans and Moës, 1990; Timmermans, 1991; Kaus, 1987; Sangekar *et al.,* 1987; Lippold and Günther, 1991).

(*iii*) *Polymer mucoadhesion.* This involves attachment or encapsulation of the drug with a polymer which interacts with either the mucosal epithelia/glycocalyx lining of the gastrointestinal tract (this is called 'direct' mucoadhesion) or mucous surfaces (the gel and the sloughed mucus in the lumen) lining the gastrointestinal tract, hence providing a macromolecular 'brake' to the movement of the drug. A good challenge for mucoadhesion is the delivery of orally administered polar drugs (and possibly peptides and proteins). These materials have low absorption characteristics (and for peptides and proteins have a stability problem; enzymatic degradation and biotransformation). A mucoadhesive alternative route to parenteral administration would be highly desirable (Junginger and Verhoef, 1992).

If the polymer carrier can access and interact *directly* with the surface mucosal epithelium or *glycocalyx*, the decrease in diffusion path from the oral DDS to the absorbing biological membrane could be an additional advantage for improving absorption, particularly in intestinal delivery of peptide drugs, at the same time minimizing dilution and possible degradation in the luminal fluids (Hayton, 1980). The further addition of penetration enhancers to an adhering dosage form could enable alteration of membrane permeability and inclusion of specific enzyme inhibitors could prevent early degradation of the peptide (Wearly, 1991; Junginger and Verhoef, 1992) and consequently increase bioavailability. However, the epithelium may not be accessible: instead the *indirect route* of interaction with the ~40–450 μm thick mucosal surface/gel lining the gastrointestinal tract provides the most likely strategy (*Figure 3*). It is also worth noting that mucus is not a major barrier to absorption.

The adhesion of gastrointestinal retention dosage forms to the mucosa has been studied for over a decade, mainly by *in vitro* or *ex vivo* test with few *in situ* or *in vivo* studies and even fewer trials in man. Despite the fact that bioadhesion, or more specifically mucoadhesion, has led to some success in drug delivery for ocular, buccal, nasal, vaginal and cervical applications (Chen and Cyr, 1970; Schor *et al.*, 1983; Nagai *et al.*, 1984; Nagai, 1986; Duchêne *et al.*, 1988; Greaves and Wilson, 1993; Smart, 1993; Bouckaert, *et al.*, 1994), gastrointestinal mucoadhesive drug delivery systems have yet to be succesfully established (see, e.g., Helliwell, 1993; Fiebrig *et al.*, 1995a).

The mucosal substrate: mucus and mucin

The last ten years has also seen a tremendous advance in our understanding of the structure and molecular biology of mucus, and in particular its major macromolecular component, *mucin*. Mucus is a viscoelastic substance with a characteristic stickiness and *Spinnbarkeit* (the ability to stretch into strands). By weight mucus is mostly water (95%–99.5%) and exists in gel or a viscous solution form. Its most important polymeric, gel-forming component is the mucus glycoprotein or mucin (0.5%–5%) (Harding, 1989; Carlstedt and Sheehan, 1988; Neutra and Forstner, 1987; Gibbons, 1972). The adherent mucus layer in the gastrointestinal tract is secreted by specialised cells. They are surface epithelial cells found mostly in the stomach but also in other parts of the gut and the goblet cells of the small and large intestine, as well as Brunner's glands in the duodenum (Neutra and Forstner, 1987; Allen, 1989; Ito, 1981). Unlike other gastrointestinal secretions, mucus adheres to the mucosal epithelial surfaces as a water insoluble gel (*Figure 3a*) until degradation and erosion takes place (Allen, 1989) leaving a mucin solution or slough on the lumen side of the gel.

For monitoring the thickness of the mucin a novel method has been developed which for the first time has enabled the preservation and visualization of the full thickness of the adherent gastric mucus layer and the underlying mucosa (Jordan *et al*, 1998): this involves a modified periodic acid Schiff/Alcian Blue staining technique for

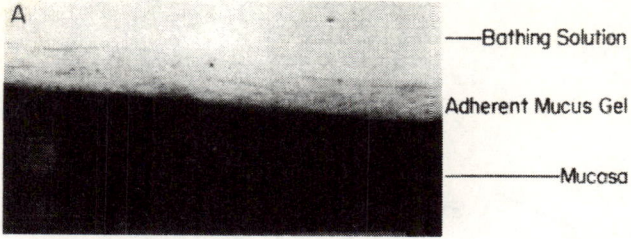

Figure 3a.   Adherent gastric mucus viewed under bright field microscopy on a transverse section (1.6 mm thick) of rat gastric mucosa. Three distinct phases are seen: mucosa, mucus gel layer and the 'bathing' solution of free mucus. Reprinted with permission from Allen (1989) and Kerss *et al.* (1982).

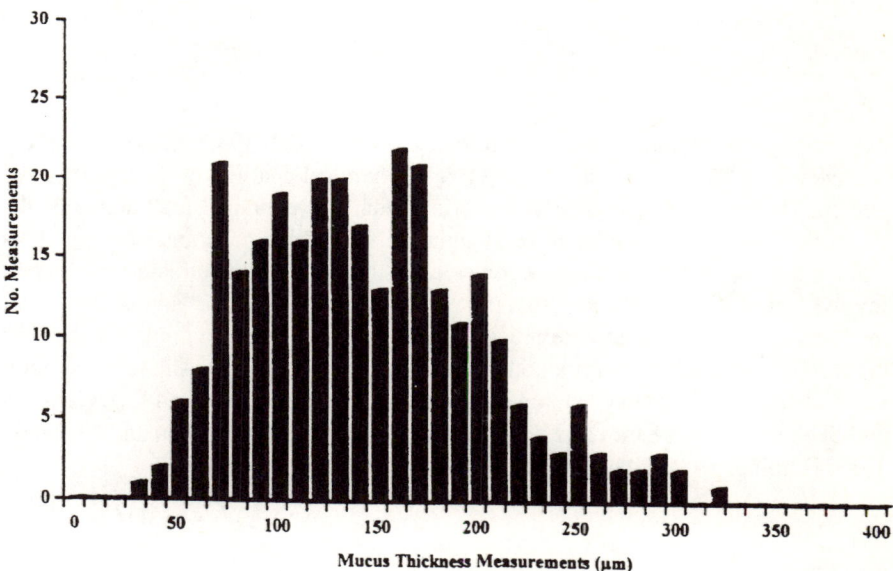

Figure 3b.   Histogram of the mucus thickness from the antrum region of human stomachs. Overall mean (from 300 measurements) = 144 µm, standard deviation = 29 µm and the median = 160 µm. The Schiff/Alcian Blue statining technique on cryostat sections of gastric mucosa was used. Reprinted with permission from Jordan *et al.* (1998).

use on cryostat sections of gastric mucosa, and an example of the distribution of mucus layer thicknesses obtained for human gastric antrum mucus is shown in *Figure 3b*.

It is believed that the adherent mucus layer plays a major role in protection of the delicate underlying epithelium against the various endogenous and exogenous insults, such as acidic pH (providing an 'unstirred boundary layer'), digestive enzymes (pepsin), pathogens (bacteria) and abrasion, while the soluble mucus may play an important role in acting as a lubricant for ingested food. The requirement for such a protective adherent gel layer becomes obvious since from a physiological point of view the luminal side of the gastrointestinal tract can still be considered as the outer side of the body. These and other aspects regarding the function of mucus have been

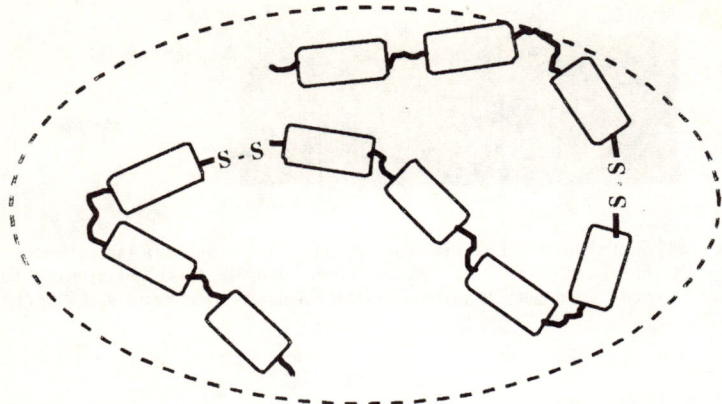

Figure 3c.   Structural hierarchy of a mucin from the g.i. tract: a colonic mucin of the MUC2 type (from Jumel *et al.*, 1997). The general structural scheme of mucins from a variety of sources is similar. See text.

extensively described by various authors e.g. Allen (1981, 1983, 1989), Silberberg and Meyer (1982) and Bhaskar *et al.* (1992). Chemical analysis of the mucus gives evidence of a rather heterogeneous material which also contains small amounts of a variety of proteins, lipids, bacteria, sloughed-off epithelial cells and in some cases nucleic acids (Creeth, 1978). It becomes clear that mucoadhesion is a process that involves large amounts of water, or more vividly, it could be seen as 'adhesion to water in a semisolid form' where the mucins play a key role in maintaining the gel-like properties of the substrate for a potential drug delivery platform. The mucins themselves display considerable heterogeneity that has been well described (e.g. Carlstedt and Sheehan, 1984; Neutra and Forstner, 1987; Allen, 1989; Sheehan and Carlstedt, 1989; Harding, 1984, 1989).

Mucin

Mucins are large molecules with molecular weights ranging from 0.5×10^6 to over 20×10^6 g/mol. They contain large amounts of carbohydrate (for gastrointestinal mucins 70%–80% carbohydrate, 12%–25% protein and up to ~5% ester sulphate). Undegraded mucins are made up of multiples of a basic unit (M ~400,000–500,000), linked together into the macroscopic mucin. Although originally thought to be arranged in a windmill type of structure (Allen, 1978), this model was shown to be incorrect: Instead the molecule is linked into linear arrays as shown by Creeth, Harding and coworkers (Harding *et al.*, 1983a,b) and by Carlstedt, Sheehan and coworkers (Carlstedt and Sheehan, 1984). Although linear, the mucin molecule in solution is loosely/randomly coiled into a spheroidal, highly swollen (by water imbibement) domain as confirmed by molecular hydrodynamics. Examples from electron microscopy clearly showing both these features are presented in *Figure 4*: *Figure 4a*, the linear 'secondary structure' and *Figure 4b* the overall spheroidal domain. The total architecture seems to be very similar for mucins from a variety of

(i)   **(ii)**

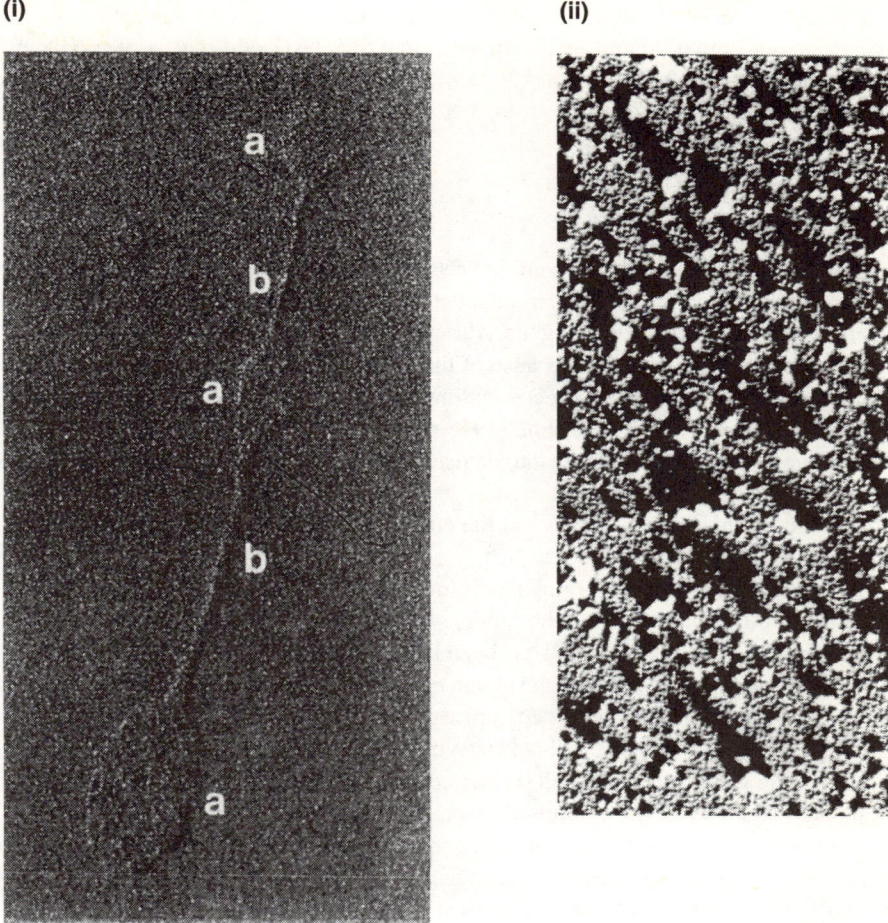

Figure 4. Electron microscopy of a bronchial mucins of the 'single subunit' type (molecular weight (~2–2.5 × 10⁶ Da) prepared by air drying onto mica (i) or critical point drying (ii) and then linear shadowing with platinum. With (a) the mucins have been strung out by the large shearing and surface tension forces experienced during air drying onto mica, leaving clear flattened (a) and linear (b) regions visible. The flattened regions are likely regions of high glycosylation (Harding *et al.*, 1983). (i) and images similar to these clearly demonstrate a linear assembly for mucin molecules and not 'windmill' or 'star-like' structures as had been proposed earlier. With (ii) the critical point drying method has preserved the overall spheroidal domain (dark spherical regions) occupied by the randomly coiled molecule in solution (Hallet *et al.*, 1984). Magnification: (i) ×161000, (ii) ×75000.

sources (for example gastric, respiratory or cervical). The basic units are linked together by regions of low or no glycosylation (*Figure 4*) which are subject to trypsin digestion: the ~400–500 kDa digestion products are thus commonly referred to as 'T-domains' (see Sheehan and Carlstedt, 1989). Every third or fourth T-domain is linked by a disulphide bridge; itself susceptible to reductive disruption by thiols. The thiol reduction products (of molecular weight between 1.5 and 2.5 MDa) are commonly referred to as 'subunits'. One of the most recent examples of such architecture in a mucin is that of colonic mucin (*Figure 3c*) (Jumel *et al.*, 1997). Even mucins produced externally by cell-lines appear to adopt this architecture, although they appear to be only up to one or two subunits in length (mol. wt < 5 MDa) (Dodd *et al.*, 1998). Mucins which are different are the submaxillary mucins, with a lower carbohydrate content and different structure, but these are not so relevant in terms of gastrointestinal adhesion strategies.

PRIMARY STRUCTURE OF MUCINS

These advances in our understanding of the gross structure of mucins have been matched by similar advances that have occurred in the last ten years in our understanding of the primary structure of mucins. Although direct sequencing of the protein chain has been virtually impossible because of the insolubility of mucins stripped of their carbohydrate, eight different genes coding for mucin production have now been sequenced (see e.g., Hounsell *et al.*, 1997 and references therein). These are called 'MUC' genes and the ones known to date and the sources of mucin they code for are given in *Table 1*. It is seen that the key gene products as far as mucoadhesion are concerned are MUC2 and MUC3 in the small intestine and colon, and MUC5AC, MUC5B and MUC6 from the stomach.

The protein sequences emerging from elucidating these genes confirm the presence of large amounts of serine and threonine, sites for the O-glycosylation, and also the large amounts of proline – which has been known for years (Harding *et al.*, 1983a,b) to be assisting with the coiling of the mucin molecule. This knowledge of the genes has also revealed the concept of a tandem repeat of sequences of amino acid throughout the linear polypeptide backbone. What is also interesting has been the identification that certain of the selectin group of cell surface adhesion molecules have mucin-like repeat sequences (such as CD43) (Hounsell *et al.*, 1997).

The O-linked carbohydrate chains may contain up to five different monosaccharides; namely D-galactose, L-fucose, N-acetylglucosamine, N-acetylgalactosamine and sialic acid (*Figure 5*). As multi-branched oligosaccharides they are covalently attached via O-glycosidic linkages from N-acetylgalactosamine to serine and threonine residues of the protein core. The absence of uronic acid and only trace amounts of mannose (<1%) distinguish mucin glycoproteins from the proteoglycans of connective tissue and serum glycoproteins, respectively. Sialic acid residues, which belong to a family of acidic sugars (Schauer, 1992) (in gastrointestinal mucins usually either N-acetyl or

Table 1.   Characterized MUC genes (from Hounsell *et al.*, 1997)

MUC gene	Location
MUC 1	Breast and colon cell surface episialin
MUC 2	Colon and small intestine goblet cell secretion
MUC 3	Intestinal tissue
MUC 4	Tracheobronchial tract
MUC 5AC	Respiratory tract and goblet cell secretion
MUC 5B	Submaxillary gland secretion
MUC 6	Gastric gland secretion
MUC 7	Salivary gland secretion
MUC 8	Respiratory tract

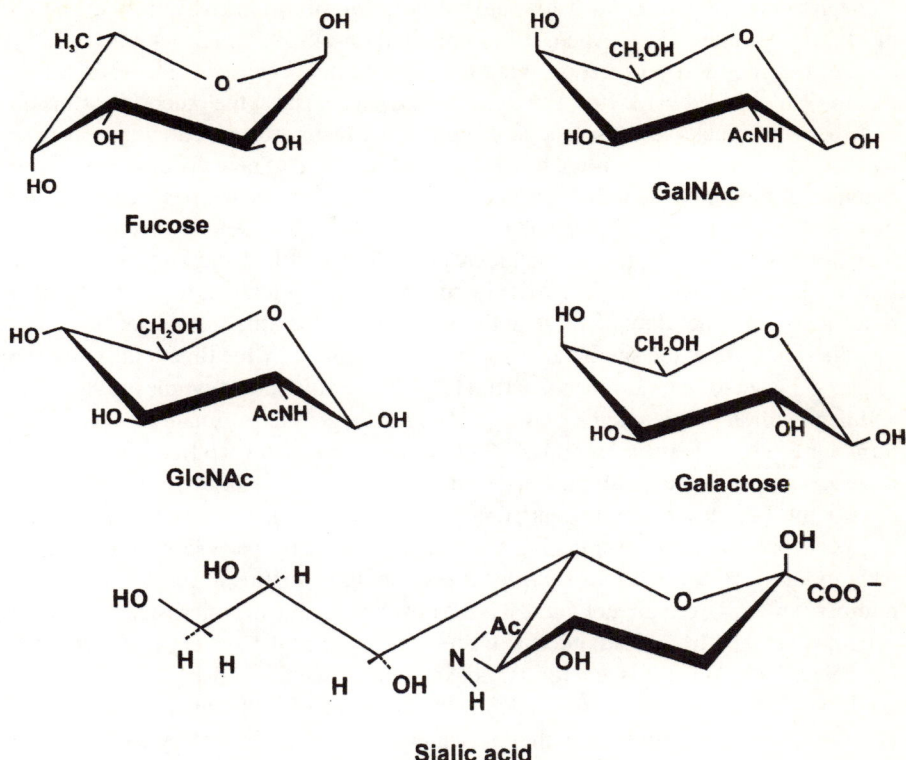

Figure 5.   The principle sugars of gastrointestinal (also bronchial and cervical) mucins. The key ones, in terms of possible interaction sites for mucoadhesives are sialic acid (–COO⁻ group for electrostatic interaction), N-acetyl glucosamine and N-acetyl galactosamine (–COCH$_3$ group, with the carbonyl for H-bonding and the hydrophobic methyl residue, and fucose (–CH$_3$ group).

N-glycollyl-neuraminic acid), are usually in a terminal position on the carbohydrate chain, whereas ester sulphate residues occur in a more internal position, e.g. as N-acetylglucosamine-6-sulphate in pig gastric mucus (Allen, 1978; Slomiany and Meyer, 1972). They both contribute in giving the molecule a net negative charge, thought to be of importance in interactions with polycationic materials (Lehr *et al.*, 1992b; Fiebrig *et al.*, 1994a, b). Other potential residues for mucoadhesive interaction are the carbonyl (hydrogen bonding) and methyl (hydrophobic bonding) groups on the N-acetyl residues (GalNAc, GlcNAc, sialic acid) and another methyl group in fucose.

Is mucus an appropriate target?

There are three physiological aspects which remain critical for the concept of gastrointestinal mucoadhesion: (i) turnover of the adherent mucus layer, (ii) interactions of the formulation with soluble, i.e. non-adherent mucus prior to adhesion and (iii) gastrointestinal motility.

TURNOVER OF THE ADHERENT MUCUS LAYER

Adherent mucus is continuously lost into the gastrointestinal lumen by proteolysis and mechanical sloughing (e.g. Allen, 1981; Allen and Caroll, 1985). The latter, caused by the ingested food and its digestion, is thought to be the major cause of the loss of gastric mucin (Waldron-Edward, 1977). A dynamic balance exists at the mucosal surface *in vivo*, between mucus secretion and mucus erosion. Mucus erosion, either by pepsin or by abrasion, must be replenished by the mucosal secretion of new material in order to maintain a protective function (Allen *et al.*, 1993). The difficulties in measuring mucus secretions *in vivo* have been outlined by Allen (1989). Studies on the turnover time of intestinal mucus gel layer in the rat *in situ* loop (Poelma and Tukker, 1987) by Lehr *et al.* (1991) have attempted to shed some light on the limitations to gastrointestinal mucoadhesion. The view of these authors is that the maximal residence time of a bioadhesive DDS at the site of adhesion would be limited by the time it takes for the mucus gel layer to be renewed as determined by the steady state of synthesis, secretion and degradation of the mucins (Allen, 1981). Although their estimate for the mucus turnover time is relatively crude (47– 270 min), it is interesting to find that this time scale is similar to the mean residence time found for mucoadhesive microspheres (94±18 min) in earlier experiments using the same animal model (rat). Furthermore it has been observed that stimulating the mucus output, by perfusion with 10 mM sodium taurocholate, led to a significant shortening of the mean residence time of microspheres. Of even greater interest is the observation that the microspheres did not become detached from dead mucosal tissue *in vitro* when the system was stirred for more than 18 h. This leads to a further consideration; that of choosing an appropriate model system. This will be discussed in more detail below. Although mucus turnover in an *in situ* isolated gut loop in the rat (which has undergone surgery and has been removed from its normal function) may be different from mucus turnover in healthy humans or patients, this physiological factor will limit potential adhesion to the adherent mucus in the gastrointestinal tract.

COMPETITIVE INHIBITORY INTERACTIONS WITH SOLUBLE MUCIN

Any formulation entering the gastrointestinal tract interacting with the mucus gel is likely also to interact with soluble mucins of the 'slough' or luminal material. This is an unavoidable complication which will reduce the efficiency of any adhesive system: that is any adhesive system targetted for groups on the mucus gel will also have the possibility to interact with the soluble mucus. Even if the epithelial cells are targetted, a 'competitive inhibition' for the mucoadhesive will recur as has been shown recently by Lehr *et al.* (1992d): Those authors used tomato lectin, a material that specifically binds to isolated pig enterocytes and monolayers of human Caco-2 cell cultures, and was proposed as a favourable candidate for specific bioadhesion to epithelial cells of the gastrointestinal tract. However, binding also occurred with crude pig gastric mucus. Thus, no mucoadhesive strategy can be 100% efficient! Other competitive inhibitors for mucoadhesion may also derive from other soluble components within the gastrointestinal tract, such as bile salts (Anderson, 1991).

GASTROINTESTINAL MOTILITY

Gastrointestinal motility patterns and in particular the so called 'housekeeper wave' which involves strong gastrointestinal contractions, serves as a cleaning mechanism to clear all indigestible materials, including non-disintegrating dosage forms, from the stomach or proximal intestine (Code and Marlett, 1975; Grundy, 1985; Leung and Robinson, 1988). Thus, a good oral mucoadhesive drug delivery system also needs to resist the cleaning action of the 'housekeeper wave' and remain in the stomach or proximal small intestine.

TARGET FOR MUCOADHESIVES

Although the most appropriate target phase that would appear to give the best efficiency for a mucoadhesive system (if it were accessible) is the underlying mucosal glycocalyx, the target phase (in the stomach, small intestine and colon) most relevant to the concept of mucoadhesion is the water insoluble mucus gel lining the mucosa of the gastrointestinal tract. This mucus layer has a variable thickness, 50–450 µm, in man and about half that in the rat (Allen, 1978; Kerss *et al.*, 1982), with regional differences: For example (A. Allen, personal communication) mucus thickness in the stomach is variable but between a mean of 100–150 µm for the firm layer of adherent gel and another 100 µm of viscous mobile mucus on top of that under unstirred conditions. In the colon the adherent gel has a mean thickness ~65 µm with something in the region of another 700 µm mobile viscous mucus that can be removed by suction. An important point is that in both cases the adherent gel barrier is continuous.

A variety of groups on the sugar residues on mucins provide potential sites for interaction of either an electrostatic, hydrogen bond or hydrophobic nature. The next question is: which is the appropriate mucoadhesive?

The mucoadhesive

The most important requirement of a mucoadhesive is that it must be *non-toxic* with no undesirable physiological or pharmacological actions, and should *not be expensive*. To this end, biopolymers, and in particular *food grade polysaccharides* are particularly attractive candidates (see Tombs and Harding, 1998). Other important criteria are that the mucoadhesive should have *good wettability* (and *spreading* ability) and high drug loading and a suitable unloading capacity. The following molecular properties are important considerations: charge, hydrodgen-bonding, hydrophobicity, flexibility (ability to overcome steric hindrance problems) and molecular weight/ molecular weight distribution. The following molecular environmental factors are important: solubility, pH, ionic strength, presence of other salts (e.g. bile) and other macromolecules (antibodies, enzymes, polysaccharide etc.).

For bioadhesion to occur, an intimate contact between the adhesive and the substrate (mucus) is a prerequisite. Factors like good wettability as well as hydration are important (Huntsberger, 1967; Chen and Cyr, 1970; Peppas and Buri, 1985). During the establishment of the adhesive bond the total surface energy between the two materials is diminished, eliminating two free surfaces and creating a new interface. This first step is believed to be followed by physical or mechanical bond formation obtained by deposition and inclusion of the adhesive material in the crevices of the mucus and chain entanglement between polymer chains of both phases (also referred to as inter-diffusion) (Boddé, 1990; Jabbari *et al.*, 1993). Lehr *et al.* (1992c) have used electron microscopy in an attempt to visualize intermixing between a polyacrylic acid derivative (polycarbophil) and mucus. They were unable to observe intermixing in the micron range but did not exclude this phenomenon for the nanometre range. Sufficient chain flexibility is required to form secondary chemical bonds such as van der Waals forces as well as hydrogen bonding (Leung and Robinson, 1988; Duchêne *et al.*, 1988). The formation of primary (covalent) chemical bonds is important in hard tissue adhesion in orthopaedics and dentistry. However, for mucoadhesion, chemical reactions of this type have not been considered so far, since a long term attachment is not required (Peppas and Buri, 1985).

POLYANIONIC AND NEUTRAL POLYMERS

Polymers with hydroxyl or carboxyl groups on their surface had been earlier claimed as the most desirable candidates for bioadhesion, rather than polymers with other functional groups or cationic moieties (Peppas and Buri, 1985). The synthetic polyacrylic acid derivatives known as polycarbophils (Carbopol® EX-55) and carbomer (Carbopol® 934) have to date been by far the most studied mucoadhesive polymers (Table 18.3 of Fiebrig *et al.*, 1995a). Both materials are polyanionic and interaction with mucus has largely been attributed to chain entanglement of the polymer chains with mucin as a result of swelling of the polymer in water and hydrogen bonding due to the carboxyl groups being in their unionised state at low pH (Robinson *et al.*, 1987; Leung and Robinson, 1988; Ponchel *et al.*, 1987a,b; Jabbari *et al.*, 1993). Poly-carbophil is described as a water insoluble but swellable polymer of polyacrylic acid crosslinked with divinylglycol and used clinically in the treatment of diarrhoea and as a bulk laxative. Carbomer is a water soluble polymer of acrylic acid loosely crosslinked with allylsucrose. There have also been a wide range of polyanionic polysaccharides as possible biopolymer alternatives, such as alginate, pectin, carrageenan, xanthan and carboxy-methyl cellulose, but macroscopic (Lehr *et al.*, 1992b) and molecular studies (Anderson, 1991; Fiebrig, 1996) have yielded little or no mucoadhesion for these substances, possibly because both mucoadhesive and the mucin are polyanionic: the findings for polycarbophil are therefore rather surprising.

POLYCATIONS

According to Anderson *et al.* (1989), Anderson (1991) and later Lehr *et al.* (1992b), the need for hydrogen-bonding capabilities and negative charge in bioadhesive materials *should not be generalized*. These workers suggested that polycationic polymers might interact with the anionic sites on the mucins more favourably due to their opposite charges providing additional molecular attraction forces. For example, interactions between charged polymeric molecules have been employed in colloidal titration (Terayama, 1952; Senju, 1969). The method is based on the principle that positively charged macromolecules will react with negatively charged macro-molecules. The neutralisation reaction will proceed stoichiometrically, allowing an estimation of either material if a standard colloid solution is used. Katayama *et al.* (1978) used the method for the titration of heparin using polydiallyldimethyl ammonium chloride as a standard polycation. Van Damme *et al.* (1992) measured the negative charge content in cartilage using polydiallyldimethyl ammonium chloride as well. Interactions between alginates and pectins with cationic polypeptides such as poly(L-lysine) and poly(Lys-Lys-Ala) have been studied using circular dichroism (Bystricky *et al.*, 1990). Differences in interaction efficiency between the polymers were attributed to differences in conformational flexibility of the polyanionic chains in solution. Takahashi *et al.* (1990) studied the characteristics of polyion complexes of chitosan with sodium alginate and sodium polyacrylate using viscometry and Fourier transform infra-red spectroscopy (FT-IR). They found that chitosan and alginate reacted with a defined binding ratio which was found to be relatively constant in media of various pH values. In contrast, for polyacrylate–chitosan interactions the unit molecular binding ratio was greatly affected by the pH. (*n.b.* chitosans are generally poorly soluble above a pH ~6).

CHITOSANS

Chitosan appears to be an ideal candidate as a mucoadhesive polycationic polymer – it is produced on a large scale (Jeuniaux *et al.*, 1989; Alimuniar and Zainuddin, 1992). Although chitosan has not yet received regulatory approval by the Food and Drug Administration (FDA) for pharmaceutical use, chitosan containing material obtained from the treatment of the waste streams of food processing plants may be used as livestock feed in the USA so long as the level of chitosan does not exceed 0.1% (Weiner, 1992). It is known to interact with molecules containing N-acetyl-glucosamine, such as lysozyme (Cölfen *et al.*, 1996). Its properties are quite different from polyanionic chitin derivatives, such as carboxy-methyl chitin (Korneeva *et al.*, 1996).

Chitosan (*Figure 6*) has been approved as a food additive in Japan since 1983 (and also apparently in some European countries) and has been placed on the '*Japanese Natural Additive List*'. It is used as a thickener and stabilizer (Weiner, 1992). It is a food ingredient in some dietary cookies and noodles from Hihon Kayaku Inc. and Tanami Foods Inc. as well as in vinegars of Nakano Inc., making use of its hypocholesterolaemic properties (Hirano, 1989). The food industry has also exploited the chelating properties of chitosan for the clarification of beverages such as apple and carrot juices (Imeri and Knorr, 1988; Soto Peralta *et al.*, 1989).

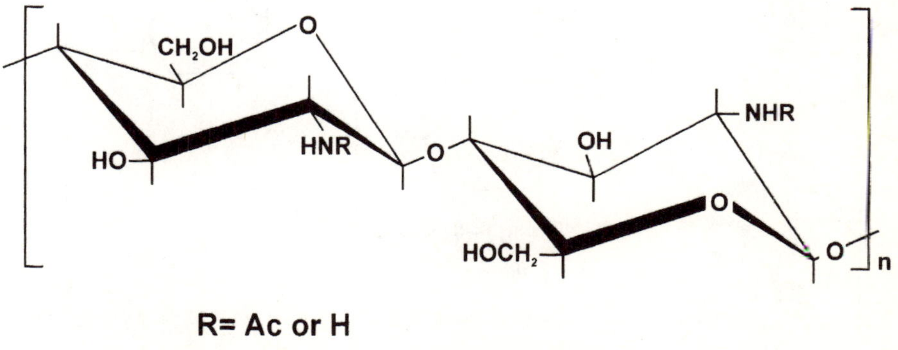

R= Ac or H

Figure 6. Chitosan. Pure poly-N-acetyl β-D(1→4) glucosamine is chitin and is insoluble. In chitosan, a proportion of the N-acetyl groups are deacetylated leaving a positive charge except at high pH (>5.5, where the molecule is again insoluble).

The lack of acute oral toxicity of chitosan has been supported by experiments in mice (Arai *et al.*, 1968) who determined an LD50 of > 10 g/kg. However the literature lacks adequate scientific studies on long term and widespread human exposure through food and pharmaceutical products (McCurdy, 1992).

Chitosan is a derivative of chitin; the insoluble structural exoskeletal polysaccharide of the shells of crabs and lobsters and can be harvested very cheaply (see Tombs and Harding, 1998); the chief producers are Norway, Japan, China and Russia. Like cellulose it is a $\beta(1\rightarrow4)$-D-glucan. Unlike cellulose the residue on the number 2 carbon atom in the ring is N-acetylated (*Figure 6*). In native chitin these residues are fully acetylated. However, after extraction the chitin molecule can be deacetylated to varying degrees to give a polycationic molecule. The degree of acetylation is represented by the parameter F_A, with $F_A = 1$ (fully acetylated) corresponding to pure chitin and $F_A = 0$ to fully deacetylated chitosan.

Variations in molecular weight and degree of deacetylation together with the ability to form gels and films allow flexibility in formulation design (Acatürk, 1989; Miyayaki *et al.*, 1990; Errington *et al.*, 1993).

DEAE DEXTRAN

DEAE-dextran (*Figure 7*) is a polycationic derivative of dextran, obtained by reaction with 2-chlorotriethyl-amin-hydrochloride or chloroethyl-diethyl-aminochloride (see

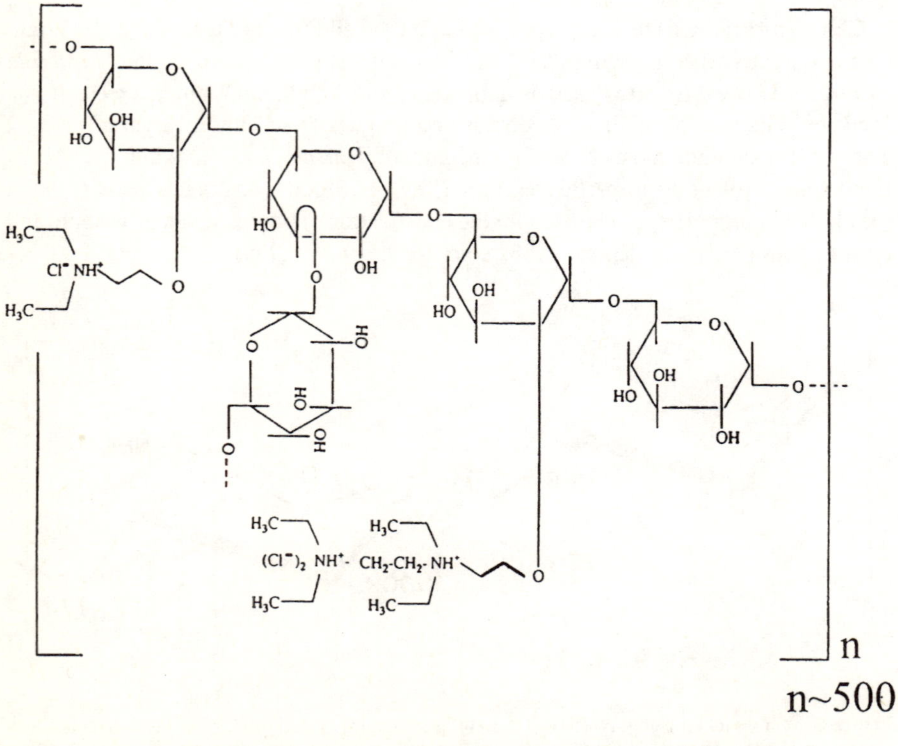

$$n\sim500$$

Figure 7.   DEAE-dextran. DEAE: di-ethylaminoethyl. From Soldani *et al.* (1987).

Soldani *et al.*, 1987). Dextran itself is an α(1→6) linked bacterial polysaccharide from *Leuconostoc mesenteroides*, with many branches of either an α(1→2), α(1→3) or α(1→4) type. In most cases the length of the side chains is short and branched residues vary between 5 and 33 per cent. Partial hydrolysis and subsequent fractionation leads to polysaccharides of a particular desired range of molecular weight. DEAE dextran is used as a weak ion-exchange material for ion-exchange columns.

MUSSEL GLUE PROTEIN

The protein *mefp-1* is one of the major adhesive proteins used by marine mussels to bind strongly to underwater surfaces. This behaviour has been related to its strong surface active and adsorptive nature (Notter 1988; Olivieri *et al.*, 1992; Hansen *et al.*, 1994). This protein and related mussel adhesive proteins, are characterised by having high lysine contents and hydroxylated amino acids: *mefp-1* for example, consists of tandemly repeated decapeptides each containing two residues of lysine, one to two residues of Dopa (Waite, 1983; Laursen, 1992) one or two residues of *trans*-4-hydroxyproline and a single residue of *trans*-2,3,*cis*- 3,4-dihydroxyproline (Taylor *et al.*, 1994). Several attempts have been made to make biomedical and commercial use of the adhesive properties of these substances (Baty *et al.*, 1997), for example, in experimental epikeroplasty and for cellular attachment (Robin *et al.*, 1988; Olivieri *et al.*, 1992) such as in the attachment of osteoblasts and epiphyseal cartilage cells to substrata (Fulkerson *et al.*, 1990). The strong adhesive properties have recently inspired a proposed use for these proteins as mucoadhesives for drug delivery

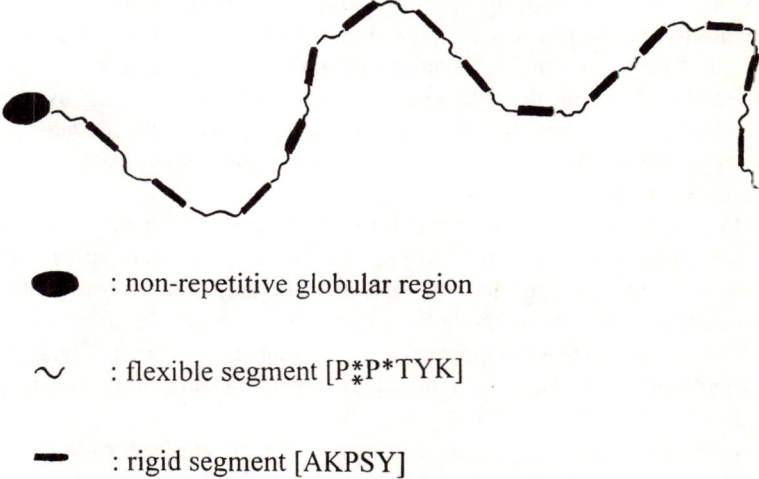

● : non-repetitive globular region

∿ : flexible segment [P*P*TYK]

▬ : rigid segment [AKPSY]

Figure 8. Consensus semi-flexible rod model for the mussel foot glue protein *mefp-1* from *Myetilus edulis*. This model takes into account the linear flexible properties consistent with molecular hydrodynamics by Deacon *et al.* (1998), earlier CD studies (Laursen, 1992) and the ability to adhere to surfaces (Baty *et al.*, 1997). The model consists of a globular region with a non-repetitive amino acid sequence and an extended region consisting of repeat sequences of amino acid with alternating stiff and flexible segments. Except at high pH (>7) and ionic strength, the chain will be relatively stiff due to electrostatic repulsion of segments. From Deacon *et al.* (1998).

(Schnurrer and Lehr, 1996) and these substances may offer an alternative to substances such as chitosans. However, applications have been hampered by the lack of knowledge on the solution structure and adhesive nature of the purified protein (Baty *et al.*, 1997) and until recently very little has been known of the oligomeric state or overall morphology of these molecules in solution (dilute or otherwise). Two recent studies have confirmed that the protein is monomeric in solution of molecular weight ~110 kDa (Deacon *et al.*, 1996) and has an extended conformation in solution (Deacon *et al.*, 1998a) as illustrated in *Figure 8*.

Direct and molecular strategies for studying mucoadhesion

Direct methods involve a study of a macroscopic interaction, usually involving whole mucus, whereas the molecular methods focus on the interactions involving the purified mucin component. Both strategies are highly complementary and should not be used in isolation.

The assay methods can either employ freshly excised tissue from various animals (frog, rat, rabbit, pig, cow, etc.), used either immediately as live or dead tissue or stored frozen and defrosted prior to use, or they use mucus or mucin at various degrees of degradation and purity either solubilised or as gel (usually from pig stomach or bovine submaxillary glands). Whatever model material is used, its relevance to the human mucus, whether in health or disease state, has to be considered (MacAdam, 1993). Dead mucosal tissue may well not produce any new mucus, while degradation of existing mucus will still take place. This will have a marked effect on the rheological characteristics of the substrate, considered to be highly relevant to adhesional phenomena. Mucus thickness may vary from species to species and intersubject, as well as intrasubject, variability of the mucosal tissue poses problems in terms of reproducibility. For the mucin based procedures, mucins, once extracted are subject to degradation by enzymes and mechanical disruption: they have to be handled with extreme care, and enzyme degradation must be kept to a minimum (e.g. by extraction in guanidine hydrochloride (Sheehan and Carlstedt, 1984) or with adequate protease inhibitors present). Mucin carbohydrate composition also varies within the gastrointestinal tract (Allen, 1989).

Small intestinal mucin is very difficult to solubilise and available usually in only small quantities. Gastric mucin from pigs appears to be an alternative since it is available in larger quantities and although its sialic acid content is low, its carbohydrate composition is comparable to human gastric mucin. Purification methods allow the removal of other components present in mucus in order to obtain purified mucin which still shows the gel-forming characteristics of native mucus (Sheehan and Carlstedt, 1989; Bell *et al.,* 1985; Allen, 1989).

Commercially available pig gastric mucins or mucus are somewhat different in the detail of their composition when compared with freshly prepared and purified material. They may be rather degraded or the freeze drying procedure may have altered the structure in such a way that it becomes difficult to redissolve them completely. Commercially available 'submaxillary' mucins are quite different from the mucins secreted in the gastrointestinal tract. They are secreted in a viscous soluble form rather than as water-insoluble gels (for a discussion of these differences see Gottschalk *et al.*, 1972). Nevertheless, highly purified mucins can give more accurate information on the actual nature of the interaction of a putative mucoadhesive with the main mucin-forming component. The use of dilute mucin solutions also allows the study of mucin–bioadhesive polymer interactions on a fundamental level.

It has been recognized that the degree of hydration of the bioadhesive DDS, as well as the amount of water available, plays an important role in determining the strength of adhesion or whether adhesion can take place at all (Leung and Robinson, 1988; Chen and Cyr, 1970). The hydration aspect can be controlled in local applications such as mouth or vagina by drying excess water in the area immediately prior to application (Deasy and O'Neill, 1989). In the gastrointestinal tract, however, excess water at the site of adhesion as well as excess in the amount of surrounding liquid cannot be controlled. Lehr *et al*. (1992b) pointed out that numerous so-called mucoadhesive polymers adhere only under conditions where the amount of interstitial liquid is limited. This kind of dry-to-wet adhesion or 'blotting adhesion' is due to the capillary forces drawing liquid from the mucus into the delivery system (Huntsberger, 1967; Lehr *et al.,* 1992b; Mortazavi and Smart, 1993). If the polymer involved offers no intrinsic ability to form a bond with the substrate (e.g. some cellulose derivatives), the initial adhesive forces, although high at the beginning, may become negligible as soon as the material is fully hydrated (Junginger and Lehr, 1990). Therefore, adhesion measurements in fully hydrated systems and over a period of time are necessary to avoid attributing a high adhesive force erroneously to intrinsic mucoadhesive properties. The adhesion mechanism of capillary attraction between a dry, water-absorbing polymer and a wet, mucosal surface being dehydrated is quite different to the interactions between two hydrogels (polymer and mucus) in equilibrium with a third liquid phase (Mortazavi and Smart, 1993).

Direct assay methods for mucoadhesion

By direct we mean 'whole mucus' assay procedures, and one of the simplest and most effective methods is tensiometry, which uses the force required to detach two surfaces, one coated with the mucus substrate, the other with mucoadhesive, as an index of mucoadhesion. Other direct methods (*Table 2*) include the 'flow through' technique, colloidal gold staining (adhesion number) and the *in-vivo* methods (endoscopy, radioisotope imaging)

TENSIOMETRY

The method employs putative mucoadhesive polymers that are usually in the form of tablets made by direct compression of the polymer or polymer coated surfaces from casting of polymer solutions. These are consequently put in contact with a mucus surface usually with a given force applied to the system for a given period of time after which the adhesive joint is destroyed by applying a vertical force in the opposite direction or a shear force in the horizontal direction. The force required to destroy the bond is taken as a qualitative and quantitative parameter for adhesion. If the experiment is done under full hydration of the polymer and in an aqueous environment

Table 2. Mucoadhesive assay methods

	Comment	References
'Direct' assays:		
Tensiometry	Force F required to dislodge two surfaces (one coated with mucus, the other mucoadhesive)	Lehr et al. (1992c)
Flow-through	Flow rate, dV/dt required to dislodge two surfaces	Mikas and Peppas (1990); Junginger et al. (1990)
Colloidal gold staining	Measures 'adhesion number'	Park (1989)
In vivo	Endoscopy, radioisotope imaging	Anderson (1991); Aoyagiet al. (1992)
Molecular mucin based assays:		
Viscometry and rheology	Intrinsic viscosity [η] can be related to complex size via MHKS 'a' coefficient	Harding (1995, 1997a)
Dynamic light scattering	Diffusion coefficient, D, can be related to complex size via MHKS 'ε' coefficient	Harding (1995, 1997a)
Turbidity/light scattering	'SEC MALLS' particularly useful for determining mol. wt. distribution of mucin. Turbidity: semi-quantitative indicator	Jumel et al. (1996, 1997); Fiebrig (1995)
Analytical ultracentrifugation	Change in mol. wt (sedimentation equilibrium) Sedimentation coefficient ratio of complex to mucin Schlieren 'fingerprinting'	Harding (1995); Cölfen and Harding (1997) Deacon et al. (1998a); Harding (1997b) Deacon et al. (1998b)
Surface Plasmon Resonance	Needs mobile and immobile phase	see, e.g. Silkowski et al. (1997)
Imaging methods	Atomic Force Micrsocopy Transmission electron microscopy (conventional and gold labelled) Scanning Tunneling Microscopy	Deacon et al. (1999) Fiebrig et al. (1995b, 1997) Roberts et al. (1995)

the likelihood of mimicking *in vivo* conditions is higher, given that a potential formulation, which is usually swallowed with liquid would not arrive at the target site in a totally dry state. However, this method neglects the fact that a swallowed formulation does not make intimate contact with the mucus gel spontaneously. Furthermore, the cohesion of mucus is also related to its thickness and rheological features. The method does not distinguish adhesion and cohesion (Huntsberger, 1967; Lehr *et al.*, 1992c). Although information on the screening of polymers can easily be obtained, the method appears unsuitable for assessing adhesive behaviour of formulations intended for gastrointestinal application. Similar comments apply to the related 'Wilhelmy Plate' method used by e.g. Smart *et al.* (1984). Tensiometry seems more useful for buccal, vaginal or other applications where liquid is controllable and more limited.

Despite these limitations, Lehr and coworkers found a very interesting spectrum of results in terms of the mucoadhesive performance of a range of biopolymers (Lehr *et al.*, 1992b) and their results are summarised in *Table 3*. The disappointing performance of the neutral and anionic polysaccharides is clearly shown. By contrast the polycationic chitosans appear to give a very favourable interaction strongly indicative of the importance of electrostatic interactions, although polycationic dextran derivatives showed little adhesive potential. A possible explanation for this is the branching of the dextran which may shield off and provide steric blockage of any interaction with anionic groups on the mucin glycoprotein substrate. *Table 3* also shows the potential of the mussel foot glue protein (Schnurrer and Lehr, 1996). Lehr *et al.* (1992a) and Schnurrer and Lehr (1996) also obtained the rather surprising result of a very favourable interaction for the synthetic polyanionic polymer, polycarbophil. We will consider these results again later when we have dealt with the molecular hydrodynamics.

FLOW THROUGH SYSTEMS

Here mucoadhesive coated spheres are placed on a mucus gel surface and the shearing flow of fluid required to dislodge them is used as an index of adhesion: this seems to model the situation in the gastrointestinal tract better than tensiometry. A flow channel device was first described by Mikos and Peppas (1990). The channel had a length of approximately 30 cm, a width of 4 cm and a height of approximately 0.5 cm, and it was thermostatted by a jacket connected to a constant temperature water bath. A cavity inside the channel allowed placement of a mucin gel or the mucosal of a tissue and the placement of a single polymer microparticle on top of it. The channel was connected through a set of valves to a gas cylinder. The volumetric flow rate was gradually increased until the particle, which was observed by an optical microscope, was detached from the mucous surface. This particular system could be suitable as a model for studying nasal mucoadhesion. For gastrointestinal models a fluid has to be substituted for the air and live tissue has been used to monitor intestinal drug absorption at the same time (Junginger, *et al.*, 1990). The observation of adhesion of a formulation (which ought to be insoluble so as to avoid dilution and rapid wash-off as well as rapid drug leaching) from a flow of solution directly onto a mucus tissue would be most desirable. A swellable, but insoluble formulation is usually achieved by crosslinking of the polymer chains. This would lead to chain rigidity which in turn

Table 3.   Mucoadhesive performance: Tensiometric analysis

Biopolymer	F (mN/cm^2)
Neutral polysaccharides:	
Hydroxy-propyl cellulose	~0 (2.8±2.8)
Hydroxy-ethyl starch	~0 (0.6±0.8)
Scleroglucan	~0
Anionic polysaccharides:	
Pectin	~0
Xanthan	~0
CMC (low viscosity)	1.8±1.1
CMC (medium)	~0 (0.3±0.3)
CMC (high viscosity)	1.3±1.0
Chitosans:	
Wella low viscosity	3.9±1.2
Wella high viscosity	6.7±0.7
Knapezyk	5.7±1.1
Daichitosan-H	8.0±5.7
Daichitosan-VH	9.5±2.4
Sea-cure 240	4.1±2.9
Sea-cure 210+	9.5±2.5
Sigma	6.6±3.0
Cationic dextrans:	
DEAE-dextran	~0
Amino-dextran	~0
Proteins:	
Mussel glue protein *mefp-1*	~9

Adapted from Lehr *et al.* (1992) and Schnurrer and Lehr (1996)
F: Force required for detachment of two surfaces in contact (mucoadhesive and mucus)

could limit mucoadhesion since the proposed 'interpenetration' or 'interdiffusion' mechanism would be restricted.

COLLOIDAL GOLD STAINING

Instead of measuring the adhesion strength or the duration of adhesion, the 'adhesion number' can be determined as a direct function of adhesion. In the method described by Park (1989) the adherent material (or substrate) consisted of colloidal gold particles of an approximate diameter of 18 nm with mucin adsorbed onto their surface (the particular mucin used was a solution of bovine submaxillary mucin Type I). Colloidal gold (cAu) sols were prepared by reducing $HAuCl_4$ with reducing agents like sodium citrate. Particle sizes varied depending on the reducing agent as well as on the preparation procedure.

The bioadhesive material used by Park was a copolymer made from acrylic acid and acrylamide cross-linked with N,N'-methylene-bis-acrylamide [P(AA-co-AM)]. The transparent hydrogels made of this material were cut into rectangular shapes of varying thickness. The polymer strips were incubated with the cAu-mucin conjugates and after a rinsing procedure, the absorbance of the strip was measured at a wavelength of 525 nm with a spectrophotometer using a transparent control polymer strip as a blank. The values obtained were a function of the amount of cAu adsorbed onto the surface. Alternatively the absorbance of the cAu-mucin preparation was measured before and after incubation. In this case, the magnitude of the decrease in the absorbance value from an initial value was used as a quantitative parameter indicating an interaction between cAu-mucin and hydrogel. An image analyser was also used to quantify the intensity of red colour on the polymer surfaces. This approach was necessary for mucoadhesive polymers which were not transparent.

The cAu-mucin conjugates prepared by Park required the addition of albumin to stabilise the preparation further. Albumin molecules are believed to adsorb onto small bare spots on the cAu particle where mucin molecules do not cover (Horisberger and Rosset, 1977; De Mey, 1984). However, it can be argued whether any interaction phenomenon observed is due solely to the properties of the mucin. Moreover, if the affinity of albumin to the cAu is higher than that of mucin, a displacement of mucin from the cAu is also possible.

As outlined by Park (1989) an alternative approach using the colloidal gold staining technique is that of developing cAu-(bioadhesive polymer) conjugates instead of cAu-mucin conjugates. The polymer coated cAu particles acting as the adhesive this time could be directly applied to the surface of target tissues. In this case the cAu-polymer conjugate would act as a model drug delivery system. Chitosan-stabilised cAu has been successfully prepared by Horisberger and Clerc (1988) to use as a marker for anionic sites on various micro-organisms and by Fiebrig *et al.* (1994b, 1997) to visualize the sites of interaction of chitosan in a mucin-chitosan complex.

IN VIVO METHODS

These involve animal models, human volunteers or patients (see Table 18.3 of Fiebrig *et al.*, 1995a for an extensive comparison). For buccal, vaginal, cervical or nasal applications the residence time of the device can be inspected visually, while the subject can give direct information on aspects of tolerance (discomfort, usefulness, etc.) (Nagai, 1986; Bottenberg, *et al.*, 1991; Smid-Corbar *et al.*, 1991). Plasma levels of drug or pharmacodynamic effect (for e.g. delivery of insulin) can give indirect evidence. Aspects of intersubject variability and disease condition will need to be taken into account.

With regards to gastrointestinal bioadhesive DDS we are faced with major experimental difficulties. Once swallowed, the device has to reach the adherent mucus layer. As has been outlined earlier, the process of bioadhesion requires *intimate contact* in its first step. There is little experimental evidence for this prerequisite actually taking place in gastrointestinal bioadhesion. Lodged on the target surface a delivery system has to resist the dislodging forces of gastrointestinal motility. In humans this particular aspect has been examined for the first time in a double blind study by Anderson (1991) using coloured tablets made from DEAE-dextran and ethylcellulose (1:1) with ethylcellulose tablets as controls. Seven patients undergoing routine gastroscopy examination swallowed both tablets with approximately 20 ml of water immediately before endoscopy. Mucoadhesion, or the lack of it, was assessed using a finger controlled water jet attachment. The clinician sprayed the tablet with water, for a fixed time period and at a constant rate, and assessed adhesion in terms of the number of sprays required to dislodge the tablet. The results did not suggest any significant adhesion of the test formulation to the gastric mucosa as compared to the control. In 50% of the patients examined at longer time intervals (up until 65 min post dose) neither the control nor the test tablet could be found in the stomach. For those patients where tablets could be observed, no significant differences in adhesive behaviour between control and test tablet with the gastric mucosa could be detected when judged using the finger controlled water jet attachment. There was no actual measurement of the water spray properties. However, the results illustrate the probable lack of adhesion for both tablet formulations as well as the intersubject variation in gastric emptying times for tablet formulations in the fasted state as observed by other authors.

The rat as an *in vivo* model has a mucus layer about half the thickness of that in man (Allen, 1978; Kerss*et al.*, 1982), while little is known about mucus turnover compared to man. It has been suggested that in the rat there is very little soluble gastric mucus when compared with the dog where there are considerable amounts of this material (Robinson *et al.*, 1987). Although the mucin of the pig gastrointestinal tract is similar to that of humans with regards to its carbohydrate and protein composition (Allen, 1989), gastric emptying in pigs has been shown to be slower than in man (Aoyagi *et al.*, 1992) and consequently this animal may not be the appropriate *in vivo* model. It is however possible to deliver directly to the intestine if required.

Molecular mucin-based assay methods

These methods focus on the interaction between the key macromolecular component of mucus, namely the mucin, and the mucoadhesive material (*Table 2*). The use of a standardised material throughout the experiments allows a comparison of results and avoids inter-sample variations. Such methods also allow for the study of the factors that may influence the interactions (ionic strength, bile salts, temperature, proteins etc.) and hence the elucidation of interaction mechanisms. The concentrations employed can be very low, as in the case of analytical ultracentrifugation and dynamic light scattering.

VISCOMETRY AND RHEOLOGY

These approaches involve the measurement of either the intrinsic viscosity [η] of a dilute solution using for example simple capillary viscometers with proper thermal control (see, Harding, 1997a), or for more concentrated dispersions and gels the rheological parameters G' (loss modulus) and G' (storage modulus) representing characteristic viscoelastic behaviour using e.g. cone-and plate type of viscometers (see Ross-Murphy, 1995). If the size of the complex is to be used as an index of mucoadhesive potential, the intrinsic viscosity can be related to the size of a complex between mucoadhesive and mucin via the Mark-Houwink-Kuhn-Sakurada (MHKS) 'a' coefficient (see Harding, 1995, 1997a) if an assumption is made about conformation (namely the conformation of the complex is the same as that of the mucin):

$$[\eta] \sim M^a \tag{1}$$

($a = 0, 0.4–0.5, 1.8$ for respectively a sphere, coil and rigid rod). However this method is complicated by the presence of unreacted mucin or mucoadhesive, and that measurement of [η] normally requires extrapolation to zero concentration to avoid complications of non-ideality. In addition, if complexation is a reversible process, lowering the concentration may also cause dissociation.

DYNAMIC LIGHT SCATTERING (PHOTON CORRELATION SPECTROSCOPY)

This involves measuring the translational diffusion coefficient, D, and, where possible, its distribution, $g(D)$, using an autocorrelator, which correlates time fluctuations of scattered light through Brownian motion of the scattering macromolecules/complex. Like the intrinsic viscosity, the diffusion coefficient or $g(D)$ can be related to the size of a complex if assumptions are made about conformation. For example, in terms of molecular weight by another MHKS relation (see Harding, 1995):

$$D \sim M^{-\varepsilon} \qquad (2)$$

($\varepsilon = 0.33, 0.5–0.6, 0.85$ for a sphere, coil and rod respectively), or (better) D is combined with the sedimentation coefficient from analytical ultracentrifugation (see below) to give M *via* a relation known as the Svedberg relation. Although the non-ideality complications are not as critical as for intrinsic viscosity, for non-spherical particles the autocorrelation function has to be extrapolated to zero angle because of complications of rotational diffusion, and like viscometry, the results will be complicated by the presence of unreacted mucin or mucoadhesive: although software can provide a distribution $g(D)$ and hence in principle resolve components of different D, results have to be considered very tentatively, since such results are from mathematical manipulation of data rather than a genuine mechanical separation as would be provided by chromatographic and sedimentation based procedures. The lack of mechanical separation also means samples have to be scrupulously clean from dust and large particulates that do not directly arise from the mucoadhesive-mucin complexation process.

TURBIDITY/LIGHT SCATTERING

Turbidmetric methods, although approximate, have been succesfully applied to large supramolecular assemblies such as the T-even bacteriophages (see Bahls and Bloomfield, 1977; Harding, 1986) and have been recently used to study mucin-mucoadhesive complexes (see Fiebrig, 1997). A good quality spectrophotometer is required which measures only the loss of intensity of an incident beam as it passes through a suspension, and does not record appreciable amounts of scattered light. The loss of intensity has to be due to scattering and not absorption, so a wavelength is chosen away from any absorption bands.

Although simple, turbidimetry is only an approximate way of sizing a macromolecular assembly. More useful information is found if a light scattering photometer is used, which records the scattered intensity envelope away from the incident angle. From this and an accurate knowledge of the concentration of the scattering particles the molecular weight and radius of gyration can be measured (see Tanford, 1961). Care has to be taken with the analysis of very large particles (M> ~50 million Da), since application of the simpler 'Rayleigh-Gans-Debye' theory ceases to be valid. As with dynamic light scattering, samples and scattering vessels have to be scrupulously free of dust and other large contaminating particulates.

A revolutionary development has been the coupling on-line of light scattering

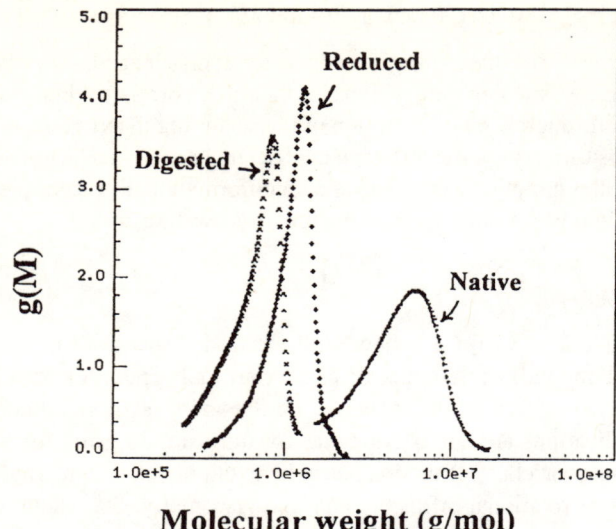

Molecular weight (g/mol)

Figure 9. Molecular weight distributions evaluated from the technique of size-exclusion chromatography coupled to multi-angle laser light scattering, 'SEC-MALLS' for pig colonic mucin, in its native, thiol reduced (or 'subunit') and papain digested ('T-domain') forms. From Jumel *et al.* (1997).

photometers with size exclusion chromatography: this (i) provides an on-line clarification system, and (ii) provides a physical (as opposed to mathematical) separation of polydisperse systems (Wyatt, 1992). The method is ideal for checking the molecular integrity of a mucoadhesive and a mucin prior to mixing. The usefulness of size-exclusion chromatogarphy coupled to multi-angle laser light scattering (SEC/MALLS) as a rapid assay for mucin molecular weight was first demonstrated by Jumel *et al.* (1995, 1997): *Figure 9* shows an example of a characterisation of pig gastric mucin (Jumel *et al.,* 1995) and colonic mucin, its subunits and T-domains (Jumel *et al.,* 1997). Beyond molecular weights of ~10 million, SEC systems generally fail to separate. However, other separation systems are now available, such as 'field flow fractionation' (FFF) and work is currently in progress to monitor the effectiveness of these systems for the characterisation of mucoadhesives (Deacon, 1999; Deacon *et al.,* 1999)

ANALYTICAL ULTRACENTRIFUGATION

In analytical ultracentrifugation the sedimentating boundary (sedimentation velocity) or equilibrium concentration distribution (sedimentation equilibrium) of a macromolecule or macromolecular complex in solution or suspension, under the influence of a centrifugal field, is recorded using absorption (uv or visible) or refractometric (Schlieren or Rayleigh interference) optics. Although the method is nearly 80 years old since its inception by T. Svedberg and coworkers, this technique has undergone a startling renaissance over the last decade, particularly in the protein biochemistry/ molecular biology fields, where its power in characterising the stoichiometries, strengths and conformations of protein-protein interacting systems is now widely recognised (see Harding and Winzor, 1999). It has enormous potential for the study of mucoadhesion. However, because of the generally much larger size and polydispersity of mucin, mucoadhesive (if it is not a protein) and complex, we have to apply ultracentrifugation as an assay for mucoadhesion in a slighty different way than used for the analysis or protein-protein interaction phenomena:

- *Change in molecular weight, M, as a measure of mucoadhesion.* Provided the molecular weight of the complex is within the the range ~1000<M<50 million Da we can apply the absolute molecular weight probe of sedimentation equilibrium in the ultracentrifuge. The ratio of the molecular weight of the complex to that of the largest of the reactants (normally the mucin, of molecular weight typically between 2 and 10 million Da) can then be used as a measure of mucoadhesive properties.

- *Change in sedimentation coefficient, s.* For complex sizes within the range 10000<M<10^9 Da we can apply sedimentation velocity in the ultracentrifuge to measure the sedimentation coefficient (sedimentation rate per unit centrifugal field). This will also be a measure of molecular weight and will also be influenced by conformation and hydration effects (see Pavlov *et al.*, 1997). We can either use the ratio of the sedimentation coefficients of the complex to the fastest sedimenting reactant {normally the mucin, of *s* typically between 20 and 60 Svedbergs (S)} as a measure of mucoadhesive potential, or we can assume a conformation for the complex (mucin appears to adopt a random coil conformation in solution) and convert the ratio of sedimentation coefficients to a ratio of molecular weights using the MHKS 'b' coefficient (see Harding, 1995, 1997b):

$$s \sim M^b \tag{3}$$

where b = 0.667, 0.4–0.5 and ~0.15 for a sphere, coil and rod, respectively.

- *Change in sedimentation concentration of reactants.* If the complexes are too large even for sedimentation velocity to pick up we can use a 'fingerprinting' assay whereby we determine the mucin or mucoadhesive loss (compared to a control) caused by complexation as a measure of mucoadhesive properties. This is particularly useful with the Schlieren refractometric optical system on the ultracentrifuge.

SURFACE PLASMON RESONANCE

This method which has had a significant impact in the study of protein-protein systems also has considerable potential for the study of mucoadhesion. It involves immobilising one phase and flowing over the second phase: in many ways it mimicks the 'flow through' technique described above in the 'Direct' assay procedures section. However, it has a serious disadvantage over the ultracentrifuge in that the method is not 'clean' in the sense that a third 'immobilising' phase is necessary. This is conventionally dextran, so assumptions over inertness have to be made. The greater the amount of material deposited from the mobile phase onto the immobile phase will affect the 'evanescent wave'. After certain assumptions, these affects can be related to a molar dissociation constant. If possible, experiments should be repeated with the mobile and immobile phases reversed, and the best application of this method is when used in conjunction with the analytical ultracentrifuge (see Silkowski *et al.*, 1997).

IMAGING METHODS

Another, but relatively new surface probe which 'images the surface' is Atomic Force Microscopy. The atomic force microscope can be used to investigate the interaction between mucin and mucoadhesive polymers. There are various modes of atomic force microscopy which include contact, non-contact and tapping, all of which may be performed in a liquid or a dry environment. In tapping mode the tip is brought into contact with the surface in a rapid intermittent fashion with the probe making very little contact with the sample. For standard forms of imaging (contact and non-contact) the picture is built up due to the collected topographical information caused by the deflection of a laser beam, targeted onto the reverse side of the cantilever, as it raster scans the surface. As tapping mode has intermittent contact with the surface it is ideal for soft samples, nearly achieving the resolution of contact mode with the non-invasive nature of non-contact modes of imaging. Using tapping mode it is possible to visualise mucin and polymers and study whether they are interacting without the harsh preparations needed in other methods. Two recent papers have indicated the usefulness of combining these measurements with those from surface plasmon resonance (Shakesheff *et al.*, 1995; Chen *et al.,* 1996).

Transmission electron microscopy has provided valuable information about the structure of mucins noted above (*Figure 4*). It has also been used to probe mucoadhesive complexes, in terms of conventional rotary shadowing of unlabelled material, and also on complexes where the mucoadhesive (chitosan) has been specifically labelled with colloidal gold or with gold tagged wheat germ agglutinin (Fiebrig *et al.*, 1997).

Case study: interaction of polycationic biopolymers with gastric mucin

We now illustrate the study of the molecular interactions between mucin and mucoadhesives by considering some of the work conducted by ourselves and others at Nottingham in conjunction with colleagues at Bristol, Lund, Trondheim and Oslo on the behaviour of mixtures of gastric mucin with DEAE-dextran, chitosan and mussel glue protein. The techniques that have been the cornerstone of our 'molecular' based approach have been primarily sedimentation velocity and electron microscopy. The sedimentation coefficient and molecular weight/molecular weight distributions (measured either by sedimentation equilibrium in the ultracentrifuge or using SEC-MALLS) has been routinely used by us as a measure of the structural integrity of the mucins (Jumel *et al.*, 1995, 1997).

1. DEAE DEXTRAN – GASTRIC MUCIN

We use the criterion of ratio of sedimentation coefficients, s, as an index of mucoadhesion between gastric mucin from pig stomachs. Two forms of the pig gastric mucin were used: a low-molecular weight 'single subunit-type' form (M ~2 million Da, s (at 20°C) ~17 Svedbergs, S, where $1S = 10^{-13}$ sec) and a 'whole mucin' form (M ~8 million, s ~42S). A Beckman (Palo Alto, USA) XL-A analytical ultracentrifuge has been used (Giebeler, 1992) with scanning uv-absorption optics. At 280 nm the mucin shows an absorption maximum whereas the DEAE-dextran is transparent and cannot be detected. Additionally, the DEAE dextran has a much smaller sedimentation

Table 4. Sedimentation velocity assay: pig gastric mucin and *DEAE-dextran*

mucin:DEAE-dextran ratio	Buffer and temperature	s_{mucin} (S) control	s_{mix} (S) complex	s_{mix}/s_{mucin}
mucin subunits (M ~2 million Da):				
2.0:1.9 (mg/ml)	pH6.8, I=0.1, 20°C	17	19	1.1
	pH6.8, I=0.1, 37°C	17	20	1.2
1.8:3.2	pH6.8, I=0.1, 37°C	18	25	1.4
whole mucins (M ~8 million Da):				
0.2:1.0	pH6.8, I=0.1, 20°C	35	65	1.9
	pH7.0, I=0.1, 20°C	42	55	1.3

UV absorption optics (using the MSE Centriscan and Beckman XL-A analytical ultracentrifuges). DEAE dextran control: sedimentation coefficient, s ~2 Svedbergs (S). For the subunit data, s values at 37°C have been normalised (for fluid density and viscosity) to 20°C (Anderson, 1991)

Table 5. Sedimentation velocity assay: pig gastric mucin and *chitosans*

chitosan	Buffer and temperature	s_{mucin} (S) control	s_{mix} (S) complex	s_{mix}/s_{mucin}
sea-cure +210	pH4.5, I=0.1, 20°C	52	780	15
	pH4.5, I=0.1, 37°C	53	1990	38
KN50 Trondheim	pH4.5, I=0.1, 20°C	52	1630	31
	pH4.5, I=0.1, 37°C	53	2340	44

UV absorption optics (using the Beckman XL-A analytical ultracentrifuge).
Chitosan controls: sedimentation coefficient, s ~1.5 Svedbergs (S)
Mucin:chitosan ratio, 0.2 mg/ml: 1.0 mg/ml
Mucin M ~11 million Da
Sea-Cure +210 (Pro-Nova, Drammen, Norway): degree of acetylation, F_A ~0.11
KN50 (NTH-Trondheim): degree of acetylation, F_A ~0.42

coefficient (~2S, as measured by a different analytical ultracentrifuge, the Beckman Model E with refractometric Schlieren optics) compared to the mucins (17-42S), so the ratio of s for the mixture to that of the mucin control can be used. An inspection of *Table 4* shows clearly that the interaction between the mucin and DEAE dextran is very modest for both 'subunits' and 'whole mucins' alike, with the maximum value of s_{mix}/s_{mucin} being only 1.9. This is consistent with the results from macroscopic tensiometry analyses of Lehr *et al.* (1992c) (*Table 3*) and is probably a manifestation of steric shielding of the charged residues by the $\alpha(1\rightarrow3)$ or $\alpha(1\rightarrow4)$ branches in the dextran chain. This finding is also consistent with the lack of bioadhesive effect for coated tablets of DEAE-dextran in *in-vivo* trials (Anderson, 1991).

2. CHITOSANS – GASTRIC MUCIN

The very modest results for DEAE-dextran contrast dramatically with those for chitosans which demonstrate a huge interaction with pig gastric mucin. *Table 5* shows the results for two chitosans of different properties. Sea-Cure 210+ is a commercial high quality chitosan available from Pronova Ltd. (Drammen, Norway), and is highly positively charged with a low degree of acetylation of the C2 N-groups (F_A ~0.11): *Table 5* shows the formation of large complexes of s_{mix}/s_{mucin} ~15-38, with no residual unreacted mucin left, corresponding (if we assume a coiled conformation for the complex, i.e. $s \sim M^{0.5}$) to particles of mol wt ~10^9-10^{10} Da, and consistent with a strong electrostatic interaction. The observations are consistent with the macroscopic

(a)

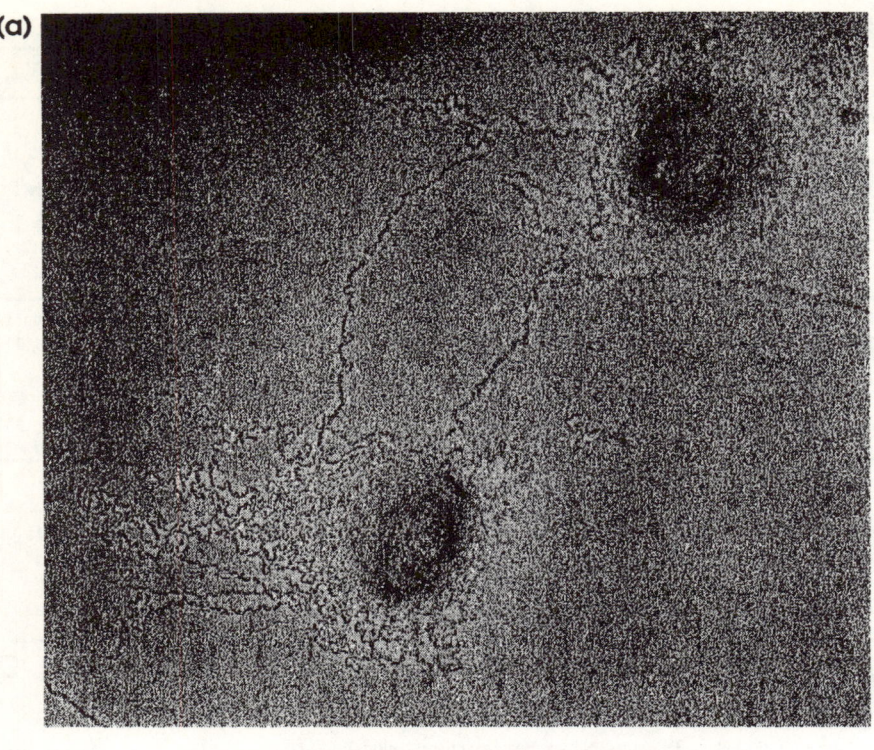

(b)

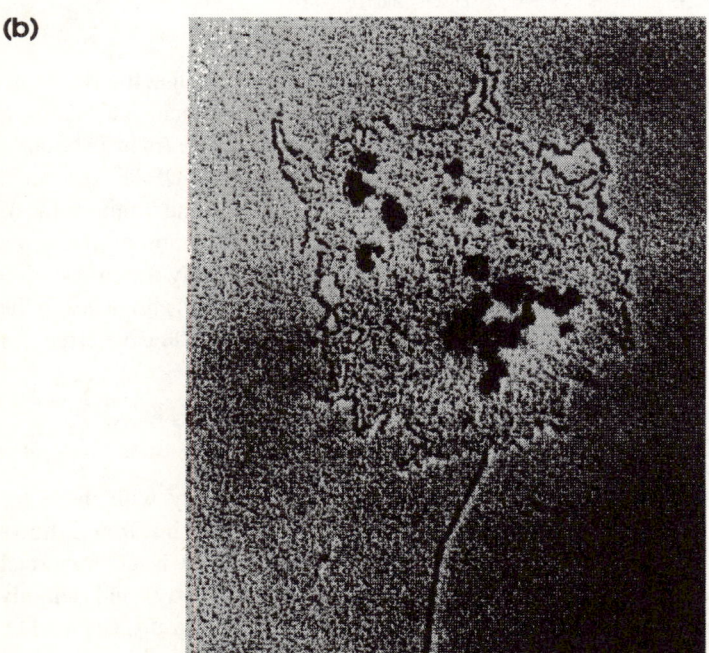

Figure 10. Transmission electron microscopy (both magnification ×75000) of a chitosan-mucin complex. Chitosan: Sea Cure +210 (F_A = 0.11). Mucin: pig gastric mucin. (a) conventional rotary shadowing (b) complex formed after the chitosan had been conjugated with colloidal gold. From Fiebrig *et al.* (1997).

tensiometry results of Lehr *et al.* (1992c) and also from electron microscopy studies (Fiebrig *et al.*, 1995b, 1997) using rotary shadowing of the complexes conventionally prepared by air drying onto mica and where the chitosan had been labelled directly by colloidal gold (*Figure 10*) and by wheat-germ-agglutinin linked gold. With this labelling strategy, the chitosan is seen to be distributed throughout the complex with 'hotspots' clearly evident.

Significantly however, the chitosan of much lower charge (KN50 of F_A ~0.42) also shows a very strong interaction of comparable magnitude (*Table 5*). This would appear to indicate that substitution of a charged group on C2 with an acetyl group

Table 6. Sedimentation velocity assay: pig gastric mucin and chitosan (Sea Cure 210+) *and bile salt*

sodium tauro-cholate conc.	Temp	s_{mucin} (S) control	s_{mix} (S) complex	s_{mix}/s_{mucin}
0mM	20°C	52	1085	21
	37°C	53	936	18
3mM	20°C	52	1375	26
	37°C	53	1877	35
6mM	20°C	52	973	19
	37°C	53	736	14

UV absorption optics (Beckman XL-A analytical ultracentrifuge).
Chitosan control: sedimentation coefficient, s ~1.5 Svedbergs (S)
Mucin:chitosan ratio, 0.2 mg/ml: 1.0 mg/ml. Bile salt: sodium taurocholate
Mucin M ~11 million Da

Table 7. Sedimentation velocity assay: pig gastric mucin and chitosan (Sea Cure 210+). *Effect of pH.*

pH	Temp	s_{mucin} (S) control	s_{mix} (S) complex	s_{mix}/s_{mucin}
2.0	20°C	45	980	22
	37°C	132	1626	12
4.5	20°C	52	780	15
	37°C	53	1990	38
6.5	20°C	32	1524	48
	37°C	46	1580	34

UV absorption optics (Beckman XL-A analytical ultracentrifuge).
Chitosan controls: sedimentation coefficient, s ~1.5 Svedbergs (S)
Mucin:chitosan ratio, 0.2 mg/ml: 1.0 mg/ml.
Mucin M ~11 million Da

Table 8. Sedimentation velocity assay: chitosan (Sea Cure 210+) *and different gastric mucins*

mucin	mucin:chitosan ratio	s_{mucin} (S) control	s_{mix} (S) complex	s_{mix}/s_{mucin}
PGM	0.2: 1.0 (mg/ml)	53	780	15
HGM	0.3: 1.0	11	222	20
OGM	1.0: 1.0	7	7*	1

UV absorption optics (Beckman XL-A analytical ultracentrifuge).
Chitosan control: sedimentation coefficient, s ~ 1.5 Svedbergs (S)
PGM: pig gastric mucin
HGM: human gastric mucin(Dr A. Corfield, University of Bristol)
OGM: Orthana (sialic acid free) pig gastric mucin (Orthana Ltd., Kastrup)
* A trace amount sedimenting at 415S was just visible

Figure 11.   Sedimentation Schlieren 'fingerprinting' of mucoadhesive interaction. Top image: Schlieren image for residual chitosan (Sea Cure +210) left after the rest had complexed with pig gastric mucin at 0.1M ionic strength. Bottom image: chitosan control under the same conditions. Area under top Schlieren boundary = 247 pixels. Area under bottom Schlieren boundary = 1017 pixels. % chitosan interacted = 75.7%. Rotor speed = 35000 rev/min, temperature = 20°C. Schlieren data captured via a CCD camera onto a PowerMac computer. From Deacon *et al.* (1999).

(containing a carbonyl group for possible H-bonding and a hydrophobic methyl group) does not compromise the interaction. This suggests that non-electrostatic interactions may also be significant, or it may suggest there was still enough charge on the chitosan to interact sufficiently with all the mucin.

The addition of bile salts (*Table 6*) does not appear to compromise the interaction significantly, except at ~6mM, but still significant complex sizes were present (s_{mix}/s_{mucin} ~14-18). Lowering the pH from 6.5 towards and below the pKa of the sialic acid residues on the mucin appears to reduce the complex size (s_{mix}/s_{mucin} drops from ~48 to ~22 at 20°C and from ~34 to ~12 at 37°C), but the interaction is still significant (*Table 7*). The observations of *Tables 6* and *7* are therefore supportive of the view that there may be a significant contribution from non-electrostatic as well as electrostatic types of interaction in the chitosan-mucin system. *Table 8* compares different types of gastric mucin, and although the preparation of the human gastric mucin is of much smaller molecular weight (s ~11S corresponds to a molecular weight between that of a T-domain and subunit), the interaction is not compromised. However, a much smaller pig gastric mucin preparation that is sialic acid free (from Orthana Ltd., Kastrup, Denmark) appears to give very little interaction.

Building on this investigation of the performance of mucins from different sources, Deacon *et al.* (1998b) have very recently investigated the mucoadhesive perfomance of chitosan with highly specific mucins from different regions of the stomach: corpus,

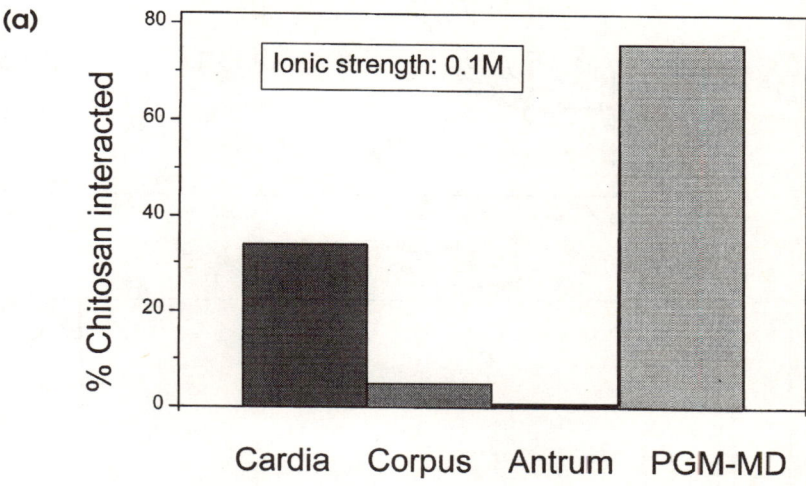

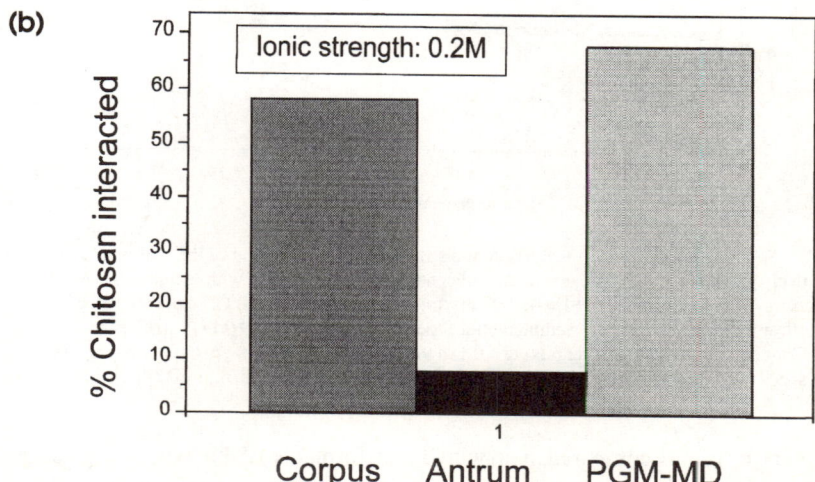

Figure 12. Comparison of amount of chitosan (Sea-Cure +210) interacted with pig gastric mucin for native pig gastric mucin prepared at Nottingham (PGM-MD) with three highly specific mucins from different regions of the stomach prepared at Lund (Cardia, Corpus and Antrum), using the Schlieren fingerprinting procedure (a) ionic strength 0.1M; (b) ionic strength = 0.2M. Other conditions as *Figure 11*.

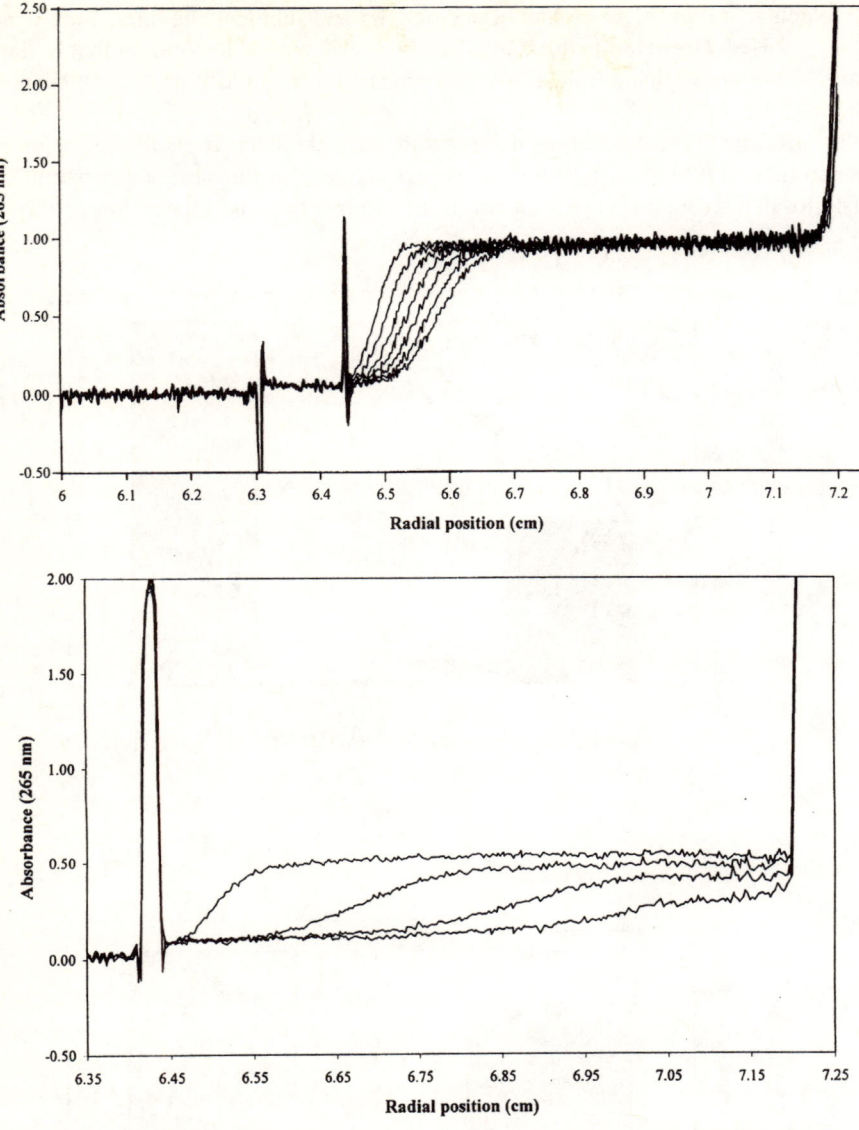

Figure 13. Sedimentation velocity analysis demonstration of mucoadhesion of the mussel glue protein *mefp-1* with pig gastric mucin. XL-A analytical ultracentrifuge used equipped with scanning uv-absorption optics (a) *mefp-1* control. Rotor speed = 40,000 rev/min, temperature = 20.0°C, scan interval = 10 min, loading concentration = 0.8mg/ml. Sedimentation coefficient, s_{20} = (2.34 ± 0.17)S. (b) *mefp-1*: mucin complex (Concentration of mucin after mixing = 0.1 mg/ml; concentration of *mefp-1* after mixing = 0.4 mg/ml. Rotor speed = 2,000 rev/min, temperature = 20.0°C, scan interval = 10 min, s_{20} ~ 7000S. (From Deacon *et al.*, 1998).

antrum and cardia, and compared them with the performance of the 'whole' pig gastric mucin. Because of scarcity of material, the sedimentation 'fingerprinting' technique (*Figure 11*) had to be employed (see page 67). This involved the use of the criterion of 'amount of chitosan interacted', as determined from the areas under refractive index gradient curves (Schlieren optics on the analytical ultracentrifuge, linked on-line via a CCD camera to a PowerMac computer). *Figure 12* shows a clear difference (under 0.1M ionic strength solvent conditions) of the interaction strength in the order cardia>corpus>antrum. Although at 0.2M insufficient cardia material was available, interestingly the corpus and antrum showed an increase in interaction strength, again consistent with the hypothesis that non-electrostatic types of interaction play a role in mucin-chitosan binding.

3. MUSSEL GLUE PROTEIN – GASTRIC MUCIN

The highly positively charged 'glue protein' *mefp-1* from the feet of *Mytilus edulis*, has to date given the strongest demonstration of an interaction with mucin, at least on the basis of molecular hydrodynamics (Deacon *et al.*, 1998). *Figure 13* compares the sedimentation velocity profile for native *mefp-1* and the mixture of *mefp-1* with mucin. The sedimentation coefficient for the native *mefp-1*, $s_{20,w}$ of 2.3S is consistent with an extended conformation for this protein (model shown in *Figure 8*) of molecular weight M ~110,000 Da which has been shown by sedimentation equilibrium in the ultracentrifuge to be monomeric in solution. However, when mixed with mucin, the *mefp-1* is completely reacted, and moves down at an s ~7000S, and corresponds to an s_{mix}/s_{mucin} of ~130 (*Table 9*).

From molecular hydrodynamics to dosage form

The next problem is constructing an adequate delivery system out of the mucoadhesive polymer. The best strategy would appear to be encapsulation *via* microspheres, and indeed similar encapsulation strategies have been invoked also for intravenous systems. The delivery system could be typically microspheres using e.g. alginate beads for the encapsulation of insulin producing islet of langerhans cells for the treatment of diabetics (Soon-Shiong *et al.*, 1994), or for mimicking secretory granule encapsulation systems (Siegel, 1998)

He *et al.* (1998a) have recently investigated the use of chitosan microspheres for gastroretention. Non-crosslinked and crosslinked chitosan microspheres were prepared by a spray drying method. The microspheres so prepared were spherical in shape and positively charged. The particle sizes ranged from 2 to 10 μm. The size and zeta potential of the particles were influenced by the level of crosslinking. With decreasing amount of crosslinking agent (either glutaraldehyde or formaldehyde), both particle size and zeta potential were increased. He *et al.* (1998a), found that the preparation conditions also had an influence on the particle size. DSC studies revealed that drugs incorporated in the microspheres (H2 antagonist drugs cimetidine, and famotidine) were molecularly in the form of a solid solution. The release of model drugs

Table 9. Sedimentation velocity assay: pig gastric mucin and mussel glue protein *mefp-1*

mucin:*mefp-1* ratio	Buffer and temperature	s_{mucin} (S) control	s_{mix} (S) complex	s_{mix}/s_{mucin}
0.1 mg/ml: 0.4 mg/ml	pH6.8, I=0.1M 20°C	53	7000	130

UV absorption optics (Beckman XL-A analytical ultracentrifuge).
Mefp-1 controls: sedimentation coefficient, s ~ 2.3 S
Mucin M ~11 million Da
Mefp-1 M ~110000 Da

(cimetidine, famotidine and nizatidine) from these microspheres was fast, and accompanied by a burst effect.

A modified spray drying method in the form of a novel w/o/w emulsion-spray drying technique was developed by He *et al.* (1998 b) to prepare chitosan microspheres with a sustained drug release pattern. The release of the model drugs (cimetidine, famotidine), from the microspheres prepared by the emulsion-spray drying method, was greatly retarded with release lasting for several hours, compared with drug loaded microspheres prepared by conventional-spray drying or emulsion methods where drug release was almost instantaneous. The slow release of the drug was partly due to the poor wetting ability of the microspheres, which floated on the surface of the dissolution medium. The addition of a wetting agent increased the release rate significantly. The coating of the microspheres with gelatin decreased the rate of release of drug in the presence of wetting agents.

The mucoadhesive properties of chitosan and chitosan microspheres has been evaluated by He *et al.* (1998c) studying the interaction between mucin and chitosan in aqueous solution by turbidimetric measurements and the measurement of mucin adsorbed on the microspheres. A strong interaction between chitosan microspheres and mucin was found. Adsorption studies were carried out for the adsorption of mucin to chitosan microspheres with different crosslinking levels. The adsorption of type III mucin (1% sialic acid content), to chitosan microspheres followed Freundlich or Langmuir adsorption isotherms. When the content of sialic acid was increased (i.e. type I-S mucin, 12% sialic acid content), the adsorption type followed more closely an electrostatic attraction type of isotherm. The heat of the adsorption was found to be 13–23 kJ/mol. A salt-bridge interaction was proposed for the interaction of the negatively charged mucus glycoprotien with chitosan microspheres. The level of mucin adsorption was found to be proportional to the absolute values of the positive

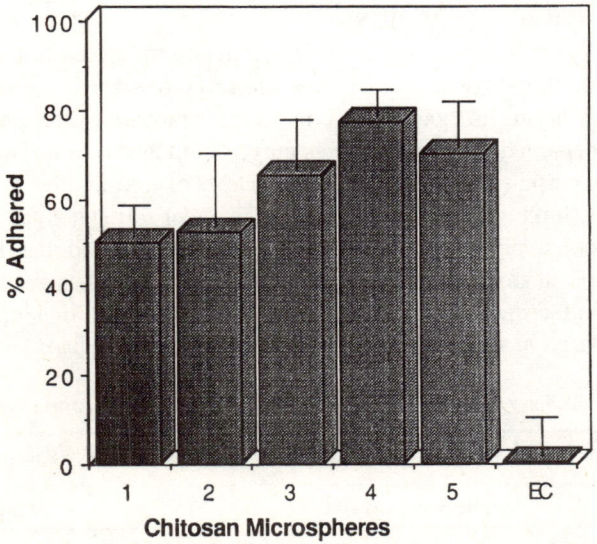

Figure 14. Mucoadhesive measurement of chitosan microspheres on rat small intestine by a particle counting technique. (EC – Ethyl cellulose microsphere).

zeta potential of chitosan microspheres and negative 'zeta potential' of mucin glyco-protein. Factors leading to a reduction or a reversal of these absolute values (e.g. different crosslinking levels of chitosan microspheres, different types of mucin, different pH, or ionic strength of the medium used) led to a reduction in the amount absorbed.

Biological studies have shown that chitosan microspheres were retained by a biological tissue; rat small intestine (*Figure 14*). Non-adhesive microsphere of the same size prepared from ethyl cellulose (EC) were used as a control. More than 50% of the administered chitosan microspheres were absorbed to the tissue, whereas but few of the EC microspheres were bound.

Futher experiments are now in progress to include phase 1 clinical studies in human subjects, where the retention of the microspheres can be followed in a non-invasive manner through a process of labelling of the microspheres with a gamma emitting radionuclide (Indium-111)

Processes involving the coating of chitosan onto delivery systems can also be used to achieve mucoadhesion. Takeuchi *et al.* (1996) have developed a chitosan coated liposome system. The liposomes were loaded with insulin. Preferential adsorption of the coated liposomes in the rat small intestine was demonstrated. The blood glucose levels were reported to be reduced significantly after one administration of the system to rats.

Acknowledgement

We thank Professor A. Allen and Dr Peter Dettmar for their helpful comments.

References

ACATÜRK, F. (1989). Preparation of a prolonged-release tablet formulation of diclofenac sodium. Part 1: Using chitosan. *Pharmazie* **44**, 547–549.

AL-DUJAILI, H., SALOLE, E.G. AND FLORENCE, A.T. (1983). Drug formulation and oesophageal injury. *Adverse Drug Reactions and Acute Poisoning Reviews* **2**, 235–256.

ALIMUNIAR, A. AND ZAINUDDIN, R. (1992). An economical technique for producing chitosan. In *Advances in Chitin and Chitosan*. Eds. C.J. Brine, P.A. Sandford and J.P. Zikakis, pp 627–632. London: Elsevier Applied Science.

ALLEN, A. (1978). Structure of gastrointestinal mucus glycoproteins and the viscous and gel-forming properties of mucus. *British Medical Bulletin* **34**, 28–33.

ALLEN, A. (1981). Structure and function of gastrointestinal mucus. In *Physiology of the Gastrointestinal Tract*. Ed. L.R. Johnson, pp 617-639. New York: Raven Press.

ALLEN, A. (1983). Mucus – a protective secretion of complexity. *Trends in Biochemical Sciences* **8**, 169–173.

ALLEN, A. (1989). Gastrointestinal mucus. In *Handbook of Physiology – The Gastrointestinal Physiology. Salivary, Gastric and Hepatobiliary Secretions*. Ed. S.G. Schultz and G.M. Makhlout, Section 6, Vol. III, pp 359–382. USA: American Physiological Society.

ALLEN, A. AND CAROLL, N.J.H. (1985). Adherent and soluble mucus in the stomach and duodenum. *Digestive Diseases and Sciences* **30**, 55S–62S.

ALLEN, A., FLEMSTRÖM, G., GARNER, A. AND KIVILAAKSO, E. (1993). Gastroduodenal mucosal protection. *Physiological Reviews* **73**, 823–857.

ANDERSON, M.T., HARDING, S.E. AND DAVIS, S.S. (1989). On the interaction in solution of a candidate mucoadhesive polymer, diethylaminoethyl-dextran, with pig gastric mucus glycoprotein. *Biochemical Society Transactions*, 631st meeting Guildford, **17**, pp 1101–1102.

ANDERSON, M.T. (1991). The interaction of mucous glycoproteins with polymeric materials. *Ph.D. thesis*, University of Nottingham, UK.

AOYAGI, N., OGATA, H., KANIWA, N., UCHIYAMA, M., YASUDA, Y. AND TANIOKA, Y. (1992). Gastric emptying of tablets and granules in humans, dogs, pigs, and stomach-emptying-controlled rabbits. *Journal of Pharmaceutical Sciences* **81**, 1170–1174.

ARAI, K., KINUMAKI, T. AND FUJITA, T. (1968). Toxicity of chitosan. *Bulletin of the Tokai Regional Fisheries Research Laboratory* **56**, 89–94.

BAHLS, D.M. AND BLOOMFIELD, V.A. (1977). Turbidmetric determination of bacteriophage molecular weights. *Biopolymers* **16**, 2797-2799.

BATY, A.M., LEAVITT, P.K., SIEDLECKI, C.A., TYLER, B.J., SUCI, P.A., MARCHANT, R.E. AND GEESEY, G.G. (1997). Adsorption of adhesive proteins from the marine mussel, *Mytilus edulis*, on polymer films in the hydrated state using angle dependent X-ray photoelectron spectroscopy and atomic force microscopy. *Langmuir* **13**, 5702-5710.

BEACHEY, E.H. (1980). *Bacterial Adherence*, Series B, Volume 6, London and New York: Chapman and Hall.

BECHGAARD, H. AND LADEFOGED, K. (1978). Distribution of pellets in the gastrointestinal tract. The influence on transit time exerted by the density and diameter of pellets. *Journal of Pharmacy and Pharmacology* **30**, 690–692.

BEERMANN, B., GROSCHINSKY-GRIND, M. AND ROSEN, A. (1976). Absorption, metabolism and excretion of hydrochlorothiazide. *Clinical Pharmacology and Therapeutics* **19**, 718–721.

BEERMAN, B. AND GROSCHINSKY-GRIND, M. (1978). Enhancement of the gastrointestinal absorption of hydrochlorothiazide by propantheline. *European Journal of Clinical Pharmacology* **13**, 385–387.

BELL, A.E., SELLERS, L.A., ALLEN, A., MORRIS, E., ROSS-MURPHY, S. (1985). Properties of gastric and duodenal mucus: effects of proteolysis, disulphide reduction, bile, acid, ethanol and hypertonicity on mucus gel structure. *Gastroenterology* **88**, 269–280.

BHASKAR, K.R., GARIK, P., TURNER, B.S., BRADLEY, J.D., BANSIL, R., STANLEY, H.E. AND LAMONT, J.T. (1992). Viscous fingering of HCl through gastric mucin. *Nature* **360**, 458–461.

BODDÉ, H.E. (1990). Principles of Bioadhesion. In *Bioadhesion – Possibilities and Future Trends*. Eds. R. Gurny and H.E. Junginger, Wissenschaftliche Verlagsgesellschaft mbH Stuttgart, APV, Vol. 25, pp 44–64.

BOEDECKER, E.C. (1984). *Attachment of Organisms to the Gut Mucosa*. Volumes I and II. Boca Raton, Florida: CRC Press.

BOOTH, C.C. (1967). Sites of absorption in the small intestine. *Federation Proceedings* **28**, 1583–1588.

BOTTENBERG, P., CLEYMAET, R., DE MUYNCK, C., REMON, J.P., COOMANS, D., MICHOTTE, Y. AND SLOP, D. (1991). Development and testing of bioadhesive, fluoride-containing slow-release tablets for oral use. *Journal of Pharmacy and Pharmacology* **43**, 457–464.

BOUCKAERT, S., TEMMERMAN, M., DHONT, M. AND REMON, J.P. (1994). The treatment of bacterial vaginosis with a bioadhesive vaginal slow-release tablet with metronidazole. *Proceedings of the International Symposium on Controlled Release of Bioactive Materials* **21**, 585–586.

BOWMAN, W.C. AND RAND, M.J. (1980). *Textbook of Pharmacology*. Oxford: Blackwell Scientific Publications, Chapter 25.

BRAHAMS, D. (1984). Death of a patient participating in trial of oral morphine for relief of postoperative pain. *Lancet* **1**, 1083–1084.

BYSTRICKY, S., MALOVÍKOVÁ, A. AND STICZAY, T. (1990). Interaction of alginates and pectins with cationic polypeptides. *Carbohydrate Polymers* **13**, 283–294.

CARGILL, R., CALDWELL, I.J., ENGLE, K., FIX, J.A., PORTER, P.A. AND GARDNER, C.R. (1988). Controlled gastric emptying. I. Effects of physical properties on gastric residence times of non-disintegrating geometric shapes in beagle dogs. *Pharmaceutical Research* **5**, 533–536.

CARGILL, R., ENGLE, K., GARDNER, C.R., PORTER, P., SPARER, R.V. AND FIX, J.A. (1989). Controlled gastric emptying. II. *In vitro* erosion and gastric residence times of an erodible device in beagle dogs. *Pharmaceutical Research* **6**, 506–509.

CARLSTEDT, I. AND SHEEHAN, J.K. (1984). Is the macromolecular architecture of cervical, respiratory and gastric mucins the same? *Biochemical Society Transactions* **12**, 615–617.

CARLSTEDT, I. AND SHEEHAN, J.K. (1988). Structure and macromolecular properties of mucus glycoproteins. *Monographs in Allergy*, Basel, **24**, 16–24.

CHEN, J.L. AND CYR, G.N. (1970). Compositions producing adhesion through hydration. In *Adhesion in Biological Systems*. Ed. R.S. Manly, pp 163-181. New York: Academic Press.

CHEN, X., DAVIES, M.C., ROBERTS, C.J., SHAKESHEFF, K.M., TENDLER, S.J.B. AND WILLIAMS, P.M. (1996). Combined surface plasmon resonance and scanning force microscope instrument. *Journal of Vacuum Science Technology* **B14**, 1582-1586.

CODE, C.F. AND MARLETT, J.A. (1975). The interdigestive myo-electric complex of the stomach and small bowel of dogs. *Journal of Physiology* **246**, 289–309.

CÖLFEN, H. AND HARDING, S.E. (1997). MSTARA and MSTARI: interactive PC algorithms for simple, model independent evaluation of sedimentation equilibrium data. *Eur. Biophys. J.* **25**, 333-346.

CÖLFEN, H., HARDING, S.E., VÅRUM, K.M., AND WINZOR, D.J. (1996). A study by analytical ultracentrifugation on the interaction between lysozyme and extensively deacetylated chitin (chitosan). *Carbohyd. Polym.* **30**, 45-53.

CREETH, J.M. (1978). Constituents of mucus and their separation. *British Medical Bulletin* **34**, 17–24.

DAVIS, S.S. (1985). The design and evaluation of controlled release systems for the gastro-intestinal tract. *Journal of Controlled Release* **2**, 27–38.

DAVIS, S.S., STOCKWELL, A.F., TAYLOR, M.J., HARDY, J.G., WHALLEY, D.R., WILSON, C.G., BECHGAARD, H. AND CHRISTENSEN, F.N. (1986). The effect of density on gastric emptying of single- and multiple-unit dosage forms. *Pharmaceutical Research* **3**, 208–213.

DAVIS, S.S. (1989). Small intestine transit. In *Drug Delivery to the Gastrointestinal Tract*. Eds. J.G. Hardy, S.S. Davis and C.G. Wilson, pp 49–61. Chichester: Ellis Horwood Ltd.

DEACON, M.P. (1999). PhD Dissertation, University of Nottingham, England.

DEACON, M.P., DAVIS, S.S., WHITE, R., WAITE, J.H. AND HARDING, S.E. (1996). Monomeric behaviour of *Mytilus edulis* (mussel) glue protein in dilute solution. *Biochem. Soc. Trans.* **25**, 422S.

DEACON, M.P., DAVIS, S.S., WAITE, J.H. AND HARDING, S.E. (1998). Structure and muco-adhesion of mussel glue protein in dilute solution. *Biochemistry* (in press).

DEACON, M.P., DAVIS, S.S., WHITE, R.J., NORDMAN, H., CARLSTEDT, I., ERRINGTON, N., ROWE, A.J. AND HARDING, S.E. (1999). Are chitosan-mucin interactions specific to different regions of the stomach? Sedimentation velocity in the ultracentrifuge can give us a clue. *Carbohydrate Polymers* (in press).

DEASY, P.B. AND O'NEILL, C.T. (1989). Bioadhesive dosage form for peroral administration of timolol base. *Pharmaceutica Acta Helvetiae* **64**, 231–235.

DE MEY, J. (1984). Colloidal gold as marker and tracer in light and electron microscopy. *Electron Microscopy Society of America* **14**, 54–66.

DODD, S., PLACE, G.A., HALL, R.L. AND HARDING, S.E. (1998). Hydrodynamic properties of mucins secreted by primary cultures of Guinea-pig tracheal epithelial cells: determination of diffusion coefficients by analytical ultracentrifugation and kinetic analysis of mucus gel hydration and dissolution. *European Biophysical Journal* **28**, 38–47.

DUCHÊNE, D., TOUCHARD, F. AND PEPPAS, N.A. (1988). Pharmaceutical and medical aspects of bioadhesive systems for drug administration. *Drug Development and Industrial Pharmacy* **14**, 283–381.

ERRINGTON, N., HARDING, S.E., VÅRUM, K.M., ILLUM, L. (1993). Hydrodynamic characteri-zation of chitosans varying in degree of acetylation. *International Journal of Biological Macromolecules* **15**, 113–117.

FIEBRIG, I., HARDING, S.E. AND DAVIS, S.S. (1994a). Sedimentation analysis of potential interactions between mucins and a putative bioadhesive polymer. *Progress in Colloid and Polymer Science* **93**, 66–73.

FIEBRIG, I., HARDING, S.E., STOKKE, B.T., VÅRUM, K.M., JORDAN, D. AND DAVIS, S.S. (1994b). The potential of chitosan as mucoadhesive drug carrier: studies on its interaction with pig gastric mucin on a molecular level. *European Journal of Pharmaceutical Sciences* **2**, 185.

FIEBRIG, I., DAVIS, S.S. AND HARDING, S.E. (1995a). Methods used to develop mucoadhesive drug delivery systems: bioadhesion in the gasterointestinal tract. In *Biopolymer Mixtures*. Eds. S.E. Harding, S.E. Hill and J.R. Mitchell, Chapter 18. Nottingham: Nottingham University Press.

FIEBRIG, I., HARDING, S.E., ROWE, A.J., HYMAN, S.C. AND DAVIS, S.S. (1995b). Transmission electron microscopy studies on pig gastric mucin and its interactions with chitosan. *Carbohyd. Polymers* **28**, 239-244.

FIEBRIG, I., VÅRUM, K.M., HARDING, S.E., DAVIS, S.S. AND STOKKE, B.T. (1997). Colloidal gold and colloidal gold labelled wheat germ agglutinin as molecular probes for identification in mucin/chitosan complexes. *Carbohydrate Polymers* **33**, 91-99.

FÖRSTER, H. AND LIPPOLD, B.C. (1982). Conception of peroral sustained-release dosage forms: calculation of initial and maintenance dose with consideration of accumulation. *Pharmaceutica Acta Helvetiae* **57**, 345–349.

FULKERSON, J.P., NORTON, L.A., GRONOWICZ, G., PICCIANO, P., MASSICOTTE, J.M. AND NISSEN, C.W. (1990). Attachment of epiphyseal cartilage cells and 17/28 rat osteosarcoma osteoblasts using mussle adhesive protein. *Journal of Orthopaedic Research* **8**, 793-798.

GIBBONS, R.A. (1972). Physico-chemical methods for the determination of the purity, molecular size and shape of glycoproteins. In *Glycoproteins: Their Composition, Structure and Function*. Ed. A. Gottschalk, pp 31-141. Amsterdam: Elsevier Publishing Company.

GIEBELER, R. (1992). The Optima XL-A: A new analytical ultracentrifuge with a novel precision absorption optical system. In *Analytical Ultracentrifugation in Biochemistry and Polymer Science*. Eds. S.E. Harding, A.J. Rowe and J.C. Horton, pp 16-25. Cambridge: Royal Society of Chemistry.

GOTTSCHALK, A., BHARGAVA, A.S. AND MURTY, V.L.N. (1972). Submaxillary gland glycoproteins. In *Glycoproteins: Their Composition, Structure and Function*. Ed. A. Gottschalk, pp 810-829. Amsterdam: Elsevier Publishing Company.

GREAVES, J.L. AND WILSON, C.G. (1993). Treatment of diseases of the eye with mucoadhesive delivery systems. *Advanced Drug Delivery Reviews* **11**, 349–383.

GRÖNING, R. AND HEUN, G. (1984). Oral dosage forms with controlled gastrointestinal transit. *Drug Development and Industrial Pharmacy* **10**, 527–539.

GRÖNING, R. AND HEUN, G. (1989). Dosage forms with controlled gastrointestinal passage – studies on the absorption of nitrofurantoin. *International Journal of Pharmaceutics* **56**, 111–116.

GRUBER, P., LONGER, M.A. AND ROBINSON, J.R. (1987). Some biological issues in oral controlled drug delivery. *Advanced Drug Delivery Reviews* **1**, 1–18.

GRUNDY, D. (1985). *Gastrointestinal Motility*. Lancaster: MTP Press Ltd.

GU, J.-M., ROBINSON, J.R. AND LEUNG, S.-H.S. (1988). Binding of acrylic polymers to mucin/epithelial surfaces: structure-property relationships. *Critical Reviews in Therapeutic Drug Carrier Systems* **5**, 21–67.

HALLETT, P., ROWE, A.J. AND HARDING, S.E. (1984). A highly expanded spheroidal conformation for a mucin from a cystic fibrosis patient: new evidence from electron microscopy. *Transactions of the Biochemical Society* **12**, 878-879.

HANSEN, D.C., LUTHER, G.W. AND WAITE, J.H. (1994). The adsorption of the adhesive protein of the blue mussel *Mytilus-edulis-L* onto type 304L stainless steel. *Journal of Colloid and Interface Science* **168**, 206-216.

HARDING, S.E. (1989). The macrostructure of mucus glycoproteins in solution. *Advances in Carbohydrate Chemistry and Biochemistry* **47**, 345–381.

HARDING, S.E. (1984). An analysis of the heterogeneity of mucins. No evidence for a self-association. *Biochemical Journal* **219**, 1061-1064.

HARDING, S.E. (1986). Applications of light scattering in microbiology. *Biotechnology and Applied Biochemistry* **8**, 489-509.

HARDING, S.E. (1989). The macrostructure of mucus glycoproteins in solution. *Advances in Carbohydrate Chemistry and Biochemistry* **47**, 345-381.

HARDING, S.E. (1995). On the hydrodynamic analysis of macromolecular conformation. *Biophysical Chemistry* **55**, 69-93.

HARDING, S.E. (1997a). The intrinsic viscosity of biological macromolecules. Progress in measurement, interpretation and application to structure in dilute solution. *Progress in Biophysics and Molecular Biology*. Ed. T.L. Blundell, **68**, 207–262.

HARDING, S.E. (1997b). Chitosan-Mucin Complexes. Characterisation by sedimentation velocity analytical ultracentrifugation. In *Chitin Handbook*. Eds. R.A.A. Muzzarelli and M.G. Peter, pp 457–466, European Chitin Society.

HARDING, S.E., ROWE, A.J., AND CREETH, J.M. (1983a). Further evidence for a flexible and highly expanded model for mucus glycoproteins in solution. *Biochemical Journal* **209**, 893–896.

HARDING, S.E., CREETH, J.M. AND ROWE, A.J. (1983b). Modelling the conformation of mucus glycoproteins in solution. In *Proceedings of the 7th International Glucoconjugates Conference*. Eds. A. Chester, D. Heinegard, A. Lundblad and S. Svensson, pp 558–559. Sweden: Olsson-Reklambyra.

HARDING, S.E. AND WINZOR, D.J. (1999). Analytical Ultracentrifugation. In *Protein-Ligand Interactions: A Practical Approach*. Eds. S.E. Harding and P.Z. Chowdhry, Vol. 1, Chap. 8. Oxford: Oxford University Press.

HARPER, C.M. AND RALSTON, M. (1983). Isobutyl 2-cyanoacrylate as an osseous adhesive in the repair of osteochondral fractures. *Journal of Biomedical Materials Research* **17**, 167–177.

HARRIS, D., FELL, J.T., SHARMA, H.L. AND TAYLOR, D.C. (1990a). GI transit of potential bioadhesive formulations in man: a scintigraphic study. *Journal of Controlled Release* **12**, 45–53.

HARRIS, D., FELL, J.T., TAYLOR, D.C., LYNCH, J. AND SHARMA, H.L. (1990b). GI transit of potential bioadhesive systems in the rat. *Journal of Controlled Release* **12**, 55–65.

HAYTON, W.L. (1980). Rate-limiting barriers to intestinal drug absorption: A review. *Journal of Pharmacology and Biopharmaceutics* **8**, 321–334.

HE, P., DAVIS, S.S. AND ILLUM, L. (1998a). Chitosan microspheres prepared by spray drying. *International Journal of Pharmaceuticals* (in press).

HE, P., DAVIS, S.S. AND ILLUM, L.(1998b). Sustained release chitosan microspheres prepared by novel spray-drying methods. *Journal of Microencapsulation* (in press).

HE, P., DAVIS, S.S. AND ILLUM, L. (1998c). *In vitro* evaluation of the microadhesive properties of chitosan microspheres. *International Journal of Pharmaceuticals* **166**, 75–68.

HELLIWELL, M. (1993). The use of bioadhesives in targeted delivery within the gastrointestinal tract. *Advanced Drug Delivery Reviews* **11**, 221–251.

HIRANO, S. (1989). Production and application of chitin and chitosan in Japan. In *Chitin and Chitosan*. Ed. G. Skjåk-Bræk, T. Anthonsen, T. and P. Sandford, pp 37–43. London: Elsevier Applied Science.

HOLLINGSBEE, D.A. AND TIMMINS, P. (1990). Topical adhesive systems. In *Bioadhesion – Possibilities and Future Trends*. Eds. R. Gurny and H.E. Junginger, Wissenschaftliche Verlagsgesellschaft mbH Stuttgart, APV, Vol. 25, pp 140–164.

HORISBERGER, H. AND ROSSET, J. (1977). Colloidal gold, a useful marker for transmission and scanning electron microscopy. *Journal of Histochemistry and Cytochemistry* **25**, 295–305.

HORISBERGER, M. AND CLERC, M.-F. (1988). Chitosan-colloidal gold complexes as poly-cationic probes for the detection of anionic sites by transmission and scanning electron microscopy. *Histochemistry* **90**, 165–175.

HÖRSTEDT, P., DANIELSSON, A., NYHLIN, H., STENLING, R. AND SUHR, O. (1989). Adhesion of bacteria to the human small-intestinal mucosa. *Scandinavian Journal of Gastroenterology* **24**, 877–885.

HOUNSELL, YOUNG, M. AND DAVIES, M.J. (1997). Glycoprotein changes in tumours: a renaissance in clinical applications. *Clinical Science* **93**, 287–293.

HUNTSBERGER, J.R. (1967). Mechanisms of adhesion. *Journal of Paint Technology* **36**, 199–211.

HWANG, S.-J., PARK, H. AND PARK, K. (1998). Gastric retentive drug-delivery systems. *Critical Reviews in Therapeutic Drug Carrier Systems* **15**, 243–284.

IMERI, A.G. AND KNORR, D. (1988). Effects of chitosan on yield and compositional data of carrot and apple juice. *Journal of Food Science* **53**, 1707–1709.

ITO, S. (1981). Functional gastric morphology. In *Physiology of the Gastrointestinal Tract*. Ed. L.R. Johnson, Vol. 1, pp 517–555. New York: Raven Press.

JABBARI, E., WISNIEWSKI, N. AND PEPPAS, N.A. (1993). Evidence of mucoadhesion by chain interpenetration at a poly(acrylic acid)/mucin interface using ATR-FTIR spectroscopy. *Journal of Controlled Release* **26**, 99–108.

JEUNIAUX, C., VOSS-FOUCHART, M.-F., POULICEK, M. AND BUSSERS, J.-C. (1989). Sources of chitin, estimated from new data on chitin biomass and production. In *Chitin and Chitosan*. Eds. G. Skjåk-Bræk, T. Anthonsen and P. Sandford, pp 3–11. London: Elsevier Applied Science.

JORDAN, N., NEWTON, J., PEARSON, J. AND ALLEN, A. (1998). A novel method for the visualization of the *in situ* mucius layer in rat and man. *Clinical Science* **95**, 97–106.

JUMEL, K., FIEBRIG, I. AND HARDING, S.E. (1995). Rapid size distribution and purity analysis of gastric mucus glycoproteins by size exclusion chromatography/ multi angle laser light scattering. *International Journal of Biological Macromolecules* **18**, 133–139.

JUMEL, K., FOGG, F.J.J., HUTTON, D.A., PEARSON, J.P., ALLEN, A. AND HARDING, S.E. (1997). A polydisperse linear random coil model for the quaternary structure of pig colonic. *European Biophysical Journal* **25**, 477–480

JUNGINGER, H.E., LEHR, C.-M., BOUWSTRA, J.A., TUKKER, J.J. AND VERHOEF, J. (1990). Site specific intestinal absorption using bioadhesives: Improved oral delivery of peptide drugs by means of bioadhesive polymers. In *Bioadhesion – Possibilities and Future Trends*. Eds. R. Gurny and H.E. Junginger, Wissenschaftliche Verlagsgesellschaft mbH Stuttgart, APV, Vol. 25, pp 177–190.

JUNGINGER, H.E. AND LEHR, C.-M. (1990). Bioadhäsive Arzneistoffabgabesysteme und Arzneiformen für perorale und rektale Anwendung. *Deutsche Apotheker Zeitung* **130**, 791–801.

JUNGINGER, H.E. AND VERHOEF, J.C. (1992). Perorale Applikation von Peptiden und Proteinen. *Deutsche Apotheker Zeitung* **132**, 1279–1290.

KAELBE, D.H. AND MOACANIN, J. (1977). A surface energy analysis of bioadhesion. *Polymer* **18**, 475–481.

KATAYAMA, T., TAKAI, K.-I., KARIYAMA, R. AND KANEMASA, Y. (1978). Colloid titration of heparin using cat-floc (polydiallyldimethyl ammonium chloride) as standard polycation. *Analytical Biochemistry* **88**, 382–387.

KAUS, L.C. (1987). The effect of density on the gastric emptying and intestinal transit of solid dosage forms: comments on the article by Davis *et al*. *Pharmaceutical Research* **4**, 78.

KERSS, S., ALLEN, A., GARNER, A. (1982). A simple method for measuring thickness of the mucus gel layer adherent to rat, frog and human gastric mucosa: influence of feeding, prostaglandin, N-actylcysteine and other agents. *Clinical Science* **63**, 187–195.

KIKENDALL, J.W., FRIEDMAN, A.C., OYEWOLE, M.A., FLEISCHER, D. AND JOHNSON, L.F. (1983). Pill-induced esophageal injury. *Digestive Diseases and Sciences* **28**, 174–181.

KORNEEVA, G.A., VICHOREVA, G.A., HARDING, S.E. AND PAVLOV, G.M. (1996). Hydrodynamic study of carboxymethyl chitin. *Abstracts American Chemical Society* **212** (part 1) 75-cell.

LAURSEN, R.A. (1992). In *Results and Problems in Cell Differentiation*, vol 19, Biopolymers. Ed. S.T. Case, pp 55–74. Berlin: Springer Verlag.

LEHR, C.-M., BOUWSTRA, J.A., TUKKER, J.J. AND JUNGINGER, H.E. (1990). Intestinal transit of bioadhesive microspheres in an *in situ* loop in the rat. A comparative study with copolymers and blends based on poly(acrylic acid). *Journal of Controlled Release* **13**, 51–62.

LEHR, C.-M., POELMA, F.G.J., JUNGINGER, H.E. AND TUKKER, J. (1991). An estimate of turnover time of intestinal mucus gel layer in the rat *in situ* loop. *International Journal of Pharmaceutics* **70**, 235–240.

LEHR, C.-M., BOUWSTRA, J.A., KOK, W., DE BOER, A.G., TUKKER, J.J., VERHOEF, J.C., BREIMER, D.D. AND JUNGINGER, H.E. (1992a). Effects of the mucoadhesive polymer polycarbophil on the intestinal absorption of a peptide drug in the rat. *Journal of Pharmacy and Pharmacology* **44**, 402–407.

LEHR, C.-M., BOUWSTRA, J.A., SCHACHT, E.H. AND JUNGINGER, H.E. (1992b). *In vitro* evaluation of mucoadhesive properties of chitosan and some other natural polymers. *International Journal of Pharmaceutics* **78**, 43–48.

LEHR, C.-M., BOUWSTRA, J.A., SPEIS, F., ONDERWATER, J., VAN HET NOORDEINDE, J., VERMEIJ-KEERS, C., VAN MUNSTEREN, C.J. AND JUNGINGER, H.E. (1992c). Visualization studies of the mucoadhesive interface. *Journal of Controlled Release* **18**, 249–260.

LEHR, C.-M., BOUWSTRA, J.A., KOK, W., NOACH, A.B.J., DE BOER, A.G. AND JUNGINGER, H.E. (1992d). Bioadhesion by means of specific binding of tomato lectin. *Pharmaceutical Research* **9**, 547–553.

LEUNG, S.S.-H. AND ROBINSON, J.R. (1988). The contribution of anionic polymer structural features to mucoadhesion. *Journal of Controlled Release* **5**, 223–231.

LIPPOLD, B.C. AND GÜNTHER, J. (1991). *In vivo* Prüfung einer multipartikulären Retard-Schwimmarzneiform. *European Journal of Pharmacy and Biopharmaceutics* **37**, 254–261.

LYNCH, J., POWNALL, R.E. AND TAYLOR, D.C. (1987). Site-specific absorption of hydrochlorothiazide in the rat intestine. *Journal of Pharmacy and Pharmacology* **39**, 55P.

MACADAM, A. (1993). The effect of gastro-intestinal mucus on drug absorption. *Advanced Drug Delivery Reviews* **11**, 201–220.

MANNINEN, V., APAJALAHTI, A., MELIN, J. AND KARESOJA, M. (1973). Altered absorption of digoxin in patients given propantheline and metoclopramide. *Lancet* **1**, 398–400.

MCCURDY, J.D. (1992). FDA and the use of chitin and chitosan derivatives. In *Advances in Chitin and Chitosan*. Eds. C.J. Brine, P.A. Sandford and J.P. Zikakis, pp 659–662. London: Elsevier Applied Science.

MERGENHAGEN, S.E. AND ROSAN, B. (1985). *Molecular Basis of Oral Microbial Adhesion*. American Society for Microbiology, Washington, DC.

MIKOS, A.G. AND PEPPAS, N.A. (1990). Bioadhesive analysis of controlled-release systems. IV. An experimental method for testing the adhesion of microparticles with mucus. *Journal of Controlled Release* **12**, 31–37.

MINAMI, H. AND MCCALLUM, R.W. (1984). The physiology and pathophysiology of gastric emptying in humans. *Gastroentrology* **86**, 1592–1610.

MIYAYAKI, S., YAMAGUCHI, H., TAKADA, M., HOU, W.-M., TAKEICHI, Y. AND YASABUCHI, H. (1990). Pharmaceutical application of biomedical polymers. *Acta Pharmaceutica Nordica* **2**, 401–406.

MOËS, A.J. (1993). Gastroretentive dosage forms. *Critical Reviews in Therapeutic Drug Carrier Systems* **10**, 143–195.

MORTAZAVI, S.A., CARPENTER, B.G. AND SMART, J.D. (1993). A comparative study on the role played by mucus glycoproteins in the rheological behaviour of the mucoadhesive/mucosal interface. *International Journal of Pharmaceutics* **94**, 195–201.

MORTAZAVI, S.A. AND SMART, J.D. (1993). An investigation into the role of water movement and mucus gel dehydration in mucoadhesion. *Journal of Controlled Release* **25**, 197–203.

MÜLLER-LISSNER, S.A. AND BLUM, A.L. (1981). The effect of specific gravity and eating on gastric emptying of slow-release capsules. *New England Journal of Medicine* **304**, 1365–1366.

NAGAI, T., NISHIMOTO, Y., NAMBU, N., SUZUKI, Y., SEKINE, K. (1984). Powder dosage form of insulin for nasal administration. *Journal of Controlled Release* **1**, 15–22.

NAGAI, T. (1986). Topical mucosal adhesive dosage forms. *Medicinal Research Reviews* **6**, 227–242.

NAGAI, T., MACHIDA, Y., SUZUKI, Y. AND IKURA, H. (1983). *Japan Pat.* 1,177,734, May 13, 1983.

NEUTRA, M.R. AND FORSTNER, J.F. (1987). Gastrointestinal mucus: Synthesis, secretion and function. In *Physiology of the Gastrointestinal Tract*. Ed. L.R. Johnson, pp 975–1009. New York: Raven Press.

NOTTER, M.F.D. (1988). Selective attachment of neural cells to specific substrates including *cell-tak*, a new cellular adhesive. *Exp. Cell Res.* **177**, 237–246.

OLIVIERI, M.P., BAIER, R.E. AND LOOMIS, R.E. (1992). Surface properties of mussel adhesive protein-component films. *Biomaterials* **13**, 1000–1008.

OLIVIERI, M.P., RITTLE, K.H., TWEDEN, K.S. AND LOOMIS, R.E. (1992). Comparative biophysical study of adsorbed calf serum, fetal bovine serum and mussel adhesive protein-component films. *Biomaterials* **13**, 201–208.

PARK, K. AND PARK, H. (1987). Enzyme-digestible balloon hydrogels for long-term oral drug delivery: synthesis and characterization. *Proceedings of the International Symposium on Controlled Release of Bioactive Materials* **14**, 41–42.

PARK, K. (1989). A new approach to study mucoadhesion: colloidal gold staining. *International Journal of Pharmaceutics* **53**, 209–217.

PAVLOV, G.M., ROWE, A.J. AND HARDING, S.E. (1997). Conformation zoning of large molecules using the analytical ultracentrifuge. *Trends in Analytical Chemistry* **16**, 401–405.

PEPPAS, N.A. AND BURI, P.A. (1985). Surface, interfacial and molecular aspects of polymer bioadhesion on soft tissues. *Journal of Controlled Release* **2**, 257–275.

PIMIENTA, C., LENAERTS, V., CADIEUX, C., RAYMOND, P., JUHASZ, J., SIMARD, M.A. AND JOLICOEUR, C. (1990). Mucoadhesion of hydroxypropylmethacrylate nanoparticles to rat intestinal ileal segments *in vitro*. *Pharmaceutical Research* **7**, 49–53.

PIMIENTA, C., CHOUINARD, F., LABIB, A. AND LENAERTS, V. (1992). Effect of various poloxamer coatings on *in vitro* adhesion of isohexylacyanoacrylate nanospheres to rat ileal segments under liquid flow. *International Journal of Pharmaceutics* **80**, 1–8.

POELMA, F.G.J. AND TUKKER, J.J. (1987). Evaluation of a chronically isolated internal loop in the rat for the study of drug absorption kinetics. *Journal of Pharmaceutical Sciences* **76**, 433–436.

PONCHEL, G.F., TOUCHARD, F., DUCHÊNE, D. AND PEPPAS, N.A. (1987a). Bioadhesive analysis of controlled-release systems. I. Fracture and interpenetration analysis in poly(acrylic acid)-containing systems. *Journal of Controlled Release* **5**, 129–141.

PONCHEL, G.F., TOUCHARD, F., WOUESSIDJEWE, D., DUCHÊNE, D. AND PEPPAS, N.A. (1987b). Bioadhesive analysis of controlled-release systems. III. Bioadhesive and release behaviour of metronidazole-containing poly(acrylic acid)-hydroxypropyl methylcellulose systems. *International Journal of Pharmaceutics* **38**, 65–70.

ROBERTS, C.J., SHIVJI, A., DAVIES, M.C., DAVIS, S.S., FIEBRIG, I., HARDING, S.E., TENDLER, S.J.B., WILLIAMS, P.M. (1995). A study of highly purified pig gastric mucin by scanning tunneling microscopy. *Protein Peptide Letters* **2**, 409–414.

ROBIN, J.B., PICCIANO, P., KUSLEIKA, R.S., SALAZAR, J. AND BENEDICT, C. (1988). Preliminary evaluation of the use of mussel adhesive protein in experimental epikeratoplasty. *Archives of Opthalmology* **106**, 973–977.

ROBINSON, R., LONGER, M.A., VEILLARD, M. (1987). Bioadhesive polymers for controlled drug delivery. In *Annals New York Academy of Science*. Ed. R.L. Juliano, Vol. 507, pp 307–314.

ROBINSON, J.R. (1990). Rationale of bioadhesion/mucoadhesion. In *Bioadhesion – Possibilities and Future Trends*. Eds. R. Gurny and H.E. Junginger, Wissenschaftliche Verlagsgesellschaft mbH Stuttgart, APV, Vol. 25, pp 13–28.

ROSS-MURPHY, S.B. (1995). Small deformation rheological behaviour of biopolymer mixtures. In *Biopolymer Mixtures*. Eds. S.E. Harding, S.E. Hill and J.R. Mitchell, pp 85–98. Nottingham: Nottingham University Press.

SAETTONE, M.F., CHETONI, P., TORRACCA, M.T., BURGALASSI, S. AND GIANNACCINI, B. (1989). Evaluation of muco-adhesive properties and *in vivo* activity of ophthalmic vehicles based on hyaluronic acid. *International Journal of Pharmaceutics* **51**, 203–212.

SANGEKAR, S., VADINO, W.A., CHAUDRY, I., PARR, A., BEIHN, G. AND DIGENIS, G. (1987). Evaluation of the effect of food and specific gravity of tablets on gastric retention time. *International Journal of Pharmaceutics* **35**, 187–191.

SCHAUER, R. (1992). Sialinsäurereiche Schleime als bioaktive Schmierstoffe. *Nachrichten aus Chemie Technik und Lababoratorium* **40**, 1227–1232.

SCHNURRER, J. AND LEHR, C.-M. (1996). Mucoadhesive properties of the mussel adhesive protein. *Int. J. Pharmaceutics* **141**, 251–256.

SCHOR, J.M., DAVIS, S.S., NIGALAYE, A. AND BOLTON, S. (1983). Susadrin transmucosal tablets (nitroglycerin in synchron controlled release base). *Drug Development and Industrial Pharmacy* **9**, 1359–1377.

SENJU, R. (1969). *Koroido Tekireiho [Colloid Titration]*, Tokyo: Nankodo.

SHAKESHEFF, K.M., CHEN, X., DAVIES, M.C., DOMB, A., ROBERTS, C J., TENDLER, S.J.B. AND WILLIAMS, P.M. (1995). Relating the phase morphology of a biodegradable polymer blend to erosion kinetics using simultaneous *in-situ* atomic-force microscopy and surface-plasmon resonance analysis. *Langmuir* **11**, 3921–3927.

SHEEHAN, J.K. AND CARLSTEDT, I. (1989). Models for the macromolecular structure of mucus glycoproteins. In *Dynamic Properties of Biomolecular Assemblies*. Eds. S.E. Harding and A.J. Rowe, pp 256–275. Cambridge: Royal Society of Chemistry.

SHETH, P.R. AND TOSSOUNIAN, J. (1984). The hydrodynamically balanced system (HBS®): a novel drug delivery system for oral use. *Drug Development and Industrial Pharmacy* **10**, 313–339.

SIEGEL, R.A. (1998). A lesson from secretory granules. *Nature* **394**, 427–428.

SILBERBERG, A. AND MEYER, F.A. (1982). Structure and function of mucus. In *Mucus in Health and Disease-II*. Eds. E.N. Chantler, J.B. Elder and M. Elstein, pp 53–74. New York: Plenum Press.

SILVER, F.H., LIBRIZZI, J., PINS, G., WANG, M.-C. AND BENEDETTO, D. (1994). Physical properties of hyaluronic acid and hydroxypropylmethylcellulose is solution: Evaluation of coating ability. *Journal of Applied Biomaterials* **5**, 89–98.

SILKOWSKI, H., DAVIS, S.J., BARCLAY, A.N., ROWE, A.J., HARDING, S.E. AND BYRON, O. (1997). Characterisation of the low affinity interaction between rat cell adhesion molecules CD2 and CD48 by analytical ultracentrifugation. *European Biophysical Journal* **25**, 455–462.

SLOMIANY, B.L. AND MEYER, K. (1972). Isolation and structural studies of sulfated glyco-proteins of hog gastric mucosa. *Journal of Biological Chemistry* **247**, 5062–5070.

SMART, J.D., KELLAWAY, I.W. AND WORTHINGTON, H.E.C. (1984). An *in-vitro* investigation of mucosa-adhesive materials for use in controlled drug delivery. *Journal of Pharmacy and Pharmacology* **36**, 295–299.

SMART, J.D. (1991). An *in vitro* assessment of some mucosa-adhesive dosage forms. *International Journal of Pharmaceutics* **73**, 69–74.

SMART, J.D. (1993). Drug delivery using buccal-adhesive systems. *Advanced Drug Delivery Reviews* **11**, 253–270.

SMID-KORBAR, J., KRISTL, J., COP, L. AND GROSELJ, D. (1991). Formulation and evaluation of oral mucoadhesive films containing metronidazole. *Acta Pharmaceutica Jugoslavica* **41**, 251–258.

SOLDANI, H., MACCHERONI, M., MARTELLI, F., MENGOZZI, G. AND CARDINI, G. (1987). Biochemistry and pharmacology of Diethylaminoethyl-dextran (DEAE-D). *Int. J. of Obesity* **11**, 201–207.

SOON-SHIONG, P., HEINTZ, P., MERIDETH, N., YAO, Q.X., YAO, Z., ZENG, T., MURPHY, M., MALONY, M.K., SCHMEHL, M., HARRIS, M., MENDEZ, R. AND SANDFORD, P. (1994). Insulin independence in a Type I diabetic patient after encapsulated islet transplantation. *Lancet* **343**, 950–951.

SOTO PERALTA, N.V., MÜLLER, H. AND KNORR, D. (1989). Effects of chitosan treatments on the clarity and colour of apple juice. *Journal of Food Science* **54**, 495–496.

TAKAHASHI, T., TAKAYAMA, K., MACHIDA, Y. AND NAGAI, T. (1990). Characteristics of polyion complexes of chitosan with sodium alginate and sodium polyacrylate. *International Journal of Pharmaceutics* **61**, 35–41.

TAKEUCHI, H., YAMAMOTO, H., NIWA, T., HINO, T., KAWASHIMA, Y. (1996). Enteral absorption of insulin in rats from mucoadhesive chitosan-coated liposomes. *Pharm. Res.* **13**, 896–901.

TANFORD, C. (1961). *Physical Chemistry of Macromolecules*. Chap. 5. New York: John Wiley and Sons.

TAYLOR, S.W., WAITE, J.H., ROSS, M.M., SHABANOWITZ, J. AND HUNT, D.F. (1994). *Trans*-2,3-*cis*-3,4-dihydroxyproline, a new naturally-occurring amino-acid, is the 6th residue in the tandemly repeated consensus decapeptides of an adhesive protein from *Mytilus edulis*. *Journal of the American Chemical Society* **116**, 10803–10804.

TERAYAMA, H. (1952). Method of colloid titration (a new titration between polymer ions). *Journal of Polymer Science* **8**, 243–253.

TIMMERMANS, J. AND MOËS, A.J. (1990). How well do floating dosage forms float? *International Journal of Pharmaceutics* **62**, 207–216.

TIMMERMANS, J. (1991). Floating Hydrophilic Matrix Dosage Forms for Oral Use: Factors Controlling Their Buoyancy and Gastric Residence Capabilities. *Ph.D. thesis*, Université Libre de Bruxelles, Brussels, Belgium.

TOMBS, M.P. AND HARDING, S.E. (1998). *An Introduction to Polysaccharide Biotechnology*. London: Taylor and Francis.

VAN DAMME, M.-P.I., BLACKWELL, S.T., MURPHY, W.H. AND PRESTON, B.N. (1992). The measurement of negative charge in cartilage using a colloid titration technique. *Analytical Biochemistry* **204**, 250–257.

VERE, D. (1984). Death from sustained release morphine sulphate. *Lancet* **1**, 1477.

WAITE, J.H. (1983). Evidence for a repeating 3,4-dihydroxyphenylalanine-containing and hydroxyproline-containing decapeptide in the adhesive protein of the mussel *Mytilius edulis L. J. Biol. Chem.* **258**, 2911–2915.

WALDRON-EDWARD, D. (1977). The turnover of mucin glycoprotein in the stomach. In *Mucus in Health and Disease*. Eds. M. Elstein and D.V. Parke, pp 301–307. New York: Plenum Press.

WANG, P.Y. (1974). Surgical adhesives and coatings. In *Medical Engineering*. Ed. C.D. Ray, pp 1123–1129. Chicago, U.S.A.: Year Book Medical Publishers.

WEARLEY, L.L. (1991). Recent progress in protein and peptide delivery by noninvasive routes. *Critical Reviews in Therapeutic Drug Carrier Systems* **8**, 331–394.

WEBSTER'S ENCYCLOPEDIC UNABRIDGED DICTIONARY OF THE ENGLISH LANGUAGE (1989). Avenel (New Jersey, U.S.A.): Gramercy Books.

WEINER, M.L. (1992). An overview of the regulatory status and of the safety of chitin and chitosan as food and pharmaceutical ingredients. In *Advances in Chitin and Chitosan*. Eds. C.J. Brine, P.A. Sandford and J.P. Zikakis, pp 673–672. London: Elsevier Applied Science.

WILSON, C.G. (1990). *In vivo* testing of bioadhesion. In *Bioadhesion – Possibilities and Future Trends*. Eds. R. Gurny and H.E. Junginger, Wissenschaftliche Verlagsgesellschaft mbH Stuttgart, APV, Vol. 25, pp 93–108.

WINKLER, R. (1986). *Stoma Therapy: An Atlas and Guide for Intestinal Stomas*, Stuttgart (Germany): Georg Thieme Verlag.

WRIGHT, P.S. (1981). Composition and properties of soft lining materials for acrylic dentures. *Journal of Dentistry* **9**, 210–223.

WYATT, P.J.(1992). Combined differential light scattering with various liquid chromatograpy separation techniques. In *Laser Light Scattering in Biochemistry*. Eds. S.E. Harding, D.B. Sattelle and V.A. Bloomfield, pp 35–58. Cambridge: Royal Society of Chemistry.

4

Analytical Ultracentrifuge Technologies for the Characterization of Biopolymer Gels and Microgels

HELMUT CÖLFEN

Max-Planck Institute of Colloids and Interfaces, Colloid Chemistry Department, Kantstrasse 55, D-14513, Teltow-Seehof, Federal Republic of Germany

Introduction

Gels and microgels are an important class of substances from both scientific and economical viewpoints. A variety of analytical techniques are available for their characterization but only very few researchers have been studying gels/microgels by means of analytical ultracentrifugation in spite of the fact that this technique is a powerful tool for the determination of thermodynamic, elastic and molecular parameters and structural properties of gels. This lack of popularity might be due to experimental difficulties concerning the detection of the polymer concentration in turbid gels, adhesion problems etc. Nevertheless the potential benefit of such experiments has over the years led to several significant investigations in this field. These have resulted in the introduction of a theoretical treatment of the sedimentation of even multicomponent gels and also an improved experimental approach which permits the characterization of a gel/solvent system in a limited concentration range in a single sedimentation equilibrium experiment. Also established is the gradient method which avoids many of the adhesion or detection problems researchers have struggled with before. For microgels that have been prepared and crosslinked in emulsions, an interesting rapid sedimentation velocity technique is now available for their characterization. This review article describes the capabilities of the experimental method and what has been achieved with it in the past. Furthermore it gives an outlook of applications which may be possible in the future.

Since the introduction of a new generation of analytical ultracentrifuges namely the Optima XL-A by Beckman Instruments (Palo Alto, USA), a renaissance of analytical ultracentrifugation can be observed especially in the field of Biophysics/Biochemistry. No longer are time consuming photographic data aquisition techniques necessary anymore. Hence, it can be expected that analytical ultracentrifugation will play an important role in the characterization of biopolymer systems again. A feature of many biomolecules is self-association which can often lead to large aggregates or even gel/microgel formation. Especially if the gel/microgel quantity is small it is hard, if not impossible, to characterize their physical and structural properties by means of

Biotechnology and Genetic Engineering Reviews – Vol. 16, April 1999
0264–8725/99/15/87–140 $20.00 + $0.00 © Intercept Ltd, P.O. Box 716, Andover, Hampshire SP10 1YG, UK

common techniques. Analytical ultracentrifugation however needs only small sample amounts and can yield a large variety of information. This review will describe the capabilities of analytical ultracentrifuge technologies for the characterization of such systems.

The first experiments with gels in an ultracentrifugal field were reported in the early years of the technique by McBain and Stuewer (1936) and by the pioneers Svedberg and Pedersen (1940). For the following two decades however only two studies were published using the ultracentrifuge as a tool for the quantitative detection of microgels. In the 60's and 70's Johnson and coworkers carried out basic investigations on the behaviour of gels in the centrifugal field (Johnson and Metcalfe, 1963, 1967; Johnson, 1964, 1968, 1971, 1972; Johnson and King, 1968). After a further ten years of relative stagnation, researchers became interested once more in using the analytical ultra-centrifuge as a tool for the characterization of gel properties. In these years the characterization of gels in the analytical ultracentrifuge has been advanced with regards to both theoretical and experimental aspects. Nowadays the ultracentrifuge can be more effective than any other method known for the characterization of gels. As many as seventy samples can be characterized simultaneously in terms of thermo-dynamic, elastic, molecular and structural parameters. Another great advantage of the ultracentrifugal investigation of gels is the continuous equilibrium which can be determined by the selection of the rotational speed. Therefore the analytical ultra-centrifuge should be applied much more for gel characterization than is the case up to now, especially in the light of modern data aquisition and computer techniques. This review will cover the technical advances and applications that have been achieved so far and will look forward to what can be achieved in the future. More detailed aspects can be found in another article recently published by the author, to which the interested reader is referred (Cölfen, 1995). The present review will cover the following: a general part describing the basic principles of the technology; sedimentation velocity; sedimentation equilibrium; density gradient methodology; the gradient method, and as part of the conclusions will provide an estimate of the potential of this technology for the future.

General behaviour of a gel in an ultracentrifugal field

THEORETICAL CONSIDERATIONS

If a gel is placed in an ultracentrifugal field in the sector shaped ultracentrifuge cell, two cases can be distinguished which are schematically presented in *Figure 1*. The first case a) is the beginning of the experiment or an experiment at low rotational speed where no sedimentation of the macroscopic gel phase occurs (as sedimentation of the gel phase, the sedimentation of the gel meniscus is understood). Nevertheless, a concentration gradient of the polymer in the gel phase will occur at these lower speeds due to the sedimentation of the crosslinked polymer. The gradient indicates the locally dependent deswelling of the gel which is caused by the swelling pressure generated by the centrifugal field. The concentration gradient changes until a final equilibrium gradient is established. This concentration gradient is considered in the so-called gradient method and the determination of the sedimentation coefficient via the movement of the centre of mass.

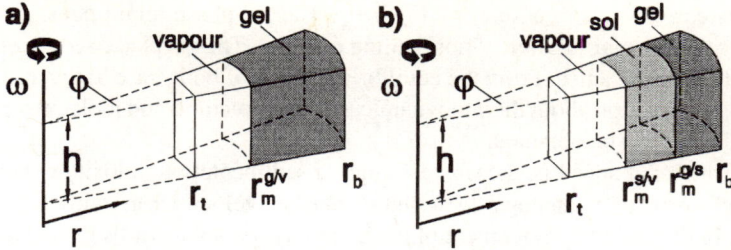

Figure 1.   Gel in an ultracentrifuge cell. a) At the beginning of the experiment or at low speeds where no sedimentation of the gel phase occurs, b) At high speeds where the gel phase has sedimented establishing a sol phase. ω is the angular velocity, h the height of the ultracentrifuge cell, φ the sector angle of the ultracentrifuge cell and *r* the distance to the axis of rotation with the indices *t* = top, *m* = meniscus, *b* = bottom, *g/v* = boundary gel/vapour, *s/v* = boundary solvent/vapour and *g/s* = gel/solvent. Redrawn from Cölfen (1993) and printed with kind permission of Dr Köster Verlag, Berlin, FRG.

The second case b) is observed at higher rotational speeds i.e. above 10,000 rpm (revolutions per minute) for the system gelatin/water. Again, the polymer concentration is increased at the cell bottom whereas it is decreased at the meniscus gel/vapour. These processes are illustrated in *Figure 2*.

At the beginning of the experiment (with the angular rotor velocity ω = 0), the polymer concentration in the gel is constant. Increase in the rotor speed ω_i after a time t leads to a decrease in the polymer concentration, c_2 at the meniscus between gel and vapour (see *Figure 2*). At a critical angular velocity ω_2 the polymer concentration drops to the value of the maximum swollen gel $c_{2,s}$. As the polymer concentration in the gel cannot be lower than $c_{2,s}$, a sol phase is introduced as soon as the polymer concentration has reached this lower limit and the meniscus gel/sol begins to sediment

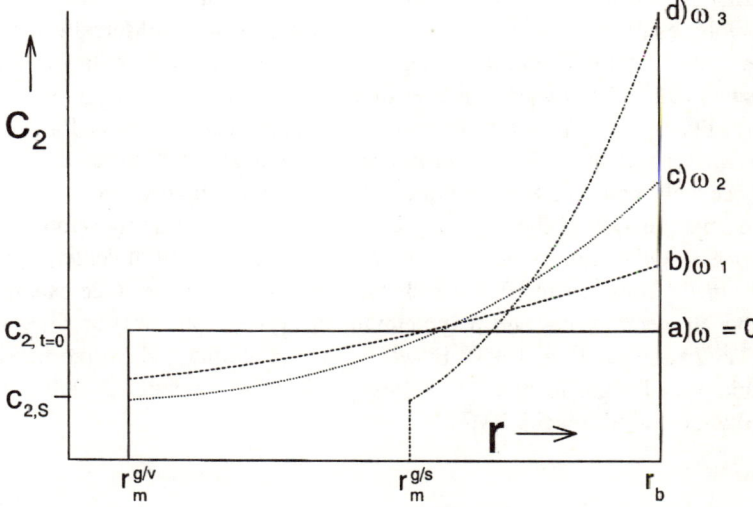

Figure 2.   Radial dependence of the local polymer concentration in the gel c_2 at different angular velocities ω_i. $c_{2,s}$ = concentration of the maximum swollen gel. Redrawn from Cölfen (1993) and printed with kind permission of Dr Köster Verlag, Berlin, FRG.

(ω_3). This corresponds to case b) of *Figure 1*. The sol phase might consist of pure solvent as well as of a solution of non-gelling material. The gel phase sediments until an equilibrium is reached. From the equilibrium states of both, case a) and case b) in *Figure 1*, information about thermodynamic, elastic, structural and molecular parameters of the gel can be obtained.

If the rotational speed is chosen very high, a sedimentation velocity experiment can be performed in analogy to a sedimentation velocity run with a polymer solution. In this case the movement of the boundary gel/sol towards the cell bottom can be measured as a function of time although recently it has been pointed out that the movement of the centre of mass has to be considered rather than that of the meniscus gel/sol (Steensgard *et al*., 1992; Borchard and Hinsken, 1997; Hinsken, 1998).

PRACTICAL PROBLEMS

Adhesion

Adhesion of the gel to the cell walls of the ultracentrifuge cell is a very significant problem when bulk gels are to be investigated. This becomes evident for the example of the system gelatin/water – the prototype of a gelling system – and also a good glue. Without any precautions, the gels stick to the centrepiece walls and windows so that elastical forces upon movement of the gel phase will distort the experimental results. One elegant way of avoiding this would be to prepare such systems in the form of a microgel and then look at their properties but this provides great practical problems of synthesis in many cases. Ways of minimising interactions between the centrepiece and window material and the gel have however been investigated.

The first person to look at the possibility of how to minimize adhesion to get reproducible results was Johnson. With Metcalfe (Johnson and Metcalfe, 1963) he reported that the same centrepiece had to be used for all sedimentation velocity experiments with gelatin/water gels to ensure reproducibility. The range of variation in the results using different centrepieces or centrepiece materials for identical samples was as high as 30%. Such errors can be minimized by impregnation of the centrepiece walls and even the windows with a thin film of highly viscous silicon oil resp. the impregnation of the centrepiece walls with a Teflon spray (Holtus, 1990; Cölfen, 1993). Other approaches consider even the construction of centrepieces and windows made from a material which shows minimized adhesion. One example are polycarbonate centrepieces and polymethylmethacrylate windows for experiments with gelatin/water (Cölfen, 1993). However, such materials reduce the maximum applicable speed significantly (40000 rpm for polycarbonate, 20000 rpm for polymethylmethacrylate acid, PMMA).

Optical detection

The choice of the optical system for a successful detection of concentration gradients inside a gel phase is very restricted as polymer concentrations are usually far above 1% (10 mg/ml) by weight and can reach concentrations of 40% (400 mg/ml) and beyond.

Whereas much effort has been spent on the development of very sensitive optical detection systems (e.g. fluorescence optics) for solution analysis down to the picomolar range of biopolymer quantities (important for example when very strong heterologeous interactions are the subject of interest), nothing has been directed at improving optical detection at very high concentrations. The only straightforward solution to the problem is the reduction of the optical pathlength from 12 mm to 6 or even 3 mm. However, shorter pathlengths are expected to cause extensive adhesion problems as in this case, the surface to volume ratio of the gel in the ultracentrifuge cell has already significantly been increased. However, even in such cases the optical signal is much higher than what some optical systems can cope with. This means that for example the Rayleigh interference optical system has no chance of providing optical records of polymer concentration inside a gel phase due to the far too high intensity differences between the interfering light beams. The ultraviolet/visible (UV/VIS) absorption optical detection system is unfortunately also not applicable due to optical saturation by the significant turbidity/light scattering: genuine UV/VIS absorption by proteins at these concentrations is in any case usually far above the linearity limit of the Lambert-Beer law and thus, the optical signal can more or less only be used at best to follow the sedimentation of the gel phase: this is also the case with Rayleigh optics. The same limitations can be expected if the gradient method is applied since even the initial gel concentrations are far above the detection limit.

Turbidity detection (which has proved to be very useful for the examination of dispersions) also unfortunately cannot be applied for gel bulk phases as in this case, the turbidity cannot be corrected for the influence of the particle size (MIE scattering). Furthermore, it is a detection system which detects time dependent changes rather than radial concentration distributions which are essential for the investigation of gels.

Hence, the only optical system which can be applied with success is the Schlieren optics: for a comparison of the optical traces for the Rayleigh interference-, Schlieren- and UV/VIS-optics, see *Figure 10* and (Cölfen, 1995). Even this system however does not detect the whole concentration gradient in every case. This holds especially for sedimentation velocity experiments, where more or less the only detectable time dependent trace is the position of the gel/sol meniscus. Therefore, assumptions about the concentration gradient or other independent local concentration gradient measurements (microtome cuts of the gel at sedimentation equilibrium with subsequent concentration determinations, etc.) have to be made. A big step forward towards a quantitative optical detection of concentration gradients has however been the low speed equilibrium gradient method (*Figure 1a*). Furthermore, sophisticated modifications like the ultrasensitive Schlieren optics (Cölfen and Borchard, 1994c) could be reversed by simply exchanging some lenses to give a very insensitive modification suitable for this very special application. Unfortunately, nobody has tried this out so far as the only commercial analytical ultracentrifuge, the Beckman XL-I does not even have the capability for a Schlieren detection, although there is the prospect of an on-line facility in the future (see, Clewlow *et al.*, 1997). However, even from the simple detection of the equilibrium position of the meniscus gel/sol (detectable by all common ultracentrifuge optical systems), a lot of information is still available.

Sedimentation velocity techniques

BULK GELS

Historically, the first investigations of gel systems used sedimentation velocity technology. However, throughout the years, it turned out that although sedimentation velocity proved to be the method of choice for microgels, *sedimentation equilibrium* proved much more effective to characterize bulk gels, and we will consider this later. The first pioneering investigations of gels in an ultracentrifugal field were reported by McBain and Stuewer (1936) on agar gels. These workers used an air driven spinner capable of rotating at speeds up to 210,000 rpm to generate centrifugal fields as high as 1,200,000g. By means of this simple device it could be shown that low concentration agar gels in the range of 0.31–1.6% by wt. and short maturation times reached swelling pressure equilibrium. When the movement of the gel/sol meniscus ($r_m^{g/s}$) was plotted against the time of sedimentation for different concentrated agar gels, the sedimentation rate of the gel was found to be constant in the beginning of the experiment but then decreasing to zero with time. A simple formula for the calculation of the swelling pressure was provided (McBain and Stuewer, 1936) which differs from that derived from thermodynamic considerations. The concentration dependence of the swelling pressure was observed to be linear with very low swelling pressures in the range of only a few millibars. The linear concentration dependence at the low polymer concentrations was seen to be analogous to the osmotic pressure behaviour of solutions. The gel concentration was assumed to be constant at different radial positions in equilibrium. This proved to be not true, as was shown in many of the later works. Theoretically it was stated that the sedimentation velocity of a gel cannot be constant. It was suggested (McBain and Stuewer, 1936) that the sedimentation rate of a gel is influenced not only by the centrifugal force but by synersis, swelling due to chemical solvation and orientation of the solvent as well as thermal molecular movements.

Svedberg was the other early worker who investigated the behaviour of gels in an ultracentrifugal field (see Svedberg and Pedersen, 1940). He also found that a gel shows a different type of behaviour from solutions in the ultracentrifuge. The latter show a constant sedimentation rate, independent of the column height whereas this is not the case for gels. This provided the verification of the considerations of (McBain and Stuewer, 1936). Furthermore, Svedberg had already stated that two cases have to be distinguished in carrying out ultracentrifuge experiments with gels: either some measureable changes in terms of the sedimentation of the gel phase occur, or they do not. This is exactly the situation shown in *Figure 1*. Svedberg also derived an equation to represent the so-called 'hydrostatic partial pressure' of a gel which is the swelling pressure at equilibrium (Svedberg and Pedersen, 1940):

$$\Pi_S = \omega^2 \int_{r_m^{g/s}}^{f} c_2 \left(1 - \tilde{v}_2 \, \rho_{01}\right) r \, dr \qquad (1)$$

with Π_S = swelling pressure, r = distance from the axis of rotation with the indices m for meniscus and g/s for the gel/solvent boundary, c_2 = polymer concentration (usually expressed as partial density of the polymer in the gel in g/ml), ω = angular velocity, $\tilde{v}_2$ = partial specific volume of the polymer and ρ_{01} = density of the pure solvent.

Equation (1) is a special case of the generalized Svedberg-Pedersen equation presented later in this review (Borchard, 1991) for binary and highly swollen gels. Svedberg pointed out that it is very often impossible to determine the polymer concentration gradient in the whole gel phase with the optical detection system of the ultracentrifuge due to the high turbidity of the gel. Therefore, he introduced an approximation method based on a mass balance and the assumption that the concentration in the middle of the gel column is equal to the average concentration of the gel column. As McBain and Stuewer before, Svedberg found for agar gels that the swelling pressure of the gel is roughly proportional to the polymer concentration in the gel.

The first basic study to be published on the behaviour of a macroscopic gel in the ultracentrifugal field came from P. Johnson in 1964 (Johnson, 1964). He investigated agar and gelatin gels in a phosphate-NaCl-buffer at temperatures between 10 and 25°C using very short maturation times of only 30 minutes for the agar gels. The concentration of the gels was determined to a low degree of accuracy by compressing the gel at 60,000 rpm and determining the mass of the compressed gel phase. After drying to constant weight, the concentration of the original gel could be obtained with the knowledge of its mass.

Nevertheless, Johnson was able to demonstrate some important properties of biopolymer gels (Johnson, 1964): he was able to show that a gel shows typical velocity characteristics in a limited range. He found a sharp gel-solution interface which was sedimenting in the direction of the applied field. Again, the sedimentation behaviour of a gel was found to be fundamentally different from that of a solution (see also Svedberg and Pedersen, 1940). In a sedimenting solution boundary the concentration is continuously decreased due to the radial dilution caused by the sector shape of the cell, whereas in a sedimenting gel the mean gel concentration increases as the gel volume is decreased with constant polymer mass. From that point of view he presented a formula to calculate the mean concentration of the gel phase at a defined time t applying a volume balance. This formula could give at least a rough estimation of the concentrations in the gels during centrifugation.

Johnson then tried to relate the initial slope of the $\log_{10} r_m^{g/s}$ plot to the square of the rotational speed (Johnson, 1964) to calculate a sedimentation coefficient of the gel phase but found only a little range where this could be accomplished. In contrast to a solution, the plot of $\log_{10} r_m^{g/s}$ against time was not linear for a gel but showing a decrease of the slope already after a few minutes. This slope was found to come to zero defining an equilibrium degree of swelling (see *Figure 3*). This equilibrium value was dependent on the rotational speed and the initial gel concentration.

Furthermore an 'induction' period at the beginning of an experiment was described where no sedimentation of the gel phase occured if the rotational speed was selected too low. This induction period was found to increase with gel concentration and lower speed. A maximum sedimentation rate could be defined in the $\log_{10} r_m^{g/s}$ plots vs. time (linear portion between rise and decrease of the slope, (see *Figure 3*) which was approximately proportional to the applied field at several concentrations and inversely proportional to the initial gel concentration between 0.5 and 2% by wt. at the same initial column length.

Nowadays, it is known that the movement of the *centre of mass* has to be considered (Borchard and Hinsken, 1997). Nevertheless Johnson used this plot to define the

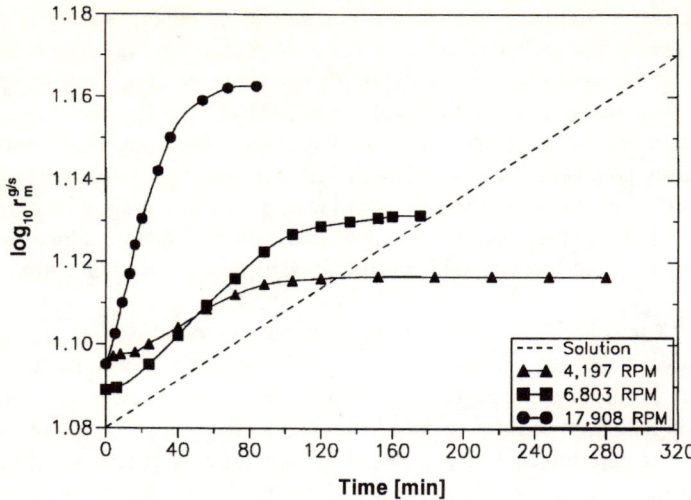

Figure 3.   Plots of $\log_{10} r_m^{g/s}$ against time from gel sedimentation diagrams for 0.5% difco-agar gel at various speeds. The dashed line represents a schematic diagram for a solution. Modified figure from Johnson (1964) printed with kind permission of the author and the Royal Society, London, U. K.

conditions under which the gel interface behaves like a solution boundary. For those cases he calculated a so called effective sedimentation coefficient s considering the movement of the boundary gel/solution towards the cell bottom in analogy to the movement of the boundary in case of a solution. Upon variation of the experimental conditions which are known to effect the gel structure like the ionic strength, pH of the buffer and the temperature on the gel sedimentation, it was found that the temperature (10–25°C) as well as a pH (6.5–7.8) and ionic strength (0.1–0.5) alteration caused little or no measurable change in the determined gel sedimentation rate which is against all expectations. It was found to be impossible to relate the effective sedimentation coefficient to a sedimenting species because the gel can be considered as a network of infinite molar mass. The motion of the gel interface was stated to be a viscous rather than an elastic type of flow as a gel showed little or no rapid recovery after the rotor has been decelerated. Johnson concluded (Johnson, 1964) that the flow of a gel must involve continuous rupture and re-formation of the junction points and explained the sedimentation of the gel phase with this concept. Consequently, the equilibrium state occurring after the sedimentation of the gel phase could be considered from the kinetic point of view as an equal rate of rupture and reformation of the junction points. From the occurrence of an induction period he concluded that the rupture of crosslinks might be a slow process. An argument against the generalization of these considerations is that it would be impossible for a gel with permanent crosslinks to sediment unless the permanent bonds are ruptured. Nevertheless, it is known that such gels also sediment (Borchard, 1975a; Cölfen and Harding, 1994). An alternative explanation is that the sedimentation of the gel can also take place as compression of the gel without rupture of crosslinks or bonds but with folding of the network chains and exclusion of solvent. This model seems to be more likely at least for chemically crosslinked gels as the energy to rupture chemical bonds is rather high. Nevertheless, up to now none of these possibilities could be clearly verified experimentally.

Johnson (1964) also observed a slower sedimenting soluble fraction of considerably less than 10% for the agar but about 30% for the gelatin gel. These so-called 'soluble parts' were found to be quite polydisperse for the case of the gelatin as deduced from the extensive spreading of the Schlieren peak. By contrast to the gel, the $\log_{10} r_m^{g/s}$ vs. time plot was found to be linear, as expected for a solution. The soluble gelatin was detected over a range of experimental conditions. Its sedimentation coefficient was similar to that of the molten gel which gives strong evidence that these soluble parts consist only of gelatin molecules which could not be incorporated into the network due to the very short maturation times of the gels or chemically damaged gelatin. Johnson himself stated that the proportion of soluble parts decreases somehow by allowing a gel to mature but still very significant soluble portions remained after long times.

Considering equilibrium aspects of the gel sedimentation, Johnson (1964) also stated that the derived increase of the $\log_{10} r_m^{g/s}$ with the rotational speed agreed qualitatively with the equation (1) of Svedberg (Svedberg and Pedersen, 1940). Considering the induction period which could be observed if the initial gel concentration was not very different from the average equilibrium concentration, Johnson also concluded from changes in the Schlieren traces of the gel phase during this period, that internal structural changes of the gel occur before any movement of the gel meniscus commences. As is now known today, this process is simply the locally dependent deswelling of the gel.

A further study from Johnson's laboratory treated the sedimentation behaviour of several dilute gelatin gels under sedimentation velocity conditions (Johnson and Metcalfe, 1963). As in the previous study (see Johnson, 1964) the separated soluble parts which occured in significant amounts behaved as real solution giving a sedimentation coefficient which equals that of gelatin in solution. It was stated that the occurence of solution components cannot be caused by the high pressures in the centrifuge cell as ΔV for the sol gel transition is negative (Flory and Garrett, 1958). Further evidence is provided by the observation that the same amount of solution component was observed over a range of centrifugal fields.

Three general characteristic features of a $\log_{10} r_m^{g/s}$ vs. time plot for a sedimenting gel were stated (Johnson and Metcalfe, 1963) which can vary widely in different gels (see *Figure 3*): a) the induction period at the beginning of the experiment, b) a period with a maximum movement of the gel boundary where the slope is constant and c) the end of the experiment where the slope is continuously decreasing to zero. It must be stated that the transition between these periods is continuous. It is important to note that the induction period is probably just due to the fact that the movement of the gel meniscus is treated as indicative for the sedimentation of the gel. In fact, sedimentation of the polymer in the gel occurs right after the application of an ultracentrifugal field which is expressed in the formation of a concentration gradient in the gel phase even if the meniscus does not sediment (Hinsken and Borchard, 1995).

Sedimentation coefficients for the gel have been calculated on the basis of the maximum movement of $r_m^{g/s}$ (Johnson and Metcalfe, 1963). These values are subsequently referred to as the *effective sedimentation coefficients* of the gel in the text that follows and can be considered to be rather inaccurate as only a small time range of the gel sedimentation (constant sedimentation velocity) can be used to calculate the sedimentation coefficient of the gel. This disadvantage can be circumvented if the movement of the centre of mass is used to calculate the sedimentation coefficient rather than that of the gel meniscus (Hinsken and Borchard, 1995). Looking at the effect of the gel maturation temperature and the run temperature on the sedimentation coefficients of the soluble component and the gel as well as the amount of soluble polymer, drastic effects were observed. Increase of the maturation temperature from 2 to 20°C at 1 h maturation time increased the amount of the soluble component from 20% to 90%. The sedimentation coefficient of the soluble component remained nearly constant whereas that of the gel was decreased by a factor of ≈ 3 from 14.5 S to 4.5 S. Below 20°C the gel sedimentation was found to be insensitive to the run temperature for a few degrees below the setting temperature, whereas it was very dependent on it above 20°C reflecting the breakdown of the gel structure up to complete gel melting. An attempt was made to monitor the gelation in a sedimenting solution by decreasing the temperature from 23°C (melting point) to 19°C over 4.5 h. Needle like sedimenting striations appeared in the Schlieren patterns after a time, which corresponded to aggregates leading to later gelation. Such aggregates have only been observed near the melting point or just before gelation. Their concentration is decreased drastically with the maturation time of the gel, much more than the concentration of the non-aggregated solution component.

The maturation time of the gels was also found to influence the sedimentation behaviour of the gel and the amount of solution component. On aging at 20°C, there was a large decrease in the gel sedimentation coefficient during the first hour whereas the changes with maturation times up to three months were small. In the entire period from a few minutes to 3 months the solution component was decreased from 40% to 25% for the system gelatin/water. If the same experiment was performed at a temperature of 2°C, the solution component (20% after 1 h) vanished after 44 h, whereas the sedimentation coefficient of the gel increased by a factor of 2.

By studying the effect of the initial gel concentration between 1 and 3% by wt. on the sedimentation behaviour and the amount of soluble component, it was found that at 20°C and an initial gel concentration of 1.5% by wt., material was still partly in solution, which was built into the network at 2% by wt. The effect was less pronounced at 18°C and vanished at 2–5°C. Some experiments on the effect of the ionic strength showed that deionized gels, representing somehow a structure of precipitated aggregates embedded in a weak network, sedimented much faster than a normal gelatin/water gel. When KCl (c = 1 mol/l) was added, the sedimentation rate was slower than that of the gel with water. This implies that the sedimentation velocity of the gel depends on its structure. Addition of KSCN (0.5 mol/l) to a 1% by wt. gelatin solution prevented the formation of aggregates and hence no gel was formed.

Separating the gelatin gel into the gel and the solution component by preparative centrifugation showed that the freeze dried solution component was far more rapidly soluble than the freeze dried gel fraction and gave no gel even at 2% by wt. At a concentration of 5% by wt. a gel was formed which showed that the gelatin molecules in the soluble fraction are at least partly able to form a network. In an experiment with a 2% by wt. gel (Johnson and Metcalfe, 1963) where the solution phase was removed three times from the ultracentrifuge cell and replaced by the same amount of water, no solution component separated from the sedimenting gel interface anymore, after the gel was allowed to mature for 40 h at 4°C. This gives evidence for the conclusion that the low molecular solution component may be removed by such a procedure. Further evidence for this was found in an experiment where the sedimentation of the first of the three extracts was compared with a diluted solution of the remaining gel phase. The extract showed a considerable tail of slower sedimenting components e.g. components with a lower molar mass.

As the sedimentation behaviour of a gel was found to be dependent on its structure it was investigated if the sedimentation coefficient of the gel can be correlated with its rigidity. It was found that the rigidity of the gelatin gels seemed to be related with the amount of solution component but not with the sedimentation coefficient of the gel. Nevertheless, it was pointed out (Johnson and Metcalfe, 1963) that the soluble parts do not directly contribute to the gel rigidity but only indirectly by lowering the concentration of molecules which build up a network. Highly rigid gelatin gels had only 20% solution component whereas gels with low rigidity had 40–50% solution component. This effect was even observed if the molar mass of the two types of gelatin molecules forming the gel was comparable.

Systematic sedimentation velocity experiments on diluted gels from commercial gelatins and gelatins from soluble collagens were described by Metcalfe (1965). This study contains far more results than could be included in the previous publication (Johnson and Metcalfe, 1963). Metcalfe compared the sedimentation behaviour of the gelatin gels over a wide range of physical conditions using mainly three quantitative measures: a) the maximum rate of gel sedimentation of the gel interface as defined above, b) the amount of solution component and c) the sedimentation coefficient of the solution component.

It could thus be shown that sedimentation studies are a sensitive method for the examination of changes in the gel structure and the interaction of the gel with other molecules. Although a qualitative relationship between the proportion of the solution component and the rigidity of the gels was found, no quantitative relation could be derived for this as it was concluded that mainly the changes within the gel itself, e. g. of its structure, are responsible for the rigidity changes. Also, no correlation was found between the gel rigidity and the maximum gel sedimentation rate which seemed to be more dependent on the gel structure. Reflecting upon the reproducibility of the experiments, considerable time dependent effects after gelling through further crosslinking of the gelatin gel were found (Metcalfe, 1965). In contrast to the previous study (Johnson and Metcalfe, 1963), it was found that adhesion of the weak gels to the ultracentrifuge centrepieces had no effect on the sedimentation coefficient of the gel whether the material of the centrepiece was changed, the surface was lubricated, the sector angle was altered or the gel was loosened from the cell walls and windows before the experiment. Nevertheless, the more rigid gels were loosened generally from the cell walls and windows as a considerable adhesion took place. The gel sedimentation coefficient was found to be proportional to the applied centrifugal field at constant gel column length just as it was found for agar gels (Johnson, 1964), whereas the induction period was inversely proportional to the field. The sedimentation coefficient of the gel was found to be proportional to the length of the gel column, independent of the sector angle of the centrepiece (e. g. the gel volume) and unchanged if a water layer was placed above the gel column or not. The independence of the sedimentation coefficient from the presence of a water layer on top of the gel column had already demonstrated that the total hydrostatic pressure has a negligible influence on the sedimentation behaviour of the gel, as was also shown theoretically later (Borchard, 1991). However, the observed dependence of the sedimentation coefficient of the gel on the column length indicates an error in the definition of gel sedimentation as the sedimentation of a gel has to be treated as a movement of its *centre of mass* rather than a movement of the meniscus gel/sol (Borchard and Hinsken, 1997). Therefore, all results based on the effective sedimentation coefficient from the movement of the meniscus gel/sol have to be treated with care, especially as the reported gel sedimentation coefficients are generally of the same order of magnitude of soluble polymers and not a factor of 10 lower, which has been found recently (Borchard and Hinsken, 1997).

As had been reported in many previous studies dealing with ultracentrifuge experiments with gels, Metcalfe (1965) had problems in detecting the concentration gradient inside the gel phase, especially in the more concentrated gels. Therefore, it was attempted to monitor the concentration distribution inside the gel phase by placing a gel consisting of alternating dyed (labelled) and undyed gelatin in the ultracentrifuge cell. This attempt was not successful as the dyed gelatin was distributed in irregular lumps throughout the gel column after the experiment implying a circulation of the polymer in the gel phase during the sedimentation. The density differences between dyed and undyed gel have been given as possible explanation as the applied dyestuff was known to promote gelation. Alternatively, this observation partly supports the explanation of Johnson – at least for the case of the physically crosslinked gelatin/water gel – that gel sedimentation occurs when crosslinks are ruptured (Johnson, 1964). It is also likely that adhesion of the gel lead to the observed rupture of the gel.

Consequently, it was attempted to establish a large density gradient inside the gel phase by layering a 2% by wt. gelatin gel above a 4% by wt. gelatin gel. The sedimentation coefficient derived from the movement of the boundary between the two gels as well as that from the movement of the meniscus gel/solvent from the 2% by wt. gel agreed closely with the values obtained for the separate gels with the same column height. From this it could be deduced that the sedimentation rate of the boundary between the two different concentrated gels was approximately independent of the presence of the gel above it. A continuous density gradient has not been established as expected, initially due to the view of rupture and reformation of the network upon sedimentation.

When the effect of the initial gel concentration on the gel sedimentation coefficient was investigated in the range of 1.5–5% by wt., a maximum of the sedimentation coefficient for 2% by wt. was found for 18°C shifting down to 1.5% by wt. at 5°C. A plot of log $1/s$ (s = sedimentation coefficient of the gel) against log c (c = average concentration in the gel at time t) yielded ranges of linear dependencies with a slope n for gelatin/water gels. Using data for dilute agar/water gels (Johnson, 1964), a similar range of linear relationship was found for agar, restricted extensively by the equilbrium approach of the agar gels. Nevertheless, it was stated that the maximum gel sedimentation rate alone is not sufficient to completely characterize the sedimentation of the gel as no relationship between this sedimentation rate and the induction period could be found. If the movement of the centre of mass is taken to calculate the sedimentation coefficient of the gel (Borchard and Hinsken, 1997), no induction period is observed anymore which shows that the definition of the gel sedimentation coefficient of the gel *via* the movement of the centre of mass is more correct.

In addition to the already published (Johnson and Metcalfe, 1963) results of the effect of the ionic strength on the gelatin gel sedimentation coefficient, the concentration of KCl in the range of 10^{-5} to 1 mol/l was found to have a dramatic effect on the sedimentation coefficient of the 2% gel investigated at 18°C (Metcalfe, 1965). Increasing the KCl concentration for example from 10^{-3} to 10^{-2} mol/l and hence the ionic strength, the sedimentation coefficient of the gel was decreased from nearly 60 S to 20 S. In contrast to the gel, the increase of the ionic strength led to an increased sedimentation coefficient of the solution component. The pH affected the gel sedimentation coefficient very strongly as well, giving a maximum near the isoelectric point of the gelatin.

The solution component was the subject of intensive study. It could be shown that the fraction of soluble material could be decreased significantly to negligible magnitude by decreasing the temperature of the ultracentrifuge experiments to 5°C or lower. Sedimentation studies of these fractionated solution components suggested a molar mass of only 10,000 g/mol in contrast to the tenfold or higher value of the gelatin. This could explain the already outlined lack of gelling ability of such a solution component. Further work, especially on the soluble parts from gelatin gels applying optical rotation measurements and amino acid analysis, has been reported by King (1967).

Further experiments were carried out to compare the sedimentation behaviour of gels from acid and alkali processed gelatin, fractionated gelatin and gelatin from soluble collagen. The sedimentation coefficients for gels from soluble collagens matured at 1 h at 18°C were found to agree with those of alkali processed commercial gelatins. But in contrast to the other gels investigated, the amount of solution component at 18°C in gels from soluble collagens was found to be much lower (e. g. < 15% by wt.) down to only 2.5% by weight. Under these conditions the amount of solution component from fractionated gelatins (the α and β components have been purified) was found not to be very much dependent upon the composition of the gelatin, e.g. the fraction. Comparing the sedimentation coefficients of the corresponding gels, a small increase of s was found for the α-fraction whereas s was slightly decreased with respect to the unfractionated gelatin for the β-fraction.

The effect of heating was studied with a gel containing only a very small amount of solution component. The amount of solution component was found to increase slightly with each heating step due to the thermal degradation of the gelatin, whereas the sedimentation coefficient of the gels increased reflecting the weaker gel structure.

Another study on the sedimentation behaviour of gelatin gels was published in 1967 by the Johnson group (Johnson and Metcalfe, 1967). This publication dealt mainly with the results in Metcalfe's PhD Dissertation (Metcalfe, 1965) but included a more detailed consideration of the dependence of the gel sedimentation coefficient on the average gel concentration. When the column length of the gel $(r_b - r_m^{gls})$ was introduced into this empirical equation, a more detailed plot of $\log s + \log (1 + r_m^{gls}/r_b)$ versus $\log (1 - (r_m^{gls}/r_b)^2)$ could be introduced which gave a straight line with the slope $n + 1$. The parameter n was suspected to be somehow related to the gel structure. Considering the ill-defined nature of the sedimentation coefficient via the movement of r_m^{gls}, the parameter n can only be a qualitative and empirical quantity.

Considering the application of Equation (1) in this review to the sedimentation velocity experiments with gels, it could be stated that this equation must apply more to equilibrium conditions than to the steady flow during the sedimentation velocity run: as an explanation it can be said that under these conditions swelling is a slow process compared to gel sedimentation.

This work was followed by a further sedimentation velocity study on gelatin gels attempting to describe the sedimentation behaviour of gels observed in the previous studies more quantitatively (Johnson and King, 1968). An important point was the attempt to measure the polymer concentration distribution inside the gel phase during sedimentation which normally could not be observed throughout the whole gel phase with any of the standard ultracentrifuge detection optics. As the attempts with layers of dyed and undyed gelatin failed (Metcalfe, 1965), a gel was set up from completely dyed gelatin. When the optical density of the photographic negatives recorded during the sedimentation of the gel were evaluated with a microdensitometer, it was found that the optical density, and hence the polymer concentration, was approximately constant throughout the gel. This is in contradiction to the prediction of Svedberg and Pedersen (Equation (1)) which would lead to a concentration gradient inside the gel phase (Svedberg and Pedersen, 1940) as well as with numerous experimental findings obtained later. Further, with increasing time an increase in the optical density was recorded. This was stated to be due to the concentration increase of the gel as its volume is decreased during the sedimentation. The constant polymer concentration in the gel phase must be an artifact of the optical densitometer readings of the Schlieren negatives or an effect of a too high concentration of the dyestuff far above the linearity range of Lambert-Beer's law.

When the concentration dependence of the gel sedimentation coefficient was investigated at temperatures of 4, 10 and 18°C, it was found that $\log s$ vs. $\log c$ gave a linear dependence above gel concentrations of 1.5% by wt. (experimentally determined sol/gel transition). With falling temperature the dependence of the gel sedimentation coefficient upon gel concentration increased. From this observation an empirical relation was deduced which was already indicated in the previous paper (Johnson and Metcalfe, 1967):

$$s = k \, \frac{r_b - r_m^{g/s}}{c^n} \tag{2}$$

Plots of $\log s + \log (1 + r_m^{gls}/r_b)$ versus $\log (1 - (r_m^{gls}/r_b)^2)$ yielded the empirical parameter n (Johnson and Metcalfe, 1967). After it had been pointed out that the shape of the $\log r_m^{gls}$ vs. time plot influences the derived n, some dependencies of n on the temperature and the maturation time were presented (see *Figure 4*).

It was stated from this behaviour that at temperatures below 18°C n increases with decreasing temperature with the maturation time indicating drastic changes in the gel structure on aging. Furthermore, it was pointed out that n was also dependent on the ionic strength and the pH of the solvent. An increase in the ionic strength caused a significant decrease of n in the isoelectric region at all considered temperatures. However, this effect was small or causing a slight increase in n at higher pH. If the pH was increased away from the isoelectric region, n decreased significantly. When urea, a structure disruption agent which breaks hydrogen bonding, was added in increasing quantities n fell rapidly down to zero.

From the experimental results it was concluded that the empirical parameter n gives an indication of the crosslinking degree and the gel structure. High values of n ($n = 3$) were found to represent extensive crosslinking whereas a low value of $n = 1$ indicated

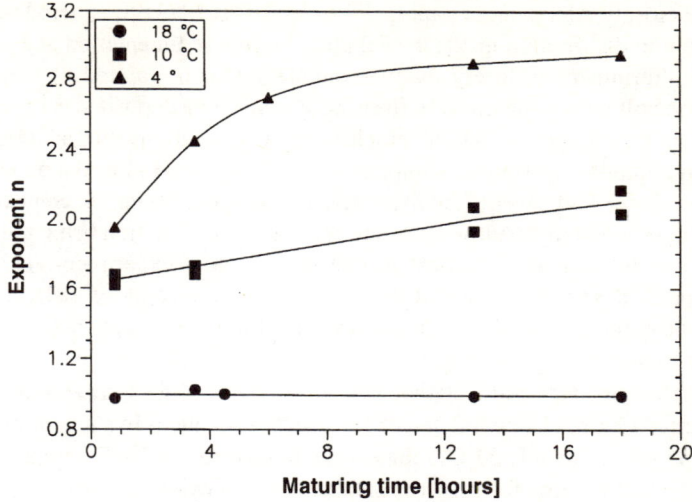

Figure 4.   Effect of temperature and maturation time on the value of *n* for aqueous lime processed ossein gelatin gels of 2% by weight. Figure redrawn from Johnson and King (1968) with kind permission of the author and the Royal Photographic Society of Great Britain, Devon, UK.

a weakly crosslinked network (see *Figure 4*). From the sedimentation behaviour of highly asymmetric molecules with high effective volumes in dilute solution compared with that of weakly interacting systems, it was concluded that *n* was suitable for interpreting solution sedimentation behaviour as well.

In a more comprehensive article, Johnson (1968) summarized the basic findings so far derived on the sedimentation behaviour of gels. On the analysis of flow under the centrifugal field, an experiment was reported where the rotor was stopped after the gelatin gel was maximally compressed. Leaving this gel for 15 h at 18°C, only very limitited reswelling was observed. This gel showed similar sedimentation behaviour to the original one with the exception that no induction period was observed anymore. From this experiment it was concluded that the sedimentation of a gelatin gel must be regarded as irreversible although numerous later results show the reversibility. Recent results show that the reported irreversible behaviour can be related to additional crosslinking of the gel by the soluble gelatin component during the gel sedimentation which leads to the formation of a gradient gel (Cölfen and Borchard, 1994; 1995).

Some more detailed considerations as well as some further experimental results concerning the structural parameter *n* were described in this work. Addition of sodiumdodecylsulfate in small amounts < 0.01 mol/l suprisingly strengthened the gel reflected in an increasing *n* which was explained by an associating effect of this long chain molecule. Higher surfactant concentrations caused a steady decrease of *n* again. Chemical crosslinking of the gelatin with glutardialdehyde decreased *n* further from an already low level, although the number of crosslinks was obviously increased. The same behaviour was found for almost completely covalently crosslinked 3% acrylamide gels which had an *n* value of nearly 0. These findings were explained in a way that *n* is low or zero for gels with a small number of covalent bonds and high for those with a large number of weak bonds. This interpretation suggests only some qualitative proportionality between *n* and the crosslinking density.

In two closely related latter papers, Johnson (1971, 1972) considered the polymer concentration distribution inside the gel phase during sedimentation and at sedimentation equilibrium more closely. For the example of agar, he pointed out that in the case of gels, the polymer is not removed from the system being deposited at the cell bottom as it occurs during sedimentation velocity runs with solutions. For agar gels without significant amounts of soluble components he showed that after the experiment the compressed gel pad swelled to its original condition when in contact with the supernatant solvent in the ultracentrifuge cell. The sedimentation behaviour observed for this swollen gel was identical to that in the initial experiment. This swelling behaviour is a very important difference to the behaviour of gelatin gels with significant amounts of soluble component as stated above and explained later (Johnson, 1968; Cölfen and Borchard, 1995).

The techniques for sedimentation velocity analysis of gels presented above have been applied to the gel-like fraction of porcine gastric mucus in a mucus dispersion at pH 3.5 and approximately 20°C (Johnson and Rainsford, 1972). The typical sigmodial plot of $\log_{10} r_m^{g/s}$ vs. time for gels was obtained for the whole mucus as well as for the mucus in which the supernatant fraction containing 3 other components had been removed by previous centrifugation. A structural parameter n of 2.5–2.6 (see Equation 2) was derived for a 2% by wt. mucus gel with the procedures described above, indicating a large amount of weak intermolecular interactions. This interaction was found to get stronger at higher concentrations with $n = 3.3$ for a 6.9% by weight of mucus gel.

At a pH of 7.3 the proportion of the gelling component was found to be much smaller than at pH 3.5. The same effect up to a vanishing gel content was observed as expected when several structure disorganizing agents like 8 mol/l urea, 6 mol/l formamide, 6 mol/l guanidine hydrochloride, 10% triton X-100 and particularly 2% sodium deoxycholate with and without 5% mercaptoethanol were added.

A similar sedimentation velocity method to those presented above for the rapid characterization of gels was described by a Ukrainian group (Babskij and S'edin, 1977). They investigated Difco agar gels (matured for 24 h and 3 months) at 20°C, pH 7 and the very low concentration of 0.15% using Schlieren optics. Considering the rapid sedimentation which took place already at the acceleration of the rotor, they derived basically the same results which were already presented in the ultracentrifuge papers by Johnson's group (see for example Johnson, 1964; Johnson and Metcalfe, 1963; 1967) before without citing them. But the interpretation differs. During the induction period at the beginning of the experiment, a destruction of the gel structure was assumed to occur as discussed by Johnson in the early papers. (The notes of caution with the induction period discussed above for Johnson's papers must be applied here as well). The constant sedimentation rate for the following period is reported to represent that of the aggregates which form a gel again later induced by the increased concentration and the hydrostatic pressure leading to a decrease in the sedimentation velocity until the sedimentation equilibrium is reached. This argument would not explain the equilibrium situation reported for the sedimentation of agar gels (Johnson, 1971; 1972), because it is most unlikely that the original network structure would be built up from the sedimenting aggregates again which must be postulated by the equilibrium nature of this process. Furthermore, the figures in this article show the sedimentation of a phase with a defined phase boundary and not the typical Schlieren peak or striations to be expected if the sedimenting species would consist of aggregates at certain times. The investigated agar gels contained soluble parts which were found to sediment individually. Overall, the results presented (Babskij and S'edin, 1977) are only of qualitative nature and outline the use of sedimentation velocity experiments with gels. To derive quantitative results, it was considered as necessary to accelerate the rotor with a well defined characteristic. Furthermore, the construction of specialized centrepieces was considered to be necessary for future applications without pointing out requirements these centre-pieces have to meet.

After these studies, sedimentation velocity experiments ceased to be performed (or at least appeared to be) for the characterization of gels for more than 20 years as since the early 70's, the power of sedimentation equilibrium experiments with gels was recognized as a more powerful alternative procedure.

The sedimentation velocity of a gel was subject of a recent paper by Borchard and Hinsken (1997). In contrast to the treatment so far that the movement of the meniscus gel/sol must be used for the calculation of the sedimentation coefficient, they could show that the movement of the center of mass has to be considered. In contrast to a solution, where a relative motion of the polymer to the solvent occurs, for a gel, a relative movement of all components to each other takes place. Thus for an n-component system, $n-1$ independent fluxes have to be considered resulting in a relevant flux density under sedimentation velocity conditions. From irreversible thermodynamics, an alternative definition of the sedimentation coefficient s was derived for a binary system and early stages of sedimentation ($t \to 0$):

$$s = \frac{\alpha}{\rho_2}(1 - \tilde{v}_2 \rho) \qquad \text{with } \alpha = \frac{\rho_2 v_2}{(1 - \tilde{v}_2 \rho)\omega^2 r}$$

$$\text{equivalent to} \quad s = \frac{dr_s / dt}{\omega^2 r} = \frac{\ln(r_s / r_{s,0})}{\omega^2 t} \tag{3}$$

where α is a phenomenological coefficient, $v_2 = $ the relative velocity of the polymer and r_s is the radial position of the center of mass defined by:

$$r_s = \frac{\int_{r_m}^{r_b} \rho_2(r) r^2 dr}{\int_{r_m}^{r_b} \rho_2(r) r\, dr} \cdot \frac{\sin \varphi}{\text{arc}\varphi}, \text{ resp. } r_{s,0} \text{ at } t = 0 \text{ accessible by extrapolation.}$$

With such a treatment, induction periods are no longer observed at the beginning of the experiment because the changes of the polymer concentration gradients have already been taken into account, even if sedimentation of the meniscus gel/sol has still not occurred. This is an important improvement as it enables the determination of sedimentation coefficients of gels even at low rotational speeds, where no sedimenta-tion of the meniscus occurs and thus adhesion and detection problems are minimized.

For longer experimental times, the equilibrium is approached, thus no relative flux density occurs anymore, meaning that the flux by sedimentation has to be balanced by that of diffusion in analogy to the solution case:

$$\underbrace{\alpha(1 - \tilde{v}_2\rho)\omega^2 r}_{\text{Sedimentation}} = \underbrace{\alpha\left(\frac{\partial \tilde{\mu}_2}{\partial \rho_2}\right)_{\text{T,P}} \left(\frac{\partial \rho_2}{\partial r}\right)}_{\text{Diffusion}} \tag{4}$$

Thus, by measurement of the polymer concentration gradient inside the gel phase $\rho_2(r)$ at sedimentation equilibrium, the diffusion coefficient of the gel D can be obtained by Equation (5), if s and α are already known from the early experimental stages via Equation (3)

$$D = \alpha\left(\frac{\partial \tilde{\mu}_2}{\partial \rho_2}\right)_{\text{T,P}} = \frac{s\omega_i^2 r_s \rho_2}{(d\rho_2 / dr)} \tag{5}$$

where $\tilde{\mu}_2$ = specific chemical potential of the polymer.

For a 4.5% by wt. κ-carrageenan/water gel at 10°C, s was found to be 0.10–0.14 S and D to be $3.4–3.8\ 10^{-10}$ cm²/s. Thus, s is found lower by a factor of 10, than that for the soluble polymer whereas D is even 3 orders of magnitude lower. Due to these extremely low transport quantities of a gel, the equilibration times are very long, as has already been reported before.

The definition of s via the movement of the center of mass may for the first time provide a quantitative analysis of quite rapidly obtainable sedimentation velocity data in terms of gel structures (following the early attempts by Johnson and coworkers who used however an inappropriately defined s). However, the method is restricted to completely transparent and thus not too concentrated gels. Furthermore, the Schlieren gradients are rather broad which might cause inaccuracies in the concentration determination. This however could well be circumvented by the use of on-line Schlieren optics (Clewlow *et. al.*, 1997) and should not be regarded as a major problem of the method.

MICROGELS

The analytical ultracentrifuge had, by 1958, already been used for the determination of the amount of microgel formation resulting from emulsion polymerisations (Shaskoua and van Holde, 1958). The microgels studied consisted of styrene crosslinked with divinylbenzene, methyl acrylate crosslinked with divinylbenzene and acrylonitrile crosslinked with methylene-bisacrylamide. Afterwards, styrene and acrylonitrile had been grafted onto the polymers. The success of the grafting was studied employing the Schlieren optical system of the ultracentrifuge. In incomplete grafting reactions two components could be observed in the Schlieren patterns. The fast component was the microgel, the slow one the linear polymer. The concentrations of the individual components have been determined by measuring the areas under the Schlieren peaks. Afterwards they have been corrected taking the concentration dependence of the sedimentation rates into account. The additional fact has been taken into account, that for two chemical species separated into two components which sediment with different speeds, the faster component contains both species whereas the slower component consists of one single species. A mixture of microgel and 25% of linear polymer was analyzed with this method (Shaskoua and van Holde, 1958) yielding 26% of linear polymer. This therefore established the credentials of the ultracentrifuge as a quantitative method for analyzing linear polymer/microgel mixtures.

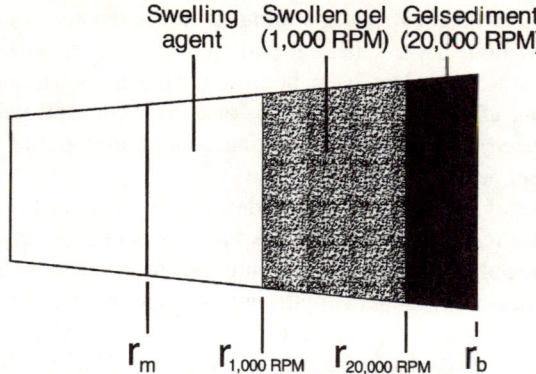

Figure 5. Sector shaped ultracentrifuge cell (single sector) for determination of the swelling of small amounts of polymer gel (schematic). *r* is the distance from the rotor axis with the indices *m* = meniscus of the swelling agent and *b* = cell bottom. The indices 1,000 rpm and 20,000 rpm refer to the gel boundary at rotor speeds of 1,000 or 20,000 rpm. Redrawn from Lange (1986) with kind permission of Steinkopff Verlag, Darmstadt, FRG.

In another study this method has been applied to characterize the crosslinking efficiency of an emulsion polymerization Shaskoua and Beaman (1958). The components for the polymerization of a microgel described in Shaskoua and van Holde (1958) have been used in different combinations and amounts to find the minimum quantity of crosslinking agent necessary to form the microgel. The ratio of the microgel and the linear polymer has been determined with the ultracentrifuge as described before (Shaskoua and van Holde, 1958).

Lange (1986) introduced the analytical ultracentrifuge for the determination of the degree of swelling and crosslinking of even extremely small gel quantities in a dispersion containing small swollen particles. Lange placed the gel with an excess of swelling agent in the ultracentrifuge cell and centrifuged at 20,000 rpm for one or two hours to compress the gel with a smaller specific volume than the swelling agent at the cell bottom (see *Figure 5*).

Afterwards he reduced the speed as much as possible (1,000 rpm) to allow the gel to swell to its maximum degree of swelling. The centrifugal field at 1,000 rpm is assumed to be so low that no deswelling of the gel due to the generated swelling pressure occurs. The swelling equilibrium was reached after only one hour. These conditions might differ from system to system. From the position of the boundary gel/ solvent which could be observed with Schlieren, absorption or Rayleigh interference optics, he could calculate the volume of the swollen gel from the known dimensions of the cell. As the volume of the polymer in the gel was known from its mass and partial specific volume, the degree of swelling could be calculated by dividing the volume of the swollen gel through the polymer volume.

These calculations can only be made if the gel contains only crosslinked molecules, the polymer is distributed largely homogeneously in the gel and no substantial swelling agent occlusions occur. These assumptions were found to be fulfilled for the investigated crosslinked polybutadienes, polychloropropenes and powdered polyurethane foams. Nevertheless, a note of caution needs to be expressed if microgels are suspected to contain soluble polymer (Mächtle *et al.*, 1995). This soluble fraction must be extracted prior to the swelling experiment.

The assumption of no substantial swelling agent occlusions may be questioned at least for rather monodisperse spherical particles which cannot be packed without solvent occlusions. Nevertheless, it can be expected that in such a case the determined degree of swelling after speed reduction is also constant (and independent of the former applied speed). Therefore, such a criterion used in this study cannot be taken to prove no solvent occlusions in every case.

From the degree of swelling the average degree of polymerization p_c and the molar mass M_c of an elastically effective network chain between two crosslinks could be determined for the polybutadiene and polychloropropene gels according to the theory of Flory and Rehner (1943) under simplifying assumptions. The method described is a rapid and effective way to derive the above mentioned parameters of gels. Unfortunately, it is restricted to uncharged polymer gels without soluble parts or with completely extractable soluble parts which means mainly chemically crosslinked gels. Furthermore, occlusions of solvent might be a problem.

The degree of swelling of only partially crosslinked microgels can be determined with a procedure introduced by Müller (see Müller et al., 1991). He measured the distribution of the sedimentation coefficients (s-distribution) of individually suspended latex particles in a thermodynamically good solvent at least at two different rotational speeds. At the low speed (2,000 rpm) the s-distribution for the swollen crosslinked particles was determined whereas at a higher speed (e.g. 40,000 rpm) the corresponding distribution of the soluble polymer could be obtained. The derived s-distributions yielded information about the portions of dissolved and crosslinked polymer, the degree of branching of the soluble polymer and the degree of swelling Q (Lange, 1986) due to the following equation:

$$Q = \frac{f \, d_{T,K}^2}{s_{swollen}} \frac{\rho_{T,K} - \rho_0}{18 \, \eta} \tag{6}$$

with s = sedimentation coefficient, $d_{T,K}$ = diameter of the compact, unswollen particles, $\rho_{T,K}$ = density of the compact, unswollen particles, ρ_0 = density of the dispersion medium and η = viscosity of the diluted dispersion. f is a factor according to $m_r = f \cdot m$ where the mass of the particle m_r reduced by the soluble part is related to the mass m of the particle consisting of soluble and insoluble parts. f could directly be obtained with the interference optics applied, whereas the diameter $d_{T,K}$ of the unswollen particle had to be determined first in a separate experiment e.g. via turbidity measurements in an analytical ultracentrifuge. In Equation (6) it is assumed that the hydrodynamic diameter of the particle is not affected by the leaching process of the particles in the dispersing medium. A further approximation has been made by neglecting the concentration dependence of s when using Equation (6) with the sedimentation coefficient at 5 g/l instead of that at zero polymer concentration. From the degree of swelling the polymerization degree p_c and the molar mass of the elastically effective network chains between the crosslinks M_c is available applying the Flory-Rehner (1943) theory.

The method was tested with styrene-butadiene latices from a batch process in cyclohexane as solvent. The results derived are presented in *Figure 6*.

It can be seen that the particle size does not increase anymore after a conversion of 70% has been reached whereas the crosslinking takes place indicated by a decreasing degree of swelling instead of the former branching of the molecules at low conversion

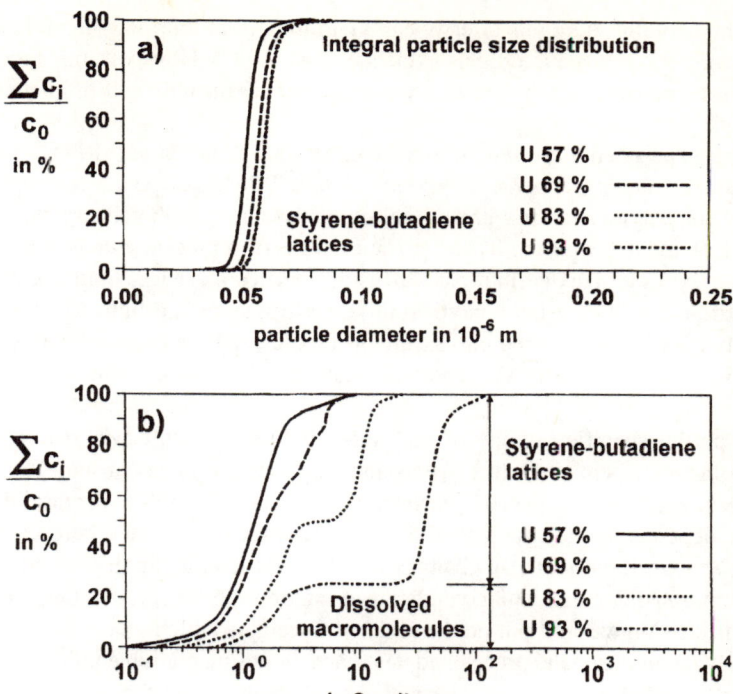

Figure 6. Particle size distribution and s-distribution of latices of the same polymerization process with increasing conversion U. s_{50} for the latices after 83% conversion is 10.5 Svedbergs whereas it is 40 Svedbergs after 93% conversion. The corresponding Q-values (Q is the degree of swelling derived using eq. (6) and the sedimentation coefficients at 5 g/l (Müller *et al.*, 1991) are 3,000 for $U = 83\%$ and 200 for $U = 93\%$. $\Sigma c_i / c_0$ reflects the sedimentation coefficient distribution. Redrawn from Müller *et al.* (1991) with kind permission of Steinkopff Verlag, Darmstadt.

degrees. With increasing conversion the s-distribution becomes bimodal at 83% conversion indicating the presence of crosslinked (high s-values) and soluble (low s-values) polymer. This makes it clear that the solution component has a much lower s-value than the gel. The s-distributions could also be used to detect small differences in latex stabilities.

Although the method described might bear some inaccuracies caused by the neglect of the concentration dependence of the sedimentation coefficients, it enables a rapid characterization of latices by the degree of swelling, the distribution of crosslinking and branching and the portions of soluble and crosslinked polymer.

A study on the characterization of microgel properties itself using analytical ultracentrifugation was published by Mächtle and coworkers (Mächtle *et al.*, 1995). The methods used were similar to those already described before for the quantitative detection of the amounts of microgel and uncrosslinked polymer, namely sedimentation velocity and density gradient centrifugation (Shaskoua and van Holde, 1958; Shaskoua and Beaman, 1958; Buchdahl *et al.*, 1963; Mächtle, 1992). The successful application of a simple step by step crosslinking theory of primary linear macromolecules as well as the agreement with results from light scattering makes this study an interesting alternative to the characterization of bulk gel properties with sedimentation equilibrium experiments as it is shown that rapid sedimentation velocity experiments with microgels already can yield important thermodynamic and structural information of the microgels (Mächtle *et al.*, 1995). However, this requires the feature that microgels of the same network structure as the bulk gels of interest can be prepared.

In the experimentally well grounded paper (Mächtle *et al.,* 1995), 14 nearly monodisperse aqueous poly-n-butylmethacrylate (PBMA) dispersions were prepared by emulsion polymerization with different amounts of methallylmethacrylate (MAMA) between 0 and 10% by wt. These particles were first precisely characterized with respect to particle size distributions, diffusion coefficient, sedimentation coefficients and particle densities using analytical ultracentrifugation and light scattering. Then the particles were transferred into tetrahydrofurane (THF), a good solvent for PBMA to allow swelling of the crosslinked molecules and dissolution of all non crosslinked ones.

The partial specific volume of the non-crosslinked sample was determined via density measurements in both solvents and was assumed to be constant for different degrees of crosslinker up to 10%. Strictly, this cannot be correct as the partial specific volume depends on the structure of the material which is changed here. But for this particular investigation, these changes which might lead to significant errors in the determination of the molecular weight are of minor importance as the emphasis of the study/discussion was not put on the molecular weights which would have then been more advantageously and precisely determined by sedimentation equilibrium experiments. Nevertheless, the molecular weights determined with the Svedberg equation from sedimentation velocity data were consistent and reasonable for both the gel *and* the solution component. Hence the differences in the partial specific volume of the different crosslinked microgels must be negligible.

The different crosslinked particles have been investigated in a density gradient as well as in aqueous as in THF dispersion/solution. In the aqueous dispersion, the density of the particles was found to be nearly constant, whereas in THF dispersion/solution a transition between the dissolved completely uncrosslinked molecules and the totally crosslinked microgel was observed. As it could be shown that the density gradient technique is able to resolve small structural differences between linear, branched and crosslinked molecules (Buchdahl *et al.*, 1963), from the observation of no more than two bands in the density gradient it seems to be the case that linear molecules can co-exist with the microgel without a significant amount in the transitional state (the branched large molecule in this example). From the density gradient experiments, the change from 0.1 to 0.2% of the crosslinker MAMA increased the microgel amount from 5 to 50% (compare also *Figure 7*).

The sedimentation behaviour of the different THF solutions/dispersions has been systematically investigated at different concentrations. In dependence of the amount of crosslinking agent, the transition between uncrosslinked soluble polymer and microgel formation could clearly be observed. In agreement with the density gradient technique the transition was between 0.1 and 0.2% of crosslinking agent. An attempt was made to calculate the critical amount of crosslinker by applying a theory of Flory for the stepwise crosslinking of a macroscopic gel phase but this yielded a value of only 0.02% of crosslinker. This difference was explained by a lower reactivity of MAMA in copolymerization as well as by side reactions.

Figure 7 is a good example for the transition between a pure solution and a microgel

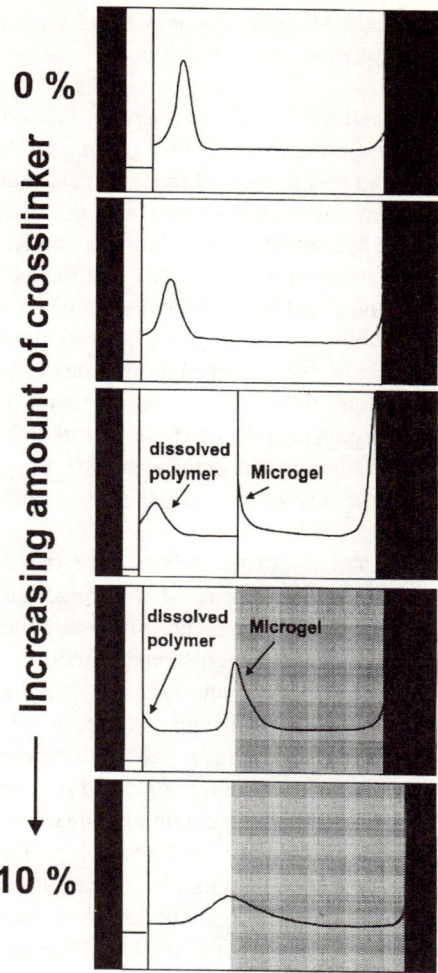

Figure 7. Schematical Schlieren-photos of sedimentation runs in an analytical ultracentrifuge for five microgel dispersions with different concentrations of the crosslinker. The photos show slow Schlieren-peaks of macromolecules and fast Schlieren peaks of microgel particles. Redrawn from Mächtle *et al.* (1995) with kind permission of Steinkopff Verlag, Darmstadt, FRG.

for showing that the Schlieren pattern of a microgel has more solution characteristics (Schlieren peak) rather than the characteristics of a separate gel phase (compare for example *Figure 7* with *Figure 10*).

In *Figure 7* it can be seen that at low amounts of crosslinking agent, pure solution behaviour is observed. But if the sedimentation coefficients of the solution components are compared for 0, 0.1 and 0.2% of MAMA, a steady increase can be observed hinting at the presence of branched polymer next to the linear uncrosslinked molecules. Although these transition states are not detected by density gradient experiments (due to the low overall concentration of a species with constant density), evidence for their presence can be seen in the increase of the solution peak sedimentation coefficient *s* with increasing degree of MAMA. If the sedimentation rate of the microgel is compared for different degrees of MAMA, a clear increase of the crosslinking density resulting in a decrease in the swelling ratio of the microgels: hence increasing s-values are observed.

If the bottom region of the cell is observed for microgel samples, it can be seen that according to the sample and experimental conditions, a more or less broad dark zone is observed which restricts the observation of the complete polymer concentration gradient as well known from experiments with bulk gel phases (Svedberg and Pedersen, 1940; Johnson, 1964; Metcalfe, 1965; Johnson and King, 1968) but much less pronounced. This observation supports the view that the dark zone near the cell bottom is caused by light deviated out of the optical system. This problem should be possible to solve by use of a Schlieren optical system with additional focussing lenses. However, as the investigation of the sedimentation behaviour of a microgel is a sedimentation velocity technique rather than a sedimentation equilibrium method established for bulk gels, the region near the cell bottom is of no interest because only the relative movement of the Schlieren peak or the area under the peak is of interest for sedimentation velocity experiments rather than the observation of the complete concentration gradient.

The amount of microgel and dissolved polymer for the different ratios was determined by integration of the Schlieren peaks of both the solution and the microgel. The results from both components agree well and show that the transition between linear and crosslinked polymer occurs for crosslinker amounts of 0.1–0.5% and is continuous. This agrees with the results from light scattering. If the molar masses determined with the analytical ultracentrifuge (Svedberg relation) for different crosslinker ratios are compared with those from light scattering and the theoretically expected ones on the basis of the amounts of microgel and soluble polymer with their molar masses, the results do not show the continuous increase of the molar mass expected theoretically and confirmed by light scattering. This showed that the amount of branched molecules must be negligible so that only microgel and linear polymer are present. The molar masses for both the soluble polymer and the microgel are constant as it can be expected because in the analytical ultracentrifuge (AUC), the fractionated components are considered and not their mixture. Thus, the continuous increase of the molar mass observed with light scattering is caused by the increased amount of microgel with increasing crosslinker density and thus an increase of the detected average molar mass.

From the sedimentation coefficient of the microgel, the volume swelling ratio Q was calculated from the particle diameters applying Stokes law. The values were in agreement with those from light scattering. The swelling ratio decreases strongly with the amount of crosslinker < 1% from more than 30 to 5. For crosslinker contents > 1% the swelling ratio is still decreasing. Even for 10% crosslinker, the microgel is weakly swollen. From the swelling ratios of the microgels, the Flory-Huggins interaction parameter χ was calculated to be 0.44 applying the Flory-Huggins theory for the crosslinking of a macroscopic gel. This value and its consensus with the results of other authors show that THF is a good solvent for the microgel and furthermore, that the Flory-Huggins theory for macroscopic gels can be applied to microgels down to 60 nm as well.

From the concentration dependence of the sedimentation coefficients for the different crosslinked samples, it could be deduced that only the microgel crosslinked with 10% of MAMA behaves as hard sphere e.g. shows no concentration dependence of s. The concentration dependence was more pronounced the lower the crosslinking density, showing a systematical dependence. From that, a soft transition from the linear polymer coil via a 'hairy ball' to the crosslinked hard sphere with increasing crosslinker amount could be concluded.

Sedimentation equilibrium techniques

BULK GELS

The basis for the application of sedimentation equilibrium techniques for the characterization of gels had already been supplied by the observations of the sedimentation velocity characteristics by Johnson's group at Cambridge (see above). The most important finding in that respect was the demonstration that true equilibria are obtained (Johnson, 1971; 1972). Johnson proved for agar gels that these equilibrium gel column lengths are real equilibrium values with the thermodynamic principle of path independence of an equilibrium. The same column length was reached independently if this value was achieved by deswelling (approaching equilibrium from lower speeds) or swelling of the gel (approaching equilibrium from higher speeds).

Discussions on the polymer concentration distribution inside the sedimenting gel and the comparison with a solution had already been published before (Johnson, 1964; Metcalfe, 1965; Johnson and King, 1968). As Svedberg and Metcalfe, Johnson stated that the gel concentration cannot be observed throughout the whole gel phase with the Schlieren optics of the ultracentrifuge. As the optical detection system could not be applied to derive the polymer concentration inside the gel phase, a microtome was used to cut the gel column into thin slices after it was removed from the ultracentrifuge cell once the equilibrium had been reached. The gel density of agar was found to be linearly dependent on the polymer concentration as was derived for gelatin/water and κ-carrageenan/water gels by other workers as well (Cölfen and Borchard, 1994d).

Combination of Svedbergs Equation (1) (Svedberg and Pedersen, 1940) with that of Freundlich and Posnjak $\Pi_s = \Pi_0 \cdot c_2^k$ yielded a k-value of approximately 2 for agar from the sedimentation equilibrium concentration distribution. The Π_0 of $2.52 \cdot 10^5$ dynes/cm^2 was found to be in agreement with one for gelatin of $2.7 \cdot 10^5$ dynes/cm^2 given by Freundlich. Nevertheless, it is known that the relation of Freundlich and Posnjak is only empirical and not all workers found that this equation describes the concentration behaviour of the swelling pressure (see for example Svedberg and Pedersen, 1940). This is also valid for some later studies which although not cited here can be found in Cölfen (1995).

Borchard was the first to apply the analytical ultracentrifuge as pressure generator for the determination of the swelling pressure and thermodynamic properties of a chemically crosslinked gel in a larger temperature range from 25–70°C (Borchard, 1975a). Polystyrene-cyclohexane gels with different crosslinking densities and removed soluble components have been investigated in two specially constructed types of ultracentrifuge cells. Here a sintered metal plate was placed onto the gel so that this plate caused the pressure. The equilibrium state was reached from lower as well as from higher rotational speeds. Because the buoyancy term was found to be very small for the system investigated and hence no concentration gradient could be detected inside the gel phase, the polymer concentration in the gel was calculated from the shift of the sinter metal plate and the initial degree of swelling of the gel.

The calculated equilibrium pressure tensions were related to the corresponding volume fractions at different temperatures. The swelling pressure was the pressure needed to keep the gel volume constant at different temperatures with respect to an appropriately chosen reference temperature. The polymer volume fraction in the gel corresponding to the constant gel volume could be calculated assuming volume additivity. The intersections between the curve for the polymer volume fractions at different temperatures and the isotherms in the volume fraction *vs.* equilibrium pressure tension plots gave the swelling pressures at the different temperatures.

From the swelling pressures, the change in the chemical potential of the solvent $\Delta\mu_1$ and from that the differential entropy of dilution ΔS_1 and the differential dilution enthalpy ΔH_1 were calculated in good agreement with the values determined by other authors. It was found that the evaluation of the results with the statistical theories for swollen networks led to physically meaningless results giving some evidence that some of the basic assumptions of the statistical theories are not fulfilled anymore for the investigated gels.

An idealized theoretical treatment of an elastic swollen gel which is compressed by an ultracentrifugal field leading to an equilibrium degree of swelling was given by Bloomfield (1976). The model used and fitted to the experimental values was able to describe the degree of swelling of a gel under ultracentrifugal force satisfactorily but assuming a distinct Flory-Huggins interaction parameter χ. Therefore, the applied model could not be used to describe the swelling behaviour of the gel completely just from the experimental parameters and results. Assuming volume additivity of mixing and considering the free energy of the gel as a sum of ultracentrifugal, elastic and mixing terms, Bloomfield derived an expression containing the unknown Flory-Huggins interaction parameter χ and experimental parameters which depends upon the degree of swelling of the gel. For simplicity the gel has been assumed to be of rectangular shape. For the experimental values which had to be inserted into this equation, it was found that χ could not exceed the value of 0.5. This is physically meaningful as the gel is to be considered as molecule with an infinite high molar mass which has to demix if χ exceeds 0.5.

Assuming χ-parameters of 0, 0.25 and 0.5 to cover the whole possible range and using the experimental values for an experiment with casein/water in a preparative ultracentrifuge as well as those for the free swelling of casein/water in absence of the ultracentrifugal field, Bloomfield (1976) derived three pairs of curves. From these plots he was able to determine the degree of swelling under the ultracentrifugal force from a given degree of free swelling. A basic finding from these considerations was that the effect of the ultracentrifugal field on the degree of swelling increased considerably with increasing χ. This is conclusive as a low χ represents high polymer-solvent interactions whereas for the formation of a compressible network, a considerable amount of polymer-polymer interactions is required to build up the network junctions represented by a $\chi \cong 0.5$. The best agreement with data from intrinsic viscosity measurements for the casein micelles investigated was found for a $\chi \cong 0.5$. The same χ has been found in ultracentrifugal studies of gelatin/water gels (Borchard and Cölfen, 1992).

The compression effect was observed to rise with increasing degree of swelling, increasing rotational speed and the increased gel column length which is to be expected. It must be noticed that the gel piece did not swell to its original amount anymore after the ultracentrifugal field was removed. As one explanation it was postulated that additional crosslinks have been formed during compression. The same was found for gelatin/water gels with soluble components which can act as crosslinking agents (Cölfen and Borchard, 1995).

The first improvements in the experimental set-up for sedimentation equilibrium experiments with gels were published by Holtus and Borchard using a Beckman Model E ultracentrifuge (Holtus and Borchard, 1989; Holtus, 1990). These improvements concern the modification of the Schlieren optical system to an off-line data capture system with a modulable laser light source to enable the application of 6-hole rotors due to the extreme experimental durations of about *three weeks*. Further improvements were made with the temperature measurement and control system and the vacuum system.

An equation based on irreversible thermodynamics formally similar to the equation of Svedberg (Svedberg and Pedersen, 1940) (Equation (1) above) was provided to facilitate the calculation of the swelling pressure of the isotropic binary gel as a function of the radial displacement from the centre of the rotor. This equation contained the locally dependent partial density of the polymer $\rho_2(r)$ as concentration variable. As this quantity was not directly accessible if the polymer concentration gradient could not be detected in the whole gel phase (see also Svedberg and Pedersen, 1940; Johnson, 1964; Metcalfe, 1965; Johnson and King, 1968), an approximation had to be used relating $\rho_2(r)$ to the locally dependent density of the gel assuming volume additivity of mixing and applying a mass balance assuming a linear polymer concentration gradient in the gel phase. This procedure required the determination of the dependence of the gel density on the polymer concentration – a very tedious and time consuming measurement (Cölfen and Borchard, 1994d). Holtus suspected turbidity of the gel causing the optical detection problems. But this would not explain the observation that the dark zone near the cell bottom in the Schlieren patterns is very well defined (at large ranges of initial polymer concentrations being investigated) and not continuous as it has to be expected for a continuous turbidity increase with the polymer concentration. Overall, the ultracentrifuge had been applied as a pressure generator here, generating radially increasing pressures inside the gel phase which led to a locally dependent deswelling of the gel and finally to a continuous swelling pressure equilibrium. With this method, the continuous dependence of the swelling pressure of a gel upon the polymer concentration could be determined in a larger concentration interval. This is a very important advantage as no other method for the determination of the swelling pressure with this capability exists. Furthermore, the swelling pressure curves of up to five samples could be obtained simultaneously under exactly the same experimental conditions which is a further very important advantage of this procedure.

This new method was tested on dialyzed gelatin/water gels in the concentration range of 2–8% by wt. without any detectable soluble components. It could be shown that the equilibria could be reached from higher and lower speeds as introduced as a proof for equilibria of agar gels by Johnson (1971; 1972). The induction period reported by Johnson for sedimentation velocity experiments (Johnson, 1964) was found in this study as well during the approach to equilibrium when the rotor speed selected has been too low.

The determined dependence of the swelling pressure on the polymer concentration (subsequently referred to as '*swelling pressure curve*' in the text that follows) suffered from an error. It had been assumed that all gels were swollen initially to their maximum degree of swelling and hence the concentration at the meniscus $r_m^{g/s}$ had to be the initial polymer concentration. Gelatin/water gels are able to swell much more as corresponds to their initial concentration. Therefore, the concentration of the maximum swollen gel (to be determined separately) has to be known for the applied mass balance, rather than the initial concentration of the gel. Although the errors in the swelling pressure curves caused by this wrong assumption are significant (Cölfen, 1993), the general order of magnitude of the swelling pressure curves as well as their relation to each other is not changed basically.

It was found that the swelling pressure curves – although the swelling pressure equilibrium could be proved – intersected for gelatin gels with initial concentrations lower than 7% by wt. at 20°C (example given in *Figure 9*). At 10°C, all swelling pressure curves in the investigated concentration range intersected. This finding was in contrast to the Flory-Huggins theory of a homogeneously swollen gel with a concentration independent interaction parameter. It was concluded that the lower concentrated gelatin networks must be inhomogeneous. Furthermore, the swelling pressure curves showed a significant dependence upon the gelation time of the gel (three days and seven days) although the selected times were large and the gelatin gel should be considered as matured in both cases. Nevertheless, the swelling pressure curves for the longer maturation times were steeper, indicating an increased crosslinking density with respect to the shorter matured gel. These findings are in qualitative agreement with observations on the empirical parameter n by other workers (Johnson and King, 1968). In this study n was increasing with the maturing time, indicating an increased crosslinking upon maturing in the time interval up to 20 hours. This increase was more pronounced at 4° and vanishes at 18°C.

As an alternative explanation of the gelation time dependence of the swelling pressure curves for gelatin/water gels, a stiffening of the chains due to further helication with time and hence an altered stress-strain behaviour was given by Holtus (1990). Furthermore, an effect of the gelation temperature on the swelling pressure curves was observed, as could be expected due to the changes in the network crosslinking density. These results showed again that swelling pressure curves may be used as a sensitive measure of structural changes in the gel network. The reproducibility of the swelling pressure curves was found to be good.

An attempt was made to calculate the concentration independent Flory-Huggins interaction parameter χ and the crosslinking density by fitting of the swelling pressure curves to a modified Flory-Huggins equation of the type of Equation (9). This gave physically meaningless negative crosslinking densities and showed that at least the model with a concentration independent interaction parameter could not be applied.

In 1991 a trilogy of papers was published treating the swelling pressure equilibria of swollen crosslinked systems in an ultracentrifugal field in detail. The first paper dealt with theoretical considerations for a binary gel (Borchard, 1991). After pointing out the general behaviour of a sedimenting gel and the nature of the swelling pressure equilibrium, irreversible thermodynamics has been applied to derive the 'generalized

Svedberg-Pedersen equation' for the swelling pressure of a binary gel:

$$\Pi_S = \omega^2 \int_{r_m^{g/s}}^{r} \frac{\rho_2}{\rho_1 \tilde{v}_1} \left(1 - \tilde{v}_2 \rho\right) r \, dr \tag{7}$$

where ρ is the gel density, ρ_1 the partial density of the solvent and ρ_2 the partial density of the polymer. All densities depend on the radial distance r. Basic assumptions for the application of this equation were:

- The deformation of the binary gel leads to a continuous isothermal equilibrium,
- The gel remains isotropic during the deformation (which is certainly valid for small deformations),
- Gel and solvent are incompressible e. g. the partial specific volume is pressure- and location independent,
- Volume additivity of the pure components,
- In the equilibrium case, the swelling equilibrium is reached at the meniscus gel/sol $(r_m^{g/s})$.

This equation had been used already in earlier works (Holtus and Borchard, 1989; Holtus, 1990). As the theoretical considerations are given in more detail in this work, they were not mentioned before. A new but very important assumption in this study is the assumption of the swelling equilibrium at $r_m^{g/s}$. This means that the concentration at this meniscus is not the initial gel concentration as assumed before but that of the maximum swollen gel which needs to be determined separately.

Equation (7) is not only formally similar to the Equation (1) above of Svedberg it can actually be shown that Equation (1) is a special case of Equation (7) for highly swollen gels which is expressed in the term 'generalized Svedberg-Pedersen equation'. The application of Equation (7) for gels where the concentration gradient in the gel cannot be detected in the whole gel phase suffers from the required knowledge of the locally dependent partial density of the polymer in the gel which is not experimentally accessible in these cases. Therefore Equation (7) has been rearranged substituting the partial specific volume of the solvent $\tilde{v}_1$ in the gel by the much easier accessible specific volume of the pure solvent $\overline{V}_{01}$ which is fulfilled for not too concentrated gels. The resulting equation now only contained parameters which could be derived experimentally:

$$\Pi_S = \omega^2 \int_{r_m^{g/s}}^{r} \left(\rho - \frac{1}{\overline{V}_{01}}\right) r \, dr \tag{8}$$

Considering the influence of the hydrostatic pressure on the gel concentrations, e.g. an effect of the hydrostatic pressure on the obtained swelling pressures, an expression was derived to calculate the polymer concentration in the gel at the meniscus gel/sol $(r_m^{g/s})$ under the influence of the hydrostatic pressure of the solvent column. The effect was estimated to be small in agreement with earlier findings of Johnson (1964).

The second part of the 'trilogy' deals with the determination of molecular parameters from the swelling pressure curves (Holtus *et al.*, 1991). Applying Flory-Huggins statistical theory of polymer solutions and the theory of rubber elasticity to the swelling of a nonelectrolyte polymer system, a modified semi-empirical Flory-Huggins equation was obtained. Substituting the difference in the chemical potentials of the solvent $\Delta\mu_1$ by $-\tilde{v}_1 \Pi_S$ with $\tilde{v}_1 =$ partial molar volume of the solvent, an equation has been derived which semiempirically relates the swelling pressure of the gel at a known concentration to molecular parameters of the gel:

$$-\frac{\Pi_S \tilde{v}_1}{RT} = \underbrace{\ln(1-w_2) + w_2 + \chi_{w,0}\, w_2^2 + \chi_{w,1}\, w_2^3}_{\text{Mixing Term}} + \underbrace{C_w\, w_2^{1/3}}_{\text{Network Term}} \tag{9}$$

with $w_2 =$ polymer concentration in the mass fraction scale (index 0 refers to the initial concentration), $\chi_w =$ linearly concentration dependent interaction parameter $\chi_w = \chi_{w,0} + \chi_{w,1} w_2$ which is an apparent value due to the semiempirical nature of the Flory-Huggins equation applied, $C_w =$ apparent network constant (explained in more detail in the original article). Both the interaction parameter and network constant are not explicitly stated as *apparent values* in the following discussion, simply to avoid repetitions, although a strict consideration would, of course, require this. Furthermore, it must be stressed that Equation (9) is only strictly valid for nonionic gels. For polyelectrolyte gels, the mixing and network contributions must be extended by a third term which accounts for charge contributions. Nevertheless, Equation (9) was found to describe the behaviour of gelatin/water and κ-carrageenan/water well, even if no additional salt is added.

It is not just the interaction parameters of the polymer-solvent system that can be derived. For example, the network constant C_w allows us to calculate the elastic modulus, the shear modulus and the molar mass of the network chains if the functionality of the crosslinking points is known. C_w and both terms of χ_w could be derived by means of a nonlinear iteration procedure from the swelling pressure curves. This iteration delivers good and unique results.

The generalized Svedberg-Pedersen equation for a binary system (Equation (8)) as well as the modified Flory-Huggins equation (Equation (9)) have been applied to ultracentrifuge experiments with a photographic gelatin in the concentration range of 2–10% by wt. The sample has been dialyzed to meet the assumption of a binary system as close as possible. In order to enable the calculation of the swelling pressure curves via a mass balance which is given in the third part of the trilogy (Cölfen and Borchard, 1991), the maximum degree of swelling of the physically crosslinked networks had to be determined separately which is very tedious, time consuming, and in the case of physically crosslinked gels, sometimes even incorrect due to a dissolution of the gel.

A basic conclusion from the generalized Svedberg-Pedersen equation is that the swelling pressure of a gel must be constant at a certain polymer concentration, independent of the applied rotational speed. This is clearly fulfilled for the investigated dialyzed gelatin/water gels as shown in *Figure 8*.

The interaction parameters derived for the gelatin/water gels gave further insight into the gel structure. From the concentration dependencies of both terms of the linearly concentration dependent interaction parameter, it could be derived that the degree of branching increased with increasing polymer concentration of the gels. Extrapolation of the two linearly concentration dependent parts of the interaction parameter to zero concentration would even allow the estimation of the interaction parameter of a completely uncrosslinked and unbranched polymer chain, a quantity

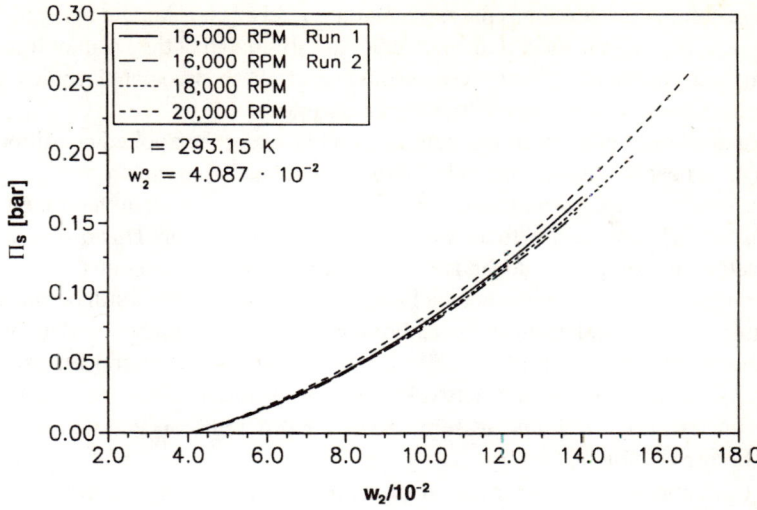

Figure 8.   Swelling pressure Π_s of the system gelatin/water vs. concentration of gelatin for different rotational speeds. w_2 is the weight fraction of the polymer. Reproduced from Holtus *et al.* (1991) with kind permission of Steinkopff Verlag, Darmstadt, FRG.

which is difficult to measure directly because of the self association of the gelatin. Both parts of χ_w were found to be temperature independent at 10 and 20°C.

The χ_w-parameters which have been calculated for the initial gelatin gel concentration were found to be nearly constant between 10 and 20°C and 2 and 10% by wt. The value derived of 0.497 was found to be very close to 0.5. Hence the gelatin/water gels must be very close to a miscibility gap as a gel is a system with an infinite molar mass which demixes at $\chi_w > 0.5$ if the influence of the network term in Equation (9) can be neglected. Furthermore, the χ_w-parameter of 0.497 was in very good agreement with results of other workers derived for gelatin solutions with osmosis and light scattering in the range of 0.491–0.499. The network constant C_w at 10°C was found to be nearly twice as large as that at 20°C. This finding could be confirmed by measurements of the complex shear modulus. Overall, the agreement with the results of other independent methods shows that the ultracentrifuge can be used not only to derive the swelling pressure curves of gels but, furthermore, some of their molecular parameters. It is remarkable that such good agreement with the results of other methods could be achieved, although one of the basic assumptions for the application of the generalized Svedberg-Pedersen equation is that the system is binary which is not strictly fulfilled. Furthermore, the investigated system is clearly no nonelectrolyte system, as assumed in the modified Flory-Huggins equation (Equation (9)) and no additional salt has been added. A possible explanation for the good agreement of the results is that gelatin can contain salt/ash from the manufacturing process which might shield the charges of the polyelectrolyte. But although the results agree rather well with those from other techniques, the applied Flory Huggins equation can only be semi-empirical.

From the network constant, the number average molar mass of the chains between the crosslinks could be calculated assuming a certain functionality and endlinked chains. For gelatin/water it was found that the network chains had a higher molar mass than the primary chain for some gels. This could only be possible if gelatin molecules associate before crosslinking. This explanation is likely as it is known that gelatin is a self-associating system, even if the molecules are so small that they cannot form a gel (Cölfen and Borchard, 1994). A crosslinking of these associated chains would furthermore lead to the observed branched networks.

As gelatin/water gels are in the state of a rubber under the applied conditions, the Young and shear modulus of the gels could be calculated assuming a poisson ratio of 0.5. The derived values were found to be about ten times lower than the storage shear modulus at 1 Hz obtained with a torsion oscillation viscometer. This difference was explained by the frequency dependence of the complex shear modulus.

The question arose as to whether the Flory-Huggins theory could still be applied for the highly branched gelatin/water gels. For a gel concentration above 6% by wt. this was considered to be no problem (coil overlap) whereas at lower concentrations the local polymer densities in the network might be different. The latter case would explain intersecting swelling pressure curves for different concentrated gels as reported before (Holtus, 1990).

The third part of the paper trilogy dealt with remaining unsolved problems of the ultracentrifugation of gelatin/water and κ-carrageenan/water. These problems were mainly caused by the presence of soluble parts in the gel which was assumed to be a binary mixture (Cölfen and Borchard, 1991). After a description of the sedimentation of a gel in an ultracentrifugal field, a mass balance based on the cell geometry and the equilibrium position of the gel/sol meniscus ($r_m^{g/s}$) is given which enables calculation of the polymer distribution in the ultracentrifuge cell if the concentration gradient is not visible throughout the gel phase. An assumption for this mass balance was a linear concentration gradient as has been proved before (Johnson, 1971;1972; Holtus, 1990). Furthermore, the relation between the gel density and the polymer concentration had to be determined as well as the maximum degree of swelling of the gel.

For the example of a gelatin/water gel, it was shown that the swelling pressure curves no longer superimpose for different rotational speeds (see *Figure 9*) once soluble parts are present in the gel. This hints at the occurrence of an irreversible process. But in this study it was supposed that true equilibria have not been reached caused by the presence of soluble polymer material. A rough estimation of the time to reach an equilibrium state was given. The influence of the soluble parts was supposed to increase with increasing rotational speed. Overall the swelling pressure equilibrium of the gel was thought to be accompanied by a sedimentation-diffusion equilibrium of the soluble parts. This assumption neglected the self-association of the soluble parts which implies an association of soluble parts to network chains if certain concentration limits would be exceeded (Cölfen and Borchard, 1994). The altered solvent activity caused by the superimposition of the gradient of the soluble parts was given as explanation for the observed additional deswelling of the gel at increased rotational speeds.

It was observed that the intersection of the swelling pressure curves described before (Holtus, 1990; Holtus *et al.*, 1991) could be altered by changing the rotational speed. In conclusion, it was pointed out that extreme care is necessary when interpreting intersecting swelling pressure concentration curves. The quantity of soluble parts present in the gelatin gels was found to be independent of the gelation time. The finding was explained by the low molar mass of the soluble parts and their lack of gelling ability. This was confirmed in a later study (Cölfen and Borchard, 1994).

Despite the obvious problems with the presence of soluble parts, it could be shown that the reproducibility of the experiments was good even under such circumstances.

For the system κ-carrageenan/water equilibria could be proved by the method used by Johnson and Holtus (Johnson, 1971; Holtus, 1990), although soluble parts could be detected. It was stated that the soluble parts did not influence the swelling pressure equilibria of the gel in that case. This finding supports the view that the phenomena described for gelatin/water before are not only caused by a superimposition of the swelling pressure equilibrium of the gel with a sedimentation-diffusion equilibrium of the soluble parts. There is further evidence for an irreversible process such as the change of the crosslinking density of gelatin/water gels upon increase of the rotational speed (see also Johnson, 1968). Discontinuous steps of the refractive index gradient have been observed in the Schlieren patterns of the sol phase formed by deswelling of the κ-carrageenan gel. This was explained by the formation of aggregates and microparticles which are not able to move into the gel network. These aggregates could be formed by those parts of a carrageenan sample which are generally not able to gel.

In a publication of Borchard and Cölfen (1992), the findings of the previous 'trilogy of papers' have been presented in a comprehensive way with the addition of some interesting new results which will be presented here:

It could for example be shown that a linearly concentration dependent χ_w-parameter in Equation (9) is sufficient for the description of the polymer-solvent interactions of the system gelatin/water. Higher χ_w-terms proved to be of negligible magnitude. A more detailed consideration of the previously reported gelatin network formation of associated chains was given. Taking the overlap region (assumed to be similar to that of collagen) of two associating or crosslinking molecules into account, the average functionality of the network chains could be estimated. For a 3% by wt. gelatin gel at 20°C an average functionality of 2.9 was derived which clearly shows the network formation of associated polymer chains. It was pointed out that this functionality < 3 clearly shows that not all chains can be endlinked to the network which leads to the previously observed highly branched structure of gelatin gels (Holtus *et al.,* 1991). A model for the gelatin gel network was given, taking these special findings into account.

As further reason for the previously reported intersection of the swelling pressure curves for gelatin/water gels (see *Figure 9*), the concentration dependence of the χ_w-parameter was given. This concentration dependence of χ_w can be due to the fact that a large concentration range is covered in the experiments. Nevertheless, a clear reason for the intersection could still not be given due to the lack of model calculations with varying χ_w-parameters.

The concentration dependence of the static shear modulus calculated from the network constant C_w has been compared with the directly measured real part of the complex shear modulus at 1 Hz. Both concentration dependencies have been found to be of a linear type in the double logarithmic scale. For different gelatin/water gels, it could be shown that at low gel concentration, the real part of the complex shear modulus is about ten times higher than the static value from the ultracentrifuge measurements. This discrepancy was found to decrease at increasing polymer concentration due to the decreasing viscous properties of the gel. Therefore it should, in principle, be possible to determine the intersection point between the lines in the double logarithmic plot for the static shear modulus (from ultracentrifuge measurements) and the dynamic shear modulus (real part of the complex shear modulus determined in a torsional oscillation experiment). This point should define the gel concentration at which all viscous properties of the gel disappear. As the gel transforms into the glassy state, this characteristic point is the *glass transition*.

A comprehensive description of the modifications of the experimental ultracentrifuge set-up for gels (Cölfen and Borchard, 1994a,b) as well as experimental results (Holtus *et al.*, 1991) has been given by Cölfen (1993). The theory of the sedimentation of a gel derived by Borchard for binary (Borchard, 1991) and ternary gels (Borchard, 1994) was extended to an N-component system at constant temperature and pressure. The change of the specific chemical potential of the component i (1 = solvent, 2 = crosslinked polymer, 3 - N = non-gelling components)$\Delta\tilde{\mu}_i$ is given by:

$$\left[\Delta\tilde{\mu}_i\left(c_1,c_2,c_3.....c_{N-1}\right)\right]_{T,P} = \omega^2 \int_{r=r_m^{g/s}}^{r}\left(1-\tilde{v}_i\,\rho\right)r\,dr \;\; ;i=1,2,3,....,N \tag{10}$$

This equation is the generalized Svedberg-Pedersen equation for an N-component system. Up to now, the relation between the difference of the osmotic pressure and the change of the chemical potential is only defined for the binary case which certainly restricts the application of Equation (10). But even if it would be defined for N-component systems, a better treatment of a multicomponent gel would not be possible because up to now it is not yet experimentally possible to detect the radial concentration gradients of all individual components with the present analytical ultracentrifuge detection optics. The approach to determine the individual concentration gradients of all components in a ternary gel by extracting the soluble component, staining it and then performing an ultracentrifuge experiment with the combined application of Schlieren (sum gradient) and uv-absorption optics (stained soluble component) is already very difficult to realize (Cölfen and Borchard, 1995). Furthermore, it has to be taken into account that a gel will deswell if it is in contact with a solution instead of a pure solvent. Therefore, the approach in Equation (10) is of a theoretical nature in the general form. Nevertheless, it is the most general description of the sedimentation equilibrium of a gel available up to now, containing all cases treated before.

It could be shown experimentally that a significant error in the determination of the swelling pressure curves occurs if the gel is considered to be swollen to its maximum degree of swelling at its initial concentration as assumed by Holtus (1990). Therefore, a precise measurement of the polymer concentration in the maximum swollen gel had to be introduced, which proved to be quite difficult in the case of dissolving samples as gelatin/water at 20°C.

As has been reported before (see for example Holtus and Borchard, 1989), the swelling pressure curves for gelatin/water gels intersected beyond certain limits of the initial gel concentration (*Figure 9*).

Some variations of the two terms of the concentration dependent χ_w-parameter and the network constant C_w in Equation (9) for a given swelling pressure curve showed that initially intersecting swelling pressure curves did not intersect anymore if both terms of the χ_w-parameter were taken as constant. This shows that the concentration dependence of the χ_w-parameter is responsible for the intersection of the swelling pressure curves for gelatin/water and not an inhomogeneity of the gel.

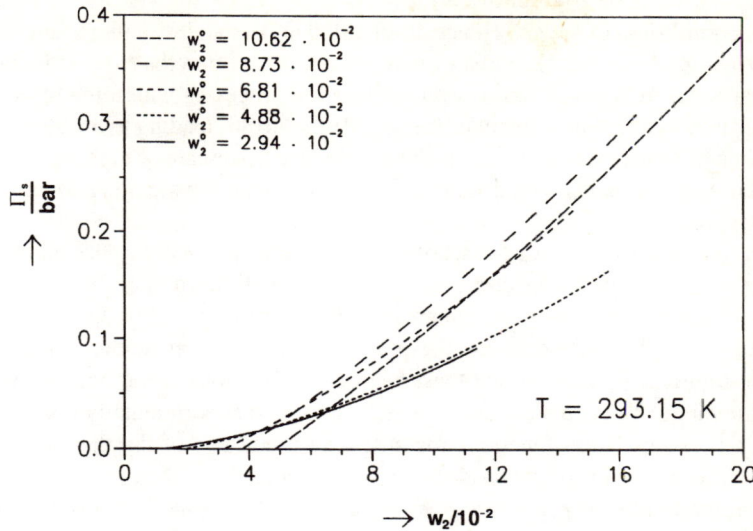

Figure 9. Swelling pressure curves for the system gelatin/water. Π_s = swelling pressure and w_2 = polymer concentration in the mass fraction scale with the index 0 for the initial gel concentration. Figure reproduced from Cölfen (1993) with kind permission of Dr Köster Verlag, Berlin, FRG.

As gelatin gels containing soluble parts cause intense experimental difficulties (Cölfen and Borchard, 1991) for the derivation of correct swelling pressure curves, a first attempt was reported to investigate the behaviour of stained soluble parts inside the gel phase during a sedimentation equilibrium experiment. Soluble parts have been extracted from gelatin/water gels as described in (Cölfen and Borchard, 1994) and stained with fluorescein-isothiocyanate. The distribution of the soluble parts during a sedimentation equilibrium experiment could be observed using UV-absorption optics. It was found that the gradient of the soluble parts did not reach its supposed equilibrium if the rotational speed was lowered from a higher one, not even after months. Nevertheless, it was still concluded that the swelling pressure equilibrium of the gel was superimposed by a sedimentation-diffusion equilibrium of the soluble parts leading to additional deswelling of the gel although this would be a reversible and not in fact the observed irreversible process. Therefore, it was concluded that the soluble parts (which have been stained) could not have a significant influence on the swelling pressure equilibria of the gel due to their observed low concentration of about 2.5 mg/ml. The fact that a gel which contains soluble parts does not swell up to its original degree of swelling once it has been exposed to a higher rotational speed was interpreted to be caused by higher molecular weight soluble parts (e.g. the presence of soluble parts with molecular weights above 4,000 g/mol which are normally only weakly attached to the gel network and hence readily soluble in an excess of solvent). These soluble parts should have a very low back-diffusion so that their concentration gradient leads to the observed overall deswelling of the gel.

This interpretation that a gel containing soluble parts does not swell up to its original degree of swelling once it has been exposed to a higher rotational speed cannot be upheld from the present point of view. A sedimentation-diffusion equilibrium of the soluble parts must be (by definition) a reversible process as well as the swelling pressure equilibrium of the gel. Hence, it is hard to understand why there is no sign of back diffusion of the soluble parts, even after a period of months from reducing the ultracentrifugal field. Even for very slow diffusion processes which might occur in gels, at least a minimal movement of the soluble molecules leading to an alteration of their concentration gradient should be detectable on a time scale of months. As this is not the case, the influence of the soluble parts on the gelatin/water gel structure must be irreversible.

An attempt was made to use the set of thermodynamic parameters derived from the swelling-pressure equilibrium curves to predict stability limits for gels (Cölfen, 1993). This was done by extrapolating the regression functions for $\chi_{w,0}$, $\chi_{w,1}$ and C_w (derived from Equation (9)) dependent on the polymer concentration to the whole concentration range. It was found that these regression functions could only be applied within the range of gel concentrations which have been experimentally observed.

With the extrapolation functions for the radial prediction of the thermodynamic parameters an attempt was also made to find out the nature of a dark zone near the cell bottom in the Schlieren pattern of gels, especially gelatin/water. A correlation was found between the mixing and the network term in Equation (9) for the investigated gelatin/water gels at that radial position where the dark zone in the Schlieren pattern starts. A closer investigation of the change of these terms with the polymer concentration showed that both terms are equal at the meniscus gel/sol. But with increasing polymer concentration (i. e. with increasing radius inside the gel phase), the mixing term increases much faster than the network term (Flory and Rehner, 1943). As the χ_w-parameters for this concentration region are higher than 0.5, a demixing of the gel with its infinite high molar mass should occur if only the mixing term is considered. But the network term of the chemical potential seems to prevent this. At a certain ratio between network and mixing term, which has been found constant for different concentrated gelatin gels, it seems that the network term cannot prevent a demixing anymore. At exactly this location, the start of the dark zone in the Schlieren pattern is observed. Due to this interpretation, the dark zone in the Schlieren pattern should be caused by a demixed gel. This demixing of gelatin gels could not yet be proved by other methods and must hence be treated with caution as several other possibilities can be responsible for the dark zone as well. Nevertheless, it was reported that during the sedimentation of an 18.25% by weight gelatin gel at 10°C, a dark zone in the Schlieren pattern suddenly formed within the gel phase. At this zone, a discontinuity of the polymer concentration gradient inside the gel phase was observed. This behaviour was explained by a demixing of the gel into two gels with the same crosslinking density but a different distribution of the network chains (Cölfen, 1993). This could lead to light refraction which are beyond the limits of registration by the optical system. Nevertheless, this gel would still remain clear as it could be observed after the ultracentrifuge experiment in contrast to a demixing solution.

It could be shown that the swelling pressure–concentration curves which are derived from ultracentrifuge experiments are very reproducible (Cölfen, 1993). Furthermore, it was demonstrated that the application of the circular sample channels in the newly designed 10-hole centrepieces yields the same results as those derived in sector shaped cells. This justified the application of the 10-hole centrepieces. Summarizing the modified ultracentrifugal technique (Cölfen and Borchard, 1994a,b) for the investigation of gels, it was pointed out that no other method is able to characterize 70 samples (e. g. a complete gel/solvent system) in only one experiment which lasts a few days. Such measurements require highly sophisticated devices for the experimental set up and the data acquisition which are quite expensive. As further advantage of the ultracentrifuge investigation of gels, the continuous equilibrium which can be determined by the selection of the rotational speed was pointed out. As a rather large concentration range is covered within the gel phase, unstable regions could be detected. With the set of thermodynamic parameters derived from Equation (9). the prediction of these regions is principally possible.

It was pointed out that it should be possible to apply the generalized Svedberg-Pedersen equation to the solution phase as well as yielding the concentration dependence of the osmotic pressure which could be used to determine M_n of the soluble parts as well as their second osmotic virial coefficient. It was stated that the greatest disadvantage of the technique was the necessity for supplementary measurements, such as the determination of the maximum degree of swelling or the concentration dependence of the gel densities. The present limitations of the ultracentrifuge technique were seen in the restriction to clear gels due to the optical detection systems of the ultracentrifuge. The application of a mass balance for the calculation of the non-detectable polymer concentration inside the gel phase was suggested. Furthermore, it was stressed that gels cannot be investigated, if the adhesion cannot be succesfully suppressed. Also, gels with a vanishing buoyancy term (e. g. $(1 - \tilde{v}_2 \rho) = 0$) are not accessible due to the lack of sedimentation. A further limitation was seen in the fact that it is not possible to cover the complete concentration range in the gel phase in the ultracentrifugal field. The reason for this is the limitation of the applicable swelling pressure by the maximum rotational speed of the ultracentrifuge.

Borchard extended the theory of the sedimentation of a binary gel (Borchard, 1991) to the ternary system (Borchard, 1994). The theory treats a gel with one inert soluble component (Borchard, 1994). Basically the same approach as for the binary system (Borchard, 1991) was used, applying irreversible thermodynamics. The difference to the binary system was made in distinguishing between the solution and the gel phase taking into account that the gel phase consists of three components (crosslinked polymer, soluble polymer and solvent) whereas the solution phase only consists of solvent and soluble component. Finally, equations were obtained to represent the change of the specific chemical potential of each component in each of the two phases.

This general solution allowed several cases to be distinguished. The case that the soluble component is so high molecular that it does not enter the gel phase was discussed as well as that of a free mobility of the soluble component within the network. The latter case would lead to an additional deswelling of the gel. Experimentally, it becomes quite obvious that already the ternary gel is difficult to handle as the radial concentration distribution of the soluble component as well as that of the crosslinked polymer have to be determined independently. Hence the superposition of the swelling-pressure equilibrium of a gel with a sedimentation-diffusion equilibrium of soluble parts could not yet be fully quantified experimentally.

The heterogeneous swelling equilibrium has been described with a ternary state diagram (Borchard, 1994) to illustrate the situation if different amounts of an inert soluble component are added to the elastic binary mixture. Furthermore, it was discussed what happens if an ultracentrifugal field is applied to such a ternary mixture.

It was pointed out that the gel phase needs to move towards the cell bottom due to mass conservation. Furthermore, the addition of a soluble component to a binary gel has been discussed in terms of mass conservation. It was stressed that the total amount of soluble component in the gel can be determined by dialyzing the gel. But this implies different assumptions: The gel is insoluble under the conditions studied and the soluble component is completely inert to the network and can thus be dialyzed, which can be a very slow process. The latter assumption is at least not fullfilled for physically crosslinked gelatin/water gels (Cölfen and Borchard, 1994; 1995).

The amount of the soluble component in the solution phase was suggested to be obtainable via the application of the interference optics. Here, it must be taken into account that meniscus depletion methods for the determination of the fringe shift at the meniscus solution/vapour cannot be applied in every case because 1) this would decrease the accuracy of the evaluation for the gel due to the decrease of the gel column height unless additional experimental time is invested for the overspeeding after the equilibrium run and 2) the soluble parts may be of such low molecular mass that the meniscus depletion does not work anymore (Cölfen and Borchard, 1994). The suggestion to stain the soluble component selectively with an UV-marker to determine its sedimentation equilibrium concentration profile is difficult to realize, although possible (Cölfen and Borchard, 1995). It becomes quite obvious that the ternary system – which considers the soluble components with their molar mass distribution as one single component – is difficult to handle experimentally due to the limitations of the present optical ultracentrifuge detection systems.

In 1994, two papers were published which were dedicated to the development of a more effective experimental set-up for the hitherto very time consuming sedimentation equilibrium experiments with gels (Cölfen and Borchard, 1994a,b). In the first part, the basic improved instrumentation applied to the Beckman Model E was described. Modifications involved a modulation system with a pulsed laser light source for Schlieren optics which has already partly been described by Holtus (Holtus and Borchard, 1989; Holtus, 1990). By means of a modified sophisticated Schlieren optical system (Cölfen and Borchard, 1994b) which generates a very small band of light illuminating the cell, it is possible to resolve even a 0.3°C sector on the spinning rotor at its maximum speed of 60,000 rpm. This is the most important prerequisite for the application of multichannel centrepieces. A further optimization concerned the photographic system for the Schlieren optics which was replaced by a fully automatic picture digitization system based on a video camera.

The second paper dealt with several technical improvements for an efficiency increase of the sedimentation equilibrium experiments (Cölfen and Borchard, 1994b).

Figure 10.   A sedimenting κ-carrageenan/water gel (2% by wt. at 20°C, 20,000 rpm after at least 1 day) observed with different optical detection systems. (a) Schlieren optics (the original Schlieren optics of the Beckman Model E with the mercury lamp has been used. The light intensity in the gel phase can be much increased if a laser is applied as light source), (b) Rayleigh interference optics with a He-Ne Laser as light source and water as reference solvent, (c) Beckman Optima XL-A ultra-violet absorption optics at three different wavelengths (curves 1 and 2 = 230 nm, curve 3 = 280 nm, curve 4 = 350 nm). The absorbance is defined as log (I_0/I) with I = Intensity. The time between the scans for the different wavelengths was 18 min (99 averages). Water was used as reference solvent. KEL-F (Beckman, Palo Alto, USA) centrepieces were used to avoid adhesion of the gel. All pictures do not correspond to the state of sedimentation equilibrium. Reproduced from Cölfen and Borchard (1994b) with kind permission of Academic Press, Inc., Orlando, Florida, USA.

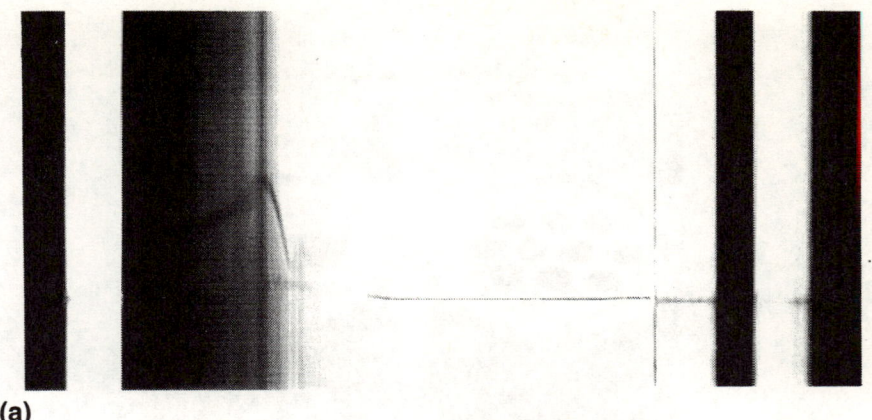

(a)

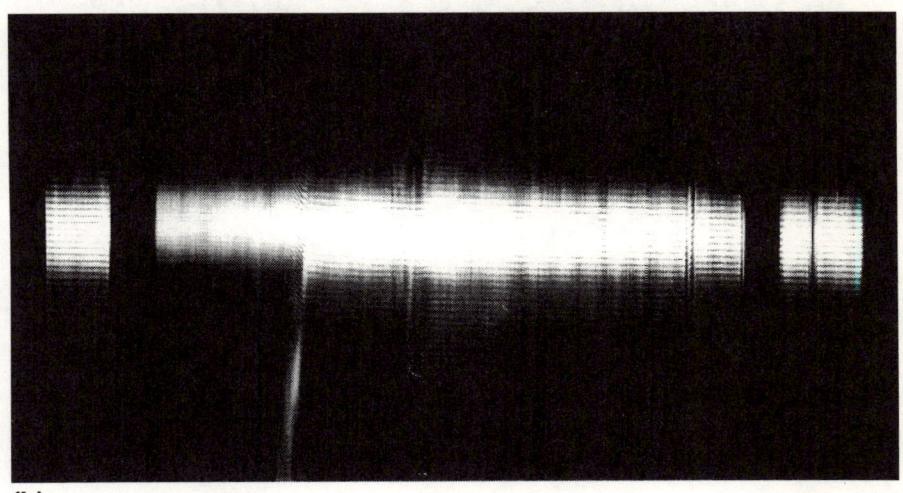

(b)

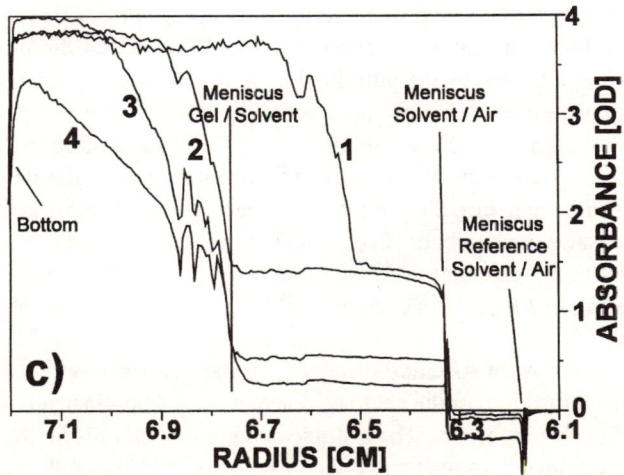

c)

Figure 11. 10-hole short column multichannel centrepiece (1) and a conventional 4° monosector centrepiece (2).

It was pointed out that the Schlieren optical system is the detection system of choice if concentrated and turbid gels need to be investigated (*Figure 10a*). Also, it was shown that no fringes can be detected in the gel phase if the Rayleigh interference optics were used due to the intensity differences of the interfering light beams (*Figure 10b*). The high turbidity of the gels also restricted the detection with the UV-absorption optics due to the significant light scattering phenomena (*Figure 10c*).

Modifications of the original Schlieren optical system of the Model E ultracentrifuge were described in detail. These modifications were necessary because only a very small band of light is needed to illuminate the measuring cell if multiplace rotors are to be used with multichannel centrepieces (resolution 0.3° on the spinning rotor). An important increase in the efficiency of the ultracentrifugal experiments was achieved by the introduction of 10-hole short column multichannel centrepieces with circular sample compartments (*Figure 11*). Together with the application of the 8-hole rotor, a simultaneous investigation of up to 70 samples in one equilibrium experiment was possible for the first time. If the new 10-channel centrepieces with their circular sample chambers are used, the mass balance for sector shaped sample chambers (Cölfen and Borchard, 1991) has to be modified. Therefore, a volume calculation for the various phases in the ultracentrifuge cell was presented in order to enable the use of the mass balance for the new centrepieces as well.

The centrepieces were specially designed for the Schlieren optics and allow the sedimentation equilibrium of the gel to be reached in two days instead of one week as before for the 1 cm gel columns. The restricted choice of materials for the centrepieces for experiments with gelatin/water gels was discussed. Polycarbonate was found to be the best material for the centrepieces for this particular system whereas Polymethyl-methacrylate (PMMA) was used for the manufacturing of the cell windows. PMMA was tested to be applicable up to 20,000 rpm, whereas the polycarbonate centrepieces could be used up to 40,000 rpm. It was found that even with polycarbonate centre-pieces and PMMA windows, the adhesion of gelatin at the cell walls of the centrepieces and at the cell windows could not be completely inhibited, especially not for higher concentrated gels. As this adhesion becomes obvious in a broadening of the meniscus gel/sol, the experimental accuracy, especially in the case of short gel columns is decreased. Therefore, a simple correction was presented to locate the meniscus in case of its broadening. The benefit of this correction could be demonstrated for gelatin/water gels.

After the procedure to determine molecular, thermodynamic and elastic properties of gels from sedimentation equilibria had been established, another study was carried out investigating the sedimentation equilibria of κ-carrageenan/water gels more closely as it had been done in previous studies (Hinsken, 1995). The gels have been investigated in the concentration range between 1 and 5% by wt. due to the high turbidities of the gels and the restricted carrageenan solubility. Again, an equilibrium could be verified by reaching it from different rotational speeds (Johnson, 1971; 1972; Holtus, 1990; Cölfen and Borchard, 1991). Nevertheless, this proof of attainment of equilibrium was only possible if the higher speed was applied for shorter times than that needed to reach a new equilibrium. In the latter case the superposition of the sedimentation-diffusion equilibrium of the soluble parts (0.3–0.6 mg/ml) with the swelling pressure equilibrium of the gel was given as an explanation for the failure of the proof of equilibrium. Back diffusion of the soluble parts was considered to be very slow which causes an additional deswelling of the gel with respect to the state reached from lower speeds. As the soluble parts are not characterized yet, this interpretation may be doubtful as it is not clear if the soluble parts are low molecular which would not explain their sedimentation at the applied relatively low speeds or they are polymers and just not able to gel (for example other non-gelling carrageenan types). In the latter case they might at least be able to associate under the conditions of the κ-carrageenan network formation restricting their free mobility in the network which has been postulated.

As a further reason for a very slow back-diffusion of the soluble parts, the reversible association of the soluble parts to the gel network as well as a self association of the soluble parts is presented without experimental proof of the suspected self-association of soluble parts as given in Cölfen and Borchard (1994) for gelatin/water. It was argued that the soluble parts would then be trapped by their association reactions and back-diffusion should be decreased considerably.

As a third possibility of the observed path dependence reaching the equilibrium, partial crystallinity of the junction points was considered. The crystalline junction points were assumed to be surrounded by a highly amorphous phase and not the pure solvent which should cause anisotropy of the gels. The soluble parts were considered to be trapped in the crystal junction points leading to a gradient of the crosslinking density depending on the radius. The last possibility discussed in terms of a crosslinking density gradient by trapped soluble parts is similar to that found as explanation for irreversible structural changes in gelatin/water gels (Cölfen and Borchard, 1995) but with the difference that the junction points are treated as crystals for κ-carrageenan here. The crystalline regions should be detectable by X-ray techniques. Other possibilities provided to explain the path dependence were the high deformations of the gels leading to an unstable network structure with regrouping of junction points and the prevention of reswelling of the gel by friction due to a high gel strength.

These explanations all differ from each other and no experimental evidence has as yet been provided to validate any of these over the others. In addition, the amount of the soluble proportion (0.3–0.6 mg/ml) was stated to be of negligible magnitude as the results for the gels were found in excellent agreement with the scaling theory of de Gennes (1975). This is somewhat contradictory as, from the discussion given above, a clear influence of the soluble parts must be expected unless only the alteration of the network structure by the centrifugal force or a prevention of swelling by the strength of the gel are responsible for the observed path dependence of the swelling pressure equilibria. What seems to be clear is that an irreversible process takes place, as observed in other studies (Johnson, 1968; Cölfen and Borchard, 1991).

Certainly, further experiments are needed to clear the role of the soluble parts on the swelling pressure equilibria or even the network structure of κ-carrageenan. It would be particularly interesting to investigate the long term behaviour (e.g. if the original equilibrium state can be reached, even after months), the suspected self association of soluble parts (although some evidence for a reversible self-association has been given) and their behaviour during sedimentation of the gel which is accessible in principle via stained soluble parts.

However, for speeds below a critical value, reversible swelling pressure equilibria could be proved from superposition of swelling pressure – concentration curves for κ-carrageenan/water. The theoretical dependence of the concentration of the maximum swollen gel on the initial gel concentration was treated. It was considered that for κ-carrageenan the chains are in the partially helicated state prior to gelation rather than random coils and the influence of partial helication on the quadratic mean end-to-end distance of the chains was discussed. This is a much more realistic treatment of the terms involved in the network constant C_w in Equation (9) than that used before and gives another argument for the semi-empirical character of the modified Flory-Huggins equation (Equation (9)) at least for the system κ-carrageenan/water and all other systems where the polymer chains at least partly helicate prior to gelation.

A linear relationship was found between the initial gel concentration and the concentration in the maximum swollen κ-carrageenan/water gel. This relation could be supported by theoretical considerations under certain assumptions and was also found for different gelatin/water gels (Cölfen, 1993).

Hinsken observed so called 'double peaks' in the Schlieren patterns of the sol phase as reported before (Cölfen and Borchard, 1991) and interpreted them in terms of association of the κ-carrageenan. The results were found in qualitative agreement with the Gilbert theory for reversible associating polymers in terms of the boundary shape. The question was raised why a self association of the soluble parts could be observed although they were not incorporated into the network, and suggesting as the answer the presence of different non-gelling carrageenan components which are still able to associate.

As for gelatin/water an intersection of the swelling pressure–concentration curves below certain initial polymer concentrations in the gel was observed (see *Figure 9* for the gelatin example). The Flory-Huggins interaction parameter was found to be linearly concentration dependent in analogy to the system gelatin/water. But in contrast to that system, χ_w was found to be around 0.44 increasing very slightly with the polymer concentration. From that it was concluded that water is a good solvent for κ-carrageenan/water. Via the linear concentration dependence of both χ_w-terms in Equation (9) the crosslinking and branching degree was found to increase with the polymer concentration. Nevertheless, simulation calculations using the presented datasets to investigate the influence of the concentration dependence of the interaction parameter on the swelling pressure–concentration curves were not performed, although such calculations helped to understand a similar behaviour for the system gelatin/water (Cölfen, 1993).

A quadratic concentration dependence was supposed for the network constant C_w in the small concentration range investigated: this finally led to a good agreement with the predictions of de Gennes following the C* theorem. Extrapolation to a vanishing C_w yielded a value of 0.99% by weight which was identical with the critical concentration of gel formation.

With a rather high assumed molar mass of about 500,000 g/mol for the primary chain of κ-carrageenan, the number of crosslinking points per polymer chain could be estimated to be at least *five*, probably much higher. This result is clearly different from that for gelatin gels where for some gels less than two crosslinking points have been observed per chain (Borchard and Cölfen, 1992).

The determined static shear moduli for κ-carrageenan in the double logarithmic log G* vs. log M plot showed a linear concentration dependence in analogy to the results summarized in (Borchard and Cölfen, 1992). The slope of both lines was 2.26 which was in good agreement with the value of 2.25 predicted theoretically by the scaling theory of de Gennes (1975). Therefore, it was concluded that the exponent of 2.25 in the double logarithmic shear modulus *vs.* concentration plot is not only restricted to chemically crosslinked gels but is in a certain sense universal for gelling systems. This statement should be treated with extreme caution as no consideration has been included as to why all other results for gelatin/water gels (except the particular one presented) – and independent of whether static or dynamic shear moduli for gelatin/water – clearly disagreed with an exponent of 2.25 (Borchard and Cölfen, 1992). The static shear moduli yielded too high slopes compared to a value of 2.25 whereas the slope for the dynamic values was too low. As the concentrations in these investigations were partly as low as those investigated by Hinsken for κ-carrageenan and furthermore the concentration dependence was considered over a much higher concentration interval, the concentration differences certainly cannot account for the differences in the scaling exponent. So it cannot be concluded without doubt from these results that an exponent of 2.25 as predicted by the C*-theorem must be obtained for low polymer concentrations. The curves published by other authors for higher concentrated gelatin/water gels as summarized in the paper by Borchard and Cölfen (1992) should have been taken into account as well. From this, the slope of 2.25 seems to be the exception, independent of the method of measurement used.

Furthermore, it could be established from the previous studies that the shear modulus as determined by the ultracentrifuge is a *static* value due to its equilibrium nature whereas for example in torsional oscillation experiments, *dynamic values* are obtained. For a given system, the dynamic values obtianed have been always higher {indeed, for the system gelatin/water at low polymer concentrations even by factor of ten (Borchard and Cölfen, 1992) due to the viscous part of the dynamic shear modulus}. In a creep experiment, it could be stated that the shear modulus decreased by a factor of two after 24 hours experimental duration (Cölfen and Borchard, 1992). Furthermore, it is known that the slope of the regression lines in double logarithmic shear modulus *vs.* polymer concentration plots are always higher for the static shear moduli than they are for the dynamic ones for concentrations beyond the glass transition (Cölfen, 1993).

Nevertheless, from the good agreement of the experimental results with the C*-theorem of de Gennes (1975) within the concentration range investigated, it was concluded that the influence of the soluble parts on the swelling pressure equilibria is negligible for κ-carrageenan/water. Most of the primary chains have been assumed to become a network chain (the amount of soluble parts was determined to be <0.056% by wt.) (Hinsken and Borchard, 1995) which certainly justifies this conclusion.

The influence of soluble parts on the swelling pressure–concentration curves e. g. the entire network structure of the gel was the subject of a further study (Cölfen and Borchard, 1995). An alkaline treated gelatin of $M_n = 68,000$ g/mol containing soluble parts with a M_w of 2,900 g/mol was investigated. It could be shown from this study that these soluble parts possess no gelling abilities but are at least partly able to associate/aggregate to the network and form new crosslinks upon compression. For such a gel, it was found, that the swelling pressure–concentration curves for the same gel at different rotational speeds did not coincide as it is required as proof of equilibrium (Holtus *et al.*, 1991). It was found that the slope of the swelling pressure curves decreased with increasing rotational speed after a certain critical speed limit has been exceeded, hinting at an increase in the crosslinking density and assuming that the semi-empirical Flory-Huggins theory for homogeneous networks could be applied. Nevertheless, the maximum swelling pressure at the cell bottom was constant for a given rotational speed as would be expected. It was stated that the process observed is irreversible because even after nine weeks the gel did not swell up to its original degree of swelling at a certain rotational speed, once the speed has been increased. It was pointed out that the initial interpretation of this phenomenon as a superposition of a sedimentation-diffusion equilibrium of the soluble parts with the swelling pressure equilibrium of the gel leading to additional deswelling (Cölfen, 1993) cannot be upheld because this would be a reversible process as discussed above. If an irreversible process takes place, it is no longer possible to compare the swelling pressure curves nor any of the parameters derived by application of Equation (9) because they correspond to different network structures of the gel. But the apparent thermodynamic parameters derived for such a case hint at the increase of the crosslinking density upon increase of the rotational speed.

Nevertheless, it could be shown that it is still possible to prove swelling pressure equilibria by means of coinciding swelling pressure curves if the network structure is kept constant. This can be maintained by reaching the equilibrium at a certain speed and then awaiting the equilibrium at a lower one, where the gel network structure is still the same but different from the network structure at the initial gel concentration as soon as a critical value of the rotational speed has been exceeded. Also, it could be shown that the reproducibility of the swelling pressure curves is good despite the alterations of the network structure after an increase of the rotational speed (Cölfen and Borchard, 1995). This gives evidence for a good reproducibility of the network structure even if the crosslinking density has been changed by additional crosslinking of soluble parts leading to a gradient gel.

As an irreversible process upon increase of the rotational speed is observed only if the gel contains soluble parts (compare for example with the observations of Holtus (1990)), the irreversibility must therefore be related to the presence of this low molecular weight component. Because of this, the results from the investigation of the concentration distribution of fluorescein-isothiocyanate stained soluble parts in gelatin/water gels at different rotational speeds (Cölfen, 1993) have been reconsidered (*Figure 12*).

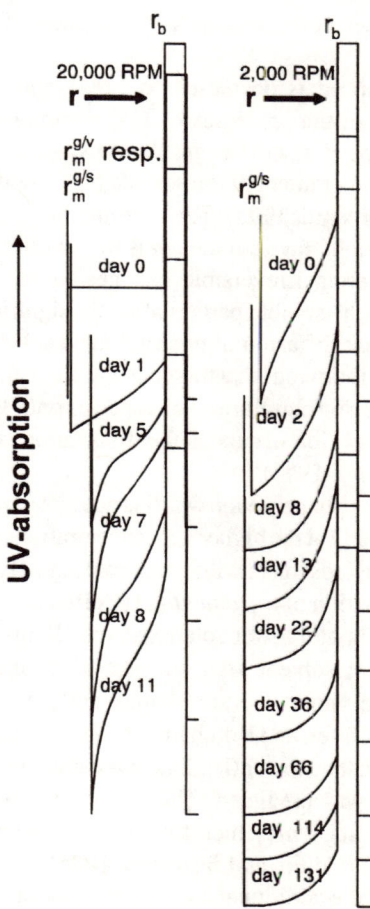

Figure 12. Ultra-violet absorption as a function of radial displacement from the centre of rotation r for a sedimentation equilibrium experiment with an 8.5% by weight gelatin/water gel 20°C. The gel contains 0.26% by weight (related to the solution) FITC-stained soluble parts ($M_w = 2,900$ g/mol). The filling height of the ultracentrifuge cell was 1 cm. The index b refers to the cell bottom, m to the meniscus with the phase boundaries explained in *Figure 1*. The scanning wavelength was 390 nm. The left series of traces shows the deswelling of the gel after the rotational speed of 20,000 rpm is applied, whereas the right series shows the swelling of the previously compressed gel at a lower speed. Figure reproduced from Cölfen (1993) with kind permission of Dr Köster Verlag, Berlin, FRG.

It can be seen that the swelling pressure equilibrium of the gel at 20,000 rpm (constant $r_m^{g/s}$) is reached after approximately seven days and this agrees with previous experiments (Holtus *et al.*, 1991). From the changes in the gradient of the soluble parts ($M_w = 2,900$ g/mol) it was concluded that they must be associated or aggregated to the gelatin network, as the soluble parts are not only self-associating but furthermore similar to the gelatin in their chemical composition (Cölfen and Borchard, 1994). The formation of the observed steep gradient cannot be explained by the sedimentation of a low molecular weight species at moderate rotational speeds. Further evidence for the association of the soluble parts into the gelatin network can be provided from the observation that the concentration gradient of the soluble parts still becomes steeper after the swelling pressure equilibrium of the gel has been attained. As the association is favoured at higher polymer concentrations, it should preferably take place at the cell bottom where the polymer concentration is the highest. This is precisely what is observed.

When the rotational speed is decreased to 2,000 rpm (the minimum speed to maintain the observation of the processes by UV-absorption optics in a sufficiently stable running rotor), a speed where the gel should swell up to its original degree of swelling, the concentration gradient of the soluble parts should vanish. But this is *not* observed (Cölfen and Borchard, 1995). The swelling pressure equilibrium of the gel is roughly reached after eight days but the gel is not swollen to its original degree of swelling anymore, indicating irreversible changes in the network structure. The concentration gradient of the soluble parts is also still significant. Slowly, it becomes flatter due to the slow back diffusion of uncrosslinked soluble parts, until it remains constant after 114 days. It is obvious that this concentration gradient cannot correspond to the sedimentation-diffusion equilibrium of a species with $M_w = 2,900$ g/mol at 2,000 rpm even when self-association occurs as the association constant was found to be small (Cölfen and Borchard, 1994).

The interpretation of this phenomenon was that soluble parts, which partly associate or even aggregate, preferably at the higher gel concentrations of the deswollen gel near the cell bottom, form new crosslinks with neighbouring polymer network chains and hence leading to the formation of a *gradient gel* (Cölfen and Borchard, 1995). If such an altered gel swells again at a lower rotational speed, this gel should not swell up uniformly in an excess of solvent after such an ultracentrifuge experiment. This feature could be verified when a sector shaped gel piece swelled up to a nearly rectangular shaped one (Cölfen and Borchard, 1995). Hence, the crosslinking density at the meniscus gel/sol must be lower (leading to the uptake of more solvent) than that at the cell bottom, as it was predicted. The observed increase of the crosslinking density leading to a gradient gel must therefore be closely related to those soluble parts present in the gelatin gels (Cölfen and Borchard, 1995).

It was pointed out that the additional crosslinking does not allow the investigation of the original gel structure anymore (Cölfen and Borchard, 1995). As this is the aim of the ultracentrifuge experiments, future investigations of gels with soluble parts causing additional crosslinking have to be carried out at rotational speeds below the critical value without the sedimentation of the gel phase (this is the *'Gradient method'*). Under such conditions, the crosslinking of associated/aggregated soluble parts is unlikely due to the much smaller concentration gradients inside the gel, as we consider below.

Density gradient techniques

MICROGELS

The density gradient technique has been applied to detect a microgel in an acrylonitrile-vinylacetate copolymer using a dimethylformamide – bromoform density gradient in the ultracentrifuge cell (Buchdahl *et al.*, 1963). The copolymer solution could be separated into three fractions. One of the fractions was assumed to be the linear copolymer, the second one to be highly branched and the third to be weakly crosslinked. The structural differences between the highly branched and the crosslinked fraction were found to be rather small causing a difference in the apparent partial specific volumes of only about 0.0005 ml/g which could still be resolved with the density gradient in the analytical ultracentrifuge.

Mächtle looked at the behaviour of styrene/acrylonitrile-copolymer (SAN) grafted onto polybutylacrylate particles (PBA) in a tetrahydrofuran-diiodomethane density gradient to study, if the grafted molecules were completely bound covalently or not (Mächtle, 1992) by analogy with the studies of Shaskoua and van Holde (1958) and Shaskoua and Beaman (1958). About 20% of the SAN was found not to be covalently bound to PBA. With a sedimentation velocity run, the sedimentation coefficient and the molar mass of the dissolved SAN was determined as well as its concentration via integration of the Schlieren peak.

The gradient method

BULK GELS

In a recent paper, Hinsken *et al.* (1995) presented a new approach to evaluate sedimentation equilibrium experiments with gels. This so-called '*gradient method*' evaluates the local polymer concentration via the detectable concentration gradient in the Schlieren pattern. As this is not possible when high rotational speeds are applied because of the occurrence of a dark zone near the cell bottom (Svedberg and Pedersen, 1940; Johnson, 1964; Metcalfe, 1965; Johnson and King, 1968; Holtus, 1990; Cölfen, 1993), low speeds are used so that no sedimentation of the macroscopic gel phase occurs (see case a) in *Figure 1*). The local slopes of the swelling pressure curves are calculated using a differentiated form of the generalized Svedberg-Pedersen equation (Equations 7 and 8) for a binary system (Borchard, 1991).

As the concentration range between the meniscus gel/vapour and the cell bottom is much smaller (< 33.5% concentration change from the initial gel concentration for the carrageenan example) than for the method used before, only a little part of the swelling pressure–concentration curve can be determined. But in contrast to those derived via a mass balance assuming a linear concentration gradient before, these swelling pressure curves are calculated for the real concentration gradients, whatever they are like. A further large advantage of the gradient method is that irreversible changes in the network which were observed for gelatin/water (Cölfen and Borchard, 1995) are not likely to occur at the selected low speeds beyond the critical speed where irreversible changes of the network structure are observed. Hence, the network can be treated as unaltered so that all thermodynamic, molecular and structural parameters derived from the swelling pressure–concentration curves are corresponding to the original network and are not apparent ones (Cölfen and Borchard, 1995). Therefore, it has been suggested that the gradient method can be used as a test for statistical theories which describe the gelation process.

By means of the path independence reaching an equilibrium, equilibria could be proved by constant equilibrium Schlieren gradients inside the gel phase (within the accuracy of the measurement) independent of whether they have been reached from either higher or lower rotational speeds. The Schlieren patterns presented were evaluated quantitatively by integration of the Schlieren curve. As the gels were investigated in a monosector cell, no baseline is present which is normally required for the proper integration procedure. Additionally, the Schlieren gradients are very broad so that the concentration determination from these pictures can be inaccurate. The polymer concentration in the gel was approximately found to be a linear function of the radius although the gradients in the evaluated Schlieren patterns were bending upwards in the region of the cell bottom. This reflects a potential error of the gradient method: the quantitative evaluation of broad Schlieren gradients.

However, the polymer concentration gradients evaluated were used to calculate $(d\Pi_s /d\rho_2)$ which is the change of the swelling pressure with the polymer concentration. These values were plotted against the polymer concentration yielding linear functions for all evaluated Schlieren patterns. But the curves corresponding to the Schlieren patterns stated to be the equilibrium Schlieren pictures before did not superpose which has to be expected for superposing swelling pressure–concentration curves indicating an equilibrium. Instead, curves for an equilibrium at 8,000 and 6,000 rpm were found to be similar. These results indicate present inaccuracies of the gradient method.

As no further experiments, particularly those using doublesector cells (baseline), were performed in this study, it it is impossible to assess what difficulties or limitations arise when the gradient method is applied even with good quality Schlieren patterns with baselines. The promising advantages expected for its application thus remains to be confirmed. The potential source of error, the concentration determination from the integration beyond the rather broad gradients in the Schlieren patterns, especially at low phase plate angles, might still be a problem. But as long as no extremely low gel concentrations are used, the Schlieren optical system is the only realistic choice for a concentration determination in the gel phase as both the Rayleigh interference and the UV-absorption optics fail (Cölfen and Borchard, 1994b): see also *Figure 10*.

To derive a complete swelling pressure–concentration curve for a gel with a certain network structure, it has been suggested that it could be constructed from the parts of the swelling pressure curves derived for gels with different initial concentration but the same network structure. Such gels can only be prepared by drying a given gel to different concentrations. A proof that the network structure of physically crosslinked gels is not altered upon drying has not to date been given so that it is not yet clear if a swelling pressure–concentration curve can be constructed from different parts. Nevertheless, the small concentration range of a swelling pressure–concentration curve derived with the gradient method should already be sufficient to calculate thermodynamical or structural parameters using the Flory-Huggins theory.

Conclusions

The results from more than 60 years of ultracentrifugal investigations of gels permit the following conclusions to be drawn:

- Thermodynamic, elastic and structural parameters of bulk gels can be characterized in an effective and elegant manner by sedimentation equilibrium since the ultracentrifugal field causes a continuous radial dependent deswelling of the gel. Usually, these experiments require several weeks, but short column multichannel centrepieces have been suggested, which reduce this time to a few days and allow the possibility of investigating up to 70 samples simultaneously in one experiment if an 8-hole rotor is used.
- Sedimentation velocity experiments can rapidly characterize the same parameters as mentioned above for microgels as well as the sedimentation coefficient of bulk gels which can possibly be related to structural parameters
- Sedimentation velocity experiments allow the quantitative determination of soluble components in bulk and microgels
- Density gradient centrifugation of microgels is a very sensitive method for the investigation of small structural changes or the effects of grafting reactions etc.
- With bulk gels, one may encounter extensive experimental difficulties due to adhesion of the gel and the high turbidity and polymer concentration which restricts optical detection
- It would be of considerable benefit if the on-line Schlieren system (Clewlow *et al.*, 1997) available for older ultracentrifuges could be adapted to the commercially available Beckman Optima XL-I as quickly as possible: the Schlieren optical system is the only optical system for the ultracentrifuge which will allow the application of all of the technology described in this review.

OUTLOOK FOR THE FUTURE

Due to the rapidly increasing technical possibilities, especially in the computer and electronics sector, it is expected that the efficiency of the application of ultracentrifuge technologies to the characterisation of biopolymer gels can be further increased on a large scale with the incorporation of on-line recording techniques. This will certainly also inspire new applications.

Some trends can already be observed. It has become clear for example that the investigation of gel properties via a sedimenting gel meniscus, although exclusively applied in the past, has several disadvantages. There are adhesion problems, effects of the hydrostatic pressure of the sol column (although small) and anisotropic deformation in the ultracentrifuge cell, which have not yet been exactly theoretically treated. The general thermodynamics of anisotropic deformation has however been treated by Borchard (1975) but this has not yet been applied to the ultracentrifugation of gels. Further problems arise from the failure of the Schlieren optical system in the region of the cell base and, most important, additional crosslinking of the gel by soluble components. All these problems can probably be avoided if the rotational speed is chosen low enough that no sedimentation of the macroscopic gel phase occurs. In such a case, the polymer concentration gradient can be detected in the whole gel phase, the hydrostatic pressure of a sol column has no effect because no sol is present, the concentration gradients in the gel are so moderate that the concentration limit for additional crosslinking *via* soluble components is not exceeded and the deformations are so small that the gel can be treated as isotropic. Therefore, it is expected that future work will be carried out using the gradient method and investigating the swelling pressure equilibrium in a single gel phase where no sol phase has yet been introduced (Hinsken *et al.*, 1995). The disadvantage that a smaller polymer concentration range is covered by the swelling pressure curves here is not important for deriving the thermodynamic and elastic properies of the gel. Only if the stability of a gel system should be investigated, the two phase (gel + sol) method is more advantageous as the polymer concentration range covered is much larger. Here, one simply uses the ultracentrifuge to create a desired concentration gradient and to detect if a suspected instability occurs at the predicted polymer concentration or not.

As many technologically important gels are polyelectrolytes, it is expected that the theoretical treatment of the sedimentation behaviour of polyelectrolyte gels will be brought forward, resulting in a proper characterization of such systems by analytical ultracentrifugation.

The determination of the sedimentation coefficient s via the movement of the centre of mass raises hopes that it will be possible to relate s to the structure of the gel (via a parameter like the empirical parameter *n* of Johnson's group (see Johnson and Metcalfe, 1967). If the above mentioned disadvantages of the two phase method can either be avoided or suppressed, this would be the most rapid structural characterization of gels with the analytical ultracentrifuge.

Another rapid and precise characterization of gels based on sedimentation velocity experiments is achieved by studying the microgel properties rather than that of the macroscopic gel phase with all its difficulties and pitfalls (Shaskoua and van Holde, 1958; Shaskoua and Beaman, 1958; Lange, 1986; Mächtle *et al.*, 1995). As this characterization method avoids the main disadvantages of the investigation of a bulk gel phase – namely anisotropic deformation and adhesion – the calculation of thermodynamic parameters from the swelling behaviour might have future potential for gel characterization, although the ultracentrifugal investigation of microgels has, up to now, been more or less applied to studies of the efficiency of a grafting reaction (Shaskoua and van Holde, 1958; Shaskoua and Beaman, 1958; Mächtle, 1992). As long as parameters like the Flory-Huggins interaction parameter or the molar mass of the crosslinked chains are desired, a sedimentation velocity experiment with microgels should be the method of choice. For this, a significant potential of applications is evident for microgel particles prepared by chemical crosslinking in emulsions (the wide field of polymer dispersions). For most physically crosslinked gels, this method cannot be applied due to the dissolution in an excess of solvent.

For physically crosslinked gels, the investigation of the bulk gel phase is still the method of choice because it is very often desired to study the properties of gels which will dissolve in an excess of solvent. For these gels, the 'gradient method' described in Hinsken *et al.* (1995) has the advantage that it produces swelling pressure curves *via* the exact radial polymer concentration and not via a mass balance assuming a certain concentration gradient. As the polymer concentration is detectable throughout the whole gel phase, the 'gradient method' is suitable for on-line techniques which have recently been developed for the Schlieren optical system (Clewlow *et al.*, 1997). With such on-line systems, exact kinetic studies of the gel sedimentation are for the first time possible and the accuracy of the determination of s via the movement of the centre of mass should be much better than possible at the moment.

An application of a real time optical system can also be seen in monitoring the kinetics of swelling of extremely low amounts of latices or microgels. If the gels are highly compressed and deswollen at a high speed (say 60,000 rpm), the kinetics of swelling could be readily monitored as soon as the speed is lowered to 1,000 rpm to allow swelling of the particles.

Another interesting application is the investigation of the diffusion of small protein molecules (which are detectable via the uv-absorption optics) in polysaccharide gels: the gel itself will be invisible for uv-absorption optics so long as the gel concentration is so low that light scattering of the gel does not disturb the observation of the protein. This would be an interesting development and may yield important information, especially for the food industry if one thinks of flavour release or related fields. It should be possible to create a synthetic boundary in the gel phase in analogy to the solution case. The diffusion of the protein should take place in reasonable time. One could then think of more complicated situations where the radial concentration of the gel is extensively varied by application of a high centrifugal field so that in one experiment the influence of the gel concentration on the protein diffusion can be monitored.

Another application for ultracentrifuge techniques would be the characterization of hybrid gels with inorganic components, a topic which appears to be gaining in industrial importance in recent years. For example, if an inorganic compound is synthesized in a gel phase, it is of interest how well the inorganic particle sticks to the functionalized polymer. In most cases, the inorganic matter has a higher density than the gel so that sedimentation of the inorganic particles should be expected if they are not (or weakly) complexed by the polymer. If the inorganic particles absorb in the UV/VIS, their concentration can independently be determined by the absorption optics, whereas Schlieren optics would deliver the complete concentration gradient. Also, possible changes in the gel structure by the inorganic component can be characterized using the discussed techniques.

To summarize, the analytical ultracentrifuge could very well play an important role in future characterization of biopolymer gels in the following areas:

- Simultaneous structural, thermodynamic and elastic characterization of a large number of gels (up to 70 at the present), which is not possible by any other method known up to now
- Rapid characterization of microgels and monitoring of the efficiency of crosslinking or grafting reactions
- Characterization of polyelectrolyte gels
- Kinetic swelling studies as well as kinetic studies of the gel sedimentation applying real time detection systems
- Diffusion studies of proteins in polysaccharide gels
- Characterization of organic/inorganic hybrid gels
- Stability investigations of gel systems *via* the accessible set of thermodynamic parameters and possible experimental verification by sedimentation velocity

Because of the importance of gel technologies in the food and increasingly the pharmaceutical and health-care industries, the potential of ultracentrifugation methodologies as reliable and complementary gel characterisation technologies to traditional rheological approaches should be taken very seriously.

Acknowledgements

All authors and editors are acknowledged for their kind permission to reproduce the figures used in this review. I would especially like to thank Dr P. Johnson (Cambridge, UK) for making available the PhD theses of his former students to me.

References

BABSKIJ, V. G. AND S'EDIN, A.A. (1977). Investigation of gel forming structures by means of Analytical Ultracentrifugation. *Khim. Biol. Nauki.* (Dopov AN Ukr. Ser. B) **10**, 926–929.

BLOOMFIELD, V.A. (1976). Ultracentrifugal compression of gels. *Biopolymers* **15**, 1243–1249.

BORCHARD, W. (1975). Zur Thermodynamik von elastischen Mischphasen. *Habilitation*, Clausthal, 108.

BORCHARD, W. (1975a). Über das Quellungsverhalten von Polystyrol verschiedener Netzwerkdichte in Cyclohexan. *Progress in Colloid & Polymer Science* **57**, 39–47.

BORCHARD, W. (1991). Swelling pressure equilibrium of swollen crosslinked systems in an external field. I: Theory. *Progress in Colloid & Polymer Science* **86**, 84–91.

BORCHARD, W. AND CÖLFEN, H. (1992). Characterization of thermoreversible gels by means of sedimentation equilibria. *Macromoecular Chemistry, Macromolecular Symposium* **61**, 143–164.

BORCHARD, W. (1994). The sedimentation diffusion equilibrium of a ternary gel. *Progress in Colloid & Polymer Science* **94**, 82–89.

BORCHARD, W. AND HINSKEN, H. (1997). The sedimentation velocity of a gelled polymer. *Progress in Colloid & Polymer Science* **107**, 172–179.

BUCHDAHL, R., ENDE, H. A. AND PEEBLES, L.H. (1963). Detection of structural differences in polymers by density gradient ultracentrifugation II: Detection of Microgel. *Journal of Polymer Science Part C* **No. 1**, 143–152.

CLEWLOW, A.C., ERRINGTON, N. AND ROWE, A.J. (1997). Analysis of data captured by an online image capture system from an analytical ultracentrifuge using Schlieren optics. *European Biophysical Journal* **25**, 311–317.

CÖLFEN, H. (1993). Bestimmung thermodynamischer und elastischer Eigenschaften von Gelen mit Hilfe von Sedimentationsgleichgewichten in einer Analytischen Ultrazentrifuge am Beispiel des Systems Gelatine/Wasser. *PhD thesis*, Duisburg; 1. Auflage Verlag Köster, Berlin 1994.

CÖLFEN, H. (1995). Analytical ultracentrifugation of gels. *Colloid & Polymer Science* **273**. 1101–1137.

CÖLFEN, H. AND BORCHARD, W. (1991). Swelling pressure equilibrium of swollen crosslinked systems in an external field. III: Unsolved problems concerning the systems gelatin/water and κ-carrageenan/water. *Progress in Colloid & Polymer Science* **86**, 102–110.

CÖLFEN, H. AND BORCHARD, W. (1994). Soluble parts in gelatin/water gels. *Acta Polymerica* **45**, 325–329.

CÖLFEN, H. AND BORCHARD, W. (1994a). A modified experimental set-up for sedimentation equilibrium experiments with gels. Part 1: The instrumentation. *Progress in Colloid & Polymer Science* **94**, 90–101.

CÖLFEN, H. AND BORCHARD, W. (1994b). A modified experimental set-up for sedimentation equilibrium experiments with gels. Part 2: Technical developments. *Analytical Biochemistry* **219**, 321–334.

CÖLFEN, H. AND BORCHARD, W. (1994c). Ultrasensitive Schlieren optical system. In *Biochemical Diagnostic Instrumentation*. Eds. R.F. Bonner, G.E. Cohn, T.M. Laue, A.V. Priezzhev. Proceedings, SPIE **2136**, 307–314.

CÖLFEN, H. AND BORCHARD, W. (1994d). Determination of the partial specific volumes of thermoreversible gelatin/water and κ-carrageenan/water gels. *Macromolecular Chemistry and Physics* **195**, 1165–1175.

CÖLFEN, H. AND BORCHARD, W. (1995). Influence of soluble parts in gelatin/water gels on their network structure in an ultracentrifugal field. *Macromolecular Chemistry and Physics* **196**, 3469–3485.

CÖLFEN, H. AND HARDING, S.E. (1994). Unpublished observations.

DE GENNES, P.G. (1975). *Scaling concepts in Polymer Physics*. London: Cornell University Press.

FLORY, P.J. AND REHNER, J.R. (1943). Statistical mechanics of cross-linked polymer networks II: Swelling. *Journal of Chemical Physics* **11**, 521–526.

FLORY, P.J. AND GARRETT, R.R. (1958). Phase transitions in collagen and gelatin systems. *Journal of the American Chemical Society* **80**, 4836–4845.

HINSKEN, H. AND BORCHARD, W. (1995). Continuous swelling pressure equilibria of the system κ-carrageenan/water. *Colloid & Polymer Science* **273**, 913–925.

HINSKEN, H., SELIC, E. AND BORCHARD W. (1995). Formation of reversible concentration gradients during the centrifugation of gels. *Progress in Colloid & Polymer Science* **99**, 154–161.

HINSKEN, H. (1998). Eine neue Methode zur Bestimmung Kinetischer und thermodynamischer Großen thermoreversibler Gele mit Hilfe der analytischen Ultrazentrifuge: Die Gradientenmethode am Beispiel des Systems κ-Carrageenan/Wasser. *PhD thesis*, Duisburg.

HOLTUS, G. AND BORCHARD, W. (1989). Swelling pressure equilibrium of physical networks in the field of an analytical ultracentrifuge. *Colloid & Polymer Science* **267**, 1133–1138.

HOLTUS, G. (1990). Untersuchung der Quellungsdruckgleichgewichte von wäßrigen Gelatine-Gelen in einer Analytischen Ultrazentrifuge. *PhD thesis*, Duisburg.

HOLTUS, G., CÖLFEN, H. AND BORCHARD, W. (1991). Swelling pressure equilibrium of swollen crosslinked systems in an external field. II: The determination of molecular parameters of gelatin/water gels from the swelling pressure–concentration curves. *Progress in Colloid & Polymer Science* **86**, 92–101.

JOHNSON, P. AND METCALFE, J.C. (1963). Sedimentation studies of gelatin gels. *Journal of Photographic Science* **11**, 214–224.

JOHNSON, P. (1964). A sedimentation study on gel systems. *Proceedings of the Royal Society of London* **A278**, 527–542.

JOHNSON, P. AND METCALFE, J.C. (1967). Physico-chemical studies on gelatin gels from soluble and insoluble collagens. *European Polymer Journal* **3**, 423–447.

JOHNSON, P. AND KING, R.W. (1968). Sedimentation studies on gelatin gels. *Journal of Photographic Science* **16**, 82–88.

JOHNSON, P. (1968). Physicochemical studies on strongly interacting systems. In *Solution properties of natural polymers*. Special publication No. **23**, pp 243–262. London: Royal Society of Chemistry.

JOHNSON, P. (1971). Velocity and equilibrium aspects of the sedimentation of agar gels. *Journal of Photographic Science* **19**, 49–54.

JOHNSON, P. (1972). Velocity and equilibrium aspects of the sedimentation of agar gels. In *Photographic Gelatin*. Ed. R.J. Cox, pp 13–27. London, New York: Academic Press

JOHNSON, P. AND RAINSFORD, K.D. (1972). The physical properties of mucus: Preliminary observations on the sedimentation behaviour of porcine gastric mucus. *Biochimica et Biophysica Acta* **286**, 72–78.

KING, R.W. (1967). Physical and Chemical Studies on Gelatin Gels and Sols. *PhD thesis*, Cambridge.

LANGE, H. (1986). Determination of the degree of swelling and crosslinking of extremely small polymer gel quantities by Analytical Ultracentrifugation. *Colloid & Polymer Science* **264**, 488–493.

MÄCHTLE, W. (1992). Analysis of Polymer Dispersions with an Eight-Cell-AUC-Multiplexer: High Resolution Particle Size Distribution and Density Gradient Techniques. In *Analytical Ultracentrifugation in Biochemistry and Polymer Science*. Eds. S.E. Harding, A.J. Rowe and J.C. Horton, pp 147–175. Cambridge: Royal Society of Chemistry.

MÄCHTLE, W., LEY, G. AND STREIB, J. (1995). Studies of microgel formation in aqueous and organic solvents by light scattering and analytical ultracentrifuge. *Progress in Colloid & Polymer Science* **99**, 144–153.

MCBAIN, J.W. AND STUEWER, R.F. (1936). Anwendungen des einfachen luftgetriebenen Zentrifugenkreisels auf kolloidchemische Probleme. *Kolloid-Zeitschrift* **74** Heft 1; 10–16.

METCALFE, J.C. (1965). A physico-chemical study of gelatin gels. *PhD thesis*, Cambridge.

MÜLLER, H.G., SCHMIDT, A. AND KRANZ, D. (1991). Determination of the degree of swelling and crosslinking of latex particles by Analytical Ultracentrifugation. *Progress in Colloid & Polymer Science* **86**, 70–75.

SHASKOUA, V.E. AND VAN HOLDE, K.E. (1958). Graft Copolymers: Synthesis and Characterization. *Journal of Polymer Science* **28**, 395–411.

SHASKOUA, V.E. AND BEAMAN, R.G. (1958). Microgel: An Idealized Polymer Molecule. *Journal of Polymer Science* **33**, 101–117.

STEENSGAARD, J., HUMPHRIES, S. AND SPRAGG, P. (1992). Measurement of sedimentation coefficients. In *Preparative Centrifugation, A practical Approach*. Ed. D. Rickwood, p 193. Oxford: Oxford University Press.

SVEDBERG, T. AND PEDERSEN, K.O. (1940). *The Ultracentrifuge*. pp 29–33. Oxford: Oxford University Press, or Svedberg, T. and Pedersen, K. O. (1940). *Die Ultrazentrifuge*. pp 26–29. Dresden: Steinkopff Verlag.

5
Biotechnological Strategies for the Modification of Food Lipids

WENDY M. WILLIS AND ALEJANDRO G. MARANGONI*

Department of Food Science, University of Guelph, Guelph, Ontario, N1G 2W1, Canada

Introduction

Now that the nutritional requirements of infants, adults and patients are becoming more clearly defined, biotechnology is moving to the forefront of lipid modification strategies. Medium chain fatty acids are an important source of rapid energy for preterm infants and for patients with fat malabsorption-related diseases. Polyunsaturated fatty acids (PUFAs) are important both for infant development and disease prevention in adults. Lipase-catalysed interesterification has been used to produce structured lipids containing medium chain fatty acids and PUFAs. These techniques have also been used to increase PUFA concentrations in fish oils, and to introduce these fatty acids into vegetable oils. Genetic engineering of oilseed plants can be used to develop plants which produce medium chain fatty acids and PUFAs. By manipulating growth conditions, high-PUFA oils can also be obtained from algae and fungi. Even though biotechnological processes applied to the modification of the structure and properties of fats and oils are expensive relative to chemical processes, their use may be justified due to their greater lipid tailoring potential.

Biotechnology, as it applies to fats and oils, involves the production of nutritionally improved products through enzymatic modifications, biotransformations and genetic engineering. Recent trends in health-related concerns about fats and oils have included decreasing consumption of fats high in saturated and *trans* unsaturated fatty acids as well as increasing consumption of essential and polyunsaturated fatty acids. Lipase-catalysed reactions, biotransformations and genetic engineering of plants are all strategies being used to develop fats and oils with increased concentrations of medium chain, essential and polyunsaturated fatty acids. The products manufactured by these processes have been used to meet infant fat nutritional requirements and for disease treatment and prevention in adults. These technologies have also been used to modify the physical properties of certain lipids, including their melting profiles and textural properties.

*To whom correspondence may be addressed. E-mail: amarango@uoguelph.ca

Biotechnology and Genetic Engineering Reviews – Vol. 16, April 1999
0264–8725/99/15/141–175 $20.00 + $0.00 © Intercept Ltd, P.O. Box 716, Andover, Hampshire SP10 1YG, UK

141

Medium chain fatty acids (MCFAs) have been used in the treatment of fat malabsorption related diseases and as a significant source of energy for preterm infants. Consumption of fats and oils containing polyunsaturated fatty acids (PUFAs) such as eicosapentaenoic acid (EPA) and docosahexaenoic acid (DHA), is increasing due to their role in preventing atherosclerosis in adults. The concentration of these fatty acids as well as their positional distribution within a triacylglycerol molecule are important factors.

Typically, chemical methods of lipid modification have been used in industrial applications due to their lower cost. However, biotechnology allows production of fats and oils with specific compositions, both in terms of positional distribution of specific fatty acids and fatty acid composition.

The aim of this paper is to provide an overview of how biotechnology, in the form of lipase-catalysed reactions, biotransformations and genetic engineering of oilseeds allow both nutritional and physical modification of fats and oils.

Metabolism and absorption of fats and oils

The positional distribution and fatty acid composition of fats and oils can have an impact on their digestion and absorption (*Figure 1*). Differences in the requirements for specific structures and composition of fats and oils are due to differences between adult and infant digestive systems. The less mature infant digestive system possesses reduced levels of pancreatic lipase and bile salts, and greater lingual lipase activity (Zoppi *et al.*, 1972; Fredrikzon and Olivecrona, 1978; Watkins, 1988).

Lingual lipase, which accounts for 50 to 70% of lipid hydrolysis in infants (Watkins, 1988), but has minimal importance in adults' digestion, is active in the upper intestinal tract, hydrolysing triacylglycerols (TAGs) to monoacylglycerols (MAGs), diacylglycerols (DAGs), and free fatty acids (FFAs). It is more specific towards short and medium chain fatty acids and fatty acids in the sn-3 position, producing mainly 1,2-DAGs and FFAs (Christensen *et al.*, 1995; Small, 1991). Gastric lipase in the stomach is also more specific towards short and medium chain fatty acids and continues to hydrolyse positions 1 and 3 of the TAGs to produce FFAs, MAGs and DAGs. Short and medium chain fatty acids, including butyric, caproic, caprylic and capric acid are more soluble in aqueous media and can be absorbed through the stomach directly into venous circulation, travelling via the portal vein to the liver where they are oxidized and used as a rapid source of energy (Borum, 1992; Nelson, 1992). This attribute is particularly important for treatment of fat malabsorption related diseases and in the provision of a rapid source of energy for premature infants. Medium chain fatty acids also provide a rapid source of energy in muscle since their transport into the mitochondria for β-oxidation is not carnitine dependent (Bruckner, 1992).

Pancreatic lipase is present in the small intestine and provides the final hydrolysis of TAGs. It is 1,3-specific, with a slight preference towards fatty acids in the sn-1 position (Small, 1991; Berdanier, 1995). It is more active towards short and medium chain fatty acids and has low activity with respect to long chain PUFAs in the sn-1 and sn-3 positions (Christensen *et al.*, 1995).

After hydrolysis, the fatty acids and 2-MAGs in the form of micelles with bile salts are absorbed through the intestinal mucosa. The positional distribution of fatty acids

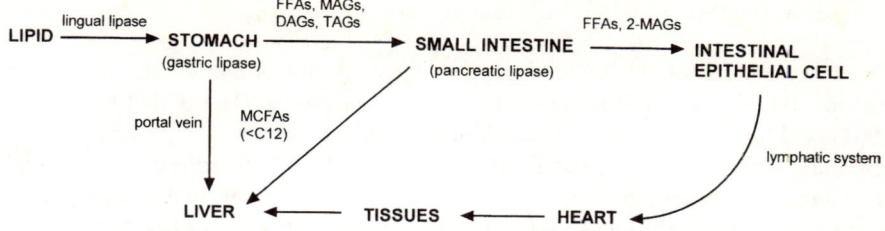

Figure 1.   Metabolism and transport of lipids and their hydrolysis components in humans.

can have a dramatic impact on their degree of absorption. Long chain saturated fatty acids such as palmitic acid are poorly absorbed in their free acid form since they are solid at body temperature and form insoluble calcium and magnesium soaps in the intestine. As a 2-MAG, palmitic acid is easily absorbed in the intestine (Innis *et al.,* 1995; Tomarelli, 1988), which is important for infants since human milkfat contains 20 to 30% palmitic acid, 70% of which is present in the sn-2 position (Innis *et al.,* 1995). Placement at the sn-2 position is also important for long chain PUFAs, which are poor substrates for pancreatic lipase and are better absorbed as 2-MAGs, requiring hydrolysis by carboxyl ester hydrolases when present in positions sn-1 and 3 (Small, 1991).

New TAGs, which are formed by reacylation of FFAs and 2-MAGs after absorption from the intestines, are combined with phospholipids and apolipoproteins to form chylomicrons for transport through the lymphatic system and into general circulation. Any short and medium chain fatty acids which were not absorbed through passive diffusion into venous circulation as well as monoacyglycerols containing these fatty acids, are also incorporated into chylomicrons and are transported through the lymphatic system (Lambert *et al.*, 1996; Bell *et al.*, 1997).

Upon transport to adipose and muscle tissue, lipoprotein lipase at the surface of capillary endothelial cells hydrolyses TAGs to 2-MAGs and FFAs, which are then transported into cells, undergoing reacylation to form TAGs for storage (Small, 1991; Tso and Wediman, 1987).

Overall, proper and efficient absorption is dependent on the composition and positional distribution of fatty acids. Therefore, any attempts to provide nutritionally improved fats and oils for applications in infant formulas, disease treatment and adult disease prevention must account for differences in metabolism as well as differences in the nutritional effects of specific fatty acids on different populations of consumers.

Metabolic significance of specific fat components

The major fatty acids associated with biotechnological modifications of fats and oils are MCFAs and long chain PUFAs. Genetic engineering of oilseed crops as well as lipase-catalysed production of structured lipids are the predominant methods of increasing the MCFA concentration. Long chain PUFAs are concentrated in fish oils and incorporated into vegetable oils by genetic engineering of oilseed crops and lipase-catalysed interesterification reactions.

MEDIUM AND LONG CHAIN SATURATED FATTY ACIDS

As previously discussed, MCFAs are important to infants and adults, for varied reasons. Infants use 25% of their caloric intake for growth (Hardy and Kleinman, 1994), and for those who are premature or have reduced biliary flow, MCFAs are an important source of energy (Borum, 1992). Medium chain fatty acids possess several advantages over other fatty acids which has increased the interest in developing new, or modifying existing fats and oils to increase their concentrations (*Table 1*). Unmodified sources of MCFAs include coconut and palm kernel oils, which lack significant concentrations of essential and polyunsaturated fatty acids, making them unsuitable for general use. Modified lipids containing increased concentrations of MCFAs have been used in the treatment of diseases where fat malabsorption and metabolism are a problem, such as AIDS, cystic fibrosis, cirrhosis and anorexia (Fan, 1997; Sandström *et al.*, 1993; Bell *et al.*, 1997; Ulrich *et al.*, 1996). Medium chain fatty acids are also essential components of formulas for premature infants (40–50% of total fatty acids) and parenteral and enteral formulations in Europe, where coconut oil is used as a source of medium chain TAGs (Mascioli *et al.*, 1988; Ulrich *et al.*, 1996; Borum, 1992). In terms of improved nutrition, MCFAs have been used as an energy supplement for athletes (Van Zyl *et al.*, 1996).

The disadvantages associated with medium chain TAGs include the potential for acidosis or toxicity at high concentrations, their negative effect on plasma concentrations of other fatty acids and their lack of essential and polyunsaturated fatty acids (Carnielli *et al.*, 1996; Sandström *et al.*, 1993; Bell *et al.*, 1997; Ulrich *et al.*, 1996).

The use of biotechnology to modify existing fats and oils and incorporate MCFAs allows producers to avoid some of the disadvantages associated with medium chain TAGs. Lipase-catalysed interesterification reactions and genetic engineering of oilseeds allow production of TAGs containing a combination of medium chain, essential and polyunsaturated fatty acids. Both of these methods also allow some degree of control over the positional distribution of these fatty acids, making their use superior to simple blending of MCTs and long chain TAGs and chemical means of lipid modification (Sandström *et al.*, 1993; Nordenström *et al.*, 1995; Jeevanandam *et al.*, 1995).

The positive health effects of long chain saturated fatty acids are really only associated with infant development. Human milk contains 20 to 30% palmitic acid, 70% of which is present in the sn-2 position of the triacylglycerol (Innis *et al.*, 1995).

As a 2-MAG, palmitic acid is an important source of energy for infants, since it is only minimally absorbed in its free acid form, forming insoluble salts with calcium (Innis *et al.*, 1995; Small, 1991; Tomarelli, 1988). A higher degree of incorporation of palmitic acid into plasma TAGs has been observed for infants fed human milk compared to a vegetable oil based formula, where the majority of palmitic acid is in the sn-1 or sn-3 positions of the triacylglycerol (Innis *et al.*, 1994; Carnielli *et al.*, 1995). Increasing the concentration of palmitic acid in the sn-2 position has also been associated with a small but significant increase in total energy absorption and reduced intestinal length required for absorption in rats (de Fouw *et al.*, 1994).

In terms of adult nutrition, it is less desirable to have long chain saturated fatty acids in the sn-2 position due to their possible hypercholesteremic effects and increased risk of heart disease (Caggiula and Mustad, 1997). Despite some concerns, not all saturated fatty acids have such a negative impact on health. While lauric, myristic and

Table 1.   Metabolic and digestive advantages associated with the consumption of medium chain fatty acids

Advantages	Reference
Extremely resistant to oxidation and stable at high and low temperatures	Megremis (1991)
Most are not incorporated into chylomicrons, do not undergo desaturation or elongation and are therefore more likely to be used for energy	Babayan and Rosenau (1991)
As TAGs, are metabolized as quickly as glucose, yet have twice the energy density of carbohydrates	Bell *et al.* (1997)
Do not promote the synthesis of eicosanoids and are not involved in free radical formation, both of which are involved in inflammatory responses	Ulrich *et al.* (1996)
Are readily oxidized for rapid energy production	Bach *et al.* (1988); Johnson *et al.* (1990)
Are easily absorbed, are completely metabolized in the liver, do not interfere with the reticuloendothelial system and are not carnitine dependent for transport into mitochondria	Borum (1992); Sandström *et al.* (1993); Mascioli *et al.* (1988)

palmitic acid are considered to be hypercholesteremic (Denke and Grundy, 1992; Zock *et al.*, 1994), with palmitic acid being less hypercholesteremic than myristic acid, stearic acid seems to have the same or a better lowering effect on total and LDL cholesterol as oleic acid (Kris-Etherton and Yu, 1997; Pai and Yeh, 1997). Therefore, the positional distribution requirements for long chain saturated fatty acids is different for infants and adults, with a sn-2 positioning ideal for infants and positioning in the sn-1 and sn-3 positions preferable for adults.

LONG CHAIN UNSATURATED AND POLYUNSATURATED FATTY ACIDS

While the issue of *trans* fatty acids (TFAs) is important in terms of its negative health effects, it is not dealt with by biotechnologically-related lipid modifications, except in the production of low *trans* fatty acid margarines to replace those that have been chemically hydrogenated, and may contain 5 to 50% *trans* fatty acids (Ohlrogge, 1983). *Trans* fatty acids seem to be metabolized as efficiently as other fatty acids, but may impair desaturation and elongation of LA to arachidonic acid (AA), thereby affecting eicosanoid production and growth (Desci and Koletzko, 1995; Ratnayake and Chen, 1996). The negative effects of TFAs on adults are decreased concentrations of high density lipoprotein (HDL) cholesterol and increased concentrations of low density lipoprotein (LDL) cholesterol, producing a hypercholesteremic effect (Mensink and Katan, 1990; Zock and Katan, 1992). Despite concerns about the effects of TFAs in human milk and infant formulas on infant health, a causal link between TFAs and infant development has not been established, with only a possible association between TFAs and lower n-3 and n-6 long chain PUFA concentrations (Carlson *et al.*, 1997). Similar concerns about a lack of epidemiological data which support the link between TFAs and increased risk of coronary heart disease in adults have brought the true negative impact of TFAs on human health into question (Shapiro, 1997).

The only fatty acids considered essential for human growth are linoleic (LA) and linolenic (LNA) acid, which are required in a range of 0.5 to 2.0% and 0.5% of total energy respectively (Bruckner, 1992; Bell *et al.*, 1997). Essential fatty acid deficiency has been associated with growth retardation, increased membrane permeability, sterility and capillary fragility (Vergroesen, 1976). Linoleic acid and LNA are precursors of 20-carbon fatty acids which are precursors of the eicosanoids. Essential fatty acids are particularly important for infants as their levels can have an effect on DHA and AA levels, especially in preterm infants where a high ratio of LA to LNA can reduce the ratio of DHA to AA in the brain and retina (Martinez, 1992; Gibson *et al.*, 1994). Increasing the concentration of LNA in infant formula can also produce an increase in DHA levels (Innis *et al.*, 1997). Competition between LA and LNA for Δ6-desaturase seems to influence the synthesis of both AA and DHA (Innis, 1992). Excessive concentrations of LA in the adult diet can increase the risk of cancer and can predispose membrane phospholipids to free radical oxidation (Grundy, 1997).

The importance of long chain PUFAs such as EPA and DHA in the prevention of heart disease in adults was first realized in the early 1970's (Bang and Dyerberg, 1972). Since then, increased consumption of EPA and DHA, in the form of fish and fish oil capsules has been associated with a reduced risk of atherosclerosis, tumour growth, thrombosis, hypertriglyceridaemia, and high blood pressure (Braden and Carroll, 1986; Shekelle *et al.*, 1981; McGee *et al.*, 1984; Joossens *et al.*, 1989). The inhibitory effect of PUFAs on many of these conditions seems to be related to their mediation of eicosanoid precursor synthesis (Braden and Carroll, 1986). In infants, DHA is required for nervous system and retinal development as shown by reduced data processing time and increased visual acuity in infants fed human milk containing DHA compared to infants fed formula with no added DHA (Makrides *et al.*, 1996; Carlson *et al.*, 1996). An exogenous source of DHA is required by infants because LNA is not readily converted to DHA (Neuringer *et al.*, 1994).

There have been conflicting opinions about the importance of DHA in infant development and consequently about its presence in infant formulas due to conflicting reports about the extent of its effects on visual function in infants. Several authors have found that supplementation of infant formula with DHA has produced some improvement in visual function compared to non-supplemented formula-fed infants for both preterm and term infants, up to one year of age (Werkman and Carlson, 1996; Carlson and Werkman, 1996; Carlson *et al.*, 1993; Makrides *et al.*, 1996). Other authors have found no improvement in visual function, even when comparisons were made between breast-fed, DHA-supplemented and DHA-non-supplemented formula-fed infants (Carlson, 1996; Innis *et al.*, 1996). Preterm infants may have a higher requirement for DHA as well as AA because they are more susceptible to pre- and post-natal deficits of AA and DHA which may lead to neurovisual development disorders (Crawford *et al.*, 1997).

While the traditional source of EPA and DHA has been fish oils, which can contain between 10 and 25% total of these PUFAs (Haraldsson *et al.*, 1993), there is difficulty in supplementing infant formula with this source since EPA seems to compete with AA for incorporation into membranes, resulting in a reduction of membrane AA and a subsequent reduction in the 2-series eicosanoids (Martinez, 1992; Makrides *et al.*, 1995). Feeding infants a fish oil source of DHA in the presence of EPA has been associated with infants with lower normalized weights and lengths than those fed standard infant formula (Carlson *et al.*, 1992). There is controversy regarding the inclusion of EPA in infant formula since it is present in low concentrations (around

0.2% of total fatty acids) in human milk, and its levels can be affected by the mother's diet (Francois *et al.*, 1998). PUFAs in TAGs from algal sources or as phospholipids from eggs are absorbed from infant formula by preterm infants as efficiently as from human milk (Carnielli *et al.*, 1998; Boehm *et al.*, 1997).

The positional distribution of long chain PUFAs, along with chain length and degree of unsaturation are important factors affecting the degree of hydrolysis and absorption. Structured lipids, with the PUFAs in position sn-2 are ideal due to the low activity of pancreatic lipase towards these fatty acids in the sn-1 and sn-3 positions and their improved absorption as 2-MAGs (Christensen *et al.*, 1995). There is some conflict as to the best form of PUFAs for good absorption, although the degree of absorption does not seem to differ (Linko and Hayakawa, 1996; Krokan and coworkers, 1983). Nelson and Ackman (1988) found that ethyl esters of EPA are absorbed better in humans than free fatty acid forms or 2-MAG forms of EPA, while Linko and Hayakawa (1996) found that the degree of absorption of free DHA was greater than 95%, while in the TAG and ethyl ester forms, degrees of absorption were only 57% and 21%, respectively. The advantage of consuming ethyl esters over natural TAGs is that the target fatty acid can be concentrated to a greater extent during processing.

As mentioned previously, long chain PUFAs are important due to their relationship to eicosanoids such as prostaglandins, leukotrienes and thromboxanes, which are derived from three different 20-carbon fatty acids and possess hormone-like activity (Berdanier, 1995). Dihomo-gamma-linolenic acid, arachidonic acid (AA) and EPA are precursors for eicosanoid series 1, 2, and 3, respectively, with AA metabolism beginning with LA, and EPA metabolism beginning with α-LNA (*Figure 2*). Methyl-interrupted PUFAs are formed by successive elongation and desaturation, with n-3 PUFAs such as DHA and EPA formed from α-LNA and n-6 PUFAs such as AA formed from LA. While EPA and DHA are produced from LNA, they are still required in the diet because the conversion efficiency of LNA is low, and direct consumption of EPA and DHA is more effective at increasing their levels in plasma lipids (Linko and Hayakawa, 1996). Eicosanoids derived from EPA have relatively weak capabilities (weaker thromboxanes), stimulating and preventing platelet aggregation and causing smooth muscle contraction. Eicosanoids derived from AA have a strong influence, stimulating and preventing aggregation of platelets and contraction of smooth muscle (Gurr, 1992). Replacement of AA with EPA through consumption of fish oils high in EPA inhibits production of strong aggregator thromboxanes by AA, thereby reducing platelet aggregation and reducing the risk of atherosclerosis (Garg *et al.*, 1990). High ratios of LA to saturated fatty acids in the diet also inhibit desaturase activity (Spielmann *et al.*, 1988).

There are some differences in the effects of these precursor fatty acids in infant development. As previously mentioned, the displacement of AA by EPA is not beneficial in infants since it has been associated with reduced growth. Excessive or limited concentrations of LA can also reduce AA levels. The difficulty in predicting the effects of different levels of eicosanoid precursors on health and disease of infants and adults is that each of the three eicosanoid families display both anti-aggregatory and anti-inflammatory precursors. With an understanding of the lipid compositional requirements of different population segments, biotechnology in the form of biotransformations, genetic modification of oilseed crops and lipase-catalysed interesterification have been applied to meet these needs.

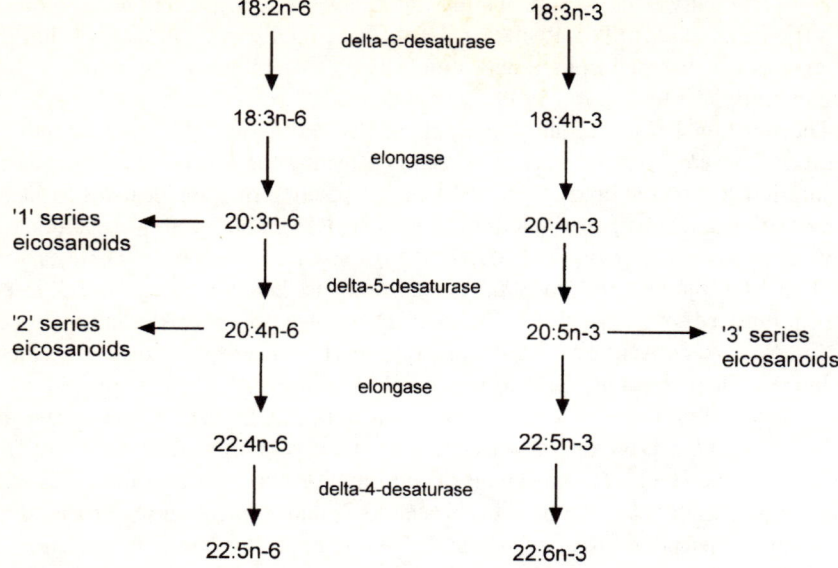

Figure 2.   Pathways for production of EPA, DHA (from linolenic acid) and AA (from linoleic acid) and their associated eicosanoids.

Biotransformations: production of PUFAs by microorganisms

SOURCES

With greater demand for high quality and less expensive sources of EPA and DHA for use as supplements in nutritionally improved fats and oils, researchers have turned to microorganisms, mainly algae and fungi, as a major source of PUFAs (*Tables 2,3,4*). The cost of deriving long chain PUFAs from microorganisms has been estimated to be half that of fish oil derived production of PUFAs (Molina Grima *et al.*, 1996). Most long chain PUFAs are derived from eukaryotes such as fungi and algae which produce TAGs in their biomass with fatty acid compositions similar to those of plant oils (Mukherjee, 1998).

Fish oil composition is more variable due to diet-related effects and fish oils are more susceptible to oxidation due to the extended periods between catching and processing, resulting in undesirable flavours and odours (Shimizu *et al.*, 1989). Microbial sources of these PUFAs are superior to fish-derived fatty acids because of improved oxidative stability, the potential to produce high concentrations of specific fatty acids and greater control of fatty acid composition (López Alonso and del Castillo, 1996; Medina *et al.*, 1995; Yongmanitchai and Ward, 1991). It is possible to use microorganisms which produce high concentrations of one specific PUFA, making it easier to isolate this PUFA from other fatty acids (Molina Grima *et al.*, 1996). Other advantages of using microorganisms for PUFA production include low energy costs since operating temperatures are close to room temperature, feasible large scale production, and the potential for the production of value-added by-products (Shimizu *et al.*, 1989; Cohen, 1990).

Table 2. Selected algae, their total lipid content and nutritionally important fatty acids

Organism	Fatty acid of interest	Maximum biomass content (g/L)	Maximum content (% w/w)	Reference
Chlorella minutissima	eicosapentaenoic acid (31.8% of FA)	0.5	14.3 (lipid)	Yongmanitchai and Ward (1991)
Phaeodactylum tricornutum	eicosapentaenoic acid (30.5% of FA)	2.5	17.4 (lipid)	Yongmanitchai and Ward (1991)
Nitzschia laevis	eicosapentaenoic acid (23.2% of FA)	–	39.3 (lipid)	Koon Tan and Johns (1996)
picoplankton strain PP301	eicosapentaenoic acid (53.2% of FA) and docosahexaenoic acid (8.9% of FA)	–	13.6 (FA)	Kawachi *et al.* (1996)
Isochrysis galbana	eicosapentaenoic acid and docosahexaenoic acid (9.5% of FA)	–	9.5 (FA)	Robles Medina *et al.* (1995)
Porphyridium cruentum (113.80)	eicosapentaenoic acid (42% of FA)	–	4.4 (FA)	Cohen (1990)
Phaeodactylum tricornutum	eicosapentaenoic acid (3.1% of biomass, 30.5% of FA)	–	–	Cartens *et al.* (1996)
Isochrysis galbana	docosahexaenoic acid (5.4% of biomass)	0.9	14.8 (FA)	Burgess *et al.* (1993)

Table 3. Selected bacteria, their total lipid content and nutritionally important fatty acids

Organism	Fatty acid(s) of interest	Maximum cell content (g/L)	Maximum content (% w/w)	Reference
Vibrio (strain T3614)	docosahexaenoic (6.7% of FA)	0.1	9.8 (lipid)	Yano *et al.* (1994)
SCRC-2738 (from mackerel intestine)	eicosapentaenoic acid (12.8% of FA)	3.7	5.7 (FA)	Akimoto *et al.* (1990)
SCRC-2738 (from mackerel intestine)	eicosapentaenoic acid (22.5% of FA)	5.3	4.3 (FA)	Akimoto *et al.* (1991)

EXTRACTION AND PURIFICATION

The major difficulty associated with the production of PUFAs using microorganisms centres around the relatively extensive purification process required to isolate and refine the oils. Some methods involve a five-step process using a combination of chloroform, methanol, and water to separate lipid classes followed by transmethylation, urea fractionation and reverse phase chromatography (Cartens *et al.*, 1996). Urea complexation is an effective method of PUFA isolation since urea preferentially complexes with saturated and monounsaturated fatty acids to form solids which can be removed from the PUFAs (Bajpai and Bajpai, 1993). Cartens and coworkers (1996)

Table 4.   Selected fungi, their total lipid content and nutritionally important fatty acids

Organism	Fatty acid of interest	Maximum biomass content (g/L)	Maximum lipid content (% w/w)	Reference
Mortierella sp.	g-linolenic (26% of lipid)	11–12	24	Hansson and Dostálek (1988)
Mortierella alpina	arachidonic (31% of lipid)	23	11	Lindberg and Molin (1993)
Mortierella alpina	arachidonic acid (31% of lipid)	22.5	44	Shinmen *et al.* (1989)
Thraustochytrium aureum	docosahexaenoic (40% of lipid)	5.7	8	Iida *et al.* (1996)
Pythium irregulare	eicosapentaenoic acid (25% of lipid)	–	10	O'Brien *et al.* (1993)
Thraustochytrium aureum	eicosapentaenoic (9% of FA) docosahexaenoic (30% of FA)	4	10	Kendrick and Ratledge (1992)
Mortierella alpina-peyron	arachidonic (5% of FA)	3.2	38	Kendrick and Ratledge (1992)
Pythium ultimum	eicosapentaenoic (0.7% of biomass w/w)	3.2	–	Wessinger *et al.* (1990)
Shiizochytrium sp.	docosahexaenoic (34% of FA)	21.0	–	Nakahara *et al.* (1996)
Thraustochytrium aureum	docosahexaenoic acid (5–7% of lipid)	1.1–5.5	1.7–25.2	Bajpai *et al.* (1991)

avoided chloroform and methanol by using ethanolic potassium hydroxide sapon-ification, followed by liquid chromatography to isolate the PUFA fraction. They obtained the same yield of EPA without the urea fractionation step. Medina and coworkers (1995) developed a two-step purification process for algal biomass involving only urea complexation and liquid chromatography, achieving a range of 94–96% purity for fractions of EPA and DHA, respectively. Molina Grima and coworkers (1996) scaled-up recovery of EPA from *P. tricornutum* using a four-step process of fatty acid extraction, saponification, lyophilization, urea fractionation and semi-preparative HPLC.

GROWTH CONDITIONS

Growth of microorganisms is highly dependent on growth media composition including temperature, pH, source, amount and ratio of carbon and nitrogen sources, and the degree of light exposure and aeration (Cohen, 1990; Bajpai and Bajpai, 1993; Rose, 1989). There are major differences in the growth conditions of different organisms, which can affect both the lipid content and fatty acid composition of the biomass (O'Brien *et al.*, 1993; Akimoto *et al.*, 1991).

In general, decreasing the temperature produces an increase in the concentration of unsaturated fatty acids, although a decrease in total lipid in the biomass can also occur (Lindberg and Molin, 1993). Membrane lipids, and in particular, phospholipids must remain in a liquid-crystalline state at lower temperatures to allow for normal membrane activity. This requires an increase in the degree of unsaturation of their component fatty acids at lower temperatures. Several authors have verified that this increase in the concentration of unsaturated fatty acids at lower temperatures occurs in both algae and fungi (Hansson and Dostálek, 1988; Lindberg and Molin, 1993; Burgess *et al.*, 1993). As previously mentioned, the optimum temperature for growth of both fungi and algae is usually between 20 and 30°C, with the temperature being modified to maximize concentrations of specific PUFAs (Kawachi *et al.*, 1996; Burgess *et al.*, 1993; Akimoto *et al.*, 1990). This temperature range provides an economical means for scale-up and production and allows production both indoors and outdoors. Molina Grima and coworkers (1995) were able to produce *I. galbana* outdoors in Spain and found that productivity was greatest in the early part of the summer when the average outdoor temperature brought the culture temperature closest to its optimum growth temperature of 20°C. Optimum pH values for individual organisms vary, although most tend to be in the range of pH 5.5 to 8 (Lindberg and Molin, 1993; Iida *et al.*, 1996).

Production of PUFAs is highly dependent on the growth stage of the micro-organism. While there is generally rapid production of PUFAs during the growth phase of the organisms, concentrations can be dramatically increased by ageing at steady state for several days at lower temperatures (Bajpai and Bajpai, 1993; Akimoto *et al.*, 1990; Shimizu *et al.*, 1989; Bajpai *et al.*, 1991). The fatty acid composition can also be affected by growth rate and storage time. Cohen (1990) found that the AA concentration produced by the microalga, *Porphyridium cruentum*, could be increased by nitrogen starvation and decreased growth, while with rapid growth, the concentration of EPA increased. Burgess and coworkers (1993) found that there was increased production of DHA by *I. galbana* by using storage conditions of low temperature and low light after the growth phase.

All organisms require a source of carbon and nitrogen for growth, although the sources can vary widely. Bajpai and coworkers (1991) found that the fungi *Thraustochytrium aureum* grew best on a carbon source of linseed oil, glucose or starch, while O'Brien and coworkers (1993) were able to use sweet whey permeate as a lactose source for fungal growth. The presence of vegetable oils such as soybean, corn, peanut and rapeseed oil, which have a high concentration of oleic and linoleic acid seems to promote the production of EPA and DHA (Shinmen *et al.*, 1989). EPA and DHA are produced from monounsaturated fatty acids, such as oleic acid which are desaturated and elongated via the n-3 pathway to produce LA then EPA and DHA, or the n-6 to produce LNA and AA (Bajpai and Bajpai, 1993). Kanisaka and coworkers (1990) also found that oleic and linoleic acid were both used rapidly by *Mortierella* sp. and desaturated to γ-linolenic acid during its growth phase. Sources of nitrogen may include potassium nitrate, corn steep liquor or yeast extract (Shinmen *et al.*, 1989; Hansson and Dostálek, 1988; Nakahara *et al.*, 1996). The concentration of nitrogen has an impact on the composition of fatty acids in the lipid fraction as shown by Cohen (1990), who found that starving the red microalga *Porhpyridium cruentum* produced a sharp increase in the concentration of AA. The ratio of carbon to nitrogen also seems to be important. Hansson and Dostálek (1988) found that a ratio of carbon to nitrogen of 80 produced the highest lipid concentration of 66% (w/w) in the biomass.

The fatty acid composition of light sensitive microorganisms can vary significantly during periods of light and dark. Storage fatty acids such as palmitic and palmitoleic acids in *I. galbana* accumulate in the presence of light and are used for energy in the absence of light (Molina Grima *et al.*, 1995). Bajpai and coworkers (1991) found that increasing the time of light exposure caused the fungi, *Thraustochytrium aureum*, to increase production of DHA, while Burgess and coworkers (1993) found that reducing light exposure caused the algae, *Isochrysis galbana*, to increase production of DHA. Photosynthetic organisms such as marine algae are superior to microbial cells because they can use sunlight for energy and CO_2 as a carbon source (Burgess *et al.*, 1993). Photosynthetic organisms also have increased concentrations of unsaturated fatty acids compared to heterotrophic organisms (Koon Tan and Johns, 1996).

SCALE-UP AND INCREASING PRODUCTIVITY

It can be difficult to scale up fungal and algal growth without experiencing difficulties. Using fungi for production has some inherent problems, including low growth rates, high viscosity and adhesion to surfaces (Hansson and Dostálek, 1988). Microalga are superior in this respect because they tend to autoflocculate, improving the harvest process (Cohen, 1990).

Fungi grow very well on solid media, but are much easier to grow in liquid media on a large scale, usually in shaker flasks (Shinmen *et al.*, 1989; Lindberg and Molin, 1993). Iida and coworkers (1996) found that a flask culture was superior to a fermenter culture of *T. aureum*, due to inhibition of growth from the stirring which occurred in the fermenter culture. Fukuda and Morikawa (1987) were able to improve the growth of fungi in fermenter cultures by immobilizing the cells with biomass support particles and by using a fluidized bed for mixing.

More recent efforts at improving the productivity of microorganisms have been concerned with designing equipment for scale up and controlling the growth environment as much as possible. As well, genetic modification has become another avenue to increase productivity of specific organisms (López Alonso and del Castillo, 1996). Productivity in terms of the total amount of specific fatty acids produced as opposed to the highest percentage of the total lipid content must be considered.

Genetic modification of oilseed crops

Genetic modification of oilseed crops to improve quality, pest and disease resistance and yield has expanded in recent years to include modification of the fatty acid composition of oils for food use as well as for oils used as lubricants and detergents (Murphy, 1996).

In general, oilseeds tend to accumulate LCFAs containing 16 or 18 carbons with one to several double bonds (Ohlrogge, 1994). Long chain fatty acid synthesis occurs in the plastids by sequential addition of two carbon units from acetyl-CoA to an acyl carrier protein (ACP) to eventually produce palmitoyl-acyl-carrier protein (palmitoyl-ACP) which is then removed from the ACP by a thioesterase (*Figure 3*). Elongation and desaturation of the fatty acid occurs in the cytoplasm to form other fatty acids (Murphy, 1994; Miquel and Browse, 1995; Slabas *et al.*, 1992). Triacylglycerols are formed by the sequential acylation of glycerol-3-phosphate to first form 1-acyl-*sn*-glycerol-3-phosphate then phosphatidic acid by lysophosphatidic acid acyltransferase. The phosphatidic acid is eventually converted to a diacylglycerol and then a

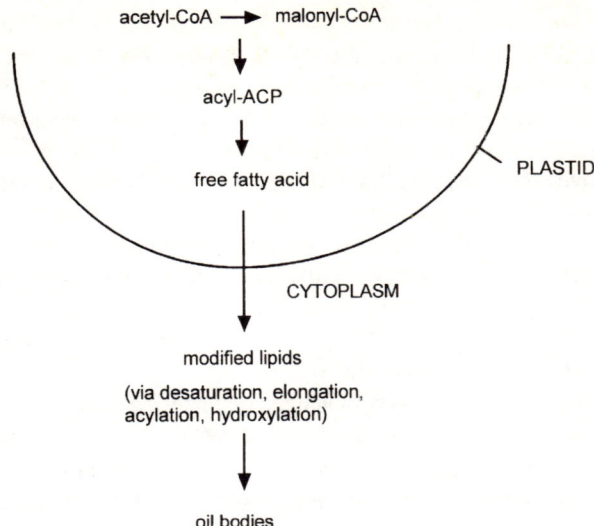

Figure 3. Simplified pathway for biosynthesis of fatty acids in plants (adapted from Jaworski *et al.*, 1992 and Slabas *et al.*, 1992).

triacylglycerol with the addition of another acyl group (Yuan and Knauf, 1998). The positional distribution of fatty acids in the triacylglycerol can be modified by the activity of lysophosphatidic acyltransferases from different sources which are specific towards fatty acids of differing chain lengths, such as for lauric acid by this acyltransferase in canola oil (Knutzon *et al.*, 1995).

The main method of fatty acid profile modification is the cloning and transfer of a gene from one plant species into another species to produce the desired levels of specific fatty acids. As well, naturally occurring enzymes can be modified or new ones can be introduced to modify the fatty acid profile of the oilseed. Once an enzyme has been identified and isolated, its associated gene can be identified and modified (Del Vecchio, 1996). Genes from bacterial, animal and yeast sources have also been incorporated into oilseeds for fatty acid modification (Miquel and Browse, 1995). Yadav and coworkers (1992) used a T-DNA tagging method to isolate the *Arabidopsis* microsomal n-3 fatty acid desaturase responsible for LNA to convert more than 75% of the LA in the seeds into LNA. A gene encoding lauric acid production from the California bay (*Umbellularia californica*) has been incorporated into *Arabidopsis thaliana*, resulting in a 70 fold increase in the 12:0-ACP thioesterase activity, and a subsequent increase in the concentration of lauric acid (Voelker *et al.*, 1992).

Previously, the main focus of transgenic modifications had been in the reduction of erucic acid in rapeseed oil. Since then, the focus has shifted towards encouraging production of medium chain fatty acids, including high lauric acid concentrations in canola oil (Del Vecchio, 1996). Some plant species produce high concentrations of MCFAs by a mechanism which seems to involve premature chain termination by acyl-ACP thioesterase to produce chain lengths of eight to fourteen carbons (Slabas *et al.*, 1992; Davies *et al.*, 1993). It is possible to further increase the medium chain fatty acid composition of oils such as the high laurate canola oil by cloning a gene which encodes for an acyltransferase which is specific for medium chain fatty acids (Knutzon *et al.*,

1995). While some seed oils, such as palm kernel oil, are not modified because they naturally contain high concentrations of MCFAs, other plants, including *Cuphea* species have been engineered to increase their medium chain fatty acid concentration (Wiberg and Bafor, 1995; Knapp *et al.*, 1991). Knapp and coworkers (1991) found that in *Cuphea viscosissima,* the total medium chain fatty acid concentration could be increased above an already high level of approximately 88% by mutational modifications. Davies and coworkers (1993) studied the ACP-thioesterases from different plant species and found that they had differing medium chain fatty acid specificities which correlated well with specific fatty acid composition in the seed. Dehesh and coworkers (1996) were able to redirect fatty acid synthesis in canola oil by performing a transgenic expression of a 8:0/10:0 specific thioesterase from *Cuphea hookeriana* to incorporate 8:0 and 10:0 into canola oil.

The disadvantages associated with genetic engineering of oilseed crops are mainly related to a lack of understanding and characterization of the processes involved in the fatty acid modification of oilseeds. Gene cloning is time consuming and expensive and there is a lack of understanding of the full implications of genetic modification on the biochemistry and regulation of oilseed metabolic processes (Miquel and Browse, 1994). The practical problems associated with crop management include pest and disease resistance and prevention of cross-pollination by non-modified plants, making the process economically unfeasible. At the present time, there are a limited number of available sequence encoding enzymes which are key in lipid metabolism, which slows down the process of gene isolation (Wolter, 1993). Despite these problems, there are several advantages associated with modifying the fatty acid profile at the level of growth and development of oilseed crops. In general, during incorporation of MCFAs into TAGs in the oilseed, there is a natural tendency for the fatty acid composition at the sn-2 position to be maintained as a long chain unsaturated fatty acid, with the MCFAs occupying positions 1 and 3 (Del Vecchio, 1996). This provides a structured lipid, which at the moment is really only obtainable at the present time by lipase-catalysed interesterification. As well, work is also being performed to encourage production of PUFAs in oilseeds, with some success already accomplished with increasing the concentration of γ-LNA through expression of the Δ6-desaturase gene (Reddy and Thomas, 1996). The ultimate goal in this case is to produce EPA and DHA in oilseeds.

Lipase-catalysed modification of fats and oils

METHODS

Traditionally, chemical interesterification has been the primary industrial means of producing modified fats and oils. It is an entropically driven reaction which, in theory, produces a complete randomization of acyl groups in TAGs (Coenen, 1974; Kuksis *et al.*, 1963; Ferrari *et al.*, 1997). Chemical interesterification is used in the manufacture of low-*trans* shortenings, margarines and spreads to improve their textural properties, modify melting behaviour and enhance stability (Nawar, 1996; Ghazali *et al.*, 1995). Chemical interesterification has the potential to be applied to the nutritional improvement of fats and oils, mainly to increase the proportion of specific fatty acids in specific positions on the glycerol backbone to improve their bioavailability. Chemical interesterification is not an effective method of producing high concentrations of MCFAs, nor structured lipids due to the randomness and lack of positional or fatty acid selectivity inherent in this method (Klemann *et al.*, 1994; Ray and Bhattacharyya, 1995). Lipase-catalysed interesterification is superior to chemical interesterification when specific positional distributions are required due to the inherent positional and fatty acid specificity possessed by lipases. Lipases catalyse the hydrolysis of TAGs, DAGs and MAGs in the presence of excess water, but under water limiting conditions, the reverse reaction, ester synthesis, can be achieved (Macrae, 1985; Jaeger *et al.*, 1994). Lipase-catalysed interesterification and hydrolysis follow a Ping Pong Bi Bi reaction for multisubstrate reactions (Malcata *et al.*, 1992; Reyes and Hill Jr, 1994).

In terms of the application of lipases to the nutritional modification of fats and oils, both the positional and fatty acid specificity of certain lipases are used (*Table 5*). Lipases which possess no specificity produce the same positional distribution as chemical interesterification with significantly greater cost and time requirements, making them unsuitable for this application (Macrae, 1983, Gunstone, 1994). Positional specificity towards positions 1 and 3 of the TAG is due to an inability of the lipase to act on position sn-2 because steric hindrance prevents access of the fatty acid in the sn-2 position to the active site (Macrae, 1983; Macrae and How, 1988). Fatty acid specificity, in terms of specificity both towards and against different fatty acids has been employed in the removal or concentrations of these fatty acids. The positional and fatty acid specificity of different lipases has been used in numerous lipase-catalysed reactions, including transesterification, acidolysis, glycerolysis and esterification to improve the nutritional quality of fats and oils. Lipases have also been used in the modification of fats and oils for the purpose of physical modifications, for use as cocoa butter substitutes and to alter melting properties.

Table 5.   Fatty acid and positional specificity of selected lipases used in the nutritional modification of fats and oils

Source of lipase	Specificity	Reference
Aspergillus niger *Aspergillus delemar*	Towards medium and short chain fatty acids	Desnuelle (1972); Stamatis *et al.* (1993)
Geotrichum candidum	Towards long chain fatty acids with *cis*-9 double bonds	Macrae (1985)
Aspergillus niger *Mucor miehei* *Rhizopus arrhizus* *Rhizopus delemar*	Towards positions sn-1 and sn-3	Macrae (1983)
Candida parapsilosis	Towards sn-2 position	Riaublanc *et al.* (1993)

Transesterification

Lipase-catalysed transesterification is defined as the exchange of acyl groups between two esters, namely two TAGs, although it can also be between ethyl or methyl esters and TAGs (*Figure 4*). Transesterification is not used as a method to transfer PUFAs from fish oils to vegetable oils due to the relatively low concentration of EPA and DHA in fish oils which does not usually exceed 25% (Ackman, 1988). It is most commonly used to produce structured lipids by the reaction between a medium chain

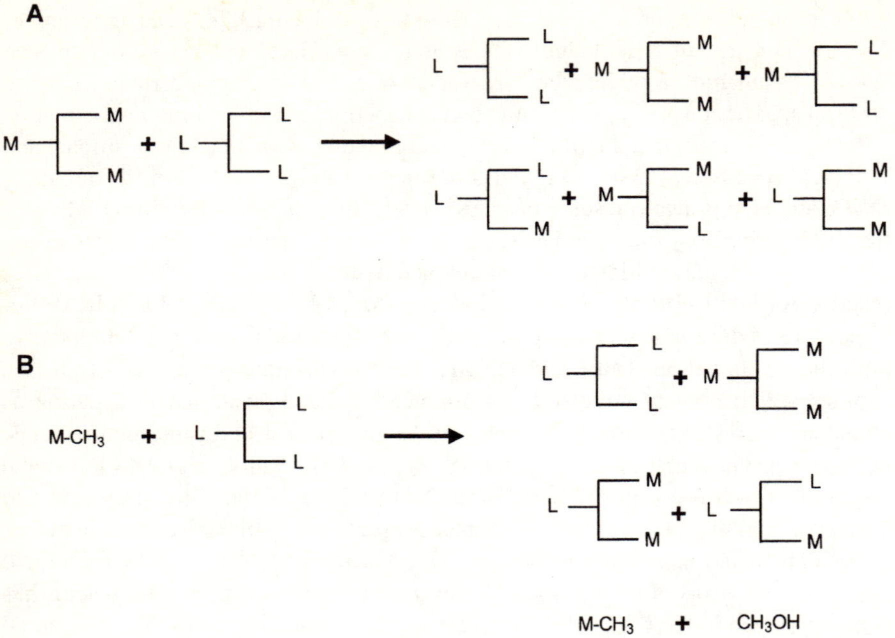

A

B

Figure 4. Possible TAG species derived from a 1,3-specific lipase-catalysed transesterification reaction between a medium chain TAG (A) or medium chain methyl ester (B) and a long chain TAG.

triacylglycerol and a vegetable oil or fish oil containing high concentrations of long chain essential and polyunsaturated fatty acids. This is not considered the ideal method for structured lipid production because there is a randomization of medium, essential and polyunsaturated fatty acids in all three positions of the triacylglycerol. Structured lipids should ideally contain medium chain fatty acids in positions sn-1 and sn-3 for rapid hydrolysis and absorption for energy production, and long chain essential and polyunsaturated fatty acids in the sn-2 position to improve their absorption. Higher concentrations of structured lipids are more easily obtained using acidolysis reactions.

Acidolysis

Acidolysis is defined as the transfer of an acyl group between an acid and an ester, and is used mainly to incorporate novel FFAs into TAGs (*Figure 5*). Acidolysis between a PUFA-rich fraction and fish oils has been a successful way of increasing the PUFA concentration in fish oils, since fish oils tend to have PUFAs in the sn-2 position, allowing more to be incorporated into positions sn-1 and sn-3.

Polyunsaturated fatty acid-enriched fish oils have been used in encapsulated form to reduce the risk of cardiovascular disease in adults. These oils are not suitable for use in infant formulas due to the high concentration of EPA which may compete with AA and affect growth. PUFA-enriched vegetable oils have been used in the prevention of cardiovascular disease in adults. The advantage of acidolysis over transesterification to produce structured lipids is a greater degree of incorporation. However, there are still difficulties in placing PUFAs in position sn-2 and MCFAs in positions 1 and 3.

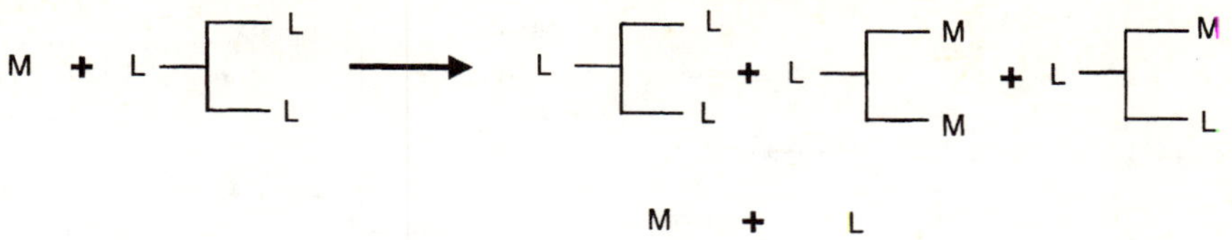

Figure 5.   Potential TAG species derived from a 1,3-specific lipase-catalysed acidolysis reaction between a long chain TAG and a medium chain fatty acid.

There are major disadvantages associated with using acidolysis as a means of improving the nutritional quality of fats and oils. Obtaining a fatty acid concentrate with high concentrations of DHA and/or EPA requires several steps, including saponification, solvent extraction, and urea inclusion or molecular distillation (Li and Ward, 1993a). Since fatty acids from the original TAG are released during the course of acidolysis, those fatty acids plus those remaining from the original substrate must be removed from the lipid. Due to the heat lability of PUFAs, traditional means of fatty acid removal such as molecular distillation have been replaced by methods such as titration with salts to precipitate fatty acids (Tanaka *et al.*, 1994).

Glycerolysis/esterification

Glycerolysis is the reaction between a TAG and glycerol, while esterification is the reaction between glycerol (or an alcohol group such as a partial glyceride) and a free fatty acid (*Figure 6*). The main application of esterification is in the production of TAGs containing all long chain PUFAs for use as adult supplements or all MCFAs for parenteral nutrition, while glycerolysis has been used to produce PUFA-containing MAGs. The advantage of glycerolysis and esterification is the high purity of TAGs containing only one fatty acid type which can be obtained, although the yield of TAGs tends to be low. The disadvantage of esterification is that the fatty acid substrates remaining must be removed. As well, esterification using ethyl esters remains economically unfeasible due to the high production cost of producing EPA and DHA concentrates (Li and Ward, 1993a; Sridhar and Lakshminarayana, 1992).

Hydrolysis/selective enrichment

As mentioned previously, some lipases are specific towards certain fatty acids, a property which can be used to concentrate these fatty acids during hydrolysis and transesterification. Lipases with decreased specificity towards DHA have been used frequently to increase the concentration of DHA in fish oil. The lower activity of some lipases towards DHA is attributed to the fact that the carbon-carbon double bond nearest to the carboxyl group is one carbon closer in DHA than in EPA which affects its ability to fit into the active site (Haraldsson *et al.*, 1993). While this selective enrichment has been used effectively in the nutritional modification of fats and oils, care must be taken to recognize the specificities of these lipases when conducting other lipase-catalysed modification methods to prevent low levels of incorporation of some fatty acids (Kosugi and Azuma, 1994; Haraldsson *et al.*, 1993; Langholz *et al*, 1989). Overall, selective enrichment is a promising method for the concentration of fatty

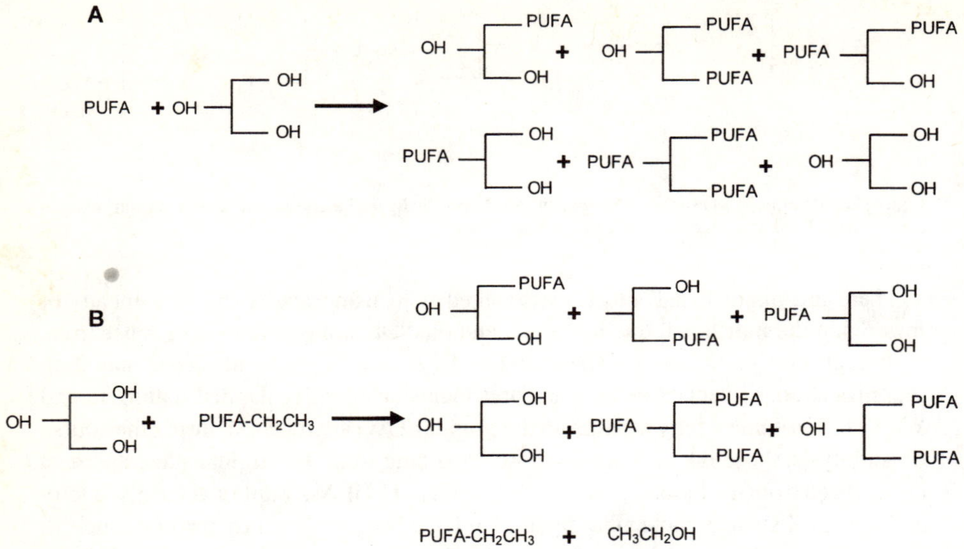

A

B

Figure 6. Potential TAG species from a non-specific lipase-catalysed esterification (A) and glycerolysis (B) reactions between glycerol and a PUFA and glycerol and a PUFA-ethyl ester, respectively.

acids in oils, specifically fish oils, since it does not require PUFA concentrates which are difficult and expensive to manufacture.

PRODUCTS

In terms of the nutritional modification of lipids, most of the lipase-catalysed methods described previously have been used to produce varied products, including medium chain triacylglycerols, structured lipids, PUFA concentrated fish oils, and PUFA containing vegetable oils. Some methods are superior for the production of specific products. As well, lipase-catalysed interesterification has been used to produce cocoa-butter substitutes and to modify the physical properties of certain lipids.

Medium chain triacylglycerols

Medium chain triacylglycerols are used predominantly in the manufacture of parenteral and enteral formulations, and in infant formulas for preterm infants since medium chain triacylglycerols are more rapidly hydrolysed and metabolised compared to long chain triacylglycerols (Mascioli *et al.*, 1988). Medium chain triacylglycerols have been blended with long chain triacylglycerols in these formulations to reduce the added cost associated with structured lipid production, although there are differing opinions as to which method is more nutritionally beneficial (Sobrado *et al.*, 1985; Christensen and Høy, 1997).

In order to produce pure medium chain triacylglycerols, esterification of glycerol and medium chain fatty acids has been the lipase-catalysed method of choice (*Table 6*). Kwon and coworkers (1996) produced MCTs from esterification of glycerol with capric acid, to obtain 47.6% (w/w) tricaprylin after 24 hours. However, lipases are not

Table 6. Production of medium-chain triacylglycerols and concentration of medium chain fatty acids by lipase-catalysed interesterification reactions

Source of lipase	Method	Substrate	Yield	Reference
Candida rugosa	esterification	capric acid glycerol	56.2 mol% tricaprin of all caprin-containing species	Kwon *et al.* (1996)
Mucor miehei	acidolysis	capric acid caprylic acid methyl esters coconut oil	capric acid increased from 6.2% to 18% carpylic acid increased from 4.9% to 17.9% (w/w)	Ghosh and Bhattacharyya (1997)

usually used to produce medium chain triacylglycerols since chemical esterification and fractionation are more cost effective. The major use of medium chain triacylglycerols is in the production of structured lipids, using 1,3-specific lipases.

Structured lipids

As previously defined, structured lipids (*Table 7*) may contain a variety of medium chain, essential and polyunsaturated fatty acids, preferably with the medium chain fatty acids in positions sn-1 and sn-3 and the long chain fatty acids in the sn-2 position.

Using a 1,3-specific lipase, acidolysis is superior to transesterification because it allows placement of medium chain fatty acids specifically in positions sn-1 and sn-3. Transesterification produces a mixture of TAGs because the starting species are both TAGs, one containing medium chain fatty acids and the other containing essential or polyunsaturated fatty acids in positions sn-1 and sn-3. Lee and Akoh (1996) transesterified tricaprin and triolein and found that the final product contained 43.3–57.7% capric acid and 42.3–56.4% LA in the sn-2 position. As well, while Soumanou and coworkers (1997) were able to produce 73%(w/w) structured lipids from transesterification, only 31% of the TAGs were disubstituted.

Despite the greater control of final product composition attainable by acidolysis, acidolysis still requires removal of residual fatty acid substrate and fatty acids produced during the reaction. As well, acyl migration may occur, resulting in placement of MCFAs in the sn-2 position. For this reason, transesterification still seems to be the reaction of choice for producing structured lipids. However, Shimada and coworkers (1996) were able to produce significant improvements in yields by running an acidolysis reaction repeatedly, for a total of three runs, to obtain a final vegetable oil product containing 100% caprylic acid in positions sn-1 and sn-3. Since this yield is significantly higher than any achievable by transesterification, the added cost of removing residual fatty acids from the acidolysis may be justified, particularly in the use of structured lipids in parenteral and enteral formulations where a high degree of substitution would be beneficial.

PUFA concentration in and from fish oils

The greatest interest in the nutritional modification of lipids in recent years has been related to increasing the n-3 PUFA concentration of fish oils above their natural

Table 7.   Production of structured lipids by lipase-catalysed interesterification reactions

Source of lipase	Method	Substrates	Yield	Reference
Rhizomucor miehei	transesterification	tricaprylin peanut oil	79 mol% SL	Soumanou *et al.* (1997)
Chromobacterium viscosum			71 mol% SL	
Candida sp.			68 mol% SL	
Mucor miehei *Candida antarctica*	transesterification tricaprin	trilinolein	78 mol% SL	Lee and Akoh (1997)
Mucor miehei	transesterification	tricaprylin EPA ethyl esters	34 mol% EPA incorporated	Lee and Akoh (1996)
Rhizopus delemar	repeated acidolysis	safflower, linseed oil caprylic acid	100% SL	Shimada *et al.* (1996)
Rhizomucor miehei	transesterification	triolein	87.7 mol%	Huang and
		caprylic acid ethyl esters	SL	Akoh (1996)
Mucor miehei	acidolysis	tricaprylin EPA	62% (w/w) EPA incorporated	Shishikura *et al.* (1994)

content of 25% EPA and DHA, because of the reduced risk of atherosclerosis related to consuming higher concentrations of EPA and DHA. Increasing concentrations of EPA and DHA has been achieved through modification of existing fish oils and through production of TAGs containing only EPA or DHA (*Table 8*). Lipases are ideal for modifying fish oils because other physical methods such as winterization, solvent crystallization and molecular distillation are not applicable due to a wide variation in fatty acid composition in the oils. Selective enrichment, through hydrolysis using lipases which are specific against DHA or EPA is a less expensive lipase-catalysed method of increasing the n-3 PUFA content in fish oils because it does not require an external source of concentrated DHA and EPA (*Table 9*). Lipases from *Mucor miehei* and *Candida cylindraceae* are specific against DHA, allowing enrichment of this fatty acid. Hills and coworkers (1990) were able to enrich DHA in cod liver oil from 9.4% to 45%, while Tanaka and coworkers (1992) were able to enrich the DHA content of tuna oil from 25.1% to 53% (w/w). Selective enrichment has also been applied to the enrichment of both EPA and DHA using lipase from *Geotrichum candidum* (Shimada *et al.*, 1994). Docosahexaenoic acid has also been concentrated in the FFA fraction by selective esterification between an alcohol such as butanol or lauryl alcohol and partial glycerides, with fatty acids other than DHA being esterified to the alcohol (Shimada *et al.*, 1997; Hills *et al.*, 1990). The main disadvantage of using this method to concentrate n-3 PUFAs in fish oils is that the fatty acids are concentrated in the form of MAGs and DAGs as a result of the selective hydrolysis process.

To avoid the production of MAGs and DAGs during enrichment, acidolysis is the only other method that has been used to modify fish oil. Adachi and coworkers (1993) used a 1,3-specific lipoprotein lipase from *Pseudomonas* sp. to enrich the total EPA and DHA concentration in sardine oil from 29% to 44.5% (w/w). A combination of acidolysis and low temperature crystallization was used by Yamane and coworkers (1993) to remove saturated and monoenoic fatty acids from the reaction mixture and

Table 8. Production of high concentrations of EPA and DHA-TAGs and concentration of EPA and DHA in fish/whale oil by lipase-catalysed interesterification reactions

Source of lipase	Method	Substrates	Yield (% w/w)	Reference
Pseudomonas sp.	glycerolysis	seal oil whale oil glycerol	42–53% MAGs	Myrnes *et al.* (1995)
Candida cylindraceae	esterification	EPA and DHA other PUFAs (18:1–18:4) glycerol	18–33% TAGs	Osada *et al.* (1990)
Candida antarctica	esterification	EPA and DHA glycerol	100% Tri-EPA and Tri-DHA	Haraldsson *et al.* (1993)
Pseudomonas sp.	esterification	EPA and DHA in a cod liver oil fatty acid concentrate glycerol	18.1% TAGs containing 36.7%	Li and Ward (1993b)
Candida antarctica	glycerolysis	EPA and DHA ethyl esters glycerol	95% Tri-DHA	Kosugi and Azuma (1994)
Mucor miehei	acidolysis	EPA, DHA enriched FA fraction cod liver oil	increased EPA and DHA content by 10%	Yamane *et al.* (1993)
Candida cylindraceae *Chromobacterium viscosum* *Pseudomonas sp.*	acidolysis	EPA, DHA enriched FA fraction sardine oil	65% EPA and DHA in oil	Adachi *et al.* (1993)
Candida antarctica	esterification	EPA glycerol	99% Tri-EPA	Haraldsson *et al.* (1995)

push the reaction equilibrium towards increased PUFA incorporation. Performing acidolysis to increase DHA concentrations in fish oil can be difficult, as found by Tanaka and coworkers (1994), who found that DHA was actually removed from a partially hydrolysed tuna oil fraction to increase the concentration of DHA in the free fatty acid fraction from 13% to 55%, thereby reducing the concentration of DHA in the actual oil. This transfer was due to the low initial concentration of DHA in the free fatty acid fraction.

Esterification between glycerol and n-3 PUFAs is an alternative method for obtaining highly concentrated sources of DHA and EPA. This method allows the production of high purity pure fractions of TAGs containing DHA and EPA. The main disadvantages of this method are the cost of obtaining a relatively pure free fatty acid substrate and the low yields of TAGs which are obtained. Kosugi and Azuma (1994) performed a batch reaction using free DHA and glycerol producing purified 95% TAG, but with yields of 66.2% DHA and 60.1% EPA. Lipases which are 1,3-specific can still be used in esterification reactions, as shown by Li and Ward (1993b) who found that lipase from *Mucor miehei* (Lipozyme IM 60) produced 92% TAGs from a PUFA concentrate and glycerol, while a non-specific lipase from *Pseudomonas sp.* (PS 30) only produced 82% TAGs. Polymerization of EPA and DHA during esterification

Table 9.   PUFA concentration in fish oil by selective hydrolysis and enrichment

Source of lipase	Target fatty acid	Oil	Yield (% w/w)	Reference
Geotrichum candidum	AA, EPA, DHA	tuna oil	85.5% TAG, with 81.5% recovery of DHA and EPA	Shimada *et al.* (1994)
Mucor miehei	DHA	cod liver oil	increase of DHA from 9.4% to 45%	Hills *et al.* (1990)
Candida cylindraceae	DHA	tuna oil	46.2% DHA in TAG fraction	Tanaka *et al.* (1992)
Rhizopus niveus	DHA	cod liver oil	29.2% DHA in MAG, 15.2% in TAG (9.6% initially)	Yadwad *et al.* (1991)
Candida rugosa *Geotrichum candidum*	DHA DHA, EPA	cod liver oil	increased EPA and DHA from 30 to 45% in partial glycerides	McNeill *et al.* (1996)
Candida cylindraceae	DHA, EPA	cod liver oil, sardine oil	50% in partial glycerides	Hoshino *et al.* (1990)
Chromobacterium viscosum	DHA	tuna oil	46.2% DHA in TAG fraction	Tanaka *et al.* (1994)
Rhizopus arrhizus	DHA	tuna oil	89% DHA in FFA fraction	Shimada *et al.* (1997)

has also been reported, requiring the addition of higher concentrations of free fatty acids or the use of the ethyl ester form of the fatty acids which do not readily polymerize, although there seems to be a lower degree of incorporation when using the ethyl ester forms (Kosugi and Azuma, 1994; Bech Pedersen and Holmer, 1995).

As of yet, any of these modifications to increase the n-3 PUFA concentration in fish oils or to produce TAGs containing only these fatty acids have had limited applications to making products geared towards disease prevention in adults. Flavour and odour problems associated with high oxidative susceptibility of PUFAs and fish oils limits their use to encapsulated oil products.

PUFA incorporation into vegetable oils

In order to increase the availability of n-3 PUFAs, acidolysis has been used to incorporate them into vegetable oils (*Table 10*), and some degree of success has been achieved. Sridhar and Lakshminarayana (1992) were able to modify the composition of groundnut oil, a staple oil in India by incorporation of a total of 17.5% EPA and DHA into positions 1 and 3. As well, Li and Ward (1993a) were able to incorporate a total of 17.7% EPA and DHA into corn oil. However, in both of these experiments, a 1,3-specific lipase was used, placing these fatty acids in positions sn-1 and sn-3. Pancreatic lipase, shows little activity towards these fatty acids so their potential for hydrolysis and absorption is reduced. Even with acyl migration or using a non-specific lipase to obtain some n-3 PUFAs in the sn-2 position, a larger proportion of these fatty acids will remain in positions sn-1 and sn-3. Therefore, while incorporating EPA and

Table 10. Incorporation of EPA and DHA into vegetable oil by lipase-catalysed interesterification reactions

Source of lipase	Method	Substrate	Yield (% w/w)	Reference
Mucor miehei *Candida antarctica*	acidolysis	EPA, EPA ethyl esters canola oil	18% EPA or 32.9% EPA ethyl esters (mol%)	Huang and Akoh (1994)
Mucor miehei	acidolysis	EPA, DHA corn oil	17.7% (W/w) EPA and DHA	Li and Ward (1993a)
Candida antarctica	acidolysis	EPA evening primrose oil	43% (mol%) EPA	Akoh *et al.* (1996)
Mucor miehei	acidolysis	EPA and DHA ground nut oil	9.5% EPA and 8.0% DHA (w/w)	Sridhar and Lakshmina-rayana (1992)
Candida antarctica	transesterification	EPA ethyl esters trilinolein	81.4 mol% EPA	Akoh *et al.* (1995)
Candida cylindraceae	selective enrichment	*B. orientalis* seed oil	up to 41% (w/w) 20:3 and 20:4 in partial glyceride fraction	Lie Ken Jie and Rahmstullah (1995)

DHA into vegetable oils would increase their use in the food industry, their concentration and availability in these oils remains low.

MODIFICATION OF PHYSICAL PROPERTIES

While the major focus of biotechnology has been on the nutritional modification of lipids, lipases have also been used in the modification of physical properties in the production of cocoa butter equivalents and modification of melting properties.

Cocoa butter equivalents

The high cost of cocoa butter for use in the confectionery industry compared to the relatively low cost of other fats and oils containing similar concentrations of the same fatty acids has resulted in attempts to manufacture cocoa butter equivalents using lipases. Cocoa butter contains 1-palmitoyl-2-oleoyl-3-stearoyl-glycerol (POS) and 1,3-distearoyl-2-oleoyl-glycerol (SOS) as its major triacylglycerol species, representing about 70% of the total (Chang *et al.*, 1990). Several inexpensive sources of lipid have been used, including palm oil midfraction or shea oil combined with stearic acid, hydrogenated cottonseed oil and olive oil, and kokum fat and palmitic acid methyl esters (Macrae and How, 1988; Sridhar*et al.*, 1991; Chang*et al.*, 1990; Bloomer*et al.*, 1990). Since stearic acid is required only in positions 1 and 3 of the triacylglycerol, a 1,3-specific lipase is ideal. Stearic acid or methyl or ethyl stearate are added by acidolysis or transesterification (respectively) to produce a combination of POS, SOS and POP. Several studies have been performed to produce cocoa butter equivalents using lipase-catalysed interesterification (*Table 11*). Macrae and How (1988) developed 1,3-specific-lipase-catalysed acidolysis processes using palm oil liquid fractions, shea oil and stearic acid to produce lipid products with compositions similar to that of

Table 11.   Production of cocoa butter equivalents using lipases

Source of lipase	Method	Substrates	Effectiveness	Reference
Rhizopus arrhizus	acidolysis	palm oil midfraction stearic acid	similar concentration of stearic, oleic, palmitic acid	Mojovic *et al.* (1993)
IM 20	acidolysis	palm olein	39.3% (w/w) cocoa butter-like TAGs	Chong *et al.* (1992)
Mucor miehei	transesterification	kokum fat methyl palmitate methyl stearate	peak melting point 32.8°C vs 32.7°C for cocoa butter	Sridhar *et al.* (1991)
Mucor miehei	transesterification	cottonseed oil olive oil	19% (2/2) cocoa butter-like TAGs	Chang *et al.* (1990)

cocoa butter. Chang and coworkers (1990) transesterified hydrogenated cottonseed oil and olive oil to obtain a mixture of POP, OSO, POS and SOS, resulting in a yield of 19% cocoa butter-like fat with a melting range of 29 to 49°C compared to a range of 29 to 43°C for cocoa butter. The cocoa butter fraction was isolated from the other fat by crystallization in acetone. Similarly, Chong and coworkers (1992) ran an acidolysis reaction between stearic acid and palm olein, and obtained a higher yield of cocoa butter-like triacylglycerols. They used steam distillation to remove free fatty acids and fractional crystallization with hexane or acetone to isolate the cocoa butter-like triacylglycerols.

The main consideration associated with using lipase-catalysed interesterification to produce cocoa butter equivalents is keeping the cost low. Major considerations include reusability of the lipase, choice of substrate to obtain maximum yields of cocoa butter-like triacylglycerols and the cost of extraction and purification.

Modification of hardness and melting properties

The physical properties of fats can be modified via chemical or enzymatic methods (Marangoni and Rousseau, 1995). The effects of blending and chemical inter-esterification on the physical and chemical properties of fats have been extensively studied (Rousseau *et al.*, 1996a,b,c; Marangoni and Rousseau, 1998a,b; Rousseau and Marangoni, 1998a,b), however, almost no information exists on the effects of enzymatic interesterification on the physical properties of fats. In order to be able to use such methods in the manufacture of new fat-containing products, it is imperative to understand the effects of enzymatic transformations on the physical properties of fats.

Rousseau and Marangoni (1998a,b) enzymatically interesterified milkfat using *Rhizopus arrhizus* lipase, which possesses long chain fatty acid and sn-1,3 specificity, in order to produce a softer, cold-spreadable butter. The hardness index, a macroscopic measure of the solid-like character of a material, decreased as a function of interesterification duration in an exponential fashion (*Figure 7*). This decrease in the solid-like character of the milkfat was also observed using more sophisticated rheological measurements, such as dynamic oscillatory controlled-stress rheometry. In these experiments, the shear storage modulus G' (or elastic component), as well as

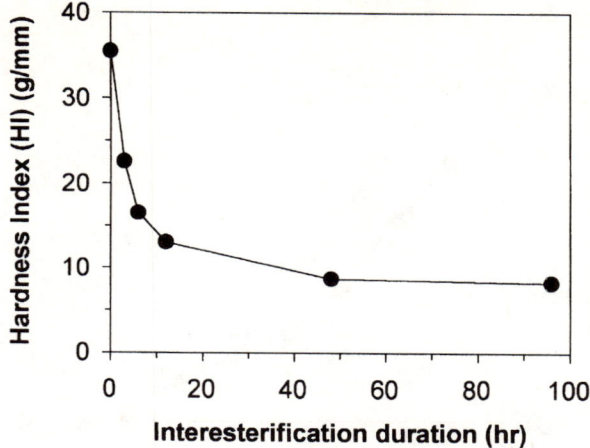

Figure 7. Hardness index as a function of time for interesterification of milkfat using lipase from *Rhizopus arrhizus.*

the shear loss modulus G" (or viscous component), decreased upon lipase-catalysed interesterification.

This decrease in the hardness of the fat was mainly attributed to a decrease in the solid-fat-content, as determined by pulsed NMR (*Figure 8*), and not to changes in the microstructure of the fat crystal network. The dropping point, a measure of the end of melt of the fat, also decreased upon enzymatic interesterification. This decrease in dropping point suggested alterations in the structure of the triglycerides, which in turn lead to changes in their ability to interact in the solid-state. This hypothesis was confirmed by X-ray diffraction studies on the milkfat where the size of the unit cell (long-spacings) increased from 4.20 nm to 4.54 nm upon enzymatic interesterification. This information, combined with the fact that the main subcell reflections (short-spacings) decreased from 0.400 and 0.433 nm to 0.392 and 0.429 nm upon enzymatic interesterification suggested that the triglycerides were crystallizing predominantly in the orthorhombic perpendicular, or β', polymorph. The packing density of TAGs in this polymorphic state is lower and the melting point is lower than for triclinic, or β, polymorph. Therefore, the decrease in the hardness of milkfat upon enzymatic interesterification was attributed mainly to a decrease in the amount of solid fat present and a slight increase in the amount of β'-crystallizing triglyceride species. The advantages of using lipases over chemical means to modify the physical properties of fats and oils are not completely clear. Further understanding of the impact of the positional distribution of different fatty acids on the physical properties of fats and oils is required.

Conclusions

The major barrier to applying biotechnology to lipid modification, in the form of genetic engineering, biotransformation and lipase-catalysed interesterification remains cost. Present chemical modification methods are relatively inexpensive, easy to run and easy to scale-up, with the major difference being a more limited ability to manufacture products of differing fatty acid composition.

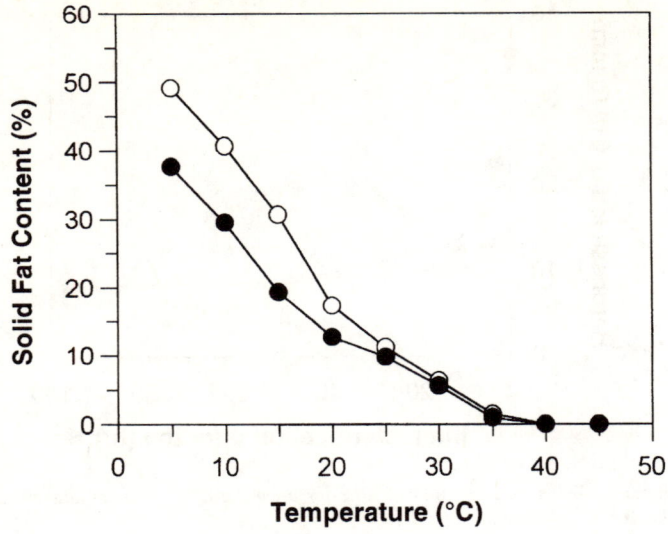

Figure 8.   Solid fat content as a function of temperature for milkfat interesterified using lipase from *Rhizopus arrhizus*. (O—O) Milkfat, (●—●) interesterified milkfat.

Genetic engineering of oilseed crops and positional distribution has the greatest start-up expense but may prove to have the greatest number of applications in the long run. Structured lipids and production of PUFAs in oilseeds are definite possibilities as greater understanding of the processes involved is acquired.

Extensive studies have been performed using lipases, however, scale-up of lipase catalysed reactions is not economical at the moment in applications towards the nutritional modification of fats and oils. The high costs of lipases, reactor systems and process control are some of the main barriers. Biotransformations to produce n-3 PUFAs have been relatively successful so far in terms of larger scale production of specialty fats and oils for nutritional applications.

Overall, biotechnology is a valuable tool in the modification of fats and oils to meet the diverse needs of consumers, however, continued cost reduction and production optimization are required.

Acknowledgements

We would like to acknowledge the financial support of the Natural Sciences and Engineering Research Council of Canada, Ontario Ministry of Agriculture, Food and Rural Affairs and Agriculture and Agrifood Canada (Science Horizons).

References

ACKMAN, R. (1988). The year of the fish oils. *Chem. Ind.* **5**,139–144.
ADACHI, S., OKUMURA, K., OTA, V. AND MANKURA, M. (1993). Acidolysis of sardine oil by lipase to concentrate eicosapentaenoic and docosahexaenoic acids in glycerides. *J. Ferment. Bioeng.* **75**, 259–264.
AKIMOTO, M., ISHII, T., YAMAGAKI, K., OHTAGUCHI, K., KOIDE, K. AND YAZAWA, K. (1990).

Production of eicosapentaenoic acid by a bacterium isolated from mackerel intestines. *J. Am. Oil Chem. Soc.* **67**, 911–915.

AKIMOTO, M., ISHII, T., YAMAGAKI, K., OHTAGUCHI, K., KOIDE, K. AND YAZAWA, K. (1991). Metal salts requisite for the production of eicosapentaenoic acid by a marine bacterium isolated from mackerel intestines. *J. Am. Oil Chem. Soc.* **68**, 504–508.

AKOH, C.C., JENNINGS, B.H. AND LILLARD, D.A. (1995). Enzymatic modification of trilinolein: incorporation of n-3 polyunsaturated fatty acids. *J. Am. Oil Chem. Soc.* **72**, 1317–1321.

AKOH, C.C., JENNINGS, B.H. AND LILLARD, D.A. (1996). Enzymatic modification of evening primrose oil: incorporation of n-3 polyunsaturated fatty acids. *J. Am. Oil Chem. Soc.* **73**, 1059–1062.

BABAYAN, V.K. AND ROSENAU, J.R. (1991). Medium-chain triglycerides in cheese. *Food Technol.* **45**, 111–114.

BACH, A.C., STORCK, D. AND MERAIHI, Z. (1988). Medium-chain triglyceride-based fat emulsions: an alternative energy supply in stress and sepsis. *J. Parent. Enteral Nutr.* **12**, 82S–87S.

BAJPAI, P. AND BAJPAI, P.K. (1993). Eicosapentaenoic acid (EPA) production from microorganisms: a review. *J. Biotech.* **30**, 161–183.

BAJPAI, P., BAJAPI, P.K. AND WARD, O.P. (1991). Optimization of production of docosahexaenoic acid (DHA) by *Thraustochytrium aureum* ATCC 34304. *J. Am. Oil Chem. Soc.* **68**, 510–514.

BANG, H.O. AND DYERBERG, J. (1972). Plasma lipids and lipoproteins in Greenlandic West Coast Eskimos. *Acta. Med. Scand.* **192**, 85–94.

BECH PEDERSEN, S. AND HOLMER, C. (1995). Studies of the fatty acid specificity of the lipase from *Rhizomucor miehei* toward 20:1n-9, 20:5n-3, 22:1n-9 and 22:6n-3. *J. Am. Oil Chem. Soc.* **72**, 239–243.

BELL, S.J., BRADLEY, D., FORSE, R.A. AND BISTRIAN, B.R. (1997). The new dietary fats in health and disease. *J. Am. Diet. Assoc.* **97**, 280–286.

BERDANIER, C.D. (1995). Lipids. In *Advanced Nutrition: Macronutrients.* pp 220–222, 234–236. Boca Raton, Florida: CRC Press, Inc.

BLOOMER, S., ADLERCREUTZ, P. AND MATTIASSON, B. (1990). Triglyceride interesterification by lipases. 1. Cocoa butter equivalents from a fraction of palm oil. *J. Am. Oil Chem. Soc.* **67**, 519–524.

BOEHM, G., MÜLLER, H., KOHN, G., MORO, G., MINOLI, I. AND BÖHLES, H.J. (1997). Docosahexaenoic and arachidonic acid absorption in preterm infants fed LCP-free or LCP-supplemented formula in comparison to infants fed fortified breast milk. *Ann. Nutr. Metab.* **41**, 235–241.

BORUM, P.R. (1992). Medium-chain triglycerides in formula for preterm neonates: implications for hepatic and extrahepatic metabolism. *J. Pediatr.* **120**, S139–S145.

BRADEN, L.M. AND CARROLL, K.K. (1986). Dietary polyunsaturated fat in relation to mammary carcinogenesis in rats. *Lipids* **21**, 285–288.

BRUCKNER, G. (1992). Biological effects of polyunsaturated fatty acids. In *Fatty Acids in Foods and Their Health Implications.* Ed. C.K. Chow, pp 631–646. New York: Marcel Dekker, Inc.

BURGESS, J.G., IWAMOTO, K., MIURA, Y., TAKANO, H., MATSUNAGA, T. (1993). An optical fibre photobioreactor for enhanced production of the marine unicellular alga *Isochrysis* aff. *Galbana* T-Iso (UTEX LB 2307) rich in docosahexaenoic acid. *Appl. Microbiol. Biotechnol.* **39**, 456–459.

CAGGIULA, A.W. AND MUSTAD, V.A. (1997). Effects of dietary fat and fatty acids on coronary artery disease risk and lipoprotein cholesterol concentrations: epidemiologic studies. *Am. J. Clin. Nutr.* **65**, 1597S–1610S.

CARLSON, S.E. (1996). Arachidonic acid status of human infants: Influence of gestational age at birth and diets with very long chain n-3 and n-6 fatty acids. *J. Nutr.* **126**, 1092S–1098S.

CARLSON, S.E., CLANDININ, M.T., COOK, H.W., EMKEN, F.A. AND FILER, L.J. JR. (1997). *trans* Fatty acids: infant and fetal development. *Am. J. Clin. Nutr.* **66**, 715S–736S.

CARLSON, S.E., COOKE. R.J., WERKMAN, S.H. AND TOLLEY, E.A. (1992). First year growth of preterm infants fed standard compared to marine-oil n-3 supplemented formula. *Lipids* **27**, 901–907.

CARLSON, S.E., WERKMAN, S.H., RHODES, P.G. AND TOLLEY, E.A. (1993). Visual-acuity development in healthy preterm infants: effect of marine-oil supplementation. *Am. J. Clin. Nutr.* **58**, 35–42.

CARLSON, S.E. AND WERKMAN, S.H. (1996). A randomized trial of visual attention of preterm infants fed docosahexaenoic until two months. *Lipids* **31**, 85–89.

CARLSON, S.E., FORD, A.J., WERKMAN, S.H., PEEPLES, J.M. AND KOO, W.W.K. (1996). Visual acuity and fatty acid status of term infants fed human milk and formulas with and without docosahexaenoate and arachidonate from egg yolk lecithin. *Pediatr. Res.* **39**, 882–888.

CARNIELLI, V.P., LUIJENDIJK, I.H.T., VAN GOUDOEVER, J.B., SULKERS, E.J., BOERLAGE, A.A., DEGENHART, H.J. AND SAUER, P.J.J. (1995). Feeding premature newborn infants palmitic acid in amounts and stereochemical position similar to that of human milk: effects on fat and mineral balance. *Am. J. Clin. Nutr.* **61**, 1037–1042.

CARNIELLI, V.P., PEDERZINI, F., VITTORANGELI, R., LUUENDIK, I.H.T, BOMAARS, W.E.M., PEDROTTI, D. AND SAUER, P.J.J. (1996). Plasma and red blood cell fatty acid of very low birth weight infants fed exclusively with expressed preterm human milk. *Pediatr. Res.* **39**, 671–679.

CARNIELLI, V.P., VERLATO, G., PEDERZINI, F., LUIJENDIJK, I., BOERLAGE, A., PEDROTTI, D. AND SAUER, P.J.J. (1998). Intestinal absorption of long-chain polyunsaturated fatty acids in preterm infants fed breast milk or formula. *Am. J. Clin. Nutr.* **67**, 97–103.

CARTENS, M., MOLINA GRIMA, E., ROBLES MEDINA, A., GIMÉNEZ GIMÉNEZ, A. AND IBÁÑEZ GONZÁLES, J. (1996). Eicosapentaenoic acid (20:5n-2) from the marine microalga *Phaeodactylum tricornutum*. *J. Am. Oil Chem. Soc.* **71**, 1025–1034.

CHANG, M.-K., ABRAHAM, G. AND JOHN, V.T. (1990). Production of cocoa butter-like fat from interesterification of vegetable oils. *J. Am. Oil Chem. Soc.* **67**, 832–834.

CHONG, C.N., HOH, Y.M. AND WANG, C.W. (1992). Fractionation procedures for obtaining cocoa-butter-like fat from enzymatically interesterified palm olein. *J. Am. Oil Chem. Soc.* **69**, 137–140.

CHRISTENSEN, M.S., HØY, C.-E., BECKER, C.C. AND T.G. REDGRAVE. (1995). Intestinal absorption and lymphatic transport of eicosapentaenoic (EPA), docosahexaenoic (DHA), and decanoic acids: dependence on intramolecular triacylglycerol structure. *Am. J. Clin. Nutr.* **61**, 56–61.

CHRISTENSEN, M.S. AND HØY, C-E. (1997). Early dietary intervention with structured triacylglycerols containing docosahexaenoic acid. Effect on brain, liver and adipose tissue lipids. *Lipids* **32**, 185–191.

COENEN, J.W.E. (1974). Fractionnement et interestérification des corps gras dans la perspective du marché mondial des matières premières et des produits finis, II – interestérification. *Rev. Franç. Corps Gras.* **21**, 403–413.

COHEN, Z. (1990). The production potential of eicosapentaenoic and arachidonic acids by the red alga *Porphyridium cruentum*. *J. Am. Oil Chem. Soc.* **67**, 916–919.

CRAWFORD, M.A., COSTELOE, K., GHEBREMESKEL, K., PHYLACTOS, A., SKIRVIN, L. AND STACEY, F. (1997). Are deficits of arachidonic and docosahexaenoic acids responsible for the neural and vascular complications of preterm babies? *Am. J. Clin. Nutr.* **66 (suppl)**, 1032S–1041S.

DAVIES, H.M. (1993). Medium chain acyl-ACP hydrolysis activities of developing oilseeds. *Phytochem.* **33**, 1353–1356.

DE FOUW, N.J., KIVITS, G.A.A., QUINLAN, P.T. AND VAN NIELEN, W.G.L. (1994). Absorption of isomeric, palmitic acid-containing triacylglycerols resembling human milk fat in the adult rat. *Lipids* **29**, 765–770.

DEHESH, K., JONES, A. KNUTZON, D.S., VOELKER, T.A. (1996). Production of high levels of 8:0 and 10:0 fatty acids in transgenic canola by overexpression of *Ch FatB2*, a thioesterase cDNA from *Cuphea hookeriana*. *Plant J.* **9**, 167–172.

DEL VECCHIO, A.J. (1996). High-laurate canola. *Inform.* **7**, 230–243.

DENKE, M.A. AND GRUNDY, S.M. (1992). Comparison of effects of lauric acid and palmitic acid on plasma lipids and lipoproteins. *Am. J. Clin. Nutr.* **56**, 895–898.

DESCI, T. AND KOLETZKO, B. (1995). Do trans fatty acids impair linoleic acid metabolism in children? *Ann. Nutr. Metab.* **39**, 36–41.

DESNUELLE, P. (1972). The Lipases, In *The Enzymes,* Volume VII, 3rd edition. Ed. P.D. Boyer, pp 575–616. New York: Academic Press.

FAN, S.T. (1997). Review: nutritional support for patients with cirrhosis. *J. Gastroenterol. Hepatol.* **12**, 282–286.

FERRARI, R. AP., ESTEVES, W. AND MUKHERJEE, K.D. (1997). Alteration of steryl ester content and positional distribution of fatty acids in triacylglycerols by chemical and enzymatic interesterification of plant oils. *J. Am. Oil Chem. Soc.* **74**, 93–96.

FITCH-HAUMANN, B. (1997). Bioengineered oilseed acreage escalating. *Inform.* **8**, 804–811.

FRANCOIS, C.A., CONNOR, S.L., WANDER, R.C. AND CONNOR, W.E. (1998). Acute effects of dietary fatty acids on the fatty acids of human milk. *Am. J. Clin. Nutr.* **67**, 301–308.

FREDRIKZON, B. AND OLIVECRONA, T. (1978). Decrease of lipase and esterase activities in intestinal contents of newborn infants during test meals. *Pediatr. Res.* **12**, 631–634.

FUKUDA, H. AND MORIKAWA, H. (1987). Enhancement of γ-linolenic acid production by *Mucor ambiguus* with non-ionic surfactants. *Appl. Microbiol. Biotechnol.* **27**, 15–20.

GARG, M.L., THOMSON, A.B.R. AND CLANDININ, M.T. (1990). Interactions of saturated, n-6 and n-3 polyunsaturated fatty acids to modulate arachidonic acid metabolism. *J. Lipid Res.* **31**, 271–277.

GHAZALI, H.M., HAMIDAH, S. AND CHE MAN, Y.B. (1995). Enzymatic transesterification of palm olein with nonspecific and 1,3-specific lipases. *J. Am. Oil Chem. Soc.* **72**, 633–639.

GHOSH, S. AND BHATTACHARYYA, D.K. (1997). Medium-chain fatty acid-rich glycerides by chemical and lipase-catalyzed polyester-monoester interchange reaction. *J. Am. Oil. Chem. Soc.* **74**, 593–595.

GIBSON, R.A., MAKRIDES, M., NEUMANN, M.A., SIMMER, K., MANTZIORIS, E. AND JAMES, M.J. (1994). Ratios of linoleic acid to alpha-linolenic acid in formulas for term infants. *J. Pediatr.* **125**, S48–S55.

GRUNDY, S.M. (1997). What is the desirable ratio of saturated, polyunsaturated, and monounsaturated fatty acids in the diet? *Am. J. Clin. Nutr.* **66 (suppl)**, 988S–990S.

GUNSTONE, F.D. (1994). Marine oils: fish and whale oils. In *The Lipid Handbook*, 2nd edition. Eds. F.D. Gunstone, J.L. Harwood, F.D. Padley, p 177. London: Chapman and Hall.

GURR, M.I. (1992). *The Role of Fats in Food and Nutrition*, pp 123, 131–133, V.

HANSSON, L. AND DOSTÁLEK, M. (1988). Effect of culture conditions on mycelial growth and production of γ-linolenic acid by the fungus *Mortiella ramanniana*. *Appl. Microbiol. Biotechnol.* **28**, 240–246.

HARDY, S.C. AND KLEINMAN, R.E. (1994). Fat and cholesterol in the diet of infants and young children: implications for growth, development and long-term health. *J. Pediatr.* **125**, S69–S77.

HARALDSSON, G.G., GUDMUNSSON, B.Ö. AND ALMARSSON, Ö. (1993). The preparation of homogeneous triglycerides of eicosapentaenoic acid and docosahexaenoic acid by lipase. *Tetrahedron Lett.* **34**, 5791–5794.

HILLS, M.J., KLEWITT, I. AND MUKHERJEE, K.D. (1990). Enzymatic fractionation of fatty acids: enrichment of gamma-linolenic acid and docosahexaenoic acid by selective esterification catalyzed by lipases. *J. Am. Oil Chem. Soc.* **67**, 561–564.

HOSHINO, T., YAMANE, T. AND SHIMIZU, S. (1990). Selective hydrolysis of fish oil by lipase to concentrate n-3 polyunsaturated fatty acids. *Agric. Biol. Chem.* **54**, 1459–1467.

HUANG, K. AND AKOH, C.C. (1996). Enzymatic synthesis of structured lipids: Transesterification of triolein and caprylic acid ethyl ester. *J. Am. Oil Chem. Soc.* **73**, 245–250.

HUANG, K. AND AKOH, C.C. (1994). Lipase-catalyzed incorporation of n-3 polyunsaturated fatty acids into vegetable oils. *J. Am. Oil Chem. Soc.* **71**, 1277–1280.

HUANG, Y.-S., LIN, X., REDDEN, P.R. AND HORROBIN, D.F. (1995). *In vitro* hydrolysis of natural and synthetic γ-linolenic acid-containing triacylglycerols by pancreatic lipase. *J. Am. Oil Chem. Soc.* **72**, 625–631.

IIDA, I., NAKAHARA, T., YOKOCHI, T., KAMISAKA, Y., YAGI, H., YAMAOKA, M. AND SUZUKI, O. (1996). Improvement of docosahexaenoic acid production in a culture of *Thraustochytrium aureum* by medium optimization. *J. Ferment. Bioeng.* **81**, 76–78.

INNIS, S.M. (1992). n-3 Fatty acid requirements of the newborn. *Lipids* **27**, 879–885.

INNIS, S.M., AKRABAWI, S.S., DIERSEN-SCHADE, D.A., DOBSON, M.V. AND GUY, D.G. (1997). Visual acuity and blood lipids in term infants fed human milk or formulae. *Lipids* **32**, 63–72.

INNIS, S.M., DYER, R., QUINLAN, P. AND DIERSEN-SCHADE, D. (1995). Palmitic acid is absorbed as sn-2 monopalmitin from milk and formula with rearranged triacylglycerols and results in increased plasma triglyceride sn-2 and cholesterol ester palmitate in piglets. *J. Nutr.* **125**, 73–81.

INNIS, S.M., NELSON, C.M., LWANGA, D., RIOUX, F.M. AND WASLEN, P. (1996). Feeding formula without arachidonic acid and docosahexaenoic acid has no effect on preferential looking acuity or recognition memory in healthy full-term infants at 9 mo of age. *J. Clin. Nutr.* **64**, 40–46.

INNIS, S.M., NELSON, C.M., RIOUX, M.F. AND KING, D.J. (1994). Development of visual acuity in relation to plasma and erythrocyte w-6 and w-3 fatty acids in healthy term gestation infants. *Am. J. Clin. Nutr.* **60**, 347–352.

JAEGER, K-E., RANSAC, S., DIJKSTRA, B.W., COLSON, C.,VAN HEUVEL, M. AND MISSET, M. (1995). Bacterial lipases. *FEMS Microbiol. Rev.* **15**, 29–63.

JEEVANANDAM, M., HOLADAY, N.J., VOSS, T., BUIER, R. AND PEDERSEN, S.R. (1995). Efficacy of a mixture of medium-chain triglyceride (75%) and long-chain triglyceride (25%) fat emulsions in the nutritional management of multiple-trauma patients. *Nutrition* **11**, 275–284.

JOHNSON, R.C., YOUNG, S.K., COTTER, R., LIN, L. AND ROWE, W.B. (1990). Medium-chain-triglyceride lipid emulsion: metabolism and tissue distribution. *Am. J. Clin. Nutr.* **52**, 502–508.

JOOSSENS, J.V., GEBOERS, J. AND KESTELOOT, H. (1989). Nutrition and cardiovascular mortality in Belgium. *Acta Cardiol.* **44**, 157–182.

KAMISAKA, Y., TOSHIHIRO, Y., NAKAHARA, T. AND SUKZUKI, O. (1990). Incorporation of Linoleic acid and its conversion to γ-linolenic acid in fungi. *Lipids* **25**, 54–60.

KAWACHI, M., KATO, M., IKEMOTO, H. AND MIYACHI, S. (1996). Fatty acid composition of a new marine picoplankton species of the Chromophyta. *J. Appl. Phycol.* **8**, 397–401.

KENDRICK, A. AND RATLEDGE, C. (1992). Lipids of selected molds grown for production of n-3 and n-6 polyunsaturated fatty acids. *Lipids* **27**, 15–20.

KLEMANN, L.P., AJI, K., CHRYSAM, M.M., D'AMELIA, R.P., HENDERSON, J.M., HUANG, A.S., OTTERBURN, M.S. AND YARGER, R.G. (1994). Random nature of triacylglycerols produced by the catalyzed interesterification of short and long-chain fatty acid triglycerides. *J. Agric. Food Chem.* **42**, 442–446.

KNAPP, S.J., TAGLIANI, L.A. AND ROATH, W.W. (1991). Fatty acid and oil diversity of *Cuphea viscosissima*: a source of medium-chain fatty acids. *J. Am. Oil Chem. Soc.* **68**, 515–517.

KNUTZON, D.S., LARDIZABAL, K.D., NELSEN, J.S., BLEIBAUM, J.L., DAVIES, H.M. AND METZ, J.G. (1995). Cloning of a coconut endosperm cDNA encoding a 1-acyl-sn-glycerol-3-phosphate acyltransferase that accepts medium chain length substrates. *Plant Physiol.* **109**, 999–1006.

KOON TAN, C. AND JOHNS, M.R. (1996). Screening of diatoms for heterotrophic eicosapentaenoic acid production. *J. Appl. Phycol.* **8**, 59–64.

KOSUGI, Y. AND AZUMA, N. (1994). Synthesis of triacylglycerol from polyunsaturated fatty acid by immobilized lipase. *J. Am. Oil Chem. Soc.* **71**, 1397–1403.

KUKSIS, A., McCARTHY, M.J. AND BEVERIDGE, J.M.R. (1963). Triglyceride composition of native and rearranged butter and coconut oils. *J. Am. Oil Chem. Soc.* **41**, 201–205.

KRIS-ETHERTON, P.M. AND YU, S. (1997). Individual fatty acid effects on plasma lipids and lipoproteins: human studies. *Am. J. Clin. Nutr.* **65**, 1628S–1644S.

KROKAN, H.E., BJERVE, K.S. AND MØRK, E. (1983). The enteral bioavailability of eicosapentaenoic acid and docosahexaenoic acid is as good from ethyl esters as from glyceryl esters in spite of lower hydrolytic rates by pancreatic lipase *in vitro*. *Biochim. Biophys. Acta* **1168**, 59–67.

KWON, D.Y., SONG, H.N. AND YOON, S.H. (1996). Synthesis of medium-chain glycerides by lipase in organic solvent. *J. Am. Oil Chem. Soc.* **73**, 1521–1525.

LAMBERT, M.S., BOTHAM, K.M. AND MAYES, P.A. (1996). Modification of the fatty acid composition of dietary oils and fats on incorporation into chylomicrons and chylomicron remnants. *British J. Nutr.* **76**, 435–445.

LANGHOLZ, P., ANDERSEN, P., FORSKOV, T. AND SCHMIDTSDORFF, W. (1989). Application of specificity of *Mucor miehei* lipase to concentrate docosahexaenoic acid (DHA). *J. Am. Oil Chem. Soc.* **66**, 1120–1123.

LEE, K-T. AND AKOH, C.C. (1996). Immobilized lipase-catalyzed production of structured lipids with eicosapentaenoic acid at specific positions. *J. Am. Oil Chem. Soc.* **73**, 611–615.

LEE, K-T. AND AKOH, C.C. (1997). Effects of selected substrate forms on the synthesis of structured lipids by two immobilized lipases. *J. Am. Oil Chem. Soc.* **74**, 579–584.

LI, Z-Y. AND WARD, O.P. (1993a). Enzyme-catalysed production of vegetable oils containing omega-3 polyunsaturated fatty acid. *Biotechnology Lett.* **15**, 185–188.

LI, Z-Y. AND WARD, O.P. (1993b). Lipase-catalyzed esterification of glycerol and n-3 polyunsaturated fatty acid concentrate in organic solvent. *J. Am. Oil Chem. Soc.* **70**, 745–748.

LIE KEN JIE, M.S.F. AND SYED RAHMATULLAH, M.S.K. (1995). Enzymatic enrichment of C_{20} *cis*-5 polyunsaturated fatty acids from *Biota orientalis* seed oil. *J. Am. Oil Chem. Soc.* **72**, 245–249.

LINDBERG, A.-M. AND MOLIN, G. (1993). Effect of temperature and glucose supply on the production of polyunsaturated fatty acids by the fungus *Mortiella alpina* CBS 343.66 in fermentor cultures. *Appl. Microbiol. Biotechnol.* **39**, 450–455.

LINKO, Y-Y. AND HAYAKAWA, K. (1996). Docosahexaenoic acid: a valuable nutraceutical? *Trends Food Sci. Technol.* **7**, 59–63.

LÓPEZ ALONSO, D. AND SEGURA DEL CASTILLO, C.I. (1996). First insights into improvement of eicosapentaenoic acid content in *Phaeodactylum tricornutum* (bacillariophyceae) by induced mutagenesis. *J. Phycol.* **32**, 339–345.

MACRAE, A.R. (1983). Lipase-catalyzed interesterification of oils and fats. *J. Am. Oil Chem. Soc.* **60**, 291–294.

MACRAE, A.R. (1985). Interesterification of fats and oils. In *Biocatalysts in Organic Syntheses*. Eds. J. Tramper, H. C. van der Plas and P. Linko, Proceedings of an International Symposium held at Noorwijkerhout, The Netherlands (14–17 April 1985). pp 195–208. Amsterdam: Elsevier Science Publishers.

MACRAE, A.R. AND HOW, P. (1988). Rearrangement Process, *United States Patent*, patent number **4**, 719, 178.

MAKRIDES, M., NEUMANN, M.A. AND GIBSON, R.A. (1996). Is dietary docosahexaenoic acid essential for term infants. *Lipids* **31**, 115–119.

MAKRIDES, M., NEUMANN, M.A., SIMMER, K. AND GIBSON, R.A. (1995). Erythrocyte fatty acids of term infants fed either breast milk, standard formula, or formula supplemented with long-chain polyunsaturates. *Lipids* **30**, 941–948.

MALCATA, F.X., REYES, H.R., GARCIA, H.S., HILL, C.G. AND AMUNDSON, C.H. (1992). Kinetics and mechanisms catalyzed by immobilized lipases. *Enzyme Microb. Technol.* **14**, 426–446.

MARANGONI, A.G. AND ROUSSEAU, D. (1995). Engineering triacylglycerols: the role of interesterification. *Trends Food Sci. Technol.* **6**, 329–335.

MARANGONI, A.G. AND ROUSSEAU, D. (1998a). The influence of chemical interesterification on the physiochemical properties of complex fat systems. I. Melting and crystallization. *J. Am. Oil Chem. Soc.* **75**, 1265–1271.

MARANGONI, A.G. AND ROUSSEAU, D. (1998b). The influence of chemical interesterification on the physiochemical properties of complex fat systems. III. Rheology and fractality of the fat crystal network. *J. Am. Oil Chem. Soc.* **75**, 1633–1636.

MARTINEZ, M. (1992). Tissue levels of polyunaturated fatty acids during early human development. *J. Pediatr.* **120 (suppl)**, S129–S138.

MASCIOLI, E.A., BABAYAN, V.K., BISTRIAN, B.R. AND BLACKBURN, G.L. (1988). Novel triglycerides for special medical purposes. *J. Parent. Enteral Nutr.* **12**, 127S–131S.

MCGEE, D.L., REED, D.M., YANO, K., KAGAN, A. AND TILLOTSON, J. (1984). Ten year incidence of coronary heart disease in Honolulu Heart Program. Relationship to nutrient intake. *Am. J. Epidemiol.* **119**, 667–676.

MCNEILL, G.P., ACKMAN, R.G. AND MOORE, S.R. (1996). Lipase-catalyzed enrichment of long-chain polyunsaturated fatty acids. *J. Am. Oil Chem. Soc.* **73**, 1403–1407.

MEDINA, A.R., GIMÉNEZ GIMÉNEZ, A., GARCÍA CAMACHO, F., SÁNCHEZ PÉREZ, J.A., MOLINA GRIMA, E. AND CONTRERAS GÓMEZ, A. (1995). Concentration and purification of stearidonic, eicosapentaenoic and docosahexaenoic acids from cod liver oil and the marine microalgae *Isochyrysis galbana. J. Am. Oil Chem. Soc.* **72**, 575–581.

MEGREMIS, C.J. (1991). Medium-chain triglycerides: a nonconventional fat. *Food Tech.* **45**, 108–110.

MENSINK, R.P. AND KATAN, M.B. (1990). Effect of dietary trans fatty acids on high-density and low-density lipoprotein cholesterol levels in healthy subjects. *New Eng. J. Med.* **323**, 439–444.

MIQUEL, M. AND BROWSE, J. (1995). Molecular biology of oilseed modification. *Inform.* **6**, 108–111.

MOLINA GRIMA, E., ROBLES MEDINA, A., GIMÉNEZ GIMÉNEZ, A. AND IBÁÑEZ GONZÁLES, M.J. (1996). Gram-scale purification of eicosapentaenoic acid (EPA, 20:5n-3) from wet *Phaeodactylum tricornutum* UTEX 640 biomass. *J. Appl. Phycol.* **8**, 359–367.

MOLINA GRIMA, E., SÁNCHEZ PÉREZ, J.A., GARCÍA CAMACHO, F., GARCÍA SÁNCHEZ, J.L. AND FERNÁNDEZ SEVILLA, J.M. (1995). Variation of fatty acid profile with solar cycle in outdoor chemostat culture of *Isochrysis galbana* ALII-4. *J. Appl. Phycol.* **7**, 129–134.

MOJOVIC, L., ŠILER-MARINKOVIC, S., KUKIC, G. AND VUNJAK-NOVAKOVIC. (1993). *Rhizopus arrhizus* lipase-catalyzed interesterification of the midfraction of palm oil to a cocoa butter equivalent fat. *Enzyme Microb. Technol.* **15**, 438–443.

MUKHERJEE, K.D. (1998). Lipid Biotechnology. In *Food Lipids: Chemistry, Nutrition and Biotechnology.* Eds. C.C. Akoh and D.B. Min, pp 589–640. New York: Marcel Dekker, Inc.

MURPHY, D.J. (1994). Manipulation of lipid metabolism in transgenic plants: biotechnological goals and biochemical realities. *Biochemical Soc. Trans.* **22**, 926–931.

MURPHY, D.J. (1996). Engineering oil production in rapeseed and other oil crops. *Tibtech.* **14**, 206–213.

MYRNES, B., BARSTAD, H., OLSEN, R.L. AND ELVEVOLL, E.O. (1995). Solvent-free enzymatic glycerolysis of marine oils. *J. Am. Oil Chem. Soc.* **72**, 1339–1344.

NAKAHARA, T., YOKOCHI, T., HIGASHIHARA, T., TANAKA, S., YAGUCHI, T. AND HONDA, D. (1996). Production of docosahexaenoic and docosapentaenoic acids by *Schizochytrium* sp. isolated from Yap Islands. *J. Am. Oil Chem. Soc.* **73**, 1421–1426.

NAWAR, W.W. (1996). Chemistry. In *Bailey's Industrial Oil and Fat Products, Volume 1: Edible Oil and Fat Products, General Applications*, 5th Edition. Ed. Y.H. Hui, pp 232, 409. New York: John Wiley and Sons, Inc.

NELSON, G.J. (1992). Dietary fatty acids and lipid metabolism, In *Fatty Acids in Foods and Their Health Implications*. Ed. C.K. Chow, pp 437–365. New York: Marcel Dekker.

NELSON, G.J. AND ACKMAN, R.G. (1988). Absorption and transport of fat in mammals with emphasis on n-3 polyunsaturated fatty acids. *Lipids* **23**, 1005–1014.

NEURINGER, M., REISBICK, S. AND JANOWSKY, J. (1994). The role of n-3 fatty acids in visual and cognitive development: current evidence and methods of assessment. *J. Pediatr.* **125**, S39–S47.

NORDENSTRÖM, J., THÖRNE, A. AND OLIVECRONA, T. (1995). Metabolic effects of infusion of a structured-triglyceride emulsion in healthy individuals. *Nutrition* **11**, 269–274.

O'BRIEN, D., KURANTZ, M.J. AND KWOCZAK, R. (1993). Production of eicosapentaenoic acid by the filamentous fungus *Pythium irregulare. Appl. Microbiol. Biotechnol.* **40**, 211–214.

OHLROGGE, J.B. (1983). Distribution in human tissues of fatty acid isomers from hydrogenated oils. In *Dietary Fats and Health*. Eds. E.G. Perkins, E.G. and W.J. Visek, p 359. Champaign, Illinois: American Oil Chemists' Society.

OHLROGGE, J.B. (1994). Design of new plant products: engineering of fatty acid metabolism. *Plant. Physiol.* **104**, 821–826.

OSADA, K., TAKAHASHI, K. AND HATANO, M. (1990). Polyunsaturated fatty glyceride syntheses by microbial lipases. *J. Am. Oil Chem. Soc.* **67**, 921–922.

PAI, T. AND YEH, Y.-Y. (1997). Stearic acid modifies very low density lipoprotein lipid composition and particle size differently from shorter chain saturated fatty acids in cultured rat hepatocytes. *Lipids* **32**, 143–149.

RATNAYAKE, W.M. AND CHEN, Z.Y. (1996). Trans, n-3 and n-6 fatty acids in Canadian human milk. *Lipids* **31**, S279–S282.

RAY, S. AND BHATTACHARYYA, D.K. (1995). Comparative nutritional study of enzymatically and chemically interesterified palm oil products. *J. Am. Oil Chem. Soc.* **72**, 327–330.

REDDY, T. AND THOMAS, T.L. (1996). Expression of a cyanobacterial D6-desaturase gene results in gamma-linolenic acid production in transgenic plants. *Nat. Biotechnol.* **14**, 639–642.

REYES, H.R. AND HILL, JR., C.G. (1994). Kinetic modeling of interesterification reactions catalyzed by immobilized lipase. *Biotechnol. Bioeng.* **43**, 171–182.

RIAUBLANC, R., RATOMAHENINA, R., GALZY, P. AND NICOLAS, M. (1993). Peculiar properties from *Candida parapsilosis* (Ashford) Langeron Talice. *J. Am. Oil Chem. Soc.* **70**, 497–500.

ROBLES-MEDINA, A., GIMÉNEZ GIMÉNEZ, A., GARCÍA CAMACHO, F., SÁNCHEZ PÉREZ, J.A., MOLINA GRIMA, E. AND CONTRERAS GÓMEZ, A. (1995). Concentration and purification of stearidonic, eicosapentaenoic and docosahexaenoic acids from cod liver oil and the marine microalga *Isochrysis galbana*. *J. Am. Oil Chem. Soc.* **72**, 575–583.

ROSE, A.H. (1989). Influence of the environment on microbial lipid composition. In *Microbial Lipids: Volume 2*. Eds. C. Ratledge and S.G. Wilkinson, pp 255–278. San Diego, USA: Academic Press.

ROUSSEAU, D. AND MARANGONI, A.G. (1998a). Tailoring the textural attributes of butterfat-canola oil blends via *Rhizopus arrhizus* lipase-catalyzed interesterification. I. Compositional modifications. *J. Agric. Food Chem.* **46**, 2368–2374.

ROUSSEAU, D. AND MARANGONI, A.G. (1998b). Tailoring the textural attributes of butterfat-canola oil blends via *Rhizopus arrhizus* lipase-catalyzed interesterification. 2. Modification of physical properties. *J. Agric. Food Chem.* **46**, 2375–2381.

ROUSSEAU, D., FORESTIERE, K., HILL, A.R. AND MARANGONI, A.G. (1996a). Restructuring butterfat through blending and chemical interesterification. I. Melting behaviour and triacylglycerol modifications. *J. Am. Oil Chem. Soc.* **73**, 963–972.

ROUSSEAU, D., HILL, A.R. AND MARANGONI, A.G. (1996b). Restructuring butterfat through blending and chemical interesterification. II. Microstructure and polymorphism. *J. Am. Oil Chem. Soc.* **73**, 973–981.

ROUSSEAU, D., HILL, A.R. AND MARANGONI, A.G. (1996c). Restructuring butterfat through blending and chemical interesterification. III. Rheology. *J. Am. Oil Chem. Soc.* **73**, 983–989.

SANDSTRÖM, R., HYLTANDER, A., KÖRNER, U. AND LUNDHOLM, K. (1993). Structured triglycerides to postoperative patients: a safety and tolerance study. *J. Parent. Enteral Nutr.* **17**, 153–157.

SHAPIRO, S. (1997). Do *trans* fatty acids increase the risk of coronary artery disease? A critique of the epidemiological evidence. *Am. J. Clin. Nutr.* **66**(suppl), 1011S–1017S.

SHEKELLE, R.B., SHRYOCK, A.M. AND PAUL, O. (1981). Diet, serum cholesterol and death from coronary heart disease. The Western electric Study. *New Engl. J. Med.* **304**, 65–70.

SHIMADA, Y., MARUYAMA, K., OKAZAKI, S., NAKAMURA, M., SUGIHARA, A. AND TOMINAGA, Y. (1994). Enrichment of polyunsaturated fatty acids with *Geotrichum candidum* lipase. *J. Am. Oil Chem. Soc.* **71**, 951–954.

SHIMADA, Y., SUGIHARA, A., NAKANO, H., KURAMOTO, T., NAGAO, T., GEMBA, M. AND TOMINAGA, Y. (1997). Purification of docosahexaenoic acid by selective esterification of fatty acids from tuna oil with *Rhizopus delemar* lipase. *J. Am. Oil Chem. Soc.* **74**, 97–101.

SHIMADA, Y., SUGIHARA, A., NAKANO, H., YOKOTA, T., NAGAO, T., KOMEMUSHI, S. AND TOMINAGA, Y. (1996). Production of structured lipids containing essential fatty acids by immobilized *Rhizopus delemar* lipase. *J. Am. Oil Chem. Soc.* **73**, 1415–1420.

SHIMIZU, S., KAWASHIMA, H., AKIMOTO, K., SHINMEN, Y. AND YAMADA, H. (1989). Microbial conversion of an oil containing alpha-linolenic acid to an oil containing eicosapentaenoic acid. *J. Am. Oil Chem. Soc.* **66**, 343–347.

SHINMEN, Y., SHIMIZU, S., AKIMOTO, K., KAWASHIMA, H. AND YAMADA, H. (1989). Production of arachidonic acid by *Mortiella* fungi. *Appl. Microbiol. Biotechnol.* **31**, 11–16.

SHISHIKURA, A., FUJIMOTO, K. SUZUKI, T. AND ARAI, K. (1994). Improved lipase-catalyzed incorporation of long-chain fatty acids into medium chain triacylglycerides assisted by supercritical carbon dioxide extraction. *J. Am. Oil Chem. Soc.* **71**, 961–967.

SLABAS, A.R., SINDEN, B.S., CHASE, D., BROWN, A.P., COLEMAN, J., MARTINEZ-RIVAS, J. AND FAWCETT, T. (1992). In *Biochemistry and Molecular Biology of Membrane and Storage Lipids of Plants*. Eds. N. Murata and C. Somerville, Proceedings US/Japan Binational Seminar, Dec 13–17, 1992, Hawaii. Current Topics in Plant Physiology: An American Society of Plant Physiologists Series, Vol. 9. USA, pp 113–120.

SMALL, D.M. (1991). The effects of glyceride structure on absorption and metabolism. *Annu. Rev. Nutr.* **11**, 413–434.

SOBRADO, J., MOLDAWER, L.L., POMPOSELLI, J.J., MASCIOLI, E.A., BABAYAN, V.K., BISTRIAN, B.R. AND BLACKBURN, G.L. (1985). Lipid emulsions and reticuloendothelial system function in healthy and burned guinea pigs. *Am. J. Clin. Nutr.* **42**, 855–863.

SOUMANOU, M.M., BORNSCHEUER, U.T., MENGE, U. AND SCHMID, R.D. (1997). Synthesis of structured triglycerides from peanut oil with immobilized lipase. *J. Am. Oil Chem. Soc.* **74**, 427–433.

SPIELMANN, D., BRACCO, U., TRAITLER, H., CROZIER, G., HOLMAN, R., WARD, M. AND COTTER, R. (1988). Alternative lipids to usual w6 PUFAS: gamma-linolenic acid, alpha-linolenic acid, stearidonic acid, EPA, etc. *J. Parent. Enteral Nutr.* **12**, 111S–123S.

SRIDHAR, R. AND LAKSHMINARAYANA, G. (1992). Incorporation of eicosapentaenoic and docosahexaenoic acids into groundnut oil by lipase-catalyzed ester interchange. *J. Am. Oil Chem. Soc.* **69**, 1041–1042.

SRIDHAR, R., LAKSHMINARAYANA, G. AND KAIMAL, T.N.B. (1991). Modification of selected Indian vegetable fats into cocoa butter substitutes by lipase-catalyzed ester interchange. *J. Am. Oil Chem. Soc.* **68**, 726–730.

STAMATIS, H., XENAKIS, A., PROVELEGIOU, M. AND KOLISIS, F.N. (1993). Esterification reactions catalyzed by lipases in microemulsions: the role of enzyme location in relation to its selectivity. *Biotechnol. Bioeng.* **42**, 103–110.

TANAKA, Y., HIRANO, J. AND FUNADA, T. (1992). Concentration of docosahexaenoic acid in glyceride by hydrolysis of fish oil with *Candida cylindracea* lipase. *J. Am. Oil Chem. Soc.* **69**, 1210–1214.

TANAKA, Y., HIRANO, J. AND FUNADA, T. (1994). Synthesis of docosahexaenoic acid-rich triglyceride with immobilized *Chromobacterium viscosum* lipase. *J. Am. Oil Chem. Soc.* **71**, 331–334.

TOMARELLI, R.M. (1988). Suitable fat formulations for infant feeding. In *Dietary Fat Requirements in Health and Development*. Ed. J. Beare-Rogers, pp 1–28. USA: American Oil Chemists' Society, Library of Congress Cataloging-in-Publication Data.

TSO, P. AND WEDIMAN, S.W. (1987). Absorption and metabolism of lipid in humans. In *Lipids in Modern Nutrition*. Eds. M. Horisberger and U. Bracco, Nestlé Nutrition Workshop Series, Volume 13, pp 1–15. New York: Raven Press.

ULRICH, H., MCCARTHY PASTORES, S., KATZ, D.P. AND KVETAN, V. (1996). Parenteral use of medium-chain triglycerides: a reappraisal. *Nutrition* **12**, 231–238.

VAN ZYL, C.G., LAMBERT, E.V., HAWLEY, J.A., NOAKES, T.D. AND DENNIS, S.C. (1996). Effects of medium-chain triglyceride ingestion on fuel metabolism and cycling performance. *J. Appl. Physiol.* **80**, 2217–2225.

VERGROESEN, A.J. (1976). Early signs of polyunsaturated fatty acid deficiency. *Biblthca. Nutr. Dieta.* **23**, 19–25.

VOELKER, T.A., WORRELL, A.C., ANDERSEN, L., BLIEBAUM, L., FAN, C., HAWKINS, D., RADKE, S.E. AND DAVIES, H.M. (1992). Fatty acid biosynthesis redirected to medium chains in transgenic oilseed plants. *Science.* **257**, 72–73.

WATKINS, J.B. (1988). Lipid digestion in the developing infant. In *Dietary Fat Requirements in Health and Development.* Ed. J. Beare-Rogers, pp 29–42. USA: American Oil Chemists Society, Library of Congress.

WERKMAN, S.H. AND CARLSON, S.E. (1996). A randomized trial for visual attention of preterm infants fed docosahexaenoic acid until nine months. *Lipids* **31**, 91–97.

WESSINGER, E.W., O'BRIEN, D. AND KURANTZ, M.J. (1990). Identification of fungi for sweet whey permeate utilization and eicosapentaenoic acid production. *J. Ind. Microbiol.* **6**, 191–197.

WIBERG, E. AND BAFOR, M. (1995). Medium chain-length fatty acids in lipids of developing oil palm kernel endosperm. *Phytochem.* **39**, 1325–1327.

WOLTER, F.P. (1993). Altering plant lipids by genetic engineering. *Inform.* **4**, 93–98.

YADAV, N., WIERZBICKI, A., KNOWLTON, S., PIERCE, J., RIPP, K., HITZ, W., AEGERTER, M. AND BROWSE, J. (1992). Genetic manipulation to alter fatty acid profiles of oilseed crops. In *Biochemistry and Molecular Biology of Membrane and Storage Lipids of Plants*. Eds. N. Murata and C. Somerville, Proceedings US/Japan Binational Seminar, Dec 13–17, 1992, Hawaii. Current Topics in Plant Physiology: An American Society of Plant Physiologists Series, Vol. 9. U.S.A., pp 60–66.

YADWAD, V.B., WARD, O.P. AND NORONHA, L.C. (1991). Application of lipase to concentrate the docosahexaenoic acid (DHA) fraction of fish oil. *Biotech. Bioeng.* **38**, 956–959.

YAMANE, T., SUZUKI, T. AND HOSHINO, T. (1993). Increasing n-3 polyunsaturated fatty acid content of fish oil by temperature control of lipase-catalyzed acidolysis. *J. Am. Oil Chem. Soc.* **70**, 1285–1287.

YANO, Y., NAKAYAMA, A., SAITO, H. AND ISHIHARA, K. (1994). Production of docosahexaenoic acid by marine bacteria isolated from deep sea fish. *Lipids* **29**, 527–528.

YONGMANITCHAI, W. AND WARD, O.P. (1991). Screening of algae for potential alternative sources of eicosapentaenoic acid. *Phytochemistry* **30**, 2963–2967.

YUAN, L. AND KNAUF, V.C. (1998). Modification of plant components. *Current Opin. Biotechnol.* **8**, 227–233.

ZOCK, P.L. AND KATAN, M.B. (1992). Hydrogenation alternatives: effects of *trans* fatty acids and stearic acid versus linoleic acid on serum lipids and lipoproteins in humans. *J. Lipid Res.* **33**, 399–410.

ZOCK, P.L., DE VRIES, J.H.M. AND KATAN, M.B. (1994). Impact of myristic acid versus palmitic acid on serum lipid and lipoprotein levels in healthy women and men. *Arterioscler. Thromb.* **14**, 567–575.

ZOPPI, G., ANDREOTTI, G., NJAI, D.M. AND GABURRO, D. (1972). Exocrine pancreas function in premature and full term neonates. *Pediatr. Res.* **6**, 880–886.

6

Genetic Manipulation of Starch Biosynthesis: Progress and Potential

H. FRANCES J. BLIGH

Division of Plant Science, School of Biological Sciences, University of Nottingham, Nottingham, NG7 2RD, UK

Introduction

Of the starch produced commercially in the western world, approximately a third is used for non-food applications, while the remaining two thirds is used in the food industry. The use of starches in non-food applications is one which is becoming of increasing importance also for consumers as they become more aware of 'green' issues such as sustainability of raw material sources and biodegradation of materials (Jane *et al.*, 1994b), especially packaging. The need of industry for starches to produce, for example, plastics, packaging and paints, has resulted in research targeted towards producing starches with specific properties required for these specialist applications, as well as towards increasing the yields of starch producing plants. In addition to non-food uses, starch is used widely in the food industry, and, with the development of more processed and pre-prepared foods, demand for specialist starches is increasing. The majority of commercial starch is sourced from potato, maize and wheat, and as a result, these are the systems used most in studies of starch biosynthesis and structure, with most emphasis being placed on potato and maize. However other organisms have also been used, notably rice, which as a simple monocotyledonous plant is a useful model for the hexaploid wheat, and pea, which like maize, exhibits useful visible starch mutant phenotypes. Barley has also been used, as there is commercial interest in this as a source of starch in countries which do not have suitable growing conditions for crops such as potatoes or maize. More recently, intensive studies have been performed on crops such as cassava, which is widely grown in tropical climates, with a view to genetic improvement (Munyikwa *et al.*, 1997).

Starch structure and synthesis

Starch is a major storage carbohydrate found in photosynthetic and storage organs of plants. As starches for industrial use come from storage organs, mainly seeds and

Biotechnology and Genetic Engineering Reviews – Vol. 16, April 1999
0264–8725/99/15/177–201 $20.00 + $0.00 © Intercept Ltd, P.O. Box 716, Andover, Hampshire SP10 1YG, UK

tubers, this review will deal with synthesis in these organs only. Starch in storage organs is synthesized within amyloplasts, plastids which can contain either a single large starch granule, as in potato, or multiple smaller granules as in rice. The granules themselves can also vary with respect to shape and structure. Rice has polyhedral starch granules which are between 3 and 8 μm in diameter, while potato has ovoid granules which can be up to 100 μm in diameter (Jane *et al.*, 1994a) (*Table 1*). Despite these differences, microscopic analysis of granule structure has shown some features which appear to be constant. All granules have been shown to exhibit a concentric sphere morphology, with alternating semi-crystalline and amorphous growth shells. Growth of these rings was thought to be linked to light and dark diurnal cycling, but is now thought to be due to day to night temperature variation (Buléon, 1997). Analysis of the semi-crystalline shells has been shown to contain further stacks of alternating crystalline and amorphous lamellae (Jenkins *et al.*, 1993). The repeat size of one layer of crystalline and amorphous lamellae appears to be approximately 9 nm, irrespective of botanical origin. However, the proportional size of crystalline and amorphous layer within the 9 nm layer has been shown to vary according to the molecular composition of the starch (Jenkins and Donald, 1995). More recently, further levels of complexity have been demonstrated showing the amylopectin lamellae to be organized into blocklets (Gallant *et al.*, 1997) (see *Figure 1*).

Table 1.   Average size and shape of starch granules from botanical sources

Botanical origin	Diameter range (microns)	Shape
Maize	5–20	spherical or polygonal
Potato	15–75	oval or spherical
Rice	3–8	polygonal
Wheat A type	22–36	disk
Wheat B type	2–3	spherical
Barley A type	10–48	oval
Barley B type	2–10	spherical
Pea	10–45	irregular with indentation
Cassava	5–40	rounded with indentation
Banana	15–45	irregular
Phaseolus	10–45	disc shaped
Maize *su* mutants	2–3	spherical

At the molecular level, starch is made up of two components, amylose, which consists of linear chains of α(1–4) linked glucose molecules, and amylopectin, which is made up of many shorter α(1–4) linked chains that also have branches in the form of α(1–6) linkages. The chains of amylopectin fall into several classes, A chains which have no substitution at the 6 position, B chains, which are branched at one or several points by α(1–6) linkages, and C chains, a single one of which runs through the centre of the amylopectin cluster and has a reducing end (Robin *et al.*, 1975) (*Figure 1*). The packing of A chains of the amylopectin are thought to be responsible for the type of crystallinity exhibited by the starch, as measured by X-ray diffraction. Shorter A chains are thought to be associated with A-type crystallinity, longer A chains associated with B-type crystallinity and C-type crystallinity being thought to be a result of a mixture of A and B types (Hizukuri, 1985). Crystallinity is also an indicator of the digestibility of a starch, with B and C-type starches, found in tubers and

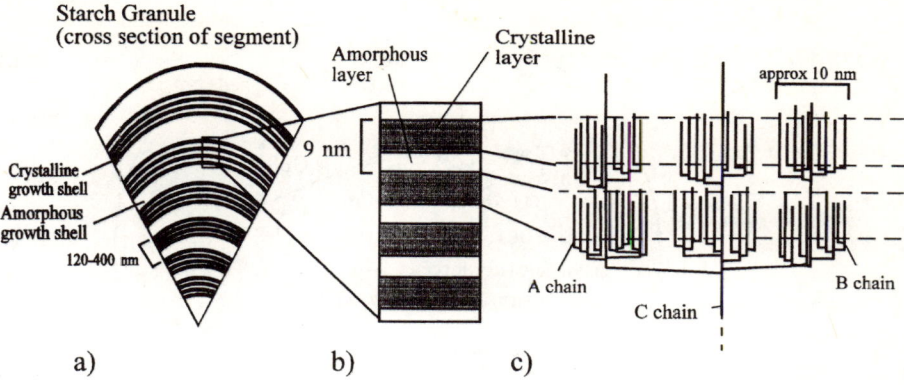

Figure 1. Schematic of starch granule structure. Cross section through a section of starch granule showing semi-crystalline and amorphous growth shells. Magnified view through one semi-crystalline shell showing amorphous and crystalline lamellae. Fine structure of crystalline and amorphous lamellae showing A, B and C chains lining up in the crystalline lamellae in blocklets with branchpoints within the amorphous layer. Based on Jenkins and Donald (1995) and Gallant *et al.* (1997).

legumes, being associated with poorer digestibility and higher enzyme resistance than A-type starches, found in cereals (Gallant *et al.*, 1972). The enzyme resistance aspect of starches is particularly relevant in the food industry where a lot of starch is processed to produce maltodextrins and corn syrup. It is also the amylopectin that provides the viscosity of starch that is required for its many food uses in applications such as thickenings and pastes. The percentages of amylose and amylopectin vary from plant to plant, and also within varieties of a plant species. For any major progress to be made in the manipulation of either yield or quality of starches within the plant, as opposed to production via post harvest modification, the fundamental processes involved in the biosynthesis of starch in storage organs, such as seeds and tubers, must be understood. So far, the enzymes involved in starch biosynthesis within the amyloplast fall into three main groups (see *Figure 2*) and a large number have now been cloned from various plant sources (*Table 2*). The first group is the ADP glucose pyrophosphorylases (AGPase), which are responsible for determining the flux of carbon into the starch biosynthesis pathway by synthesizing ADP-glucose, the substrate for starch synthases, from ATP and glucose-1-phosphate. The starch synthases transfer glucose from ADP-glucose substrate to $\alpha(1\text{--}4)$ glucan to create amylose and fall into two groups, Granule Bound Starch Synthases (GBSS) and Soluble Starch Synthases (SSS). The GBSSs are involved in the synthesis of amylose, as has been shown by the fact that null mutants of these genes, first isolated as *waxy* mutants in maize, are almost totally lacking in amylose. The SSSs are thought, by default, to be involved in the synthesis of the glucose chains that act as substrates for the Branching Enzymes, although there is little evidence to confirm this. The Branching Enzymes (BE) create the $\alpha(1\text{--}6)$ branches of the amylopectin and appear to fall into 2 groups, A and B, based on sequence homology. Mutations in some A group BE genes appear to result in what is known as the *amylose extender (ae)* mutant, which is characterized by particularly long chain amylopectin, giving the characteristics of a high amylose starch. This led to the theory that the B group of BEs had affinity for branching long glucose chains, while the A group had affinity for shorter chains, however, this

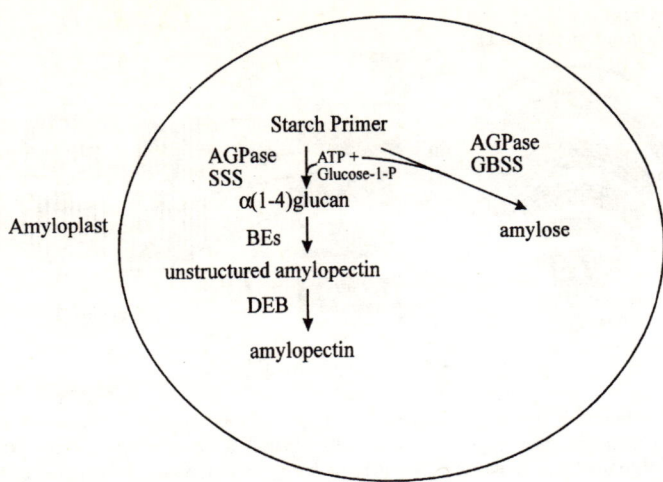

Figure 2. Schematic of enzyme groups responsible for starch biosynthesis. ATP and glucose-1-phosphate are converted to ADP glucose by ADP glucose pyrophosphorylase (AGPase), which is then used as a substrate by soluble starch synthase (SSS) or granule bound starch synthase (GBSS) to make glucan chains. Unstructured amylopectin is then thought to be created by branching enzymes (BE) and then broken down to an ordered structure by debranching enzymes (DBE).

mechanism is probably an oversimplification of a more complex interaction between both branching enzymes and starch synthases. In addition to these three main groups, there are also debranching enzymes (DBE), pullulanase and isoamylase which were initially thought to be involved in degrading of starch during germination (for review see Nakamura, 1996). Recent expression studies have suggested that these enzymes are involved more directly in starch synthesis with DBEs thought likely to be mandatory in creating the rigid structure that results in the concentric shell formation seen in starch granules (James *et al.*, 1995; Ball *et al.*, 1996). Isoamylase is also now thought to have an important role in amylopectin structure, as the *sugary-1 (su-1)* mutant of maize, a mutant with increased phytoglycogen and amylose, has now been identified as a gene with strong homology to bacterial isoamylase enzymes (James *et al.*, 1995). The importance of the two different forms of DBE, pullulanase and isoamylase, may vary between plants, as *su-1* mutants of rice have been demonstrated to have significantly low levels of pullulanase but no discernable difference in isoamylase expression (Nakamura *et al.*, 1997). Disproportionating enzyme, so far found only in potato, is another enzyme also thought initially to be involved in starch degradation that has now been shown to be highly expressed during tuber development (Takaha *et al.*, 1993). As this enzyme is thought to be able to use short chain amylose and amylopectin as substrate, it may be that this enzyme has a joint role with debranching enzymes in regulating the number of branches and chain length of amylopectin. There is also a need for a primer molecule to initiate the starch synthesis. Such a primer, named amylogenin and thought to be functionally analogous to glycogenin, was identified by Keeling *et al.* (1994) although recent data has led to debate as to whether this is the true function of this enzyme (Dhugga *et al.*, 1997). The need to understand the interactions and mode of action of enzymes involved in starch biosynthesis has been the driving force behind a lot of the

Table 2. Genes from major starch sources involved in starch biosynthesis that have been cloned

Enzyme	Source of Clones	References
AGPase small subunit	rice, potato, maize, wheat, barley	Anderson *et al.* (1989); Smith-White and Preiss (1992); Bhave *et al.* (1990); Olive *et al.* (1989); Villand *et al.* (1992)
AGPase large subunit	rice, potato, maize, wheat, barley	Satozawa *et al.* (1995); Nakata *et al.* (1991); Giroux *et al.* (1995); Olive *et al.* (1989); Villand *et al.* (1992)
GBSS	rice, potato, maize, wheat, pea, barley	Wang *et al.* (1990); Visser *et al.* (1989); Klösgen *et al* (1986); Clarke *et al.* (1991); Dry *et al.* (1992); Rohde *et al.* (1988)
SSS	rice, potato	Baba *et al.* (1993); Marshall *et al.* (1996)
Branching enzyme type A	rice, maize, pea	Mizuno *et al.* (1993); Fisher *et al.* (1993); Burton *et al.* (1995)
Branching enzyme type B	rice, maize, potato, pea	Nakamura and Yamanouchi (1992); Fisher *et al.* (1995); Poulsen and Kreiberg (1993); Burton *et al.* (1995)
Debranching enzyme (pullulanase)	rice	Nakamura *et al.* (1996)
Debranching enzyme (isoamylase)	maize	James *et al.* (1995)
Disproportionating enzyme	potato	Takaha *et al.* (1993)

experiments performed to date involving the manipulation of starch biosynthesis enzymes.

Methods for genetic manipulation

Genetic manipulation of starch falls into two main categories. The first is the more obvious use of transgenic gene expression using modern molecular biology techniques to increase or reduce the expression of specific enzymes. In common with all transgenic expression systems, there are certain requirements:

(a) The availability of a reliable transformation system for the target organism. Until recently, this has been a particular problem for cereals such as wheat and rice, although biolistics, and latterly, the development of supervirulent *Agrobacterium* strains have both played a role in the increase in the number of species of cereals that can now be transformed reliably.

(b) A highly expressing promoter system. A strong promoter is essential if a transgene is to be highly expressed, and while many model systems have successfully used the constitutive Cauliflower Mosaic Virus 35s rRNA promoter (Croy, 1993), this is not always highly expressed, and also has the disadvantage that it is expressed indiscriminately in all tissues. This problem was encountered by Stark *et al.* (1992) who expressed the *Escherichia coli* AGPase gene in potato, but found that while they could regenerate shoots and callus, the over-expression caused a lethal effect resulting in poor recovery of transgenic plants expressing the gene. This problem was overcome by using the tuber specific patatin promoter that targeted the expression almost exclusively to the tuber, and enabled the regeneration of healthy plants. Seed specific promoters such as the rice glutelin (Leisy *et al.*, 1990) and the maize GBSS (Russell and Fromm, 1997) promoters are also now available and the increasing availability of such promoters means that ideally, a promoter can be chosen that will have optimal expression in the chosen plant system.

(c) Ideally, such promoters should also be accompanied by a terminator sequence, as while not essential, these sequences have been shown to have some role in the expression and regulation of genes (Nakata and Okita, 1996). In addition to this, it is necessary to examine the expression systems of the plant in which expression is required. One example of this is in some cereals such as maize and rice, where it has been demonstrated that the presence of a large (>1kb) intron in the 5' untranslated region can lead to improved expression of transgenes (Maas *et al.*, 1991; Li *et al.*, 1995). For more comprehensive details about expression systems the reader is referred to Croy (1993).

There are three techniques that have the potential to be used in the molecular manipulation of starch enzymes.

1. The use of antisense technology to reduce, or completely inhibit a particular gene activity. This method constitutes taking the coding region of a gene, or a section of it, and placing it in the reverse orientation under the control of a suitable promoter. This is thought to result in the production of an 'antisense' mRNA molecule that is complementary to and hence can form a duplex with the mRNA of the gene whose suppression is desired. Similar effects to this have also been demonstrated when attempts have been made to over express an endogenous gene in plants. For some reason, it has been observed that over-expression of a gene can result in suppression of that gene in much the same way as antisensing can. This phenomenon has been termed sense suppression and is used with as much success as antisense suppression in the inhibition of gene activity. Both antisense and sense suppression have been used to study starch biosynthesis genes in potato (Müller-Röber *et al.*, 1992; Kuipers *et al.*, 1994; Flipse *et al.*, 1996a–d) and rice (Shimada *et al.*, 1993; Itoh *et al.*, 1997).

2. Over-expression of an endogenous gene. This method can be used to increase the expression of a gene either to increase the overall target product, such as over-expression of AGPase to increase starch production, or alternatively to increase a particular enzyme function such as BE, to increase the overall branching levels within the starch. Over-expression of endogenous genes to increase gene activity has met with little success due to the phenomenon of sense suppression. This appears to be caused by natural feedback mechanisms within the plant preventing genes from being expressed at higher levels than would be found naturally (see above) (e.g. Itoh *et al.*, 1997).

3. The third method by which starch may be modified transgenically is by the use of heterologous gene expression. This is where the transgene used is from a different genetic source from the system being transformed, for example, another plant or a completely different organism such as *E. coli*. This approach has been taken by several investigators expressing genes such as the *E. coli* AGPase or BE genes

(Stark *et al.,* 1992; Shewmaker *et al.,* 1994; Sweetlove *et al.*, 1996a,b; Kortstee *et al.,* 1996). In addition to the usual requirements for transgenic expression described above, bacterial transgene constructs have a few additional requirements for functional expression. A requirement for expression of starch enzymes in particular, is a targeting sequence to enable the expressed protein to target to the amyloplast. In the case of the *E. coli glg*C16 gene in potato, a chloroplast transit peptide from the *Arabidopsis thaliana* small subunit ribulose 1,5-bisphosphate carboxylase (Stark *et al.*, 1992). In addition to this, it is often necessary to check that the sequence environment around the start AUG codon is suitable for plant expression, as the sequence around the AUG has to have at least some homology to the consensus AACA<u>AUG</u>GC (Futterer and Hohn, 1996).

The second less obvious, but no less useful, method of manipulating gene expression, is the use of known mutants or alleles to breed for a particular starch characteristic (see *Table 3* for list of mutants). A major example of this is the use of *waxy* maize varieties that have been cultivated due to industrial uses of *waxy* maize starch. These varieties have also been used in breeding studies to produce novel starches such as the loosely branched starch obtained by Boyer *et al.* (1976) who crossed a *wx* maize with a variety carrying the *ae* mutation. Similar crosses with *wx* and *du* genes have been shown to result in an increase in shorter B chains of the amylopectin (Fuwa *et al.,* 1987). Mutagenesis followed by screening for starch mutants has also been employed more recently in peas by Hedley and colleagues who used both mutagenesis and breeding techniques to develop a series of novel starches with altered starch granule shape (Wang *et al.,* 1990, 1997; Lloyd *et al.,* 1996). Utilizing both the *rugosus* (*r* and *rb*) mutants of classical genetics and a series of others, a total of 6 alleles involved in starch biosynthesis were identified. By studying the physical nature of the starches in these mutants, combined with an analysis of starch biosynthesis enzyme activities, the enzymes associated with these loci were identified. Having identified the individual loci for these alleles, a series of further breeding experiments were performed to create double mutants, similar to those of Boyer *et al.* (1976) and Campbell (1996) for maize. The double mutants obtained had further modifications to both starch structure and content, including some interesting effects on starch granule morphology in the case of a *rb/rug-5* double mutant (Wang *et al.,* 1997). It is hoped that further study of such double mutants will extend our knowledge into the interactions of starch biosynthesis enzymes as well as allow for the breeding of novel starches for commercial use. These methods have previously been particularly slow and cumbersome, but the advent of modern genetic screening methods such as microsatellite DNA markers, has made targeted breeding programs considerably quicker and easier. A large number of the starch biosynthesis genes have now been mapped in both maize and rice, and in the case of rice, a microsatellite marker which is closely linked to the GBSS gene has been identified (Bligh *et al.*, 1995). This marker has been shown to be useful in identifying alleles of the GBSS gene which have intermediate amylose, and this in turn has the potential to be used as a breeding marker to select for intermediate or high amylose levels (Ayres *et al.,* 1997). Mapping of the BEIII gene of rice also enabled the prediction of the location of this gene in other cereals, as the maize BEII and rice BEIII genes, both thought to be group A branching enzymes, were found to be in regions within their respective genomes. These regions corresponded to homeologous regions

Table 3. Mutations known to affect structural genes for enzymes involved in starch biosynthesis

Mutation	Enzyme affected	Plant	Phenotype	Reference
waxy (wx)	GBSS	maize	no amylose	Shure *et al.* (1983)
waxy (wx)	GBSS	rice	no amylose	Sano (1984)
amylose free (amf)	GBSS	potato	no amylose	Hovenkamp-Hermelink *et al.* (1987)
low amylose (lam)	GBSS I	pea	low amylose	Denyer *et al.* (1995)
amylose extender (ae)	BE IIb	maize	high amylose	Fisher *et al.* (1996)
amylose extender (ae)	BE III	rice	high amylose	Mizuno *et al.* (1993)
brittle-2 (bt-2)	small subunit AGPase	maize	low starch	Giroux *et al.* (1994)
shrunken-2 (sh-2)	large subunit AGPase	maize	low starch	Giroux *et al.* (1994)
rugosus (r)	BE I	pea	low starch	Bhattacharya *et al.* (1990)
rugosus (rb)	small subunit AGPase	pea	low starch	Smith *et al.* (1988)
rugosus-5 (rug-5)	GBSS II	pea	altered starch granule shape	Wang *et al.* (1997)
sugary-1 (su-1)	isoamylase (debranching enzyme)	maize	high phytoglycogen	James *et al.* (1995)

in the oat and wheat genomes (Harrington *et al.*, 1997). The homeology of such genes within groups such as cereals has the potential to be utilized in breeding programs.

Targets for genetic manipulation

1. One of the major reasons for genetically manipulating starch biosynthesis enzymes is to increase our understanding of how the enzymes interact to produce starch. For many of the more targeted experiments designed specifically to construct an altered starch with a view to commercial exploitation, knowledge is often also gained regarding the mode of action of an enzyme, or its interaction with other enzymes in the pathway. Hence, this particular area, to some extent, is mixed in with experiments described in other sections below. One of the earliest starch gene antisense experiments was the antisensing of the potato small subunit AGPase gene. This resulted in a tuber with reduced dry weight, high sucrose (up to 30%) and reduced storage protein, thus demonstrating the crucial role of AGPase in overall starch biosynthesis (Müller-Röber *et al.*, 1992). In some cases, it is possible to understand the role of an enzyme because a null mutant of the enzyme has been isolated, e.g. the *waxy* mutant of maize has been demonstrated to have no, or little amylose, as this mutant is null for the GBSS. However, this is not always the case and some enzymes, such as the SSSs and some BEs have no known null mutants. An example of how transgenic plants can be used to study genes for which there are no known phenotypic mutations is in the antisensing of BE in an amylose free potato by Flipse *et al.* (1996d). This experiment would have been expected to result in the originally amylose free (*amf*) potato starch, which normally stains red with iodine, regaining at least some amylose, due to lack of branching resulting in longer glucose chains. In fact, although some of the smaller starch granules did stain blue at the core, no discernible change was found in the branching or starch levels of these potatoes compared with the controls. More interestingly, changes were found in the physico-chemical properties of this

starch, specifically in the rheological properties, which are known to be affected by amylose:amylopectin ratios, suggesting that suppression of the BE had resulted in subtle changes to the amylose levels. Despite the fact that only one gene had been isolated for BE in potato at the time, this evidence suggested that, in common with other plants, potato does indeed contain a second BE, which is responsible for the branching levels seen in the starch of these potatoes. This would appear to be confirmed by the work of Larsson *et al.* (1996) who detected a second BE isoform in potato starch, although the actual gene has still to be isolated. Another approach that has been taken is that of expressing plant starch biosynthesis enzymes in *E. coli* and observing how these enzymes affect the glycogen synthesis in this bacterium. *E. coli* itself has a well characterized glycogen synthesis pathway, and strains mutant for the three major genes of the pathway, the AGPase, glycogen synthase and the branching enzyme are all available (Damotte *et al.*, 1968; Cattanéo *et al.*, 1969). Expressing plant enzymes such as potato AGPase in *E. coli* gives us the opportunity to observe aspects of enzyme activity in a simple system which also enables large quantities of the enzyme to be cultivated for purification relatively rapidly. This system also enables assessment of enzyme activity away from isoforms and other enzymes, which may co-purify and affect results. Such a system has been used to great effect to study the allosteric regulation of potato AGPase subunits by Okita and others, who have expressed both the large and small subunits of this enzyme in *E. coli* to obtain a functional enzyme (Iglesias *et al.*, 1993). By first mutagenising potato large subunit cDNA and then expressing the potato large and small subunit genes in a *glgC⁻* strain, it was possible to identify mutants with impaired glycogen production. One particular mutant was identified as requiring higher levels of 3-phosphoglycerate for maximum activity and on sequencing, a single nucleotide change of Pro-52 to Leu was found (Greene *et al.*, 1996a). The same technique was used to identify Asp-413 as a key requirement for 3-phosphoglycerate binding in allosteric regulation of this enzyme as mutation of this residue to alanine resulted in almost complete abolition of activation by 3-phosphoglycerate (Greene *et al.*, 1996b). Similar systems have also been set up expressing maize BEI and II in *E. coli* (Guan *et al.*, 1994a,b; Guan *et al.*, 1995). The expression of both of these branching enzymes in isolation has enabled the substrate interactions for these enzymes to be studied in more detail. This has demonstrated that the purified enzymes do have specificity for different chain lengths, with BEI predominantly longer chains than BEII, as had been predicted from mutation studies (Guan *et al.*, 1997). This heterologous expression system has also enabled studies of predicted active sites, deduced from homologies between enzymes of different species, to be examined by deleting well conserved amino acids, and then assessing the activity of the enzyme. This demonstrated the importance of Asp-386, Glu-441 and Asp-509 residues, all of which were found to be essential for enzyme activity (Kuriki *et al.*, 1996). More interesting, from the point of manipulating starch structure, is the construction of BEI/BEII hybrid enzymes. While most chimeric molecules produced by this method resulted in loss of enzyme activity generally, one hybrid enzyme, which consisted of predominantly BEII, with only the carboxy-terminal part of BEI, showed a high activity with substrate affinity similar to that of BEI. This demonstrates the importance of this region of the protein in substrate specificity and hence chain length in amylopectin (Kuriki *et al.,* 1997). In the case of both BE and AGPase, it is interesting to note that the plant enzymes will complement the equivalent *E. coli* mutations, despite the fact that *E. coli* synthesizes glycogen and not starch, indicating the strong similarities between starch and glycogen biosynthesis. A similar model system for examining starch biosynthesis enzymes at the more basic level is that of *Chlamydomonas reinhardtii,* a single celled alga that can both photosynthesize and synthesize starch. The starch of this alga has been shown to have A type crystallinity, the same as that found in maize, and studies have shown the two starches to be indistinguishable, making this an ideal model organism for starch biosynthesis studies (Buléon *et al.,* 1997). Using X-ray mutagenesis, it has been possible to create a large number of starch mutations that could be identified by staining with iodine vapours. Iodine naturally stains starch black, but mutants of *Chlamydomonas* could be identified easily as they stained either yellow, red, or olive, depending on the type of mutant starch present (Ball *et al.,* 1991; Fontaine *et al.,* 1993). Using this series of mutants, the effect of mutations on genes for which no known phenotypes exist in plants have been studied and alleles corresponding to most of the major plant starch synthesis enzymes have been identified. The *st-3* mutation, which lacks one isoform of soluble starch synthase (SSII), leads to an apparent increase in amylose. It also causes an increase in the number of short chain (approx. 6 residues) and decrease in the number of intermediate chain glucans (10–40 residues) in the amylopectin (Fontaine *et al.,* 1993). The *st-2* mutants deficient in GBSS, as well as being deficient in amylose, also had altered amylopectin (Maddelein *et al.,* 1994) with a decrease in extra long (greater than 50 residues) chains. These studies have led to the conclusion that each starch synthase interacts at a different level with each branching enzyme to create a varying degree of branching (Buléon *et al.,* 1997). The role of a starch synthase in the determination of amylopectin chain length would appear to correlate with the studies on potato, where expression of *E. coli* glycogen synthase resulted in the production of a highly branched amylopectin, not dissimilar to glycogen (Shewmaker *et al.,* 1994). Studies on the *st-7* mutant in *Chlamydomonas,* which lacks an isoamylase branching enzyme, have led to the current theory that amylopectin is formed from a preamylopectin precursor by processing by debranching enzyme (Mouille *et al.,* 1996). Another interesting finding that arose from this work is the discovery that *Chlamydomonas* starch granules do not contain the characteristic ring structure associated with day/night cycles in plants. It has been speculated that this is due to the growth of this organism under controlled conditions with constant temperature and light, and that such a system would be useful in the study of the growth and pattern of these rings (Buléon *et al.,* 1997). Such work is invaluable in laying groundwork on which theories can be based regarding how various plant enzymes may be manipulated in a controlled fashion to obtain altered starches. It remains to be seen how far such studies can be extrapolated to higher plants.

2.   One of the major goals for manipulation of starch for industrial purposes is that of increasing the amount of starch synthesized in the storage organ. As the rate-limiting enzyme for starch synthesis is AGPase, this has obviously been the main target enzyme in these studies. Two groups have taken the route of over-

expressing the AGPase gene in potato (Stark *et al.*, 1992; Sweetlove *et al.*, 1996a,b). Due to the problems of sense suppression, allied with the fact that both phosphate and 3-phosphoglycerate allosterically regulate AGPase, both groups took the approach of using the *E. coli glg*C16 mutant. This has the double advantages of a) having very little homology with the plant's genes and so being unlikely to cause sense suppression, and b) the gene has a substitution of aspartic acid for glycine at position 336, which makes it resistant to allosteric inhibition (Leung *et al.*, 1986; Lee *et al.*, 1987). In both cases as well, the gene was expressed under the control of the patatin promoter and targeted to the amyloplast using the transit peptide sequence from a small subunit ribulose-bisphosphate carboxylase. Both groups claim to have increased the activity of AGPase in the transgenic potato tubers with Stark *et al.* (1992) claiming to have observed nearly 60% more starch than in controls in some experimental tubers. In comparison, Sweetlove *et al.* (1996a,b) found that while there was increased flux into starch metabolism, this was accompanied by an increased turnover rate, resulting in no overall increase in starch content. Stark *et al.* (1992) also expressed the *glg*C16 gene under the control of the 35s promoter of cauliflower mosaic virus, a highly expressing constitutive promoter. These transformants did not result in any viable transgenic plants, but the callus produced had considerably increased starch, accompanied by an increased number of starch granules, as observed under the light microscope. Another group has taken a more traditional approach to manipulation of this enzyme for increased starch content in maize. Giroux *et al.* (1996) exploited the maize *Dissociation* (*Ds*) system to target the maize gene for the large subunit of AGPase and create small insertion mutants within the gene. The *shrunken2-m1* mutation selected has a Ds transposable element already present within exon 16, an area known to be involved in allosteric regulation of this gene by phosphate. Revertants of this type of mutation are created by the excision of the *Ds* element, recreating the complete gene. However, inaccurate excision of the element can result in small in frame insertions in the gene creating an additional 1 or 2 amino acids in the protein sequence. Secondary mutants were created by reverting the *sh2-m1* mutant and sequencing the subsequent partial revertants to determine whether reversion had resulted in the production of in frame insertions, that is, insertions of 3 or 6bp that do not cause frameshift mutations, which inactivate the gene. One revertant in particular that had a 6bp insertion leading to an extra serine and tyrosine residue was noted to have an increase in seed weight of 11–18% over wild type, and this could in part be explained by an increase in the amount of starch. The authors concluded that seed weight increased as a result of increase of several seed components, not only starch, that the mutation resulted in a stronger sink, which, in turn, led to an increase in synthesis of other seed components.

3. Alteration of the amylose to amylopectin ratio has been a focus of much research, as starches with altered ratios have been shown to have many uses. This is partly because of waxy mutants, which produce starches with little or no amylose. These starches are used in both the food industry, where waxy starches have lower pasting temperatures and greater paste clarity than normal starches (Jane, 1997) and in non-food uses for, among other things, the manufacture of plastics (van Soest *et al.*, 1997), and paints (Taylor, 1997). Most waxy starches are created by

mutations in the GBSS gene, which have been selectively bred for in maize, and hence have been available for a number of years. Study of these mutants led to the initial isolation of the maize GBSS gene (Shure *et al.*, 1983; Klösgen *et al.*, 1986), which, in turn has led to the isolation of the equivalent gene in rice (Okagaki and Wessler, 1988) and potato (Visser *et al.*, 1989). This has resulted in a large number of antisense studies using the GBSS genes of several species to suppress GBSS activity and create low amylose and amylose free mutants (Visser *et al.*, 1991; Shimada *et al.*, 1993). In addition, the regulatory regions of these genes are being studied to gain an understanding of the pattern of gene expression of starch biosynthesis genes (e.g. Hirano *et al.*, 1995; Li *et al.*, 1995). Antisense experiments, especially in potato have enabled an increase in understanding of how amylose is laid down in the formation of the starch granule (Kuipers *et al.*, 1994), and also how dosage effect and amylose content are linked (Flipse *et al.*, 1996a). The advantage of using genetic manipulation for these studies is that it enables the effect of the gene suppression to be studied in a known genetic background, but eliminating the need for a lengthy breeding program with a large number of backcrosses. One crop that has no known naturally occurring phenotypic waxy mutant is wheat, where there is a strong interest in developing a low amylose variety. This would result in flour that would give extended shelf life for bread products. Waxy starches have also been shown to be superior for oriental noodles (Graybosch *et al.*, 1997). The hexaploid nature of wheat means that naturally occurring phenotypic null waxy mutants do not exist and efforts to develop waxy wheat starch have been laborious. Several groups have now developed waxy wheat lines utilizing breeding of partial waxy mutants (varieties mutant in one of the *waxy* alleles) (Nakamura *et al.*, 1995; Hoshino *et al.*, 1996) or random mutagenesis (Yasui *et al.*, 1997) and studies into the effects of the properties of the starches from these mutants are now in progress (Yasui *et al.*, 1996; Hayakawa *et al.*, 1997). The availability of a GBSS cDNA from wheat (Clark *et al.*, 1991) now means that it is theoretically possible to create a waxy wheat variety using antisense technology. This has the advantage that it would enable the waxy property to be inserted, in one generation, into a variety agronomically suitable for the required environment. Obtaining plants with a high amylose content is another goal, as high amylose starch has film formation properties particularly desirable for food uses such as fried food coating batters (Jane, 1997). High amylose starches in plastics also retain crystallinity during storage at high humidity in comparison to high amylopectin starches (van Soest *et al.*, 1997). Studies involving the expression of the GBSS gene in a normal amylose variety of rice resulted in sense suppression, suggesting that raising of amylose levels by the over-expression of the GBSS gene would not be possible due to natural feedback mechanisms (Itoh *et al.*, 1997). Maize in particular is a good source of high amylose starch, as the *ae* varieties, which are mutant in the BEII gene, have much reduced branching and result in a high apparent amylose content. No such mutants have been found in potato, although, as mentioned earlier, a BE gene in potato has been antisensed, producing an altered starch, but not one with any obvious increase in amylose (Flipse *et al.*, 1996d). Other high amylose mutants include the *ae* mutation in rice, which is thought to correspond to the BEIII gene (Mizuno *et al.*, 1993) and the original *r* mutant of pea, which encodes BEI (Bhattacharya *et al.*, 1990). It is possible that plant varieties with higher amylose levels could be engineered by suppressing more than one of the multiple BE genes within a system. Before the appropriate enzymes can be targeted, however, it is important to gain an understanding of the role of the various isoforms in the branching patterns of amylopectin. High amylose has also been found in the *sugary* mutants of maize (Shannon and Garwood, 1984), although the *sugary* mutants of rice actually have low or no detectable amylose (Kaushik and Khush, 1991). The *su-2* locus in maize is thought to encode a gene involved in gene regulation during endosperm biosynthesis (Nelson and Pan, 1995), as are the *sugary* genes from rice. This may explain the phenotypic differences between the two sets of mutants, as different regulatory genes may regulate different subsets of starch biosynthesis genes. An understanding of regulatory genes for starch biosynthesis and endosperm development may also lead to the development of varieties with altered starch. For example, while it may not be possible to over-express genes using transgenic technology directly (as studies on AGPase mentioned earlier have shown) manipulation of regulatory regions of genes may allow the normal feedback mechanisms to be overridden. Manipulation of regulatory genes may also allow for more than one gene to be influenced simultaneously, similar to the pleiotropic effects seen in the *sugary* mutants of maize and rice. For example, while it may be difficult to switch off all known branching enzymes, it may be possible to reduce branching extremely efficiently by antisensing a key regulatory gene. Drawbacks to this sort of technique at the moment include our lack of understanding of many of these genes and the fact that many of those that have been studied do seem to regulate quite a large number of quite different genes. For example, the *Flo-2* locus of rice is thought to control expression of BEs, DBE GBSS and AGPase (Kawasaki *et al.*, 1996).

4. A continuation of the concept of alteration of amylose:amylopectin ratio is to manipulate the levels of branching within the amylopectin, without actually altering its percentage of the total starch. This is particularly important as the level of branching and the length of amylopectin chains are both thought to be key components in determining physical properties of the starch. For example, an increase in the shorter B chains in starches would increase resistance to shear thinning and increase the viscosity of starch pastes, which would prevent the need for chemical crosslinking. Alternatively, shortening of A chains could be used to convert an enzyme resistant starch such as banana to a more easily digestible form. Altered branching properties would also affect the quality of plastics produced from starch. Many of the experiments into manipulating branching levels of starch have involved the over-expression of heterologous enzymes from *E. coli* and resulted in a highly branched, unstructured amylopectin, similar in structure to glycogen. While this would be expected from the expression of *E. coli* glycogen branching enzyme, which has been expressed in amylose free potato (Kortstee *et al.*, 1996), more surprising is that expression of the *E. coli* glycogen synthase enzyme also resulted in production of a highly branched starch (Shewmaker *et al.*, 1994). This is only one of many experiments that suggest that our current theories as to the role of various enzymes in starch structure are too simplified and that the reality involves complex interactions between the various synthases and branching enzymes. This is confirmed by reciprocal experiments

where maize BEs I and II have been used to complement branching enzyme mutations in *E. coli*. Despite the enzyme activity being shown to be similar to that of maize purified enzyme, the resultant polysaccharide has been shown to be highly branched and more analogous to glycogen than starch. Despite this, the two different enzymes did produce slightly different chain lengths, with BEI giving and average chain length of 14, while BEII gave an average chain length of 16 (Guan *et al.,* 1995). This is in comparison to an average chain length of 22–24 in normal maize amylopectin, and 10 for *E. coli*. Surprisingly, the result of expressing both BEs together resulted in an average chain length of 12 (Guan *et al.,* 1995). The role of the various starch synthases in controlling branching in *Chlamydomonas* has recently been studied by Buléon *et al.* (1997). This group used mutants for GBSS, SSSI or SSSII and compared the starch structure with that of the wild type. What they found was that deficiency in any one of the SSs resulted in a change in the amylopectin structure, and that each SS appeared to be involved with branching at a different level. Deficiency of GBSS resulted in not only loss of amylose, but loss of a high molecular weight portion of the amylopectin (Delrue *et al.*, 1992). Deficiency in SSSII was found to show a decrease in intermediate sized glucans, causing a B type diffraction pattern and a general loss of crystallinity (Fontaine *et al.,* 1993). Mutants deficient in both GBSS and SSSII, and hence completely dependant on SSSI for starch synthesis, were found to have only a very small amount of granular starch that was highly branched, similar to phytoglycogen (Maddelein *et al.,* 1994). Thus, these authors concluded that all the SS enzymes have an individual role in determining the structure of amylopectin either by synthesizing particular length chains, or by interaction with particular BEs. The problem with such experiments so far, is that while giving some clues into the complex nature of the regulation of branching length and number, they have not enabled any specific manipulation of branch length or number of the type that would be needed for industrial uses such as plastic manufacture.

5. Starch granule size is one of the main variables between starches of different botanical origin, and a key factor in choice of starch type for particular applications. Maximum viscosity of starch solutions is thought to be reached at maximum granule swelling during simultaneous heating and shearing, thus viscosity will be related to both size of the granule and accessibility of water to the crystalline structure. This in turn affects factors such as thickening properties and retrogradation after subsequent cooling and heating, particularly relevant with the increase in food products such as frozen meals (Lillford and Morrison, 1997). Starch in its granular form is used in applications such as dusting agents for confectionery and face powders. Applications where granule size is a key factor in choice of botanical origin include carbonless copy paper, which uses the large, smooth granules of wheat starch (Nachtergaele and Van Nuffel, 1989), and degradable plastic film, where granule size affects tensile strength and film thickness (Lim *et al.*, 1992). There is also interest in developing the use of small starch granules (e.g. rice, cassava and waxy maize) as fat mimetics, as a starch granule with a diameter of 2 microns or less will give products a similar mouthfeel and taste to fat (Daniel and Whistler, 1990). Different granule types also exhibit different digestibilities depending on their crystallinity types, with potato and banana starches (B and C type crystallinity) having poor digestibility (Jane, 1997).

Despite the importance of the starch granule size and structure with respect to functionality, very little is understood about the factors influencing this. Some mutants, notably the *sugary* (*su1* and 2) mutants of maize and the rugosus (*r, rug* and *rb*) mutants of pea (Lloyd *et al.*, 1996), are known to reduce granule size. However, in all cases, this seems to be a by product of a reduction in the total amount of starch produced. It is likely that more subtle changes would need to be studied to establish genetic factors involved in starch granule morphology, as the difference in starch granule size in varieties of maize are strongly indicative of alleles that play a role in this (Campbell *et al.*, 1996), although it is highly likely that environment is also important. The factors behind the bimodal distribution of starch granule size in cereals such as wheat and barley are another area where much research is needed as it has been shown for barley that not only do the size of the A and B granules vary with variety, but the ratio of B to A granules can vary from 9.6 to 16.8 (Oliveira *et al.*, 1994). Granule type ratio would be an important factor to control as, for example, a wheat that produced only A granules would be useful for production of carbonless paper as the A granules would not have to be separated from the unwanted B granules. The expression of the *E. coli glg*B gene for glycogen synthase mentioned earlier (Shewmaker *et al.*, 1994) as well as resulting in a starch with a much more branched structure, was noticed to have resulted in a marked increase (up to 80%) in small, round starch granules of less than 10 μm, similar to cereal starches. As this was linked with an increase in amylopectin content, it was speculated by the authors that granule size and amylopectin levels may in some ways be linked. However, the presence of a high amount of sucrose in these tubers may also suggest that this effect was, similar to the pea mutants of Lloyd *et al.* (1996), merely due to limited overall starch biosynthesis.

6.   Other constituents such as lipid, protein and phosphate are also part of the native starch structure within the plant, and these are also factors that can influence the properties of the starch. For example, reduction in lipid, especially phospholipid can result in increased clarity of starch pastes, one reason why waxy starches are often used, as they have particularly low levels of phospholipid. Lipid interactions with amylose are also thought to be a factor in determining physical properties such as swelling power and viscosity (Visser and Jacobsen, 1993). Protein content, especially around the starch granule, is thought to play a role in controlling granule swelling during processing (Lillford and Morrison, 1997) and control of this, once it is better understood, may also result in starches with better processing qualities. Protein content is also thought to play a role in starch flavour in food applications, and for non-food uses, protein removal is desirable, if somewhat difficult in the case of small granule sized starches. A large proportion of these proteins in cereals are storage proteins such as glutelins and gliadins for which genes have been cloned. While some antisensing of such genes has taken place in rice, to reduce allergic reactions caused by these proteins, this did not have any effect on the overall amount of protein produced in the seed (Nakamura and Matsuda, 1996; Tada *et al.*, 1996). This suggests that a large number of genes would have to be antisensed before significant protein reduction could be obtained. As large gene families encode most seed storage proteins, antisensing all these genes would be so complicated as to be impractical (Nakase *et al.*, 1996). Some

mutants have been shown to be deficient in a group of proteins, e.g. the *Opaque-2* gene of maize encodes a transcriptional activator and mutants in this gene have been shown to be deficient in certain zein storage proteins (Hartings *et al.*, 1989). However, such mutants also tend to have altered starch, so at present it seems unlikely that protein content could be reduced by any manipulation of regulatory proteins until the mode of action and specificity of these proteins is understood. Other factors are often produced by chemical modifications of processed starch, but have the potential to become 'value added' properties of native starch with such starches having the advantage of being more environmentally friendly to produce. Processing of starch under harsh conditions such as low pH and high temperature means that starch is often crosslinked prior to processing to prevent loss of viscosity. Starch that has been genetically modified so that it does not require chemical crosslinking prior to processing would be valuable, not only to remove the need for the chemical process, but because only some of the available methods for crosslinking starch chemically have been approved for food use, and use of, for example, crosslinked hydroxypropylate starch for frozen food applications is becoming desirable (Jane, 1997; Champagne, 1996). Phosphorylation is a modification known to have an affect on rheological properties of starch, and phosphorylation of potato starch is used to control function and appearance for food uses such as salad dressings (Poulsen *et al.,* 1997). As a certain amount of phosphorylation of amylopectin occurs naturally, and amounts vary according to potato type, it does not seem inconceivable that higher levels of natural phosphorylation could be achieved either through selective breeding for natural mutants, or by genetic modification, if the enzyme responsible for phosphorylation could be identified. Shewmaker *et al.* (1994), in the previously discussed expression of *E. coli glg*B in potato, noted that, in addition to changes in the starch granules and fine structure, overall phosphate content was significantly lower than in normal potato starch. One possibility is that, as phosphate tends to be associated with longer chains, this reduction was simply a result of the shorter chain structure of this starch. However, the authors comment that phosphate incorporation is poorly understood, and other factors, linked to the normal starch synthases in potato may be important.

7. There is also the possibility that the addition of heterologous enzymes into the plant genome could be used to make specialist carbohydrate products utilizing starch as a substrate. Cyclodextrins are cyclic oligosaccharides of 6,7 or 8 glucose molecules which can be synthesized by bacterial cyclodextrin glycosyltransferases (CGT), using starch as a substrate. Cyclodextrins are much in demand as pharmaceutical delivery systems and flavour and odour enhancement and removal systems as they have an apolar cavity which enables them to form a complex with a hydrophobic 'guest' molecule. Such complexes enable the guest molecule to have increased solubility and prolonged stability. Currently, cyclodextrins are synthesized in batch fermentors using hydrolyzed starch as a substrate, but this method is expensive for large scale use. Synthesis of cyclodextrins in plants has potential in that the one enzyme, CGT, required for conversion of starch to cyclodextrin is available. Oakes *et al.* (1991) used a CGT cloned from *Klebsiella pneumoniae* under the control of the patatin promoter and transformed this into potatoes. Although the amount of mRNA produced in the resultant transformants

was low (estimated at 0.00001–0.0001% of total mRNA) both 6 ring, and in some cases 7 ring cyclodextrins were detected in the transformed tubers at levels corresponding to conversion of 0.001–0.01% of starch. Although such levels are low, expression has the potential to be increased by uses of other bacterial CGT genes, promoters and alteration of factors such as codon usage, which may improve expression of a bacterial enzyme in plants. Understanding the factors which determine specificity of hydrolysis and synthesis of glycosidic bonds in the synthesis of carbohydrate generally in plants could also be the basis of production 'designer' carbohydrates using plants as molecular factories.

Potential new targets for starch manipulation

Future possibilities for manipulation of starch, some of which have already been considered above, are numerous in the long term, but many depend on further research into the mechanisms of starch biosynthesis, and the physical properties of altered starch structures. One area, which has been studied extensively at the physical level, is the structure of the starch granule. Current theories are based on the blocklet structure, which explains both molecular and electron microscope observations (*Figure 1*). While the dimensions of the amorphous and crystalline components of a single blocklet can vary with factors such as amylose content, the length of a single blocklet has been found to be constant, irrespective of botanical origin and mutant phenotypes, at 9 nm. Alteration of some aspect of starch metabolism that would alter this parameter would give an interesting insight into the more constant aspects of starch metabolism, as well as possibly conferring novel properties on such a starch granule. Another poorly understood area is the actual biogenesis of the starch granule. While many plants (e.g. potato) contain only one starch granule in a single amyloplast, others such as rice have amyloplasts with many starch granules. There is obvious similarity between the amyloplast and the chloroplast, and gene targeting sequences seem to work equally well for either organelle, with the chloroplast ribulose 1,5-bisphosphate carboxylase targeting sequence from *Arabidopsis* being able to target the potato amyloplast (Stark *et al.*, 1992) and the maize GBSS targeting sequence also mediating protein transport, at least *in vitro*, to the chloroplast (Klösgen *et al.*, 1989). This is not surprising, in that chloroplasts contain starch granules; however, this leaves the questions as to what the main signal initiating storage starch granule biosynthesis might be. Identification of a starch priming molecule, analogous to glycogenin in mammalian systems, could give some major ideas as to the biogenesis of the starch granule, and manipulation of the levels and sequence of such a molecule may enable control of either the number or size of starch granules in a plant. This would be particularly useful given the current use for small starch granules in both food and non-food applications, as then granule size in a high starch yielding crop such as potato could be controlled. Studies on the current candidate, amylogenin, which at least appears to perform the biochemical reaction required of a starch primer, may improve our knowledge in this area. As both wheat and barley contain starch granules of two sizes, understanding of ways in which one or other type could be selected in these crops would be extremely useful, as wheat A type granules are particularly good quality for the manufacture of carbonless copy paper, while the B type granules have potential to be used as fat mimetics. At the moment, the granule types have to be separated post harvest, which adds cost to the product, or processed mechanically to achieve a smaller granule size (Jane *et al.*, 1992).

Further understanding of the fine structure of branching levels and how these may be manipulated accurately is also required, as current manipulations in this area, while being interesting, have mainly resulted in loss of the regular structure of starch and production of phytoglycogen. One way in which some understanding may be achieved is by comparison of the branching levels of different starches and trying to relate these to differences in enzymes involved in the synthesis of these starches. One possibility would be to express enzymes from one botanical origin in an alternative host, and see whether this affects starch structure at the molecular level. Given the current theories regarding regulation of branching being controlled by DBE (Ball *et al.,* 1996), this enzyme would be a likely candidate for such studies. Further study is also needed into the functionality of such starches regarding how number and length of branches correlate to factors such as glass transition temperature and gelling and pasting qualities in engineered starches. It has also been proposed (Lillford and Morrison, 1997) that as lipid complexed within amylose is known to play a role in the reduction of starch retrogradation, that there is a possibility for manipulation of lipid levels to reduce retrogradation in high amylose starches. However, until the manner of starch/lipid interaction is better understood, this remains a goal for the future.

Given the variation in climate across the world, there is also a need to understand the mechanisms of starch biosynthesis. This would be useful not only in the major crops such as maize, potato and rice, but also in wheat, which is currently lagging behind its other cereal counterparts, mainly due to the hexaploid nature of its genetics, and barley, which has the advantage of being a particularly robust crop in colder climates such as Scandinavia. In addition to being able to manipulate crops suited to particular climates, starch structure is known to vary within a particular geographical area depending on climatic conditions within a growing season. This is particularly well documented for amylose content in rice, where one allele of the GBSS gene in particular has been shown to be particularly susceptible to temperature (Sano *et al.*, 1985). This seasonal variation of crop quality can interfere with product quality, for example, the processing of rice to produce crisped rice breakfast cereal is dependent on an ideal level of amylose in the rice to produce a product of optimum quality. Studies into the understanding of the causes of such seasonal variation, and how this may be overcome are an area where research is needed and would be useful not only in maintaining crop and product quality, but would also provide useful information regarding starch biosynthesis of gene regulation, and possibly more widespread information regarding plant gene expression.

Conclusion

In conclusion, in recent years there have been huge advances in our knowledge regarding the fine structure of starch, the enzymes involved in the biosynthesis of this fine structure and how this relates to product quality. There are still many gaps in this knowledge relating to how minor required changes may be controlled to produce starches for specific industrial uses, and also how gross changes in starch granule morphology may be effected, but our knowledge is expanding constantly. There are also more practical hurdles to be overcome before genetically engineered starches become common in the market place, not least being acceptance of the consumer, who in Europe, at least, is being slow to accept the products of plant genetic engineering on the supermarket shelves. Such problems have not however, occurred with starches from plants obtained through selective breeding of mutants. There is also the question of such plants (both mutants and transformed) producing high enough yields of required starches to make farmers consider their use economic. It is likely, though, that the needs of the marketplace will drive research into these areas forward and it will not be long before the first fruits of molecular genetic engineering are in the market place.

References

ANDERSON, J.M., HNILO, J., LARSON, R., OKITA, T.W., MORELL, M. AND PREISS, J. (1989). The encoded primary sequence of a rice seed ADP-glucose pyrophosphorylase subunit and its homology to the bacterial enzyme. *Journal of Biological Chemistry* **264**, 12238–12242.

AYRES, N.M., MCCLUNG, A.M., LARKIN, P.D., BLIGH, H.F.J., JONES, C.A. AND PARK, W.D. (1997). Microsatellites and a single-nucleotide polymorphism differentiate apparent amylose classes in an extended pedigree of US rice germ plasm. *Theoretical and Applied Genetics* **94**, 773–781.

BABA, T., NISHIHARA, M., MIZUNO, K., KAWASAKI, T., SHIMADA, H., KOBAYASHI, E., OHNISHI, S., TANAKA, K. AND ARAI, Y. (1993). Identification, cDNA cloning, and gene-expression of soluble starch synthase in rice (*Oryza-sativa* L) immature seeds. *Plant Physiology* **103**, 565–573.

BALL, S., GUAN, H.P., JAMES, M., MYERS, A., KEELING, P., MOUILLE, G., BULÉON, A., COLONNA, P. AND PREISS, J. (1996). From glycogen to amylopectin – A model for the biogenesis of the plant starch granule. *Cell* **86**, 349–352.

BALL, S., MARIANNE, T., DIRICK, L., FRESNOY, M., DELRUE, B. AND DECQ, A. (1991). A *Chlamydomonas-reinhardtii* low-starch mutant is defective for 3-phosphoglycerate activation and orthophosphate inhibition of ADP-glucose pyrophosphorylase. *Planta* **185**, 17–26.

BHATTACHARYYA, M.K., SMITH, A.M., ELLIS, T.H.N., HEDLEY, C. AND MARTIN, C. (1990). The wrinkled-seed character of pea described by Mendel is caused by a transposon-like insertion in a gene encoding starch-branching enzyme. *Cell* **60**, 115–122.

BHAVE, M.R., LAWRENCE, S., BARTON, C. AND HANNAH, L.C. (1990). Identification and molecular characterization of *shrunken-2* cDNA clones of maize. *Plant Cell* **2**, 581–588.

BLIGH, H.F.J., TILL, R.I. AND JONES, C.A. (1995). A microsatellite sequence closely linked to the *Waxy* gene of *Oryza-sativa*. *Euphytica* **86**, 83–85.

BOYER, C.D., SHANNON, J.C. AND GARWOOD, D.L. (1976). The interaction of the *amylose-extender* and *waxy* mutants of maize (*Zea mays* L.). Fine structure of *amylose-extender waxy* starch. *Starch/Starke* **28**, 405–410.

BULÉON, A., GALLANT, D.J., BOUCHET, B., MOUILLE, C., DHULST, C., KOSSMANN, J. AND BALL, S. (1997). Starches from A to C – *Chlamydomonas reinhardtii* as a model microbial system to investigate the biosynthesis of the plant amylopectin crystal. *Plant Physiology* **115**, 949–957.

BURTON, R.A., BEWLEY, J.D., SMITH, A.M., BHATTACHARYYA, M.K., TATGE, H., RING, S., BULL, V., HAMILTON, W.D.O. AND MARTIN, C. (1995). Starch branching enzymes belonging to distinct enzyme families are differentially expressed during pea embryo development. *Plant Journal* **7**, 3–15.

CAMPBELL, M.R. AND GLOVER, D.V. (1996). Interaction of two *sugary-1* alleles (*su1* and *su1(st)*) with *sugary-2* (*su2*) on characteristics of maize starch. *Starch-Starke* **48**, 391–395.

CAMPBELL, M.R., LI, J., BERKE, T.G. AND GLOVER, D.V. (1996). Variation of starch granule size in tropical maize germ plasm. *Cereal Chemistry* **73**, 536–538.

CATTANÉO, J., DAMOTTE, M., SIGAL, N., SANCHEZ-MEDINA, F. AND PUIG, J. (1969). Genetic studies of *Escherichia coli* K12 mutants with alterations in glycogenesis and properties of an altered adenosine diphosphate glucose pyrophosphorylase. *Biochemical and Biophysical Research Communications* **34**, 694–701.

CHAMPAGNE, E.T. (1996). Rice starch composition and characteristics. *Cereal Foods World* **41**, 833–838.

CLARK, J.R., ROBERTSON, M. AND AINSWORTH, C.C. (1991). Nucleotide-sequence of a wheat (*Triticum-aestivum* L) cDNA clone encoding the waxy protein. *Plant Molecular Biology* **16**, 1099–1101.

CROY, R.R.D. (1993). *Plant Molecular Biology Labfax*. Oxford, UK.: Blackwell Scientific Publications.

DAMOTTE, M., CATTANÉO, J., SIGAL, N. AND PUIG, J. (1968). Mutants of *Escherichia coli* K12 altered in their ability to store glycogen. *Biochemical and Biophysical Research Communications* **32**, 916–920.

DANIEL, J.R. AND WHISTLER, R.L. (1990). Fatty sensory qualities of polysaccharides. *Cereal Foods World* **35**, 825.

DELRUE, B., FONTAINE, T., ROUTIER, F., DECQ, A., WIERUSZESKI, J.M., VANDENKOORNHUYSE, N., MADDELEIN, M.L., FOURNET, B. AND BALL, S. (1992). Waxy *Chlamydomonas-reinhardtii* – Monocellular algal mutants defective in amylose biosynthesis and granule-bound starch synthase activity accumulate a structurally modified amylopectin. *Journal of Bacteriology* **174**, 3612–3620.

DENYER, K., BARBER, L.M., BURTON, R., HEDLEY, C.L., HYLTON, C.M., JOHNSON, S., JONES, D.A., MARSHALL, J., SMITH, A.M., TATGE, H., TOMLINSON, K. AND WANG, T.L. (1995). The isolation and characterization of novel low-amylose mutants of *Pisum-sativum* L. *Plant Cell and Environment* **18**, 1019–1026.

DHUGGA, K.S., TIWARI, S.C. AND RAY, P.M. (1997). A reversibly glycosylated polypeptide (RGP1) possibly involved in plant cell wall synthesis: Purification, gene cloning, and trans-Golgi localization. *Proceedings of the National Academy of Sciences of the United States of America* **94**, 7679–7684.

DRY, I., SMITH, A., EDWARDS, A., BHATTACHARYYA, M., DUNN, P. AND MARTIN, C. (1992). Characterization of cDNAs encoding 2 isoforms of granule-bound starch synthase which show differential expression in developing storage organs of pea and potato. *Plant Journal* **2**, 193–202.

FISHER, D.K., BOYER, C.D. AND HANNAH, L.C. (1993). Starch branching enzyme-II from maize endosperm. *Plant Physiology* **102**, 1045–1046.

FISHER, D.K., GAO, M., KIM, K.N., BOYER, C.D. AND GUILTINAN, M.J. (1996). Allelic analysis of the maize amylose-extender locus suggests that independent genes encode starch-branching enzymes IIa and IIb. *Plant Physiology* **110**, 611–619.

FISHER, D.K., KIM, K.N., GAO, M., BOYER, C.D. AND GUILTINAN, M.J. (1995). A cDNA-encoding starch branching enzyme-I from maize endosperm. *Plant Physiology* **108**, 1313–1314.

FLIPSE, E., KEETELS, C., JACOBSEN, E. AND VISSER, R.G.F. (1996a). The dosage effect of the wildtype GBSS allele is linear for GBSS activity but not for amylose content – Absence of amylose has a distinct influence on the physicochemical properties of starch. *Theoretical and Applied Genetics* **92**, 121–127.

FLIPSE, E., SCHIPPERS, M.G.M., JANSSEN, E.M., JACOBSEN, E. AND VISSER, R.G.F. (1996b). Expression of wild-type GBSS transgenes in the off-spring of partially and fully complemented amylose-free transformants of potato. *Molecular Breeding* **2**, 211–218.

FLIPSE, E., STRAATMANENGELEN, I., KUIPERS, A.G.J., JACOBSEN, E. AND VISSER, R.G.F. (1996c). GBSS T-DNA inserts giving partial complementation of the amylose-free potato mutant can also cause co-suppression of the endogenous GBSS gene in a wild-type background. *Plant Molecular Biology* **31**, 731–739.

FLIPSE, E., SUURS, L., KEETELS, C., KOSSMANN, J., JACOBSEN, E. AND VISSER, R.G.F. (1996d). Introduction of sense and antisense cDNA for branching enzyme in the amylose-free potato mutant leads to physicochemical changes in the starch. *Planta* **198**, 340–347.

FONTAINE, T., DHULST, C., MADDELEIN, M.L., ROUTIER, F., PEPIN, T.M., DECQ, A., WIERUSZESKI, J.M., DELRUE, B., VANDENKOORNHUYSE, N., BOSSU, J.P., FOURNET, B.

AND BALL, S. (1993). Toward an understanding of the biogenesis of the starch granule – Evidence that *Chlamydomonas* soluble starch synthase-II controls the synthesis of intermediate size glucans of amylopectin. *Journal of Biological Chemistry* **268**, 16223–16230.

FUTTERER, J. AND HOHN, T. (1996). Translation in plants – Rules and exceptions. *Plant Molecular Biology* **32**, 159–189.

FUWA, H., GLOVER, D.V., MIYAURA, K., INOUCHI, N., KONISHI, Y. AND SUGIMOTO, Y. (1987). Chain-length distribution of amylopectins of double-mutants and triple-mutants containing the *Waxy* gene in the inbred Oh43 maize background. *Starch-Starke* **39**, 295–298.

GALLANT, D.J., BOUCHET, B. AND BALDWIN, P.M. (1997). Microscopy of starch: Evidence of a new level of granule organization. *Carbohydrate Polymers* **32**, 177–191.

GALLANT, D.J., MERCIER, C. AND GUILBOT, A. (1972). Electron microscopy of starch granules modified by bacterial α-amylase. *Cereal Chemistry* **49** 354–365.

GIROUX, M., SMITHWHITE, B., GILMORE, V., HANNAH, L.C. AND PREISS, J. (1995). The large subunit of the embryo isoform of ADP glucose pyrophosphorylase from maize. *Plant Physiology* **108**, 1333–1334.

GIROUX, M.J. AND HANNAH, L.C. (1994). ADP-Glucose Pyrophosphorylase in *shrunken-2* and *brittle-2* mutants of maize. *Molecular & General Genetics* **243**, 400–408.

GIROUX, M.J., SHAW, J., BARRY, G., COBB, B.G., GREENE, T., OKITA, T. AND HANNAH, L.C. (1996). A single-gene mutation that increases maize seed weight. *Proceedings of the National Academy of Sciences of the United States of America* **93**, 5824–5829.

GRAYBOSCH, R.A., PETERSON, C.J. AND HANSEN, L.E. (1997). Occurence and expression of granule bound starch synthase mutants in hard winter wheat. In *Starch, Structure and Functionality*. Eds. P. J. Frazier, A. M. Donald and P. Richmond, pp 214–221. Cambridge, UK: Royal Society of Chemistry.

GREENE, T.W., CHANTLER, S.E., KAHN, M.L., BARRY, G.F., PREISS, J. AND OKITA, T.W. (1996). Mutagenesis of the potato ADP-glucose pyrophosphorylase and characterization of an allosteric mutant defective in 3-phosphoglycerate activation. *Proceedings of the National Academy of Sciences of the United States of America* **93**, 1509–1513.

GREENE, T.W., WOODBURY, R.L. AND OKITA, T.W. (1996). Aspartic-acid-413 is important for the normal allosteric functioning of ADP-glucose pyrophosphorylase. *Plant Physiology* **112**, 1315–1320.

GUAN, H.P., BABA, T. AND PREISS, J. (1994a). Expression of branching enzyme I of maize endosperm in *Escherichia coli*. *Plant Physiology* **104**, 1449–1453.

GUAN, H.P., BABA, T. AND PREISS, J. (1994b). Expression of branching enzyme-II of maize endosperm in *Escherichia-coli*. *Cellular and Molecular Biology* **40**, 981–988.

GUAN, H.P., KURIKI, T., SIVAK, M. AND PREISS, J. (1995). Maize branching enzyme catalyzes synthesis of glycogen-like polysaccharide in glgB-deficient *Escherichia-coli*. *Proceedings of the National Academy of Sciences of the United States of America* **92**, 964–967.

GUAN, H.P., LI, P., IMPARLRADOSEVICH, J., PREISS, J. AND KEELING, P. (1997). Comparing the properties of *Escherichia coli* branching enzyme and maize branching enzyme. *Archives of Biochemistry and Biophysics* **342**, 92–98.

HARRINGTON, S., BLIGH, H.F.J., PARK, W.D., JONES, C.A. AND MCCOUCH, S.R. Linkage mapping of starch branching enzyme III in rice (*Oryza sativa* L.) and prediction of location of orthologous genes in other grasses. *Theoretical and Applied Genetics* **94**, 564–568.

HARTINGS, H., MADDALONI, M., LAZZARONI, N., DIFONZO, N., MOTTO, M., SALAMINI, F. AND THOMPSON, R. (1989). The O2 gene which regulates zein deposition in maize endosperm encodes a protein with structural homologies to transcriptional activators. *EMBO Journal* **8**, 2795–2801.

HAYAKAWA, K., TANAKA, K., NAKAMURA, T., ENDO, S. AND HOSHINO, T. (1997). Quality characteristics of waxy hexaploid wheat (*Triticum aestivum* L.): Properties of starch gelatinization and retrogradation. *Cereal Chemistry* **74**, 576–580.

HIRANO, H.Y., TABAYASHI, N., MATSUMURA, T., TANIDA, M., KOMEDA, Y. AND SANO, Y. (1995). Tissue-dependent expression of the rice *Wx*(+) gene promoter in transgenic rice and petunia. *Plant and Cell Physiology* **36**, 37–44.

HIZUKURI, S. (1985). Relationship between the distribution of the chain-length of amylopectin and the crystalline-structure of starch granules. *Carbohydrate Research* **141**, 295–306.

HOSHINO, T., ITO, S., HATTA, K., NAKAMURA, T. AND YAMAMORI, M. (1996). Development of waxy common wheat by haploid breeding. *Breeding Science* **46**, 185–188.

HOVENKAMP-HERMELINK, J.H.M., JACOBSEN, E., PONSTEIN, A.S., VISSER, R.G.F., VOSS-CHEPERKEUTER, G.H., BIJMOLT, E.W., DEVRIES, J.N., WITHOLT, B. AND FEENSTRA, W.J. (1987). Isolation of an amylose-free starch mutant of the potato (*Solanum-tuberosum*-L). *Theoretical and Applied Genetics* **75**, 217–221.

IGLESIAS, A.A., BARRY, G.F., MEYER, C., BLOKSBERG, L., NAKATA, P.A., GREENE, T., LAUGHLIN, M.J., OKITA, T.W., KISHORE, G.M. AND PREISS, J. (1993). Expression of the potato-tuber ADP-glucose pyrophosphorylase in *Escherichia-coli*. *Journal of Biological Chemistry* **268**, 1081–1086.

ITOH, K., NAKAJIMA, M. AND SHIMAMOTO, K. (1997). Silencing of waxy genes in rice containing *Wx* transgenes. *Molecular & General Genetics* **255**, 351–358.

JAMES, M.G., ROBERTSON, D.S. AND MYERS, A.M. (1995). Characterization of the maize gene *sugary1*, a determinant of starch composition in kernels. *Plant Cell* **7**, 417–429.

JANE, J. (1997). Starch functionality in food processing. In *Starch, Structure and Functionality*. Eds. P. J. Frazier, A. M. Donald and P. Richmond, pp 26–35. Cambridge, UK: Royal Society of Chemistry.

JANE, J.L., KASEMSUWAN, T., LEAS, S., ZOBEL, H. AND ROBYT, J.F. (1994a). Anthology of starch granule morphology by scanning electron-microscopy. *Starch-Starke* **46**, 121–129.

JANE, J., LIM, S., PAETAU, I., SPENCE, K. AND WANG, S. (1994b). Biodegradable plastics made from agricultural biopolymers. *ACS Symposium Series* **575**, 92–100.

JANE, J., SHEN, L., WANG, L. AND MANINGAT, C.C. (1992). Preparation and properties of small-particle corn starch. *Cereal Chemistry* **69**, 280–283.

JENKINS, J.P.J., CAMERON, R.E. AND DONALD, A.M. (1993). A universal feature in the structure of starch granules from different botanical sources. *Starch-Starke* **45**, 417–420.

JENKINS, P.J. AND DONALD, A.M. (1995). The influence of amylose on starch granule structure. *International Journal of Biological Macromolecules* **17**, 315–321.

KAUSHIK, R.P. AND KHUSH, G.S. (1991). Genetic-analysis of endosperm mutants in rice *Oryza-sativa* L. *Theoretical and Applied Genetics* **83**, 146–152.

KAWASAKI, T., MIZUNO, K., SHIMADA, H., SATOH, H., KISHIMOTO, N., OKUMURA, S., ICHIKAWA, N. AND BABA, T. (1996). Coordinated regulation of the genes participating in starch biosynthesis by the rice *floury-2* locus. *Plant Physiology* **110**, 89–96.

KEELING, P.L., LOMAKO, J., GIEOWAR-SINGH, D., SINGLETARY, G.W. AND WHELAN, W.J. (1994). Novel plants and processes for obtaining them. Patent No: WO 94/04693

KLÖSGEN, R.B., GIERL, A., SCHWARZSOMMER, Z. AND SAEDLER, H. (1986). Molecular analysis of the *Waxy* locus of *Zea-mays*. *Molecular & General Genetics* **203**, 237–244.

KLÖSGEN, R.B., SAEDLER, H. AND WEIL, J.H. (1989). The amyloplast-targeting transit peptide of the waxy protein of maize also mediates protein-transport *in vitro* into chloroplasts. *Molecular & General Genetics* **217**, 155–161.

KORTSTEE, A.J., VERMEESCH, A.M.S., DEVRIES, B.J., JACOBSEN, E. AND VISSER, R.G.F. (1996). Expression of *Escherichia-coli* branching enzyme in tubers of amylose-free transgenic potato leads to an increased branching degree of the amylopectin. *Plant Journal* **10**, 83–90.

KUIPERS, A.G.J., JACOBSEN, E. AND VISSER, R.G.F. (1994). Formation and deposition of amylose in the potato-tuber starch granule are affected by the reduction of granule-bound starch synthase gene-expression. *Plant Cell* **6** 43–52.

KURIKI, T., GUAN, H.P., SIVAK, M. AND PREISS, J. (1996). Analysis of the active-center of branching-enzyme-II from maize endosperm. *Journal of Protein Chemistry* **15**, 305–313.

KURIKI, T., STEWART, D.C. AND PREISS, J. (1997). Construction of chimeric enzymes out of maize endosperm branching enzymes I and II: Activity and properties. *Journal of Biological Chemistry* **272**, 28999-29004.

LARSSON, C.T., HOFVANDER, P., KHOSHNOODI, J., EK, B., RASK, L. AND LARSSON, H. (1996). 3 Isoforms of starch synthase and 2 isoforms of branching enzyme are present in potato-tuber starch. *Plant Science* **117**, 9–16.

LEE, Y.M., KUMAR, A., TANAKA, T. AND PREISS, J. (1987). An *Escherichia-coli* ADP-glucose synthetase allosteric mutant enzyme – DNA sequencing of the structural gene and site-directed mutagenesis. *Federation Proceedings* **46**, 2045.

LEISY, D.J., HNILO, J., ZHAO, Y. AND OKITA, T.W. (1990). Expression of a rice glutelin promoter in transgenic tobacco. *Plant Molecular Biology* **14**, 41–50.

LEUNG, P., LEE, Y.M., GREENBERG, E., ESCH, K., BOYLAN, S. AND PREISS, J. (1986). Cloning and expression of the *Escherichia-coli* glgC gene from a mutant containing an ADP-glucose pyrophosphorylase with altered allosteric properties. *Journal of Bacteriology* **167**, 82–88.

LI, Y.Z., MA, H.M., ZHANG, J.L., WANG, Z.Y. AND HONG, M.M. (1995). Effects of the first intron of rice *Waxy* gene on the expression of foreign genes in rice and tobacco protoplasts. *Plant Science* **108**, 181–190.

LILLFORD, P.J. AND MORRISON, A. (1997). Structure/function relationship of starches in food. In *Starch, Structure and Functionality*. Eds. P. J. Frazier, A. M. Donald and P. Richmond, pp 1–8. Cambridge, UK: Royal Society of Chemistry.

LIM, S.T., JANE, J.L., RAJAGOPALAN, S. AND SEIB, P.A. (1992). Effect of starch granule size on physical-properties of starch-filled polyethylene film. *Biotechnology Progress* **8**, 51–57.

LLOYD, J.R., WANG, T.L. AND HEDLEY, C.L. (1996). An analysis of seed development in *Pisumsativum*.19. Effect of mutant alleles at the *r*-loci and *rb*-loci on starch grain-size and on the content and composition of starch in developing pea-seeds. *Journal of Experimental Botany* **47**, 171–180.

MAAS, C., LAUFS, J., GRANT, S., KORFHAGE, C. AND WERR, W. (1991). The combination of a novel stimulatory element in the 1st exon of the maize *shrunken*-1 gene with the following intron-1 enhances reporter gene-expression up to 1000-fold. *Plant Molecular Biology* **16**, 199–207.

MADDELEIN, M.L., LIBESSART, N., BELLANGER, F., DELRUE, B., DHULST, C., VANDENKOORNHUYSE, N., FONTAINE, T., WIERUSZESKI, J.M., DECQ, A. AND BALL, S. (1994). Toward an understanding of the biogenesis of the starch granule – Determination of granule-bound and soluble starch synthase functions in amylopectin synthesis. *Journal of Biological Chemistry* **269**, 25150–25157.

MARSHALL, J., SIDEBOTTOM, C., DEBET, M., MARTIN, C., SMITH, A.M. AND EDWARDS, A. (1996). Identification Of the Major Starch Synthase In the Soluble Fraction Of Potato-Tubers. *Plant Cell* **8**, 1121–1135.

MIZUNO, K., KAWASAKI, T., SHIMADA, H., SATOH, H., KOBAYASHI, E., OKUMURA, S., ARAI, Y. AND BABA, T. (1993). Alteration of the structural-properties of starch components by the lack of an isoform of starch branching enzyme in rice seeds. *Journal of Biological Chemistry* **268**, 19084–19091.

MOUILLE, G., MADDELEIN, M.L., LIBESSART, N., TALAGA, P., DECQ, A., DELRUE, B. AND BALL, S. (1996). Preamylopectin processing – a mandatory step for starch biosynthesis in plants. *Plant Cell* **8**, 1353–1366.

MÜLLER-RÖBER, B., SÖNNEWALD, U. AND WILLMITZER, L. (1992). Inhibition of the ADP-glucose pyrophosphorylase in transgenic potatoes leads to sugar-storing tubers and influences tuber formation and expression of tuber storage protein genes. *EMBO Journal* **11**, 1229–1238.

MUNYIKWA, T.R.I., LANGEVELD, S., SALEHUZZAMAN, S., JACOBSEN, E. AND VISSER, R.G.F. (1997). Cassava starch biosynthesis: New avenues for modifying starch quantity and quality. *Euphytica* **96**, 65–75.

NACHTERGAELE, W. AND VAN-NUFFEL, J. (1989). Starch as stilt material in carbonless copy paper – New developments. *Starch-Starke* **41**, 386–392.

NAKAMURA, R. AND MATSUDA, T. (1996). Rice allergenic protein and molecular-genetic approach for hypoallergenic rice. *Bioscience Biotechnology and Biochemistry* **60**, 1215–1221.

NAKAMURA, T., YAMAMORI, M., HIRANO, H., HIDAKA, S. AND NAGAMINE, T. (1995). Production of *Waxy* (amylose-free) wheats. *Molecular & General Genetics* **248**, 253–259.

NAKAMURA, Y. (1996). Some properties of starch debranching enzymes and their possible role in amylopectin biosynthesis. *Plant Science* **121**, 1–18.

NAKAMURA, Y., KUBO, A., SHIMAMUNE, T., MATSUDA, T., HARADA, K. AND SATOH, H. (1997).

Correlation between activities of starch debranching enzyme and alpha-polyglucan structure in endosperms of *sugary-1* mutants of rice. *Plant Journal* **12**, 143–153.

NAKAMURA, Y., UMEMOTO, T., OGATA, N., KUBOKI, Y., YANO, M. AND SASAKI, T. (1996). Starch debranching enzyme (r-enzyme or pullulanase) from developing rice endosperm – purification, cDNA and chromosomal localization of the gene. *Planta* **199**, 209–218.

NAKAMURA, Y. AND YAMANOUCHI, H. (1992). Nucleotide-sequence of a cDNA-encoding starch-branching enzyme, or Q-enzyme-I, from rice endosperm. *Plant Physiology* **99**, 1265–1266.

NAKASE, M., ADACHI, T., URISU, A., MIYASHITA, T., ALVAREZ, A.M., NAGASAKA, S., AOKI, N., NAKAMURA, R. AND MATSUDA, T. (1996). Rice (*Oryza-sativa* L) alpha-amylase inhibitors of 14–16 KDa are potential allergens and products of a multigene family. *Journal of Agricultural and Food Chemistry* **44**, 2624–2628.

NAKATA, P.A., GREENE, T.W., ANDERSON, J.M., SMITH-WHITE, B.J., OKITA, T.W. AND PREISS, J. (1991). Comparison of the primary sequences of 2 potato-tuber ADP-glucose pyrophosphorylase subunits. *Plant Molecular Biology* **17**, 1089–1093.

NAKATA, P.A. AND OKITA, T.W. (1996). Cis-elements important for the expression of the ADP-glucose pyrophosphorylase small-subunit are located both upstream and downstream from its structural gene. *Molecular & General Genetics* **250**, 581–592.

NELSON, O. AND PAN, D. (1995). Starch synthesis in maize endosperms. *Annual Review of Plant Physiology and Plant Molecular Biology* **46**, 475–496.

OAKES, J.V., SHEWMAKER, C.K. AND STALKER, D.M. (1991). Production Of Cyclodextrins, a Novel Carbohydrate, In the Tubers Of Transgenic Potato Plants. *Bio-Technology* **9**, 982–986.

OKAGAKI, R.J. AND WESSLER, S.R. (1988). Comparison of non-mutant and mutant *Waxy* genes in rice and maize. *Genetics* **120**, 1137–1143.

OLIVE, M.R., ELLIS, R.J. AND SCHUCH, W.W. (1989). Isolation and nucleotide-sequences of cDNA clones encoding ADP-glucose pyrophosphorylase polypeptides from wheat leaf and endosperm. *Plant Molecular Biology* **12**, 525–538.

OLIVEIRA, A.B., RASMUSSON, D.C. AND FULCHER, R.G. (1994). Genetic-aspects of starch granule traits in barley. *Crop Science* **34**, 1176–1180.

POULSEN, L., MUHRBECK, P. AND ADLER-NISSEN, J. (1997). Potato starch: Degree of phosphorylation related to dynamic rheological characteristics. In *Starch, Structure and Functionality.* Eds. P. J. Frazier, A. M. Donald and P. Richmond, p 258. Cambridge, UK: Royal Society of Chemistry.

POULSEN, P. AND KREIBERG, J.D. (1993). Starch Branching Enzyme cDNA From *Solanum-tuberosum.* *Plant Physiology* **102**, 1053–1054.

ROBIN, J.P., MERCIER, C., DUPRAT, F., CHARBONNIERE, R. AND GUIBOT, A. (1975). Amidons Lintnerises. *Starch/Starke* **27**, 36–45.

ROHDE, W., BECKER, D. AND SALAMINI, F. (1988). Structural-analysis of the *Waxy* locus from *Hordeum-vulgare.* *Nucleic Acids Research* **16**, 7185–7186.

RUSSELL, D.A. AND FROMM, M.E. (1997). Tissue-specific expression in transgenic maize of four endosperm promoters from maize and rice. *Transgenic Research* **6**, 157–168.

SATOZAWA T., AKAGI H., SAKAMOTO M., BABA T., SHIMADA H., FUJIMURA T. (1995). EMBL database Acc. No. D50317

SANO, Y. (1984). Differential regulation of *Waxy* gene-expression in rice endosperm. *Theoretical and Applied Genetics* **68**, 467–473.

SANO, Y., MAEKAWA, M. AND KIKUCHI, H. (1985). Temperature effects on the *wx* protein level and amylose content in the endosperm of rice. *Journal of Heredity* **76**, 221–222.

SHANNON, J.C. AND GARWOOD, D.L. (1984). Genetics and physiology of starch development. In *Starch: Chemistry and Technology.* Eds. R. L. Whistler, J. N. BeMiller and E. F. Paschall, pp 25–86. Orlando, Florida: Academic Press.

SHEWMAKER, C.K., BOYER, C.D., WIESENBORN, D.P., THOMPSON, D.B., BOERSIG, M.R., OAKES, J.V. AND STALKER, D.M. (1994). Expression of *Escherichia-coli* glycogen-synthase in the tubers of transgenic potatoes (*Solanum-tuberosum*) results in a highly branched starch. *Plant Physiology* **104**, 1159–1166.

SHIMADA, H., TADA, Y., KAWASAKI, T. AND FUJIMURA, T. (1993). Antisense regulation of the rice *Waxy* gene-expression using a PCR-amplified fragment of the rice genome reduces the amylose content in grain starch. *Theoretical and Applied Genetics* **86**, 665–672.

SHURE, M., WESSLER, S. AND FEDOROFF, N. (1983). Molecular-identification and isolation of the *Waxy* locus in maize. *Cell* **35**, 225–233.

SMITH, A. (1988). Biochemical basis of the differences between round and wrinkled pea-seeds. *Heredity* **61**, 279.

SMITH-WHITE, B.J. AND PREISS, J. (1992). Comparison of proteins of ADP-glucose pyrophosphorylase from diverse sources. *Journal of Molecular Evolution* **34**, 449–464.

STARK, D.M., TIMMERMAN, K.P., BARRY, G.F., PREISS, J. AND KISHORE, G.M. (1992). Regulation of the amount of starch in plant-tissues by ADP glucose pyrophosphorylase. *Science* **258**, 287–292.

SWEETLOVE, L.J., BURRELL, M.M. AND APREES, T. (1996a). Characterization of transgenic potato (*Solanum-tuberosum*) tubers with increased ADP-glucose pyrophosphorylase. *Biochemical Journal* **320**, 487–492.

SWEETLOVE, L.J., BURRELL, M.M. AND , T. (1996b). Starch metabolism in tubers of transgenic potato (*Solanum-tuberosum*) with increased ADP-glucose pyrophosphorylase. *Biochemical Journal* **320**, 493–498.

TADA, Y., NAKASE, M., ADACHI, T., NAKAMURA, R., SHIMADA, H., TAKAHASHI, M., FUJIMURA, T. AND MATSUDA, T. (1996). Reduction of 14–16 kda allergenic proteins in transgenic rice plants by antisense gene. *Febs Letters* **391**, 341–345.

TAKAHA, T., YANASE, M., OKADA, S. AND SMITH, S.M. (1993). Disproportionating enzyme (4-alpha-glucanotransferase – EC 2.4.1.25) of potato – purification, molecular-cloning, and potential role in starch metabolism. *Journal of Biological Chemistry* **268**, 1391–1396.

TAYLOR, P. (1997). Starch applications in surface coatings. Presented at *Gateway to Renewable Industrial Feedstocks* York, UK.

VAN-SOEST, J.J.G. AND ESSERS, P. (1997). Influence of amylose-amylopectin ratio on properties of extruded starch plastic sheets. *Journal of Macromolecular Science-Pure and Applied Chemistry* **A34**, 1665–1689.

VILLAND, P., AALEN, R., OLSEN, O.A., LUTHI, E., LONNEBORG, A. AND KLECZKOWSKI, L.A. (1992). PCR Amplification and sequences of cDNA clones for the small and large subunits of ADP-glucose pyrophosphorylase from barley tissues. *Plant Molecular Biology* **19**, 381–389.

VISSER, R.G.F., HERGERSBERG, M., VANDERLEIJ, F.R., JACOBSEN, E., WITHOLT, B. AND FEENSTRA, W.J. (1989). Molecular-cloning and partial characterization of the gene for granule-bound starch synthase from a wild-type and an amylose-free potato (*Solanum-tuberosum*-L). *Plant Science* **64**, 185–192.

VISSER, R.G.F. AND JACOBSEN, E. (1993). Towards modifying plants for altered starch content and composition. *Trends in Biotechnology* **11**, 63–68.

VISSER, R.G.F., SOMHORST, I., KUIPERS, G.J., RUYS, N.J., FEENSTRA, W.J. AND JACOBSEN, E. (1991). Inhibition of the expression of the gene for granule-bound starch synthase in potato by antisense constructs. *Molecular & General Genetics* **225**, 289–296.

WANG, T.L., BARBER, L., CRAIG, J., DENYER, K., HARRISON, C., LLOYD, J.R., MACLEOD, M., SMITH, A. AND HEDLEY, C.L. (1997). Manipulation of starch quality in peas. In *Starch, Structure and Functionality*. Eds. P. J. Frazier, A. M. Donald and P. Richmond, pp 188–195. Cambridge: Royal Society of Chemistry.

WANG, T.L., HADAVIZIDEH, A., HARWOOD, A., WELHAM, T.J., HARWOOD, W.A., FAULKS, R. AND HEDLEY, C.L. (1990). An analysis of seed development in *Pisum-sativum*.13. The chemical induction of storage product mutants. *Plant Breeding-Zeitschrift Fur Pflanzenzuchtung* **105**, 311–320.

WANG, Z.Y., WU, Z.L., XING, Y.Y., ZHENG, F.G., GUO, X.L., ZHANG, W.G. AND HONG, M.M. (1990). Nucleotide-sequence of rice *Waxy* gene. *Nucleic Acids Research* **18**, 5898.

YASUI, T., MATSUKI, J., SASAKI, T. AND YAMAMORI, M. (1996). Amylose and lipid contents, amylopectin structure, and gelatinization properties of waxy wheat (*Triticum-aestivum*) starch. *Journal of Cereal Science* **24**, 131–137.

YASUI, T., SASAKI, T., MATSUKI, J. AND YAMAMORI, M. (1997). Waxy endosperm mutants of bread wheat (*Triticum aestivum* L) and their starch properties. *Breeding Science* **47**, 161–163.

7
Polysialic Acids: Potential Role in Therapeutic Constructs

GREGORY GREGORIADIS*, ANA FERNANDES, BRENDA McCORMACK, MALINI MITAL AND XIAOQIN ZHANG

Centre for Drug Delivery Research, School of Pharmacy, University of London, 29–39 Brunswick Square, London, WC1N 1AX, UK

Introduction

The extended presence of drugs either within the vascular system or in extravascular use is often a prerequisite for their optimal use (Gregoriadis *et al.*, 1994). Many antibiotics and cytostatics for instance, as well as a variety of therapeutic peptides and proteins, are removed from the circulation prematurely and before effective concentrations in target tissues can be achieved. It follows that such drugs could be more effective, less toxic and also used in smaller quantities if their presence intravascularly or extravascularly (and hence interaction with corresponding receptors or substrates) could be prolonged (Lee *et al.*, 1995). Similarly, prolonged circulation of drug delivery systems such as liposomes (Gregoriadis, 1995), other colloidal systems (Davis *et al.*, 1984) and polymers (Domb *et al.*, 1997) would facilitate targeting of drugs to cells other than those (e.g. the reticuloendothelial system; RES) by which many of these systems are normally intercepted (Gregoriadis, 1995; Lee *et al.*, 1995).

To that end, the half-lives of a number of short-lived proteins (*e.g.* enzymes, cytokines, etc) have been successfully augmented (Nucci *et al.*, 1991) by conjugating these to low molecular weight (750–5,000) mono-methoxypoly(ethyleneglycol) (mPEG). Liposomes and polystyrene microspheres coated with mPEG or poloxamers are also known to exhibit increased half-lives (e.g. Senior *et al.*, 1991; Davis *et al.*, 1984). It appears that mPEG molecules prolong the circulation time of proteins and particles by forming a shell around their surface, thus sterically hindering interaction with factors responsible for their clearance (Torchilin and Papisov, 1994). However, because of their low molecular weight (which results in rapid excretion through the kidneys), such mPEG polymers are not suitable for prolonging the half-life of small therapeutic agents (e.g. small peptides and conventional drugs). Recently, we have reported (Gregoriadis *et al.*, 1993) on an alternative type of macromolecules which

---

*To whom correspondence may be addressed.

Biotechnology and Genetic Engineering Reviews – Vol. 16, April 1999
0264–8725/99/15/203–215 $20.00 + $0.00 © Intercept Ltd, P.O. Box 716, Andover, Hampshire SP10 1YG, UK

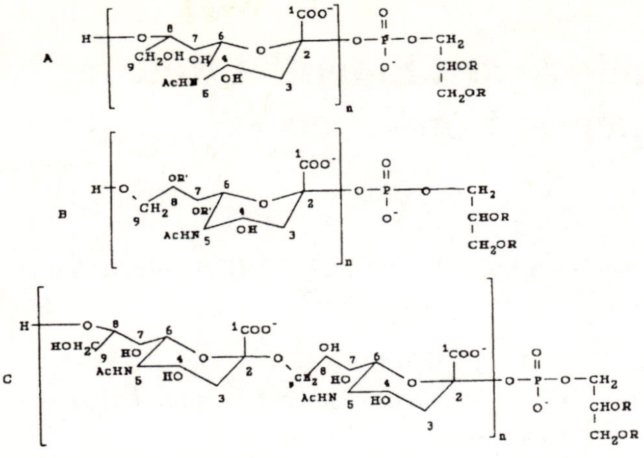

Figure 1. Structures of polysialic acids. (A) Serogroup B capsular polysialic acid B (PSB) from *N. meningitidis* or *E. coli* K1 is a homopolymer ($n = 199$) of α-(2–8)-linked *N*-acetyl neuraminic acid. (B) Serogroup C capsular polysialic acid (PSC) from *N. meningitidis* C is a homopolymer ($n = 74$) of α-(2–9)-linked *N*-acetyl neuraminic acid. (C) Polysialic acid (PSK92) from *E. coli* K92 is a heteropolymer ($n = 78$) of alternate units of α-(2–8)-α-(2–9)-linked *N*-acetyl neuraminic acid. All three polysialic acids contain a phospholipid molecule covalently linked to the reducing end of the polymers. From Gregoriadis *et al.* (1993), with permission.

may serve to increase the half-life not only of small molecules but also of large proteins, other large biopolymers, and microparticles such as liposomes. These macromolecules are naturally occurring polysaccharides, namely polymers of N-acetyl neuraminic acid (NeuNAc) (polysialic acids). They include (*Figure 1*) the serogroup B capsular polysaccharide from *Neisseria meningitidis* B and *Escherichia coli* KI, the serogroup C capsular polysaccharide C from *N. meningitidis* C, and the polysaccharide K92 from *E. coli* K92, as well as shorter chain derivatives thereof.

Clearance of polysialic acids from the circulation

It was thought (Gregoriadis *et al.*, 1993) that because of their highly hydrophilic nature and the absence of a known receptor in the body for NeuNAc, polysialic acids would be likely to circulate in the blood for prolonged periods after intravenous injection and could, thus, serve as carriers of short lived drugs or peptides. Experiments were therefore carried out in which mice were injected with a variety of polysialic acids. NeuNAc levels in blood plasma were measured by an assay modified (Gregoriadis *et al.*, 1993) to exclude the NeuNAc of plasma glycoproteins. Results showed that the clearance pattern of polysaccharide B (PSB) (see structure in *Figure 1*) from the blood circulation was biphasic with 50% of the dose removed 3 min after injection (*Figure 2*). The remainder of the dose assumed a linear rate of clearance with a half-life of 20 h. As *Figure 1* illustrates, PSB, polysaccharide C (PSC) and polysaccharide K92 (PSK92) have a phospholipid moiety covalently attached through its phosphate group to their reducing end. As a result, polysialic acids in solution exhibit micellar behaviour and form aggregates (Gotschlich *et al.*, 1981). The phospholipid moiety of the PSB used here was probably partially deacylated because

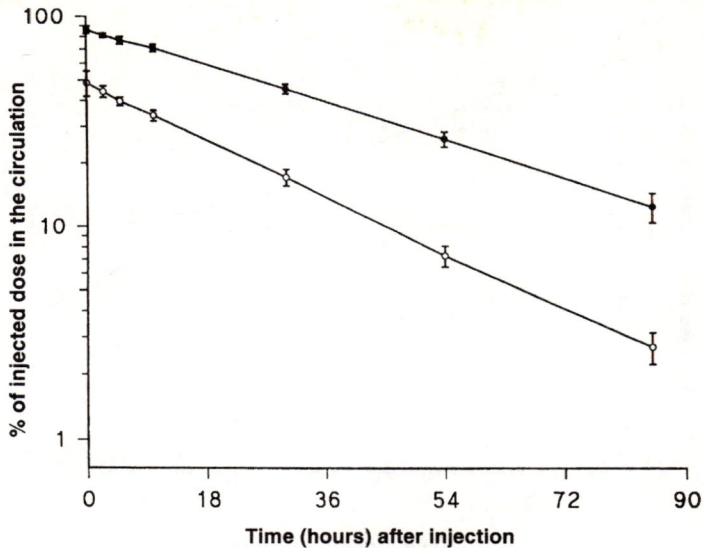

Figure 2. Clearance of PSB from the blood circulation. In six separate experiments, mice in groups of 3–4 animals were injected intravenously with 1.1–2.0 mg of intact (○) or deacylated (●) PSB and bled at time intervals. NeuNAc in the blood plasma samples was assayed as described (Gregoriadis *et al.*, 1993) and expressed as % ± S.D. of the dose in total blood. (Values from all groups treated with intact and deacylated PSB respectively, were pooled). Blood volume was estimated as 7% of the body weight. From Gregoriadis *et al.* (1993), with permission.

of long-term storage (Gregoriadis *et al.*, 1993), with only the acylated remainder expected to form aggregates. This would explain the rapid partial loss of PSB from the circulation, presumably in the form of aggregates. Results suggest that this is indeed the case: only 5–10% of the fully deacylated PSB was cleared from the circulation rapidly, the remainder exhibiting a linear rate of clearance with a half-life of 30 h (*Figure 2*). On the other hand, there was no apparent difference in the clearance patterns of PSK92 before and after deacylation (*Figure 3*). Following a relatively slow clearance during the first 6 h, patterns became linear with half-lives of 40 h (*Figure 3*).

It is therefore apparent that the rate of removal of a given polysialic acid from the circulation may be dependent on the presence or absence of phospholipid acyl groups. However, since the α-(2–8)-linked PSB is cleared more rapidly than the α-(2–8)-α-(2–9)-linked PSK92 (*Figures 2 and 3*), clearance may also depend on the structure of polysialic acids. Moreover, as the chain length of polysialic acids is an average and preparations are, therefore, polydisperse, low molecular weight polysialic acids may also contribute to the early rapid removal of some of the injected material from the circulation (as observed for PSB and, to a lesser extent, for PSK92) (*Figures 2 and 3* respectively). Indeed, experiments with a PSB of short chain length (15 NeuNAc units) have revealed that over 90% of the injected dose is removed from the circulation within 30 min (Gregoriadis *et al.*, 1993).

Clearance of a model drug bound to polysialic acid

The finding of prolonged half-lives for the polysialic acids used here was encouraging in terms of employing these as a means to extend the half-life of small drugs. Studies

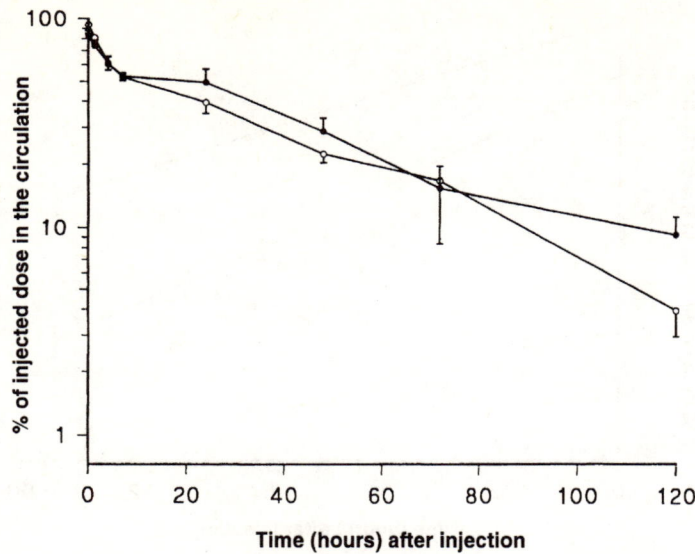

Figure 3. Clearance of PSK92 from the blood circulation. Mice in two groups of 4 were injected intravenously with 1.8 mg of intact (○) or deacylated (●) PSK92. For other details see legend to *Figure 2*. From Gregoriadis *et al.* (1993), with permission.

were therefore carried out with a model drug (fluorescein) coupled to deacylated PSB of low molecular weight (82 NeuNAc units). Data in *Figure 4* indicate that whereas fluorescein as such was removed from the circulation very rapidly, clearance of the polysialic acid-bound dye was slower and, also, independent of the dose of injected PSB for the amounts tested: following the removal of about 80% of the dose within 2.5 h, the remainder of fluorescein exhibited a half-life of 5 h, presumably that of the conjugate.

It is thus apparent that large molecular weight polysialic acids such as those described, could potentially retain rapidly cleared drugs and small peptides within the vascular and extravascular areas for extended periods of time. Also, because of the dependence of polysialic acid clearance not only on the type used and the presence or absence of the acyl groups of the phospholipid moiety but also the molecular size (Gregoriadis *et al.*, 1993), it would be possible to tailor clearance rates of drugs and peptides to satisfy specific needs. It is envisaged, for instance, that large molecular weight polysialic acids would be suitable for the delivery of one or more molecules (per molecule of polysialic acid) of low molecular weight drugs and peptides. On the other hand, shorter chain polysialic acids derived by the hydrolysis of long-chain molecules, could serve as a coat of large proteins as well as drug delivery systems such as liposomes (*Figure 5*). A variety of techniques could be used to conjugate polysialic acids to drugs and liposomes, depending on the reactive groups available on the interacting entities. Possible sites (*Figure 1*) of conjugation in polysialic acids include the non-reducing end which, on periodate oxidation, generates a reactive aldehyde, the carboxyl and hydroxyl groups, and the amino groups becoming available on deacetylation (Gregoriadis *et al.*, 1993). However, caution is required as coupling reactions could potentially damage the tertiary structure of the longer chain polysialic acids and alter their patterns of clearance.

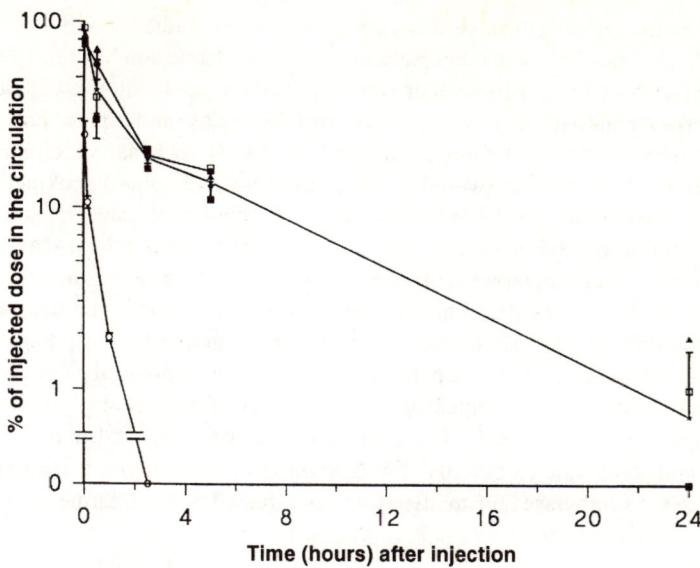

Figure 4.   Clearance of low molecular weight fluorescein-PSB conjugate from the blood circulation. Mice in groups of three were injected intravenously with 28 (▼), 102 (■), 510 (▲) and 1,528 µg (●) of PSB conjugated to [^{125}I]fluorescein or with 40 µg fluorescein only (○). Values are means ± S.D. of ^{125}I-radioactivity (closed symbols). NeuNAc (□) or fluorescein. Stars denote the mean of ^{125}I mean values for all doses at each time interval. For other details see legend to *Figure 2*. From Gregoriadis *et al.* (1993), with permission.

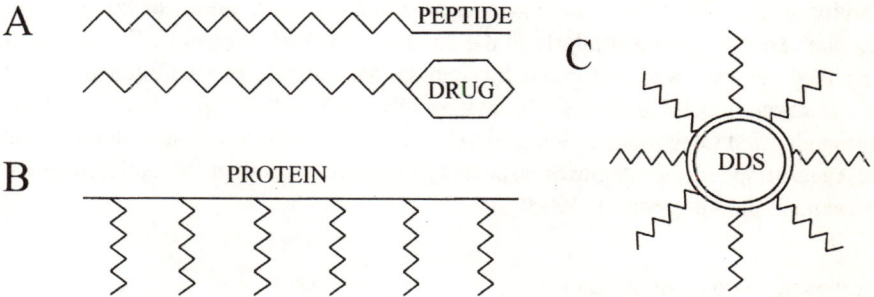

Figure 5.   Schemes of polysialic acid use in drug delivery. Long polysialic acids can be used to prolong the circulation time of small drugs and peptides.(A) Shorter polysialic acids bound to the surface of proteins (B) or liposomes (C) will render them more hydrophilic and extend their half-lives.

Polysialylated enzymes: characterization and fate *in vivo*

Therapeutic use of proteins can be hampered by proteolytic degradation and short half-lives in the circulation (Nucci *et al.*, 1991). Further, administration of large amounts of protein in order to maintain therapeutic efficacy, can often lead to toxicity and also promote adverse immune responses. As discussed elsewhere (Fernandes and Gregoriadis, 1996; Nucci *et al.*, 1991), such problems can be circumvented by the covalent coupling of proteins to hydrophilic macromolecules such as dextrans and mPEG.

The latter is by far the most successful and comprehensively studied and its use has now been extended to liposomes and other particulate systems (Lasic and Martin, 1995).

We have employed a low molecular weight polysialic acid, colominic acid (CA) (*Figure 6*), as a means to render two enzymes, namely catalase and asparaginase, more hydrophilic (Fernandes and Gregoriadis, 1996, 1997). It was anticipated that polysialylation of the proteins would not only improve their pharmacokinetics and stability, but also reduce their immunogenicity. The choice of catalase as a model therapeutic protein was based on its increasing use as an oxygen radical scavenger or in enzyme replacement therapy (Scott *et al.*, 1991). Asparaginase on the other hand, catalyses the hydrolysis of the non-essential amino-acid L-asparagine to L-aspartic acid and ammonia. As certain tumour cells that are deficient in L-asparagine synthetase (Haskell *et al.*, 1969) depend on the external supply of L-asparagine, asparaginase is currently in clinical use for the treatment of acute lymphoblastic leukaemia (Keating *et al.*, 1993). The enzyme is also active against non-Hodgkin's lymphoma and pancreatic carcinoma (Yunis *et al.*, 1977). Moreover, because normal cells produce the synthetase and are therefore not affected by the treatment, asparaginase therapy is highly selective.

POLYSIALYLATED CATALASE

Previous work (Fernandes and Gregoriadis, 1996) in which catalase was polysialylated with colominic acid following periodate oxidation of the latter at the non-reducing end (carbon 7) and subsequent coupling to the enzyme (ε amino groups) by reductive amination, showed that the extent of polysialylation was modest (3.8±0.4 moles of CA per mole of catalase). Polysialylated catalase, however, retained 70% of its initial activity at the end of the coupling reaction compared with values of 29–39% for enzyme controls treated similarly in the absence of CA and reagents. Formation of sialylated catalase was confirmed by ammonium sulphate or trichloroacetic acid precipitation, molecular sieve chromatography and SDS-PAGE electrophoresis (Fernandes and Gregoriadis, 1996). Sialylated catalase was much more stable in the presence of specific proteinases, especially chymotrypsin, than the native enzyme (Fernandes and Gregoriadis, 1996).

POLYSIALYLATED ASPARAGINASE

Earlier attempts to increase the half-life of asparaginase in the blood circulation, included entrapment into liposomes (Neerunjun and Gregoriadis, 1976) or erythrocytes (Kravtzoff *et al.*, 1990), and covalent coupling to mPEG (Cao *et al.*, 1990). The latter approach, however, led to substantial loss of enzyme activity.

Activation of colominic acid

Oxidation (activation) of CA was carried out with 0.1M sodium periodate (10 mg CA/ml periodate solution) at 20°C in the dark and in the presence of ethylene glycol to expend excess periodate. Following extensive dialysis at 4°C against ammonium carbonate solution, the preparation was freeze-dried and kept at –40°C until further use (Fernandes and Gregoriadis, 1996, 1997).

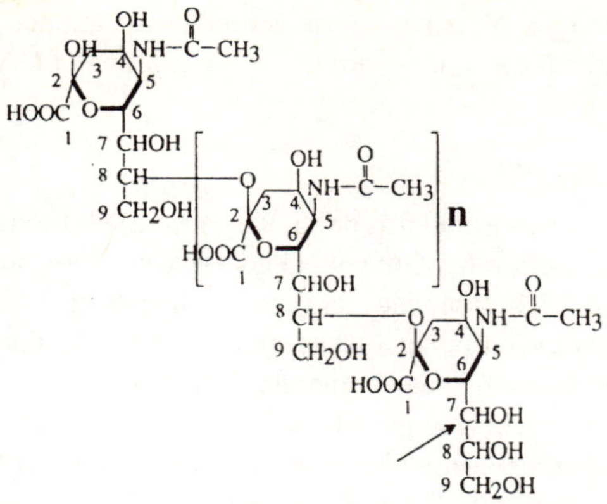

Figure 6. Structure of colominic acid. *N*-acetylneuraminic acid units are linked via α-(2→8) glycosidic linkages. Arrow indicates the carbon atom (C_7) at the non-reducing end of the sugar where periodate oxidation introduces an aldehyde group. From Fernandes and Gregoriadis (1996), with permission.

Preparation of asparaginase-colominic acid conjugates

Asparaginase (previously dialyzed to remove dextrose monohydrate) was covalently coupled to the activated CA by reductive amination in the presence of NaCNBH$_3$ (Fernandes and Gregoriadis, 1997) as previously applied for catalase (Fernandes and Gregoriadis, 1996). Three different molar ratios of CA:asparaginase (50:1, 100:1 and 250:1) were used in the coupling reaction. The sialylated asparaginase formed was isolated from non-conjugated CA by ammonium sulphate precipitation. Pellets containing the conjugate were then dissolved in 0.15M sodium phosphate buffer supplemented with 0.9% NaCl, pH 7.4 (PBS), and extensively dialysed against the same buffer. The dialysed samples were filtered to remove insoluble material, and asparaginase activity and protein concentration in the filtrates were determined spectrophotometrically. CA bound to the enzyme was measured (Fernandes and Gregoriadis, 1997) and values were expressed as moles of CA per mole of asparaginase or as percentage of polysialylated lysine residues. Solutions containing a known amount of polysialylated asparaginase (450–475 U/mg protein) were freeze-dried and kept at 4°C until further use. Results (Fernandes and Gregoriadis, 1997) on the extent of CA coupling to asparaginase and activity retention by the polysialylated enzyme revealed that polysialylation was directly dependent on the molar ratio of CA and enzyme used in the coupling reaction, with the highest degree of polysialylation (8.1 ± 1.7 moles of CA/mole asparaginase) achieved when a 250 fold excess of CA was present in the reaction mixture. This value (8.1) corresponds to an average of 11% of the available lysine ε-amino groups (Fernandes and Gregoriadis, 1997). However, as CA is polydisperse, values of degree of polysialylation are only average.

Asparaginase conjugates produced by other methods are known to lead to severe loss of enzyme activity, for instance as much as 70–90% for pegylated asparaginase (Cao *et al.*, 1990). In contrast, the coupling procedure used here led to only a modest loss (14–18%) of initial asparaginase activity in the polysialylated enzyme (Fernandes and Gregoriadis, 1997). It thus appears that polysialylation (or perhaps the presence of CA in the reaction mixture during the coupling procedure), protects the enzyme from inactivation: only 17% of asparaginase activity was retained by the enzyme when subjected to identical reaction conditions in the absence of CA.

Tritiation of asparaginase

Native asparaginase was tritiated as previously described (Fernandes and Gregoriadis, 1997). Following isolation of the labelled enzyme with ammonium sulphate or precipitation with trichloroacetic acid, more than 90% of the radioactivity was recovered with the enzyme pellet. Furthermore, polyacrylamide gel electrophoresis of the labelled enzyme confirmed (Fernandes and Gregoriadis, 1997) the absence of higher molecular mass enzyme species that could have formed through formaldehyde-induced methylene bridges. In additional experiments, asparaginase was simultaneously radiolabelled and coupled to CA as described (Fernandes and Gregoriadis, 1997), by employing $NaCNB[^3H]_3$ in the reaction mixture. The enzyme conjugate was then isolated and excess label removed by ammonium sulphate precipitation. In typical preparations, more than 90% of the radioactivity in the polysialylated asparaginase could be precipitated by trichloroacetic acid, indicating that reductive methylation ensured obligatory participation of most of the ^3H in the structure of the polysialylated enzyme.

Kinetics of polysialylated asparaginase

Plots (*Figure 7*) of the effect of substrate concentration on the activity of native and polysialylated asparaginase suggested a modest (but not significant; P>0.05) increase in the enzyme's Km value after polysialylation. Estimation of Km values according to Hanes-Woolf were 1.68×10^{-5} M for native asparaginase and 1.90×10^{-5} M, 2.15×10^{-5} M and 2.29×10^{-5} M respectively for the polysialylated constructs made with different CA to enzyme ratios (see *Figure 8*). These values are of the same order of magnitude as those reported (Howard and Carpentier, 1972) for the clinically useful asparaginases (e.g. 10^{-5} M). V_{max} values calculated from *Figure 7* and in the same order, were 0.847, 0.901, 0.910 and $0.919\mu mole\ min^{-1}\ U^{-1}$ (Fernandes and Gregoriadis, 1997). It thus appears that covalent coupling of CA to asparaginase and regardless of its degree of polysialylation, does not affect significantly the action of the enzyme on asparagine.

The effect of plasma on asparaginase activity

The results presented in *Figure 8* indicate that polysialylated asparaginase is more stable in the presence of (mouse) plasma at 37°C than the native (non-sialylated) enzyme: whereas polysialylated asparaginase retained most (65–83%) of its initial activity after exposure to plasma for 6 h, that of the native enzyme decreased to 13.5%. In addition, retention of activity by the polysialylated asparaginase was significantly higher for the preparation with the greatest number of CA molecules (preparation C; *Figure 8*). The effect of polysialylation on the stability of asparaginase (also observed for polysialylated catalase; Fernandes and Gregoriadis, 1996) has been tentatively attributed to changes in the microenvironment of the enzyme which must have occurred by the presence of the highly hydrophilic, negatively charged CA molecules.

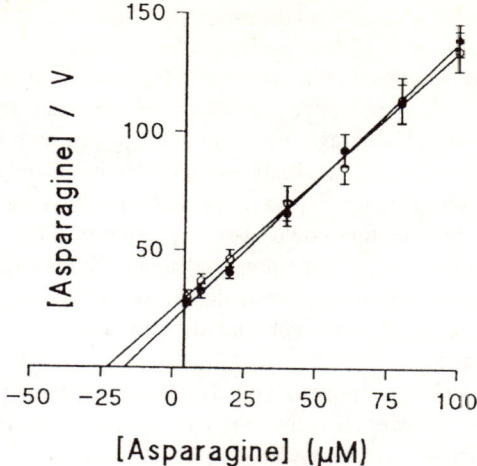

Figure 7. Hanes Woolf plots. For native (●) and polysialylated asparaginase (○). Values denote means ± S.D. (3 different experiments); V denotes velocity, expressed as μmol of liberated ammonia per min per unit (U) of enzyme. K_m values were obtained by extrapolation to the abscissa. From Fernandes and Gregoriadis (1997), with permission.

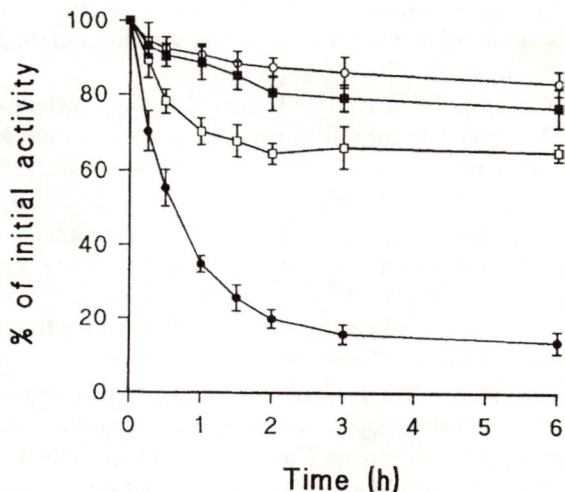

Figure 8. Retention of asparaginase activity in the presence of plasma. Native (●) and polysialylated asparaginase preparations A (□), B (■), C (○) were incubated in the presence of mouse plasma at 37°C; values denote means ± S.D. (3 different preparations). *Statistics*: results obtained at 6 h, were compared by ANOVA and *P* values corrected by the Bonferroni test. Native vs. A, B and C, *P*<0.001; A vs. B, n.s.; B vs. C, n.s.; A vs. C, *P*<0.05; n.s. = non-significant. From Fernandes and Gregoriadis (1997).

For instance, it is conceivable that a hydrophilic environment (promoted by polysialylation) combined with a shielding effect of the CA chains, contributes to a reduction in the access of plasma proteases to their target sites on the enzyme.

The effect of polysialylation on asparaginase clearance

Earlier work (results not shown) had shown that, in terms of enzyme activity, tritiation of asparaginase does not alter its pattern of clearance from the circulation. The clearance of tritiated asparaginase was therefore compared with that of the tritiated polysialylated constructs. As tritiation of the latter could only occur during the process of polysialylation, radioactivity was expected to represent polysialylated asparaginase and not intact enzyme that could also be present in the preparation. Moreover, the (radiolabelled) constructs were prepared under conditions (i.e. CA:enzyme molar ratio used in the coupling reaction) that were identical to those used for the non-radiolabelled conjugates, except that the sodium borohydride added during the reaction was tritiated.

All three constructs of polysialylated asparaginase were found to be removed from the circulation at slower rates than the native enzyme, both in terms of radioactivity (*Figure 9A*) and enzyme activity (*Figure 9B*). However, as observed with other modified enzymes (e.g. Nucci *et al.*, 1991), much of the injected dose (about 75% for native and 60–65% for polysialylated asparaginase) was removed from the circulation within 2 h after injection, the remainder exhibiting slower, linear clearance rates (*Figure 9A,B*). Moreover, on the basis of enzyme activity, terminal half-lives ($t_{1/2}\beta$) (estimated from the linear portions of asparaginase activity and ^3H radioactivity clearance patterns) were about 15 h for the native and about 38 h for the polysialylated asparaginase. Terminal half-lives were also independent of the dose injected, at least for the range of doses tested (0.5– 2.0 mg) and, for each of the polysialylated conjugates tested, similar whether derived from radioactivity or enzyme activity values. This confirmed that radioactivity measurements accurately reflected the presence of active enzyme.

Factors contributing to the clearance of injected proteins include (Delgado *et al.*, 1992) non-specific uptake by the reticuloendothelial system and receptor-mediated endocytosis by cells of the liver and other tissues where protein degradation eventually occurs. In addition, protein clearance is also influenced by molecular mass, shape and charge (Bocci, 1987; Benbough *et al.*, 1979) which determine the extent to which proteins undergo transcapillary passage or renal filtration (Bocci, 1987). In the latter case, the molecular mass cut-off is 66–70 kDa (Delgado *et al.*, 1992), i.e. well below the molecular mass (135 kDa) of asparaginase. It is likely that, at least in part, a greater resistance to plasma proteases (*Figure 8*) contributes to increases in the half-life of polysialylated asparaginase. It is also feasible that, as a result of the loss of some of the free ε-amino groups of asparaginase upon polysialylation, the modified protein is intrinsically more negatively charged. This, together with a shielding effect of the CA chains discussed above and elsewhere (Fernandes and Gregoriadis, 1996, 1997), could interfere with the interaction of the enzyme with blood and tissue components and thus curtail its recognition by tissues and removal from the circulation.

Although the reduction in the clearance rates of the polysialylated asparaginase observed in the present study is not as great as that claimed for the pegylated enzyme (e.g. Cao *et al.*, 1990), the following should be taken into consideration: (a) amounts of both native and pegylated enzyme in those studies were too low (up to 40 U per animal) compared to those (450–550 U) used in the present work and, therefore, measurements of enzyme activity may not have been as accurate, especially over

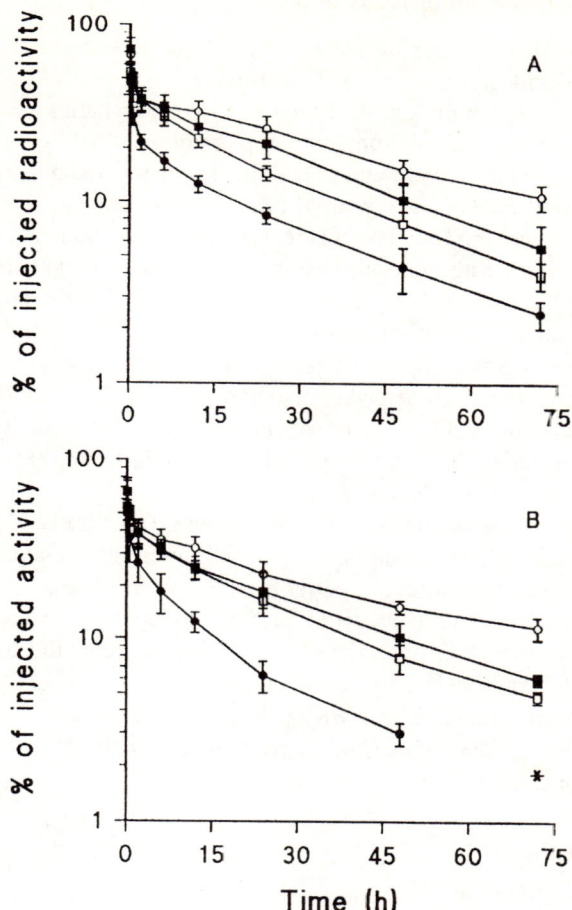

Figure 9. Clearance of asparaginase from the circulation. Mice were injected intravenously with 1 mg (550 U) tritiated native (●) and 1 mg (450-475 U) tritiated polysialylated asparaginase prepared by the use of 50:1 (□), 100:1 (■) and 250:1 (○) molar ratios of CA: asparaginase in the coupling reaction. Blood plasma obtained at time intervals was assayed for ^{3}H (A) and asparaginase activity (B). The pharmacokinetics profiles demonstrate biphasic patterns of clearance which are consistent with a two-compartment model. Values denote means ± S.D.; *n*=4 animals. *Native asparaginase activity was not detectable at 72 h. From Fernandes and Gregoriadis (1997).

extended periods of time. This could have in turn contributed to an overestimation of the half-lives for the pegylated asparaginase; (b) as pegylation of asparaginase leads to quantitative loss (Cao *et al.*, 1990; Uren and Ragin, 1979) of enzyme activity, polysialylation (which affects activity only modestly) (Fernandes and Gregoriadis, 1996, 1997) may be a preferred alternative as it would limit wastage of proteins; (c) to the authors' knowledge, there is no information on the fate and effect of the mPEG moiety of pegylated proteins subsequent to their uptake by tissues. Since mPEG is non-biodegradable, its accumulation intracellularly, especially on chronic use, may prove undesirable.

Immunogenicity and antigenicity of polysialic acids

In contrast to other hydrophilic macromolecules (e.g. dextran, mPEG), polysialic acids are biodegradable and their catabolic products (e.g. NeuNAc) are not known to be toxic. Furthermore, polysialic acids, like other polysaccharides, are T-independent antigens and do not induce immunological memory. PSB for instance, is non-immunogenic in animals and humans and this has hampered attempts to produce a vaccine against *N. meningitidis* group B or *E. coli* K1 (Moreno *et al.*, 1985). On the other hand, although PSC and PSK92 are immunogenic in humans, it is necessary to use polysaccharides with molecular weights in excess of 50.000 Da (average chain length greater than 170 NeuNAc units). However, polysialic acids coupled to proteins can become T cell dependent antigens with induction of memory, and no restriction on the size of the polymer applies. Nonetheless, immune responses are difficult to achieve, especially for PSB. An additional, perhaps more important consideration in selecting a polysialic acid for drug or enzyme delivery, is antigenicity (i.e. binding of the antigen to its antibodies). Although antibodies against some of the polysialic acid structures do exist at low levels in circulation, they are generally of low affinity, especially those against the α-(2–8) linked structures (Mandrell and Zollinger, 1982) which are present on host cell surfaces, thereby limiting any immunological response (Finne, 1982). Finally, it would be easier from the practical point of view to produce polysialic acids from non-pathogenic bacteria (as opposed to the pathogenic *N. meningitidis*). Since PSB (deacylated) and PSK92 exhibit the longest half-lives (*Figures 2* and *3*) and can be derived from the slightly pathogenic *E. coli* K1 (PSB) or the non-pathogenic *E.coli* K92 (PSK92) bacteria, these materials and their lower molecular weight products should be adopted for conjugation to therapeutic agents or to systems that could deliver such agents.

Acknowledgements

The authors thank Mrs Concha Perring for excellent secretarial assistance.

References

BENBOUGH, J.E.,WIBLIN, C.N., RAFTER, T.N.-A. AND LEE, J. (1979). The effect of chemical modification of L-asparaginase on its persistence in circulating blood of animals. *Biochemical Pharmacology* **28**, 833.

BOCCI, V. (1987). Metabolism of protein anticancer agents. *Pharmacological Therapeutics* **34**, 1.

CAO, S., ZHAO, Q., DING, Z., MA, L., YU, T., WANG, J., FENG, Y. AND CHENG, Y. (1990). Chemical modifications of enzyme molecules to improve their characteristics. *Annals of the New York Academy of Science* **613**, 460.

DAVIS, S.S., ILLUM, L., MCVIE, J.-G. AND TOMLINSON, E., EDS. (1984). *Microspheres and Drug Therapy*. Amsterdam: Elsevier.

DELGADO, C., FRANCIS, G.E. AND FISTER, D. (1992). The uses and properties of PEG-linked proteins. *Critical Reviews of Therapeutic Drug Carrier Systems* **9**, 249.

DOMB, A.J., KOST, J. AND WISEMAN, D.M., EDS. (1997). *Handbook of Biodegradable Polymers*. Amsterdam: Harwood Academic Publishers.

FERNANDES, A. AND GREGORIADIS, G. (1996). Synthesis, characterization and properties of sialylated catalase. *Biochimica et Biophysica Acta* **1293**, 92.

FERNANDES, A. AND GREGORIADIS, G. (1997). Polysialylated asparaginase: preparation, activity and pharmacokinetics. *Biochimica et Biophysica Acta* **1341**, 26.

FINNE, J. (1982). Occurrence of unique polysialosyl carbohydrate units in glycoproteins of developing brain. *Journal of Biological Chemistry* **257**, 11966.

GOTSCHLICH, E.C., FRASER, B.A., NISCHIMURA, O., ROBBINS, J.B. AND LIU, T.-Y. (1981). Lipid on capsular polysaccharides of gram-negative bacteria. *Journal of Biological Chemistry* **256**, 8915.

GREGORIADIS, G. (1995). Engineering liposomes for drug delivery: progress and problems. *Trends in Biotechnology* **13**, 527.

GREGORIADIS, G., MCCORMACK, B., WANG, Z. AND LIFELY, R. (1993). Polysialic acids: Potential in drug delivery. *FEBS Letters* **315**, 271.

GREGORIADIS, G., MCCORMACK, B. AND POSTE, G., EDS (1994). Liposomes *in vivo*: Control of behaviour. In *Targeting of Drugs: Advances in System Constructs*. New York: Plenum Press.

HASKELL, C.M., CANELLOS, G.P., LEVENTHAL, B.G., CARBONE, P.P. AND BLACK, J.B. (1969). L-asparaginase resistance in human leukemia-asparaginase synthetase. *Biochemical Pharmacology* **18**, 2578.

HOWARD, J.B. AND CARPENTIER, F.H. (1972). L-asparaginase from *Erwinia carotovora*. *Journal of Biological Chemistry* **247**, 1020.

KEATING, M.J., HOLMES, R., LERNER, S. AND HO, D.H. (1993). L-asparaginase and PEG asparaginas – past, present and future. *Leukemia Lymphoma* **10**, 153.

KRAVTZOFF, R., ROPERS, C., LAGUERRE, M., MUH, J.P. AND CHASSAIGNE, M. (1990). Erythrocytes as carriers for L-asparaginase. Methodological. *Journal of Pharmaceutics and Pharmacology* **42**, 473.

LEE, V.H.L., HASHIDA, M. AND MIZUSHINA, Y., EDS. (1995). *Trends and Future Perspectives in Peptide and Protein Drug Delivery*. Amsterdam: Harwood Academic Publishers.

MANDRELL, R.E. AND ZOLLINGER, W.D. (1982). Measurement of antibodies to meningococcal group B polysaccharide: low avidity binding and equilibrium binding constants. *Journal of Immunology* **129**, 2172.

MORENO, C., LIFELY, M.R. AND ESDAILE, J. (1985). Immunity and protection of mice against *Neisseria meningitidis* group B by vaccination using polysaccharide complexed with outer membrane proteins: a comparison with purified B polysaccharide. *Infection and Immunity* **47**, 527.

NEERUNJUN, D.E. AND GREGORIADIS, G. (1976). Tumour regression with liposome-entrapped asparaginase. Some immunological advantages. *Biochemical Society Transactions* **4**, 133.

NUCCI, M.L., SHORR, R. AND ABUCHOWSKI, A. (1991). The therapeutic value of poly(ethylene-glycol)-modified proteins. *Advanced Drug Delivery Reviews* **6**, 133.

SCOTT, M.D., LUBIN, B.H., ZUO, L. AND KUYPERS, F.A. (1991). Erythrocyte defence against hydrogen peroxide: preeminent importance of catalase. *Journal of Laboratory Clinical Medicine* **118**, 7.

TORCHILIN, V.P. AND PAPISOV, M.I. (1994). Why do polyethylene glycol-coated liposomes circulate so long? *Journal of Liposome Research* **4**, 725.

UREN, J.R. AND RAGIN, R.C. (1979). Improvement in the therapeutic, immunological and clearance properties of *Escherichia coli* and *Erwinia carotovora* L-asparaginase by attachment of poly-DL-alanyl peptides. *Cancer Research* **39**, 1927.

YUNIS, A.A., ARIMURA, G.K. AND RUSSIN, D.J. (1977). Human pancreatic carcinoma (MIA PaCa-2) in continuous culture: sensitivity to asparaginase. *International Journal of Cancer* **19**, 128.

8
Microbial Polysaccharide Products

IAN W. SUTHERLAND

Institute of Cell and Molecular Biology, Edinburgh University, Mayfield Road, Edinburgh, EH9 3JH, UK

Introduction

Exopolysaccharide-synthesising micro-organisms are of widespread occurrence. After much experimental work and examination of physical properties, a small number of microbial polysaccharides are now accepted products of biotechnology. Several others are in various stages of development. The uses and value of such polymers vary widely. Some are employed because of their unique physical properties. Others are superior to other natural or synthetic polymers. In this category are xanthan and gellan (gelrite) which, although they have found various food applications, also find diverse uses in non-food systems. Gellan has proved especially valuable in plant cell biotechnology. Xanthan is still the 'benchmark' product. It received food approval in the US many years ago; it is a relatively inexpensive product because of the very high conversion (*c.* 70%) of substrate to polymer and relative ease of processing and recovery. It is an accepted product in various industrial applications, including oil exploration and development. Because of lower yields or production and processing problems, other microbial polysaccharides are generally more expensive, some markedly so. At the top of the scale is bacterial hyaluronic acid from *Streptococcus equi* and related bacteria. Many of the major applications of microbial polysaccharides utilise their gel-forming ability or their high viscosity in aqueous solution, others use more specialised properties.

Many other exopolysaccharides (EPS) have been the subject of studies into microbial pathogenicity and in some bacteria there is a direct correlation between the presence of EPS as capsules surrounding the cell and the pathogenic state. The majority of plant-pathogenic bacteria produce copious quantities of EPS both as loose slime and capsules. Other microbial EPS play important roles in adhesion to exposed surfaces where they may be found as major constituents of 'biofilms' (Sutherland, 1997). Among these is a very unusual positively charged polysaccharide composed of D-glucosamine (*Figure 1*) (Mack *et al.*, 1996). This polymer has been demonstrated to be closely associated with the intercellular adhesion of strains of *Staphylococcus epidermidis* and with attachment to polystyrene and various types of plastic surfaces.

Abbreviations: CD, circular dichroism; EPS, exopolysaccharide; LBG, locust bean gum

Biotechnology and Genetic Engineering Reviews – Vol. 16, April 1999
0264–8725/99/15/217–229 $20.00 + $0.00 © Intercept Ltd, P.O. Box 716, Andover, Hampshire SP10 1YG, UK

EPS I

→ [6-β-D-GlcpNAc-(1→ 6)-β-D-GlcpNAc-(1→6)-β-D-Glcp-N-(1→6)-β-D-GlcpNAc-(1→ 6)-β-D-GlcpNAc-(1 →

Ratio: GlcpNAc : GlcpN c. 5:1

EPS II

→ [-6-β-D-GlcpNAc-(1→ 6)-β-D-GlcpNAc-(1→6)-β- D-Glcp N-(1→6)-β-D-GlcpNAc-(1→ 6)-β-D-GlcpNAc-(1→

Ratio: GlcpNAc : GlcpN c. 16:1   Also contains: phosphate and succinyl half esters

Figure 1.   The positively charged polysaccharides from *Staphylococcus epidermidis* strains. Results of Mack *et al.* (1996).

As a result of these diverse interests, there have now been structural studies on a very large number of microbial exopolysaccharides. We now have a better idea of the relationships which exist between structure and function in microbial exopolysaccharides (Sutherland, 1994), but it is still difficult to predict from a knowledge of polysaccharide structure which microbial polymers are likely to prove worth developing. Many reports in the literature which have suggested potential industrial usage have proved over optimistic. The series of mutants yielding xanthan with abbreviated side-chains provides an excellent example of the unpredictability of physical properties (*Figure 2*). Such mutant bacterial strains fail to yield as much polymer as the wild type and may also be less stable. The mass of the polymer produced may also be changed. At the same time, the availability of xanthan with a second acetyl group (Stankowski *et al.*, 1993) (and little if any pyruvylation) enabled us to compare such a product with the traditional mono-acetylated polysaccharide (e.g. Shatwell *et al.*, 1990). Acetan has the same cellulosic main-chain as xanthan and shows some structural similarities in its side-chain although this is a pentasaccharide as opposed to the pyruvylated trisaccharide of xanthan (Couso *et al.*, 1987). Solutions of acetan have also been shown to be highly viscous. The physical properties of acetan are indeed very similar to those of xanthan (Berth *et al.*, 1996; Ojinnaka *et al.*, 1996). Acetan, like xanthan, can be represented as forming stiff, double-stranded chains. The persistence length is *c.* 100 nm (Harding *et al.*, 1996). A similar series of altered polysaccharides with truncated side-chains, is now also being developed from acetan-synthesising *Acetobacter xylinum* (Colquhoun *et al.*, 1995).

Polysaccharide properties and applications

Many polysaccharides are used as gelling agents. This is true of starch and also of alginate, agar and carrageenan obtained from various marine algae. All these are traditional products, which have been used over long time periods. Gelation may be an inherent property of the polysaccharide or may require the presence of either monovalent or multivalent cations. Further, non-gelling polysaccharides such as xanthan may form gels in admixture with plant galacto- or gluco-mannans (synergistic gelling). The process of gel formation by microbial polysaccharides has thus come under intense scrutiny. Ross-Murphy and Shatwell (1993) divided such polymer gels and networks into three categories: systems in which junctions were formed by covalent

→ [-4-β-D-Glc*p*-(1→ 4)-β-D-Glc*p*-1-] →
$1\uparrow3$
α-D-Man*p*-O.CO.CH₃
$1\uparrow2$
β-D-Glc*p*A

'Polytetramer'

→ [-4-β-D-Glc*p*-(1→ 4)-β-D-Glc*p*-1-] →
$1\uparrow3$
α-D-Man*p*
$1\uparrow2$
β-D-Glc*p*A

'Polytetramer' (Non-acetylated)

→ [-4-β-D-Glc*p*-(1→ 4)-β-D-Glc*p*-1-] →
$1\uparrow3$
α-D-Man*p*-O.CO.CH₃

'Polytrimer'

→ [-4-β-D-Glc*p*-(1→ 4)-β-D-Glc*p*-1-] →
$1\uparrow3$
α-D-Man*p*

'Polytrimer' (Non-acetylated)

Figure 2. Mutant forms of xanthan with truncated side-chains.

cross-linking; physical junctions which could be disrupted; and entanglement net-works. The structure and properties of microbial polysaccharides including curdlan and gellan have been known for many years. Other gel-forming polymers such as mutan and a neutral polysaccharide from *Rhizobium* (Gidley *et al.*, 1987) have also been well studied. Various novel bacterial gelling polysaccharides have been reported more recently. Several of these resemble gellan in requiring deacylation prior to discovery of their ability to form gels in association with divalent actions. One of these interesting new polymers was discovered accidentally through a study of mutants of *Rhizobium meliloti*. This polysaccharide proved to be a cryptic product, synthesis of which was normally suppressed. It was a homopolysaccharide, a polymer of 1,4-β-D-glucuronic acid carrying *O*-acetyl groups on the C2 and C3 positions with molecular weight ranging from 6×10^4 to 5×10^5 (*Figure 3*) (Dantas *et al.*, 1994). As might be expected, some of its properties were very similar to those found in some algal alginates. Thus, it formed gels in the presence of monovalent, divalent or trivalent cations (Heyraud *et al.*, 1994) but these gels were thermoreversible unlike calcium or strontium alginate gels (Dantas *et al.*, 1994). A second gel-forming polymer, the galacturonic acid-containing polysaccharide 'beijeran' is produced by *Azotobacter*

-4)-β-D-Glc*p*A-(1→4)-β-D-Glc*p*A-(1→4)-β-D-Glc*p*A-(1→4)-β- D-Glc*p*A-(1→4)-β-D-Glc*p*A-(1→4)-β-D-Glc*p*A-(1-
   ↑2   ↑3   ↑2   2↑3
   O.CO.CH₃   O.CO.CH₃   O.CO.CH₃   (O.CO.CH₃)₂

Figure 3.   The structure of poly-D-glucuronic acid from *Rhizobium meliloti* M5N1. Results of Courtois *et al.* (1994).

beijerinckia strain YNM1 (Ogawa *et al.*, 1996). The repeat unit of this EPS is a linear trisaccharide carrying *O*-acetyl substituents on the C6 position of the glucose residue (*Figure 4*). Physical studies suggested that the polymer formed an extended two-fold helix and that the deacetylated macromolecule exhibited higher crystallinity (Ogawa *et al.*, 1997). Studies on each of these polymers are still at an early stage. The unique properties of exopolysaccharides and the influence of relatively minor substituents are also exemplified by the *Enterobacter* XM6 polymer (Nisbet *et al.*, 1984). In aqueous solutions XM6 yielded highly viscous solutions, as did the polymer from *Klebsiella aerogenes* type 54. At higher concentrations, XM6 formed gels in the presence of both monovalent and divalent cations and showed a very sharp transition between the ordered and disordered form at 39°C. However, the acetylated polymers with exactly the same carbohydrate structure from *K. aerogenes* type 54 failed to form gels, as also did the polymer from *Escherichia coli* K27 in which the side-chains are D-galactosyl residues replacing D-glucose (Sutherland *et al.*, 1970). This was observed both for type 54 polymers which carried *O*-acetyl groups on each tetrasaccharide repeat unit and for those which were acetylated on alternate repeat units. On deacetylation with mild alkali, all the polymers were capable of gel formation. X-ray fibre diffraction studies indicated that XM6 was highly crystalline due to strong polymer-polymer interactions, whereas the acetylated polymers were poorly crystalline with weak intermolecular interactions (Atkins *et al.*, 1987). Morris and Miles (1986) suggested that acetylation did not appear to alter the helical conformation of the polysaccharides but controlled gelation through the effects it exerted on intermolecular association and crystallisation of segments of the polymer chains.

Curdlan is a linear 1,3-β-D-glucan synthesised by several Gram negative bacteria including *Agrobacterium* spp. and is used for various food applications in Japan. Although curdlan was insoluble in water, it could be dissolved in sodium hydroxide solutions. It appeared to form triple helical structures under appropriate conditions (Deslandes *et al.*, 1980). Curdlan was also unusual in that it formed two distinct types of gel (Harada *et al.*, 1994). The first type (low-set gel) was formed when alkaline solutions of the polysaccharide were neutralised or when aqueous solutions were heated to 60°C and cooled. A second type of gel (high set gel) formed when solutions were heated above 80°C leading to a hydrophobic interaction and to stronger gels than those obtained at lower temperatures. The higher temperature followed by cooling converted the polysaccharide to the triple helical form. The polysaccharide differed from other microbial polysaccharides in forming a gel on heat treatment, with gel strength increasing with both temperature and time of heating. The high temperature gels were also much more resistant to hydrolysis by 1,3-β-D-glucanases and by acid treatment and failed to melt on further heating. This alteration in properties was ascribed to the formation of a pseudocrystalline form with greatly reduced hydration of the curdlan following high temperature treatment (Kanzawa *et al.*, 1989). Chemical introduction of *O*-acetyl groups altered the conformation and properties of curdlan and

Azotobacter beijerinkia YNM1 Exopolysaccharide (Beijeran)

$$\rightarrow [3]\text{-}\alpha\text{-}D\text{-}GalAp\text{-}(1\rightarrow3)\text{-}\beta\text{-}L\text{-}Rhap\text{-}(1\rightarrow 3)\text{-}\alpha\text{-}D\text{-}Glcp\text{-}(1] \rightarrow$$
$$\uparrow 6$$
$$O\text{-}CO.CH_3$$

Figure 4.   The structure of beijeran.

yielded a single helical structure in which there was one molecule of water per glucose residue (Okuyama *et al.*, 1996). Although it has been less studied than curdlan, the linear 1,3-α-D-glucan mutan produced by *Streptococcus mutans* also exhibits crystalline structure (Ogawa *et al.*, 1981)

Gellan has proved to be one of the most interesting recent microbial polysaccharide introductions. It is produced commercially from a strain of *Sphingomonas elodea* and has a linear tetrasaccharide repeat unit (Jansson *et al.*, 1986). The gelling properties of this polysaccharide are only fully revealed after chemical deacylation to remove the *O*-acetyl and *O*-glyceryl substituents, both of which are on the 3-linked D-glucose residue. The native polymer only formed soft, elastic gels but progressive removal of the acyl groups increased the brittleness of the gels produced in the presence of divalent cations such as Mg^{2+}. The extent of acetylation was thought to control the local crystallisation of parts of the polysaccharide chains (Morris and Miles, 1986). Chandrasekaran and Rhada (1995) indicated that monovalent ions such as K^+ promoted antiparallel alignment of the double helices of gellan. Divalent ions linked the gellan molecules causing gelation. The actual gel strength varied depending on the counterion added with the monovalent ions increasing from $Li^+ < Na^+ < K^+ < Cs^+$, while Mg^{2+}, Ca^{2+} and Sr^{2+} gels were broadly similar in strength but $< Zn^{2+} < Cu^{2+} < Pb^{2+}$. Even stronger gels were obtainable when the polysaccharide was in the proton form (Larwood *et al.*, 1996). The L-glyceryl substituents on native gellan caused significant shielding of carboxylate groups and weakened the linkage between chains. The resultant gels were therefore weak and rubbery as opposed to the hard, brittle gels of the deacylated polysaccharide. Doner and Douds (1995) demonstrated that if gellan were first freed of all multivalent cations, it could form gels in a manner similar to algal alginates on the addition of Ca^{2+} despite the very different chemical structures of the two polysaccharides. Beads of gellan were produced when solutions of the monovalent salt form were dropped into solutions of divalent cations. This extended the applications of gellan for biotechnology in the area of cell and plant culture and enzyme immobilisation. Moritika *et al.* (1992) observed that the gelling and melting temperatures of gellan increased with both increased gellan concentration and salt concentration. This was attributed to the increased number of junction zones and decreased rotational freedom of the parallel helices. Gellan is only one of a series of polysaccharides from *Sphingomonas* spp. All possessed closely related chemical structures but none of the others formed gels although most yielded highly viscous solutions with considerable thermostability.

Modified polysaccharides

As has already been mentioned, many polysaccharides require chemical modification before they exhibit useful properties. This usually entails alkali treatment to remove acyl groups. Thanks to genetical studies or to the availability of spontaneous mutants or strains which naturally produce variants on the normal EPS structure, a number of polymers are readily available in modified form. The availability of 'families' of closely related structures has also provided considerable information on the effect of specific substituents or alterations. These EPSs and products from mutant bacteria have the advantage over material chemically modified by mild acid hydrolysis, that they are likely to be more uniform in composition and molecular weight. Xanthan has been one of the most productive polymers in this respect. Betlach *et al*. (1987) prepared a series of mutant xanthans with truncated side-chains (*Table 1*) in which either the terminal sugar (β-linked D-mannose and its attached pyruvate ketal) or the terminal disaccharide were absent. Further variants lacked *O*-acetyl groups on the internal mannose residue. Another mutant product lacking the terminal mannose residue was isolated and studied by Tait *et al*. (1989). It could be further modified by treatment with β-D-glucuronidase which removed some, though not all, of the uronic acid residues. The product lacking terminal mannose residues yielded lower solution viscosity than the wild type xanthan but removal of some uronosyl residues gave a product with higher viscosity than wild type. Modelling experiments performed by Levy *et al*. (1996) have provided some possible explanations for the observed behaviour of the truncated xanthans. These included greater flexibility of the side-chains in native xanthan and an increased quantity of open helical backbone in polymer which lacked the terminal mannose but was acetylated on the internal mannose. It was suggested that this might account for the higher viscosity of this polymer than was found for either native xanthan or the deacetylated mutant product. This hypothesis is however questionable, as no account was taken of the duplex structure which is considered to be a major feature of the ordered state of 'wild type' xanthan.

Xanthan with modified side-chains in which the terminal β-D-mannosyl residues were removed by mild acid treatment, proved capable of maintaining the ordered, double-stranded state (Christensen *et al*., 1993a,b). Even removal of over 70% of the remaining aldobiouronic acid side-chains failed to affect the order-disorder transition. Thus, transition was not dependent on the terminal mannose of the side-chains. Removal of side-chain sugars did however affect both transition enthalpy and the intensity of the major peak at 204 nm in the CD spectrum of ordered xanthan. It also yielded sharper transition within a smaller temperature range than that seen with native xanthan. This may well result from production of polymer molecules more uniform in structure and in mass. Callet *et al*. (1987) demonstrated that in xanthans of the same molecular weight, neither acetyl nor pyruvate substituents influenced dilute solution viscosity. They did note that acetyl groups had a stabilising effect on the conformational transition of xanthans while pyruvate groups had the opposite effect. This was confirmed for a series of xanthans differing naturally in acylation by Shatwell *et al*. (1990). It was also clear that irrespective of the cations present, the salt concentration had a very marked effect on the transition temperature for all the xanthan variants studied.

When dilute solutions of native or deacetylated xanthan were mixed with either

Table 1. Comparison of 'gel strengths' (as indicated by the G' and tan δ values) for a range of xanthan/gluco- or galacto-mannan mixed systems (0.5% xanthan – 1% gluco- or galacto-mannan)

Xanthan	% substituent		Xanthan/LBG		Xanthan/KM	
	Acetate	Pyruvate	G'/Pa	Tan d	G'/Pa	Tan d
X646	4.5	4.4	430	0.095	300	0.106
X1128	7.7	1.7	93	0.25	–*	–*
X1128 DAC	0	1.3	330	0.049	114	0.15
X556	1.6	6.0	510	0.048	230	0.075
X556 Depyr	1.1	1.0	520	0.049	93	0.11
XBD9a	2.3		62	0.41		

- No data is shown as these systems failed to gel at ambient temperature
- DAC = deacetylated; Depyr = depyruvylated
- XBD9 is xanthan defective in terminal mannose (Tait *et al.*, 1989).

locust bean gum (LBG) or Konjac mannan, the viscosity was greatly enhanced (Goycoolea *et al.*, 1995) as a result of the interaction between these polymers. The actual viscosity depended on the proportions of the different polysaccharides present. Another marked difference from the individual solutions was the observation of significant thixotrophy. Most modifications have been achieved through loss or removal of acyl substituents or of side-chain monosaccharides. Scleroglucan or very similar polymers are produced by several fungal species. These are 1,3-β-D-glucans to which 1,6-β-D-glucose residues are attached on approximately every third main-chain glucose; they yield highly viscous aqueous solutions. It does not form gels in the presence or absence of ions such as Cr^{3+}. Stokke *et al.* (1995) modified scleroglucan through the introduction of varying amounts of carboxyl groups. The products then gelled in the presence of Cr^{3+}. The transition from viscoelastic solution to gel depended on the ion concentration and on the degree of carboxylation.

SYNERGISTIC GELLING

Xanthan and acetan show considerable structural similarities (*Figure 5*). Both also yield highly viscous aqueous solutions and undergo a thermally reversible order-disorder transition in solution, but differences are seen in synergistic gelling. The formation of synergistic gels when mixed aqueous solutions of xanthan and plant galactomannans are heated and cooled has received both study and application. Ross-Murphy *et al.* (1996) used a series of different xanthan preparations to demonstrate that the acyl groups of xanthan played a significant role in the interactions with guar gum, LBG and konjac mannan. Removal of the acetyl groups from xanthan enhanced gelation. Most xanthans formed a relatively strong gel network with LBG. An exception was a preparation in which there was a high acetate and low pyruvate content. The tan δ and *G'* values were rather higher and lower respectively than for the other mixtures. Removal of the acetate from this xanthan with mild alkali increased the *G'* value by almost 250% and the tan δ fell. Removal of pyruvate had little effect, indicating that it probably played little if any role in gelation. When mixed with konjac mannan, the xanthans showed similar behaviour to the interaction with LBG but higher polysaccharide concentrations were needed. However, the acetylated, non-pyruvylated xanthan failed to gel whereas the deacetylated material formed a strong

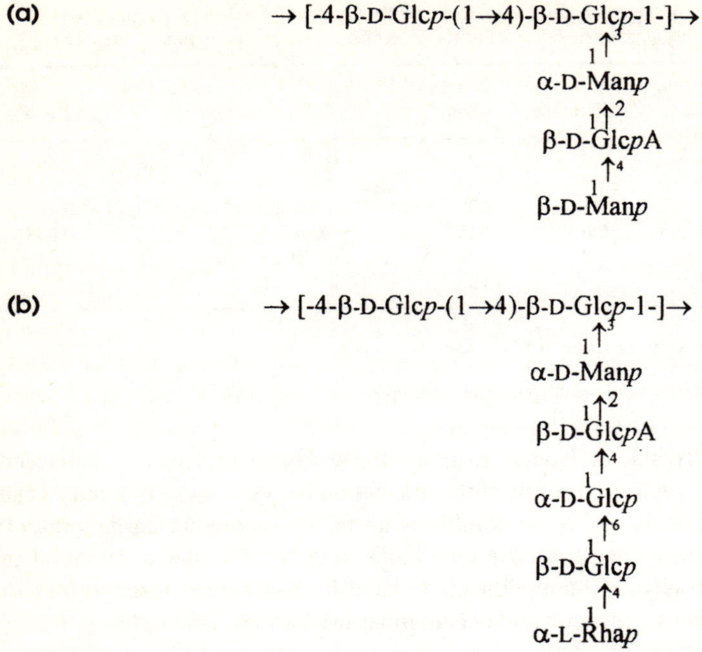

(a)

$$\rightarrow [\text{-}4\text{-}\beta\text{-}D\text{-}Glc}p\text{-}(1\rightarrow4)\text{-}\beta\text{-}D\text{-}Glc}p\text{-}1\text{-}]\rightarrow$$

$$1\uparrow^3$$

α-D-Manp

$$1\uparrow^2$$

β-D-GlcpA

$$\uparrow^4$$

β-D-Manp

(b)

$$\rightarrow [\text{-}4\text{-}\beta\text{-}D\text{-}Glc}p\text{-}(1\rightarrow4)\text{-}\beta\text{-}D\text{-}Glc}p\text{-}1\text{-}]\rightarrow$$

$$1\uparrow^3$$

α-D-Manp

$$1\uparrow^2$$

β-D-GlcpA

$$\uparrow^4$$

α-D-Glcp

$$\uparrow^6$$

β-D-Glcp

$$\uparrow^4$$

α-L-Rhap

Figure 5. Comparison of the structures of xanthan and acetan. (a) The structure of the exopolysaccharide from strains of *Xanthomonas campestris* (Xanthan). Typically the polymer carries an *O*-acetyl group on each repeating unit and 0.3 pyruvate ketals on the terminal mannose residue. (b) The structure of the exopolysaccharide from strains of *Acetobacter xylinum* (Acetan). Typically the polymer carries 2 *O*-acetyl groups on each repeating unit, one of which is possibly on a main-chain glucose residue.

gel. This again demonstrated the inhibitory effect of the *O*-acetyl groups (*Table 1*). Initially, mixtures of xanthan and LBG showed areas enriched in xanthan but, after heating above the transition temperature, these disappeared, probably due to more uniform distribution of the two component polysaccharides and disappearance of the liquid crystal mesophases (Schorsch *et al.*, 1995). Variations in the ratio of mannose to galactose in LBG also affect the properties of the mixed gels. A difference in gelation temperature of almost 13°C was observed by Lundin and Hermansson (1995) when comparing mixtures of xanthan and LBG with high and low mannose:galactose ratios.

It has now been found that although native acetan does not form gels with LBG or konjac mannan, deacetylation of the bacterial polysaccharide promotes synergistic interactions with both (Ojinnaka *et al.*, 1998). Acetan resembles xanthan in that it adopts a similar conformation in the solid state and shows the same thermally reversible transition from ordered (helical form) to disordered coil in solution. The failure of the native, acetylated acetan to form mixed gels was attributed to the solubility promoted by the presence of the *O*-acetyl groups and the resultant inhibition of intermolecular association. The role played by acyl groups and the frequent need for their removal to reveal useful properties suggest that genetic manipulation of several bacterial strains to delete the polysaccharide acetylase genes might prove a useful approach. As the relevant gene or genes have already been identified in a number of systems, such as succinoglycan production by *Rhizobium* sp. (Glucksman *et al.*, 1993) and xanthan synthesis in *X. campestris* (Betlach *et al.*, 1987), this should be feasible. Indeed, Franklin and Ohman (1996) have identified the two acetylase genes responsible for acetylation on the C2 and C3 positions of *Pseudomonas* alginate and have produced a mutant yielding non-acetylated alginate in good yield.

Biomimetics and other properties

Although high solution viscosity or the ability to form gels have resulted in commercialisation of several polysaccharides, other properties may also be very useful. These may depend on either physical or biological properties. Two most valuable exopolysaccharide products have proved to be bacterial cellulose and hyaluronic acid. The former has found several specialist applications, including use in audio-membranes. It can also be manufactured into wound dressings which show excellent retention of fluid and stimulation of healing in extensive burns or similar wounds (Joris and Vandamme, 1993). Bacterial hyaluronic acid owes its acceptance to the very high capacity for water regain and to its compatibility with the human immune system. It can be found as a replacement for hyaluronic acid in human fluids or as an effective moisturising agent in high quality cosmetics. Another example of biological properties leading to novel polysaccharide applications can be found in the range of fungal 1,3-β-D-glucans which include scleroglucan. These have proved to be potent immunomodulators, a property that is still poorly understood (Misaki *et al.*, 1993).

Possible new products

Despite examination of many bacterial isolates from different parts of the world (Dasinger *et al.*, 1994) and novel environments including deep-sea thermal vents (Guezennec *et al.*, 1994; Rougeaux *et al.*, 1996), few new polymers with interesting properties have been discovered. Among the products from five deep-sea isolates, one yielded high viscosity in aqueous solution and appeared to have some properties in common to xanthan. It also possessed high affinity for certain heavy metal ions including cadmium, lead and zinc (Loaëc *et al.*, 1997). One of the few to have been fully characterised and found to be worth further investigation is the exopolysaccharide from *Alteromonas* strain 1644 (Bozzi *et al.*, 1996a,b). This polysaccharide formed a gel in the presence of divalent cations which proved to be clear and very elastic. Exopolysaccharide from a Venezuelan soil isolate (Dasinger *et al.*, 1994) also yielded highly viscous aqueous solutions insensitive to high concentrations of NaCl or $CaCl_2$. Its composition resembled that of some Rhizobium polymers in containing D-mannose, D-glucose, D-galactose and D-glucuronic acid in the molar ratio 1:4:1:2. It also contained 10–15% acetate.

Another possible area for development lies in bacterial alginates. Earlier attempts to develop these commercially were unsuccessful, primarily because of the low molecular mass of the products. This was caused by release of alginate lyases present in the polysaccharide-synthesising bacteria. While those from *Azotobacter* spp. are in effect acylated variants of the algal material, use of the extracellular poly-D-mannuronosyl-4-epimerase enzyme (Skjåk-Bræk and Larsen, 1985) may permit tailoring of the composition and hence the physical properties of alginates of bacterial or algal origin.

The structure of the exopolysaccharide from *Lactobacillus sake* 0-1

$$\beta\text{-D-Glc}p \qquad CH_3.COO.$$
$$\downarrow 1 \qquad\qquad 2\downarrow$$
$$6$$
$$\rightarrow 4)\text{-}\beta\text{-D-Glc}p\text{-}(1\rightarrow3)\text{-}\beta\text{-L-Rha}p(1\rightarrow$$
$$1\uparrow 3$$
$$\alpha\text{-L-Rha}p \ (4\leftarrow sn\text{-glycerol 3-phosphate})$$

The structure of the exopolysaccharide from *Lactobacillus acidophilus* LMG9433

$$\beta\text{-D-Glc}p\text{NAc}$$
$$\downarrow 1$$
$$3$$
$$\rightarrow4)\text{-}\beta\text{-D-Glc}pA\text{-}(1\rightarrow6)\text{-}\alpha\text{-D-Glc}p\text{-}(1\rightarrow4)\text{-}\beta\text{-D-Gal}p\text{-}(1\rightarrow4)\text{-}\beta\text{-D-Glc}p\text{-}(1\rightarrow$$

Figure 6. Structures of some *Lactobacillus* exopolysaccharides. Robijn *et al.* (1995a,b; 1996).

Future products

What emphasis will there be in the search for new exopolysaccharide products. The search for microbial polysaccharides specifically for food use has switched to studies on bacterial species such as *Lactobacillus* spp. which are already accepted food micro-organisms. This might preclude the necessity for much of the extensive testing required to obtain approval for food use. Several of these have now been characterised (*Figure 6*) (Robijn *et al.*, 1995a,b; 1996). Production of this group of polymers is beset with difficulties – yields are low and complex media and growth conditions may be required. Apart from these, it is always possible that, given the current volume of research in this area, some new polymers will be found which possess really useful and possibly unique properties. Chemically modified polymers may also yield novel properties and applications. Perhaps more useful biological properties will also be discovered.

References

ATKINS, E.D.T., ATTWOOL, P.T., MILES, M.J., MORRIS, V.J., O'NEILL, M.A. AND SUTHERLAND, I.W. (1987). Effect of acetylation on the molecular interactions and gelling properties of a bacterial polysaccharide. *International Journal of Biological Macromolecules* **9**, 115–117.

BERTH, G., DAUTZENBERG, H., CHRISTENSEN, B.E., ROTHER, G. AND SMIDSRØD, O. (1996). Physicochemical studies on xylinan (acetan).1. Characterization by gel permeation chromatography on Sepharose CL-2B coupled with static light scattering and viscometry. *Biopolymers* **39**, 709–719.

BETLACH, M.R., CAPAGE, M.A., DOHERTY, D.H., HASSLER, R.A., HENDERSON, N.M., VANDERSLICE, R.W., MARELLI, J.D. AND WARD, M.B. (1987). Genetically engineered polymers: manipulation of xanthan biosynthesis. In *Industrial Polysaccharides*. Ed. M. Yalpani, pp 145–156. Amsterdam: Elsevier.

BOZZI, L., MILAS, M. AND RINAUDO, M. (1996a). Characterization and solution properties of a new exopolysaccharide excreted by the bacterium *Alteromonas* sp. strain 1644. *International Journal of Biological Macromolecules* **18**, 9–17.

BOZZI, L., MILAS, M. AND RINAUDO, M. (1996b). Solution and gel rheology of a new polysaccharide excreted by the bacterium *Alteromonas* sp. strain 1644. *International Journal of Biological Macromolecules* **18**, 83–91.

CALLET, F., MILAS, M. AND RINAUDO, M. (1987). Influence of acetyl and pyruvate contents on rheological properties of xanthan in dilute solution. *International Journal of Biological Macromolecules* **9**, 291–293.

CHANDRASEKARAN, R. AND RADHA, A. (1995). Molecular architectures and functional-properties of gellan gum and related polysaccharides. *Trends in Food Science and Technology* **6**, 143–148.

CHRISTENSEN, B.E., SMIDSRØD, O. AND STOKKE, B.T. (1993a). Xanthans with partially hydrolysed side chains: conformation and transitions. In *Carbohydrates and Carbohydrate Polymers*. Ed. M. Yalpani, pp 166–173. Mount Pleasant: ATL Press.

CHRISTENSEN, B.E., KNUDSEN, K.D., SMIDSRØD, O. KITAMURA, S. AND TAKEO, K. (1993b). Temperature-induced conformational transitions in xanthans with partially hydrolysed side chains. *Biopolymers* **33**, 151–162.

COLQUHOUN, I.J., DEFERNEZ, M. AND MORRIS, V.J. (1995). NMR studies of acetan and the related bacterial polysaccharide CR1/4 produced by a mutant strain of *Acetobacter xylinum*. *Carbohydrate Research* **269**, 319–331.

COURTOIS, J., SEGUIN, J.-P., ROBLOT, C., HEYRAUD, A., GEY, C., DANTAS, L., BARBOTIN, J. AND COURTOIS, B. (1994). Exopolysaccharide production by the *Rhizobium meliloti* M5N1 CS strain. Location and quantitation of the sites of *O*-acetylation. *Carbohydrate Polymers* **25**, 7–12.

COUSO, R.O., IELPI, L. AND DANKERT, M.A. (1987). A xanthan gum-like polysaccharide from *Acetobacter xylinum*. *Journal of General Microbiology* **133**, 2133–2135.

CRATER, D.L., DOUGHERTY, B.A. AND VANDERIJN, I. (1995). Molecular characterization of *hasc* from an operon required for hyaluronic acid synthesis in Group A *streptococci* – demonstration of UDP-glucose pyrophosphorylase activity. *Journal of Biological Chemistry* **270**, 28676–28680.

DANTAS, L., HEYRAUD, A., COURTOIS, B., COURTOIS, J., AND MILAS, M. (1994). Physico-chemical properties of exogel exocellular β(1→4) glucuronan from *Rhizobium meliloti* strain. *Carbohydrate Polymers* **24**, 185–192.

DASINGER, B.L., MCARTHUR, H.A.I., LENGEN, J.P., SMOGOWICK, A.A., MILLER, J.W., O'NEILL, J.J., HORTON, D. AND COSTA, J.B. (1994). Composition and rheological properties of extracellular polysaccharide 105-4 produced by *Pseudomonas* sp. strain ATCC 53923. *Applied and Environmental Microbiolology* **60**, 1364–1365.

DESLANDES, Y., MARCHESSAULT, R.H., AND SARKO, A. (1980). Triple helical structure of (1→3)-β-D-glucan. *Macromolecules* **13**, 1466–1471.

DONER, L.W. AND DOUDS, D.D. (1995). Purification of commercial gellan to monovalent cation salts results in acute modification of solution and gel-forming properties. *Carbohydrate Research* **273**, 225–233.

FRANKLIN, M.J. AND OHMAN, D.E. (1996). Identification of *algI* and *algJ* in the *Pseudomonas aeruginosa* alginate biosynthesis gene cluster which are required for alginate *O*-acetylation. *Journal of Bacteriology* **178**, 2186–2195.

GIDLEY, M.J., DEA, I.C.M., EGGLESTON, G., AND MORRIS, E.R. (1987). Structure and gelation of Rhizobium capsular polysaccharide. *Carbohydrate Research* **160**, 381–396.

GLUCKSMAN, M.A., REUBER, T.L. AND WALKER, G.C. (1993). Genes needed for the modification, polymerization, export and processing of succinoglycan by *Rhizobium meliloti*: a model for succinoglycan biosynthesis. *Journal of Bacteriology* **175**, 7045–7055.

GOYCOOLEA, F.M., MORRIS, E.R. AND GIDLEY, M.J. (1995). Screening for synergistic interactions in dilute polysaccharide solutions. *Carbohydrate Polymers* **28**, 351–358.

GUEZENNEC, J.G., PIGNET, P., RAGUENES, G., DESLANDES, E., LOJOUR, Y. AND GENTRIC, E. (1994). Preliminary chemical characterization of unusual eubacterial exopolysaccharides of deep sea origin. *Carbohydrate Polymers* **24**, 287–294.

HARADA, T., OKUYAMA, K., KONNO, A., KOREEDA, A. AND HARADA, A. (1994). Effect of heating on formation of curdlan gels. *Carbohydrate Polymers* **24**, 101–106.

HARDING, S.E., BERTH, G., HARTMANN, J., JUMEL, K., CÖLFEN, H. AND CHRISTENSEN, B.E.

(1996). Physicochemical studies on xylinan (acetan). 3. Hydrodynamic characterization by analytical ultracentrifugation and dynamic light scattering. *Biopolymers* **39**, 729–736.

HEYRAUD, A., DANTAS, L., COURTOIS, J., COURTOIS, B., HELBERT, W. AND CHANZY, H. (1994). Crystallographic data on bacterial (1→4)-β-D-glucuronan. *Carbohydrate Research* **258**, 275–279.

JANSSON, P.-E., LINDBERG, B., MAEKAWA, E. AND SANDFORD, P.A. (1986). Structural studies of a polysaccharide (S194) elaborated by *Alcaligenes* ATCC 31961. *Carbohydrate Research* **156**, 57–163.

JORIS, K. AND VANDAMME, E.J. (1993). Novel production and application aspects of bacterial cellulose. *Microbiology Europe* 27–29.

KANZAWA, Y., HARADA, A., KOREEDA, A., HARADA, T. AND OKUYAMA, K. (1989). Difference of molecular association in two types of curdlan gel. *Carbohydrate Polymers* **10**, 299–313.

LARWOOD, V.L., HOWLIN, B.J. AND WEBB, G.A. (1996). Solvation effects on the conformational behaviour of gellan and calcium ion binding to gellan double helices. *Journal of Molecular Modeling* **2**, 175–182.

LEVY, S., SCHUYLER, S.C., MAGLOTHIN, R.K. AND STAEHELIN, L.A. (1996). Dynamic simulations of the molecular conformations of wild type and mutant xanthan polymers suggest that conformational differences may contribute to observed differences in viscosity. *Biopolymers* **38**, 251–272.

LOAËC, M., OLIER, R. AND GUEZENNEC, J.G. (1997). Uptake of lead, cadmium and zinc by a novel bacterial exopolysaccharide. *Water Research* **31**, 1171–1179.

LUNDIN, L. AND HERMANSSON, A.-M. (1995). Supermolecular aspects of xanthan-lbg gels based on rheology and electron microscopy. *Carbohydrate Polymers* **26**, 29–140.

MACK, D., FISCHER, W., KOROTSCH, A., LEOPOLD, K,. HARTMANN, R., EGGE, H. AND LAUFS, R. (1996). The intercellular adhesin involved in biofilm accumulation of *Staphylococcus epidermidis* is a linear β-1,6-linked glucosaminoglycan: purification and structural analysis. *Journal of Bacteriology* **178**, 175–183.

MISAKI, A., KISHIDA, E., KAKUTA, M., AND TABATA, K. (1993). Antitumor fungal (1→3) β-glucans: In *Carbohydrates and Carbohydrate Polymers*. Ed. M. Yalpani, pp 116–129. Mount Pleasant: ATL Press.

MORITAKA, H., NISHINARI, K., NAKAHAMA, N. AND FUKUBA, H. (1992). Effects of potassium chloride and sodium chloride on the thermal properties of gellan gum gels. *Bioscience Biotechnology* and *Biochemistry* **56**, 595–599.

MORRIS, V.J. AND MILES, M.J. (1986). Effect of natural modifications on the functional properties of extracellular bacterial polysaccharides. *International Journal of Biological Macromolecules* **8**, 342–348.

NISBET, B.A., SUTHERLAND, I.W., BRADSHAW, I.J., KERR, M., MORRIS, E.R. AND SHEPPERSON, W.A. (1984). XM6 a new gel-forming bacterial polysaccharide. *Carbohydrate Polymers* **4**, 377–394.

OGAWA, K., OKAMURA, K. AND SARKO, A. (1981). Molecular and crystal structure of the regenerated form of (1→3)-α-D-glucan. *International Journal of Biological Macromolecules* **3**, 31–36.

OGAWA, K., YUI, T., NAKATA, K., NITTA, Y., KAKUTA, M. AND MISAKI, A. (1996). Chain conformation of deacetylated beijeran calcium salt. *Bioscience Biotechnology and Biochemistry* **60**, 551–553.

OGAWA, K., YUI, T., NAKATA, K., KAKUTA, M. AND MISAKI, A. (1997). X-ray study of beijeran sodium salts, a new galacturonic acid-containing exopolysaccharide. *Carbohydrate Research* **300**, 41–45.

OJINNAKA, C., JAY, A.J., COLQUHOUN, I.J., BROWNSEY, G.J., MORRIS, E.R. AND MORRIS, V.J. (1996). Structure and conformation of acetan polysaccharide. *International Journal of Biological Macromolecules* **19**, 149–156.

OJINNAKA, C., BROWNSEY, G.J., MORRIS, E.R. AND MORRIS, V.J. (1998). Effect of deacetylation on the synergistic interaction of acetan with locust bean gum or konjac mannan. *Carbohydrate Research* **305**, 101–108.

OKUYAMA, K., OBATA, Y., NOGUCHI, K., KUSABA, T. AND ITO, Y. (1996). Single helical structure of curdlan triacetate. *Biopolymers* **38**, 557–566.

ROBIJN, G.W., THOMAS, J.R., HAAS, H., VAN DEN BERG, D.J.C., KAMERLING, J.P. AND

VLIEGENHART, J.F.G. (1995a). The structure of the exopolysaccharide produced by *Lactobacillus helveticus* 766. *Carbohydrate Research* **276**, 137–154.

ROBIJN, G.W., VAN DEN BERG, D.J.C., HAAS, H., KAMERLING, J.P. AND VLIEGENHART, J.F.G. (1995b). Determination of the structure of the exopolysaccharide produced by *Lactobacillus saké* O-1. *Carbohydrate Research* **276**, 117–136.

ROBIJN, G.W., GALLEGO, R.C., VAN DEN BERG, D.J.C., HAAS, H., KAMERLING, J.P. AND VLIEGENHART, J.F.G. (1996). Structural characterization of the exopolysaccharide produced by *Lactobacillus acidophilus* LMG9433. *Carbohydrate Research* **288**, 203–218.

ROSS-MURPHY, S.B. AND SHATWELL, K.P (1993). Polysaccharide strong and weak gels. *Biorheology* **30**, 217–227.

ROSS-MURPHY, S.B., SHATWELL, K.P., SUTHERLAND, I.W. AND DEA, I.C.M. (1996). Influence of acyl substituents on the interaction of xanthans with plant polysaccharides. *Food Hydrocolloids* **10**, 117–122.

ROUGEAUX, H., PICON, R., KERVAREC, N., RAGUENES, G.H.C. AND GUEZENNEC, J.G. (1996). Novel bacterial exopolysaccharides from deep-sea hydrothermal vents. *Carbohydrate Polymers* **31**, 237–242.

SCHORSCH, C., GARNIER, C. AND DOUBLIER, J.L. (1995). Microscopy of xanthan/galactomannan mixtures. *Carbohydrate Polymers* **28**, 319–323.

SHATWELL, K.P., SUTHERLAND, I.W. AND ROSS-MURPHY, S.B. (1990). Influence of acetyl and pyruvate substituents on the solution properties of xanthan polysaccharide. *International Journal of Biological Macromolecules* **12**, 71–78.

SHATWELL, K.P. SUTHERLAND, I.W., DEA, I.C.M. AND ROSS-MURPHY, S.B. (1990). The influence of acetyl and pyruvate substituents on the helix-coil transition behaviour of xanthan. *Carbohydrate Research* **206**, 87–103.

SKJÅK-BRÆK,G. AND LARSEN, B. (1985). Biosynthesis of alginate: purification and characterisation of mannuronan C-5-epimerase from *Azotobacter vinelandii*. *Carbohydrate Research* **139**, 273–283.

STANKOWSKI, J.D., MUELLER, B.E. AND ZELLER, S.G. (1993). Location of a second *O*-acetyl group in xanthan gum by the reductive cleavage method. *Carbohydrate Research* **241**, 321–326.

STOKKE, B.T., ELGSAETER, A., SMIDSRØD, O. AND CHRISTENSEN, B.E. (1995). Carboxylation of scleroglucan for controlled crosslinking by heavy metal ions. *Carbohydrate Polymers* **27**, 5–12.

SUTHERLAND, I.W. (1994). Structure function relationships in microbial exopolysaccharides. *Biotechnology Advances* **12**, 393–448.

SUTHERLAND, I.W. (1997). Microbial Biofilm Exopolysaccharides – Superglues or Velcro? (Biofilm Exopolysaccharides) In *Biofilms: Community Interactions* and *Control*. Eds. J.W.T. Wimpenny, P. Handley, P. Gilbert, H.M. Lappin-Scott and M.V. Jones, pp 33–39. Cardiff: Bioline Publications.

SUTHERLAND, I.W., JANN, B. AND JANN, K. (1970). The isolation of *O*-acetylated fragments from the K antigen of *Escherichia coli* O8:K27(A)H by the action of phage-induced enzymes from *Klebsiella aerogenes*. *European Journal of Biochemistry* **12**, 285–288.

TAIT, M.I. AND SUTHERLAND, I.W. (1989). Synthesis and properties of a mutant type of xanthan. *Journal of Applied Bacteriology* **66**, 457–460.

THORNTON, B.P., VETVICKA, V., PITMAN, M., GOLDMAN, R.C. AND ROSS, G.D. (1996). Analysis of the sugar specificity and molecular location of the beta-glucan-binding lectin site of complement receptor-type-3 (cd11b/cd18). *Journal of Immunology* **156**, 1235–1246.

USUI, S., TOMONO, Y., SAKAI, M., KIHO, T. AND UKAI, S. (1995). Preparation and antitumour activities of beta (1→6)-branched (1→3)-beta-D-glycan derivatives. *Biological and Pharmaceutical Bulletin* **18**, 1630–1636.

9

The Production and Applications of Genetically Modified Skin Cells

STEPHEN BEVAN[1,3], ROBIN MARTIN[1*] AND IAN A. MCKAY[2]

[1]*Blond McIndoe Centre, Queen Victoria Hospital, East Grinstead, West Sussex, RH19 3DZ, UK, [2]Centre For Cutaneous Research, St Bartholomew's and The Royal London School of Medicine & Dentistry, 2 Newark Street, London, E1 2AT, UK and [3]Present address: Institute of Cancer Research, Haddow Laboratories, 15 Cotswold Road, Sutton, Surrey, SM2 5NG, UK*

Introduction

The genetic modification of skin cells brings together two of the predominant fields in biotechnology: gene therapy and tissue engineering. Two major approaches exist for accomplishing the genetic modification of skin. The *ex vivo* approach aims to culture skin cells *in vitro* and introduce genetic material before returning the cells as *in vivo* tissue. For this strategy retroviral gene transfer has been the most successful. Considerable attention has been paid to the longevity of transplanted, retrovirally modified skin cells, particularly whether gene transfer to keratinocyte stem cells has been observed. An alternative strategy is to carry out direct gene transfer and the biolistic 'gene gun' approach is straightforward and has been successful for predominantly short-term gene expression. This area of research is revealing a large number of potential applications for clinical gene transfer to the skin.

WHY MODIFY SKIN CELLS GENETICALLY?

Here we review the rationale for undertaking genetic modifications of skin for clinical or investigative purposes and assess the progress that has been made towards these goals. Whilst skin consists of several distinct cell types that can be independently cultured, this article will restrict itself to genetic modifications whose ultimate aim is to affect the property of skin as an intact tissue, rather than in isolated component cell types. We review the structure of skin and identify target cells for genetic modification before proceeding to a discussion of the methods being developed and examples of applications in each of the scenarios outlined below.

*To whom correspondence may be addressed.

Biotechnology and Genetic Engineering Reviews – Vol. 16, April 1999
0264–8725/99/16/231–256 $20.00 + $0.00 © Intercept Ltd, P.O. Box 716, Andover, Hampshire SP10 1YG, UK

There are five main reasons for attempting the genetic modification of skin cells:

- to mark individual cells or cell lineages for studies in skin development and homeostasis – *gene marking of skin cells*
- to make skin cells capable of secreting proteins for the correction of systemic disease – *systemic gene therapy*
- to model inherited or acquired skin disease through genetic modification – *modelling skin disease*
- to correct for genetic defects in skin cells which lead to disease including those that lead to uncontrolled proliferation – *gene therapy of skin disease including cancer*
- to alter the properties of cultured skin cells to accelerate their performance on subsequent grafting – *gene therapy for wound healing and tissue engineering*

THE STRUCTURE AND CELLULAR COMPONENTS OF SKIN

A generalised structure of human skin is illustrated in *Figure 1*. Roughly speaking, skin can be divided into two layers, the outer avascular epidermis consisting of keratinocytes and the underlying vascular dermis that is composed of a collagenous matrix elaborated by fibroblasts. At the epidermal-dermal junction there is a 'basement

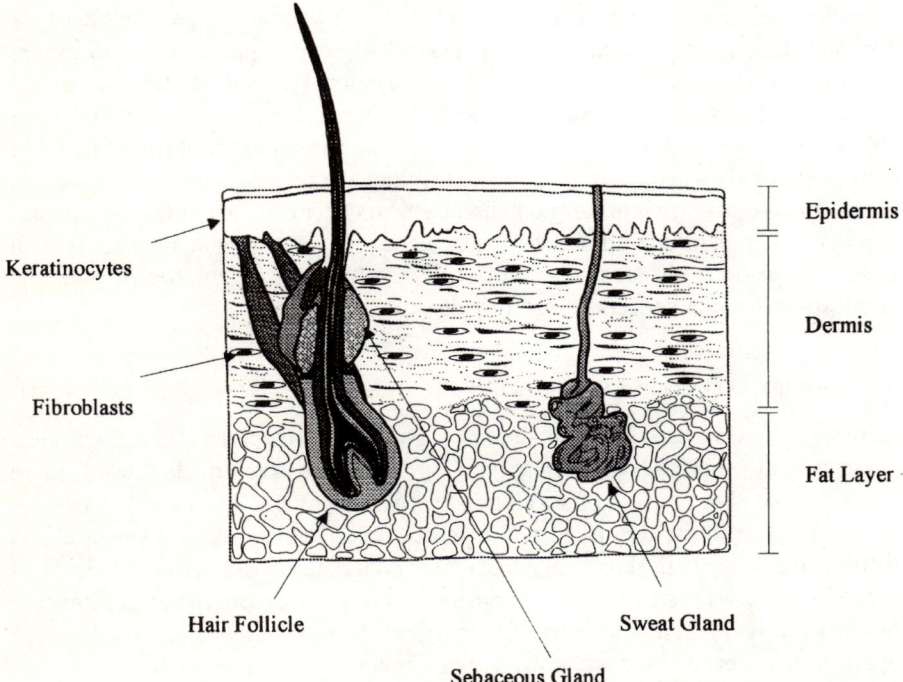

Figure 1.   The structure of human skin to show possible targets for gene modification. Keratinocytes form the epidermis that is divided from the dermis by a protein matrix structure known as the basement membrane. The dermis contains predominately fibroblasts. Blood vessels penetrate the dermis but not the epidermis. Cutaneous innervation however, extends into the epidermal layer. The hair follicle and sweat gland are also potential targets for genetic modification.

membrane'. This is a two dimensional net-like structure comprised of a number of matrix proteins: collagen IV, collagen VII and laminin, to name but a few, whose function is to provide secure anchorage for the epidermal layer above. In addition, there are skin organs or appendages such as the pilosebaceous unit and sweat glands that are spread throughout the skin and which traverse both the dermis and the epidermis.

In the epidermis, as well as keratinocytes, there are populations of melanocytes and Langerhans' cells. All have been targets for genetic modification in one study or another. There are also a number of specialised cell types such as neuroendocrine Merkel cells. The epidermis has no direct blood supply but numerous capillaries loop into the dermis near to the epithelial layer. In the dermis, the main target cells for genetic modification have been the resident fibroblasts which generate and maintain the extracellular matrix, but there are other cell types including nerve cells, dermal dendrocytes, transitory blood-borne cells and endothelial cells, all of which represent targets for genetic modification. A review of the structure of human skin that includes electronmicrographs of the principle cellular components can be found in Holbrook (1994).

ACCESS OF GENETIC MATERIAL TO SKIN CELLS *IN VIVO* AND *EX VIVO*

Given that a major function of the skin is to protect the body from infectious agents, one might suppose that it would be difficult to get viruses or nucleic acids to penetrate intact skin and this is indeed the case. However, a number of methods have been developed for circumventing the skin's defensive systems to deliver nucleic acids to target cells. These are discussed in more detail below and can be divided into viral, physical or chemical methods. These can be employed to target skin directly *in vivo* or used in *ex vivo* methods which target cultured or explanted skin cells prior to their return as grafts. One of the major advantages behind the early consideration given to gene therapy of skin was that *ex vivo* methods for isolating individual populations of skin cells and culturing them *in vitro,* before returning them to patients, were already well established (Rheinwald and Green, 1975; Green *et al.*, 1979; Hansbrough *et al.*, 1989; Langdon *et al.*, 1991). These methods remain in clinical use for wound healing and skin regeneration and many new advances are continuing to appear that are improving the delivery of cultivated cells back into functional tissues (Navsaria *et al.*, 1995). It has therefore been a natural progression to introduce new genetic material into skin cells in culture and to expand the modified cell populations in readiness for grafting back to the donor. In comparing *in vivo* with *ex vivo* methods for genetic modification one has to balance the greater accessibility of *ex vivo* gene delivery to cultured cells with the resultant problems of regenerating authentic functioning skin – a problem of tissue engineering not genetic engineering. In contrast, the benefits of direct gene delivery to intact tissue have to be balanced against the greater difficulty encountered in achieving controlled and efficient gene transfer. Along with questions of the relative efficacy of each of the different methods for delivering nucleic acids to skin, there are many issues of safety that will need to be addressed before skin cell genetic modification becomes acceptable for clinical therapeutic use. These include whether:

- integrating genetic material will activate proto-oncogenes or inactivate tumour suppressor genes;
- ligand-coupled nucleic acids or liposomes appear as foreign bodies to the immune system limiting the therapeutic intervention to a single occasion;

- expressed gene products will upset homeostatic mechanisms in the target cells and induce them to proliferate in an uncontrolled fashion;
- there is a possibility that viral vectors could recombine with endogenous cellular sequences to generate new, potentially pathogenic viruses.

The direct *in vivo* and the indirect *ex vivo* approaches are likely to differ in the spectrum of safety issues that will need to be addressed once proof of a therapeutic principle has been established. There was early recognition that potentially hazardous changes in genetically modified skin cells could be relatively easily surgically excised. This contrasts with genetic modifications to less readily accessible target cells of the blood or internal organ systems. However this belief may need to be reassessed for certain methods of direct *in vivo* gene transfer such as the use of gene guns or lipid/DNA injections, where it may be difficult to be sure that only skin cells are exposed.

EXPERIMENTAL SYSTEMS FOR GENETICALLY MODIFIED SKIN CELLS

Whilst there are a number of non-clinical applications for genetic modification of skin, many investigations have at their core an eventual clinical use. In preparation for this, combinations of *in vitro* and experimental animal systems are required before clinical trials are proposed and approved. For example, *in vitro* models of human skin have been described in which both secretion from keratinocytes (Katz and Taichman, 1994) and gene transfer into keratinocytes (Badiavas *et al.*, 1996) (Garlick and Taichman, 1992) have been studied. However, *in vitro* cultured keratinocytes may not display correct dermal-epidermal interactions or incorporate the effects of blood vessels and lymphocytes: simple *in vitro* systems may not replicate the complexity of the real *in vivo* conditions. Inevitably, studies using animal models must be considered with the understanding that they come from different species that may have different modes and efficiencies of skin cell behaviour.

Ideally, an animal chosen for an experimental model should have skin as near to human as possible, so as to provide a valid experimental system. The similarity between human and porcine skin has long been recognised, and the pig is widely used as a model for the investigation of wound healing (Compton, 1994; Dodds, 1982; Kangesu *et al.*, 1993; Bevan *et al.*, 1997; Ng *et al.*, 1997). Although the pig appears to be an ideal system for testing genetic modification of skin cells, to keep and experiment on pigs is expensive and technically challenging. Therefore some researchers have chosen to use mice, rats, rabbits or dogs instead.

The rabbit has been used extensively in studies of factor IX secretion from transduced fibroblasts, and the method of grafting the transduced cells has progressed from surgical implantation to subcutaneous injection (Zhou *et al.*, 1993).

Dogs have also been used as an animal model for gene therapy trials. Marked keratinocytes have been shown to survive after grafting to a canine full thickness wound (Stockschlader *et al.*, 1994). Trials have been conducted in dogs with human adenosine deaminase transduced canine fibroblasts, although in this instance the animals did not show sustained high level expression ten weeks post-grafting (Ramesh *et al.*, 1993).

By far the most favoured animal models are the rat and the mouse. The predominant criteria of the economy of keeping large numbers and the availability of inbred stocks, leading to reduced variability, weigh heavily against the profound differences between human and rodent skin. For example, rat fibroblasts genetically modified to express porcine growth hormone have been both injected and grafted (Chen *et al.*, 1995) and showed long term expression of the transduced gene up to 70 days post grafting (Lu *et al.*, 1996).

Use of the athymic (nude) and severe combined immunodeficient (SCID) mouse has allowed the grafting of keratinocytes from different species to mice (Morgan *et al.*, 1987). This is particularly important where naturally occurring animal models are not available for human disease, since keratinocytes from patients can then be grafted to the mouse and the proposed therapy tested *in vivo* without the risk of rejection (Fenjves *et al.*, 1997). Athymic mice also allow the secretion of human gene products from grafts of transduced cells without the risk of inactivation of those produced by the host's immune system before their effects can be studied. This has been demonstrated with keratinocytes expressing the human transferrin gene (Petersen *et al.*, 1995).

PERMANENT VERSUS TRANSIENT GENETIC MODIFICATION

When discussing gene modification it is important to be clear whether a permanent or a transient phenomenon is required. Permanent gene transfer may be required for applications such as the cure of an inherited skin disease. In contrast, a transient pulse of gene product may be more desirable for a genetic treatment to accelerate wound healing. As a rough guide, most physical and chemical gene transfer techniques into normal skin or cultured cells give predominantly transient expression. Some viruses, notably retroviruses, have very efficient mechanisms for genome integration and consequent changes in gene expression are likely to be longer lived.

In vivo methods for transfer of genetic material to skin cells

VIRAL METHODS

Many types of virus have been modified to act as vectors for the transfer of foreign nucleic acids into mammalian cells. (Viral gene delivery is known as *transduction* as opposed to *transfection* which describes the delivery of DNA to a cell assisted by chemical means). Clearly there are important safety considerations before the use of viruses *in vivo* to transduce skin tissues in animals. Even more checks will be required before the consideration of viral vectors for patients. A handful of viruses have been found to be suitable for this purpose. Often elements of the viral genome that are essential for replication or infection can be removed and replaced by the foreign nucleic acid. The missing replicative and or infective functions are supplied in *trans* by special host cell lines. Whilst many viral vectors have been employed against cultured skin cell targets, some viruses, notably adenoviruses, have been used to good effect *in vivo* (Kozarsky and Wilson, 1993; Trapnell, 1993). Adenoviruses expressing *lacZ* (the *E. coli* β galactosidase gene) or human α1 anti-trypsin were compared in *ex vivo* and *in vivo* methodologies in mice by Setoguchi *et al.* (1994). In both cases expression of the introduced gene was observed for up to 14 days. Lu *et al.*, (1996) achieved gene expression from adeno- and herpes simplex viruses, but only when the physical barrier to viral infection, the squamous layer of skin, was removed by tape-stripping. This exposed basal cells that could then be transduced with topically applied virus. There were though, significant signs of cytotoxicity following viral application. These are side effects known to limit the very highly effective gene transfer efficiencies obtainable with herpes and adenoviral vectors. Normally, components of the immune system within intact skin (Rambukkana et al., 1995) are perfectly capable of dealing with potential viral invaders. However, if the immune system is compromised by immunosuppression, opportunistic infections can occur. For example, such appears to be the case in renal transplant recipients whose skin is frequently found to be infected with papillomaviruses (Glover et al., 1993). Therefore, the clinical deployment of viral vectors has to be viewed with the potential for the exposure of compromised patient populations as well as fit and healthy individuals.

LIPID-COATS AND VESICLES APPLIED TO *IN VIVO* GENE TRANSFER

The rationale for coating nucleic acid with lipids is to allow highly negatively charged nucleic acid molecules to traverse the plasma membrane of the target cell (Cotten and Wagner, 1993). Lipid coating may also offer immunity from nucleases in phago-lysosomes. Thus, the nucleic acid is protected much as it is during viral infection. Initially different lipophilic coatings were developed for *in vitro* transfection experiments. However, researchers have attempted to exploit the lipophilic nature of these complexes in attempts to target keratinocyte stem cells *in vivo*. In hairy skin the stem cells are thought to be located close to the opening of the sebaceous gland onto the hair shaft (Lane et al., 1991). This region is known as the 'bulge' in rodent hair follicles although the bulge is not so apparent in humans. The researchers reasoned that if lipophilic complexes could diffuse in the lipid-rich sebum (the fatty lubricant secreted by sebaceous glands) to the stem cells they might transduce them. LacZ gene markers were detected three days after lipid/DNA application to mice only in the hair follicles (Li and Hoffman, 1995). LacZ expression was seen as soon as six hours post treatment in the study by Alexander and Akhurst (1995). Expression peaked at 24–48 hours and was greatly diminished by seven days. Delivery to dermis, epidermis and hair follicles was observed. The ability of topically applied lipid/nucleic acid complexes to transduce a variety of skin cell types without the associated risks of viral transduction creates an exciting avenue for further research and development.

THE 'GENE GUN' OR BIOLISTIC APPROACH

The accessibility of skin has encouraged approaches to transfection that involve blasting DNA in various forms directly at intact skin. Biolistics is the term coined to describe the impact of ballistic particles on biological systems. Perhaps surprisingly, this system can work very well and several types of gene gun have been developed for this purpose. Pellets of tungsten or gold are coated with nucleic acid and then fired from the gun at the skin surface. By adjusting parameters such as the number of pellets, the amount of nucleic acid, and the force with which the pellets are expelled, it is possible to adjust the level of gene expression and the depth to which the pellets will penetrate. Luciferase expression was detected in rat dermis 1.5 years after particle mediated transfection (Cheng et al., 1993). Transfer to the epidermis was predominant in the experiments of Lu et al., (1996) with physiological effects observed following delivery of sequences encoding TGFα (transforming growth factor α). Viral sequences have also been delivered by biolistic means. Xiao and Brandsma (1996) showed that rabbit cotton tail virus DNA could be delivered to rabbit skin. Other biolistic methods involve firing solutions of nucleic acids under high pressure (Furth *et al.*, 1995). Following high pressure jet delivery of 100–300 μl, transfected cells were detected up to 2 cm away from the skin surface. Transfection of the dermis might be anticipated to give longer term expression whilst the ability to deliver nucleic acids to the stem cell fraction will determine the longevity of the expression within epidermis. One of the most promising applications of biolistic particle mediated gene transfer may be in genetic immunisation: the delivery of an antigen to the skin by means of the DNA encoding the antigen rather than the protein itself (Nabel *et al.*, 1993; Ciernik *et al.*, 1996; Condon *et al.*, 1996). This approach side-steps the problems of purification, sensitivity and safety associated with the conventional preparation of vaccines.

SKIN PUNCTURE BY NEEDLES

As an alternative to biolistics, subcutaneous injection of nucleic acid or virus has been shown to allow gene transfer to skin cells. This method seems to target mainly dermal cells and is not as useful as the biolistic approach for targeting epidermal cells (Lu *et al.*, 1996). A refined version of this technique uses a high frequency oscillating bundle of fine metal needles to achieve skin cell transfection (Ciernik *et al.*, 1996). The results appear promising and again the deeper dermal tissues and dendritic cells are those predominantly targeted (Condon *et al.*, 1996).

ELECTRICAL METHODS

While it is feasible to use electroporation to transfect skin cells cultured *in vitro* (see below), there have been a limited number of attempts to achieve electroporation *in vivo*. One can imagine an entirely new set of safety considerations, but in one case skin-depth targeting was achieved by varying pulsed electrical fields and pressure from calliper-type electrodes on topically applied nucleic acids (Zhang *et al.*, 1996). Interestingly, in control animals, gene constructs were expressed in some hair follicles without application of the pulsed fields.

In vitro or *ex vivo* methods for transfer of genetic material into skin cells

CULTURE OF SKIN CELLS

The use of skin cells as a target for 'gene therapy' is made simpler by the fact that they can be cultured *in vitro*. Skin culture was attempted long before the advent of 'gene therapy', mainly in an attempt to replace lost or damaged skin tissue after injury. This section briefly outlines the culture of each of the most important cells in skin, and includes examples of gene transfer into each cell type.

KERATINOCYTES

Keratinocytes form the outermost layer of cells over the entire body. Early cultivation methods relied on the outgrowth of cells from explants of skin incubated in medium, although the results were insufficient for therapeutic purposes. It was not until 1975, when Rheinwald and Green managed to grow keratinocytes on a feeder layer of connective tissue cells – lethally irradiated 3T3 fibroblasts – that clinically useable grafts could be produced (Rheinwald and Green, 1975; Rheinwald, 1977; Rheinwald, 1989). Further refinements to culture conditions over the succeeding years have meant that cultures can now be maintained either as undifferentiated monolayers to maximise expansion or induced to differentiate and stratify (Boyce and Ham, 1983; Leigh and Watt, 1994).

Since it is so readily accessible and can be easily and routinely subcultured, the keratinocyte layer of the skin is a particularly attractive target for genetic manipulation (Greenhalgh *et al.*, 1994). In addition, since it covers the entire body and is so close to the circulatory system, it has been proposed to use skin as a secretory vehicle for the systemic distribution of gene products via the blood (Barra *et al.*, 1994). Keratinocytes have been tested using the entire range of gene transfer methods. Viral transfer was demonstrated by papillomavirus (Burnett and Gallimore, 1983); Epstein–Barr viral vectors (Jensen *et al.*, 1994); adenoviral vectors (Setoguchi *et al.*, 1994); transfected using poly-L-ornithine (Nead and McCance, 1995) and a variety of other chemical and physical methods (Jiang *et al.*, 1991). Earlier work was primarily as an experimental tool to study keratinocytes themselves, rather than as a precursor to gene therapy. With the exception of certain lipid/DNA complexes, most of the chemical or physical

Figure 2. *Ex vivo* retroviral gene transfer to cultured porcine keratinocytes. A section through *lacZnls* positive keratinocytes in porcine skin. Cultured autologous keratinocytes were transferred as a layer 1–2 cells thick to a de-epidermalised dermal graft in pigs. Two weeks later the graft was excised. The cultured *lacZ* marked cells have contributed to all layers of the tissue, (Ng, Bevan & Martin, unpublished; Bevan *et al.*, 1997; Ng *et al.*, 1997). Colour images similar to this appear in Ng *et al.* (1997).

gene transfer methods in routine use for laboratory tissue culture cell lines work quite inefficiently on normal diploid keratinocytes. In contrast, retroviral vectors work well and have been used extensively in gene transfer into keratinocytes (Garlick *et al.*, 1991; Garlick and Taichman, 1993; Morgan *et al.*, 1987). Since a number of retroviral producing lines are derived from 3T3 fibroblasts these can be irradiated and used as viral producing feeder layers to ensure a continual exposure of seeded keratinocytes to the relatively short-lived vectors. With high titre $> 10^6$ ml^{-1} retroviral producer lines, transduction rates can approach 100% (Carroll *et al.*, 1993). Permanent genetic modification has been anticipated since retroviral vectors integrate randomly into the genome with high efficiency. In practice some short fall is observed that has been put down to viral inactivation due to methylation (Fenjves *et al.*, 1996; Flowers *et al.*, 1990). Work from the authors' laboratory using *ex vivo* retroviral transfer into cultured porcine keratinocytes, with subsequent grafting of epidermal sheets of 1–2 cells thick, is illustrated in *Figure 2*. The β-galactosidase expression of the *E. coli lacZ* gene is detected by the X-gal histochemical stain in a section of pig skin that was harvested 2 weeks after grafting (Bevan *et al.*, 1997; Ng *et al.*, 1997). In this case, the *lacZ* gene contained a nuclear localising signal (nls) which targets the transgene to the nucleus (Ferry *et al.*, 1991). This emphasises the effect and distinguishes it from any endogenous β-galactosidase that is always cytoplasmically located.

A major discussion point with respect to keratinocyte culture is the elusive 'keratinocyte stem cell'. This has yet to be formally identified and isolated (Barrandon and Green, 1987; Barrandon, 1989; Watt, 1998). It is believed that something like stem cells must be present in culture since very large populations can be established from individual cells (Mathor *et al.*, 1996). Keratinocytes *in vitro* are not immortal though; cultures will eventually stop dividing as they reach their intrinsic 'Hayflick' limit. It is possible that the complex interactions between dermal fibroblasts and epidermal keratinocytes may not be sufficiently mimicked *in vitro* (Fusenig, 1994). Thus, cultures might be unable to provide a niche exactly comparable to the *in vivo* environment, with the consequence that stem cells may not display the same spectrum of properties *in vitro* as they do in real skin. Many groups are actively searching for markers that would enable stem cells to be specifically targeted (Li *et al.*, 1998). This is important since in many cases treatment of genetic disease will require long-lived expression of the genetic modification (De Luca and Pellegrini, 1997). The therapeutic effect will be limited if no stem cells receive the genetic modification since the modified keratinocytes will be lost through the natural turnover of cells moving from the proliferative basal layer to the non-dividing upper layers of the skin. The search for the *in vivo* location of stem cells within skin is discussed in further detail below in the section looking at the application of gene transfer as a genetic marker to track the fate of individual cells in skin.

DERMAL FIBROBLASTS

Dermal fibroblasts play an important a role in the maintenance of the skin. Soluble factors secreted by fibroblasts contribute to the basement membrane and stimulate keratinocytes to proliferate, differentiate and elaborate the keratinocyte share of basement membrane components (Marinkovich *et al.*, 1993). One of the factors secreted by fibroblasts has been identified as IGF I (insulin-like growth factor I) (Barreca *et al.*, 1992). In contrast, PDGF (platelet-derived growth factor) that stimulates fibroblast matrix production and endothelial cell mediated vascularisation of the dermis is produced by skin keratinocytes (Ansel *et al.*, 1993). Fibroblasts are often used as a 'test system' before transfer of genes into other cell types since they are abundant, well characterised and grow relatively rapidly in simple media. Calcium phosphate transfection of fibroblasts has been performed for almost 20 years (Shih *et al.*, 1979; Gorman, 1985) although more efficient viral techniques have since been developed (Dai *et al.*, 1992).

The dermis and its complement of dermal fibroblasts may be a slightly less attractive target for gene therapy since the techniques for autologous (an individual's own cells or tissues) culture and replacement are less developed than those for keratinocytes. However, as tissue engineers learn how to replace dermis as well as epidermal tissue following growth of cells in culture, *ex vivo* gene transfer to skin fibroblasts may become a very attractive option, especially since sub-cutaneous placement may avoid surface scars (Hoeben *et al.*, 1993; Ramesh *et al.*, 1993; Petersen *et al.*, 1995).

DERMAL ENDOTHELIAL CELLS

Endothelial cells are a specialised fibroblastic cell type that forms the epithelial covering of blood vessels throughout the circulatory system. The difficulty associated with targeting endothelial cells in the dermis specifically could be overcome either by modifying a viral vector so that it only transduced endothelial cells, or more simply by transducing endothelial cells *in vitro* and seeding them into the dermis before grafting on a keratinocyte sheet.

Transduction of vascular endothelial cells with retroviral vectors expressing the *lacZ* gene has been shown in one study to impair cellular proliferation *in vitro* and graft endothelialisation *in vivo* (Baer *et al.*, 1996). This may prove a major stumbling block to endothelial cell retroviral gene therapy since the decrease in proliferation associated with transduction may outweigh the benefit of the transduced gene. It remains to be seen how general an effect this is (Sackman *et al.*, 1996).

MELANOCYTES

Melanocytes are resident in the epidermis along with keratinocytes after migration from the neural crest in early development. They maintain their numbers in a constant ratio with keratinocytes by cell division and show a dendritic morphology (De Luca *et al.*, 1994). The main function of the melanocyte is to produce pigment, primarily for protection from UV radiation. It may be that the particular properties of the melanocyte might be useful for secretion of gene products since their residence time in the skin could be a solution to the problems posed by the turnover of proliferating keratinocytes. Melanocytes can be cultured *in vitro*, and several factors have been identified as stimulants for their growth, including phorbol ester, hydrocortisone and cyclic AMP. Tissue engineered skin substitutes have been constructed containing melanocytes (Boyce *et al.*, 1993). Little work on gene transfer into normal skin melanocytes seems to have been reported, although there have been many studies that have employed gene transfer to cultured melanoma cell lines (Wakeling *et al.*, 1995).

DENDRITIC CELLS

Epidermal dendritic cells, more commonly referred to as Langerhans' cells are concerned with the uptake, processing and presentation of antigen. The density of Langerhans' cells in the skin is site-specific and they are usually found suprabasally in the spinous layer. As antigen presenting cells, Langerhans' cells migrate through lymphatics to lymph nodes to present foreign antigens. This has provided a basis for the development of targeted immunisation strategies *in vivo* (Condon *et al*., 1996)

It is possible to culture Langerhans' cells *in vitro*, although like melanocytes, they do not withstand serial culture or freezing well (Abe *et al*., 1995). Indeed, the routine passage of cultured keratinocytes to be used for grafting can be performed to remove dendritic cells. This has been shown to prolong the life of cultured allogeneic keratinocytes over tissues containing Langerhans' cells (Aubock *et al*., 1988). There is an interesting debate over whether dendritic cells *in vitro* represent the *in vivo* state accurately (Tsunoda *et al*., 1997). Transduction of dendritic cells has been performed with the bacterial reporter gene *lacZ* in a retroviral vector (Aicher *et al*., 1997). Stable integration was shown for 20 days after gene transfer. The use of viral vectors to transduce dendritic cells appears to be much more successful than physical methods such as electroporation and calcium phosphate precipitation. Transduction efficiencies of 35–67% have been reported with a retroviral vector (Aicher *et al*., 1997) and 95% with an adenoviral vector (Arthur *et al*., 1997).

HAIR FOLLICLE CELLS

The culture of hair follicles can be split into two broad categories. There are attempts to grow a complete hair follicle in a 3D matrix, and there are attempts to grow specific cells associated with the hair in standard monolayer format. Hair follicle culture is complicated by the large number of different cell types associated with the follicle. These include outer root sheath cells, dermal papilla cells, dermal sheath cells and germinative epidermal cells (Reynolds and Jahoda, 1994a). Co-culture of various cell types has been attempted, usually in a collagen matrix, with varying degrees of success (Reynolds and Jahoda, 1994b; Arase *et al*., 1990; Arase *et al*., 1994), or by plating a mixed suspension of whole skin (Ihara *et al*., 1991).

Hair culture is further complicated by the close association of the sebaceous gland *in vivo*, although whole organ culture has been attempted *in vitro* (Harmon and Nevins, 1994). Culture of the human pilosebaceous unit has been achieved but results are still sub-optimal and growth could only be maintained for seven days *in vitro* (Sanders *et al*., 1994). The keratinocyte stem cells are thought to reside in the hair follicle bulge region. Whole organ culture including this region shows the most promise for the growth of hair follicles *in vitro*.

To develop effective gene therapy of hair specific cells it is necessary to target these cells, since they form only a small percentage of skin cells. Liposomes have been shown to selectively target hair follicles and this may provide the basis for future applications (Li and Hoffman, 1995). This technique is simple and non-invasive and can be used without the need for *in vitro* manipulation. Treatment of allopecia, whether by natural hair loss through ageing or as a side effect of other treatment such as chemotherapy is still some way away however. The hair follicle should be treated as an organ composed of diverse cell types rather than a single cell type, and provides a greater challenge than most gene therapy targets in skin.

Applications of genetically modified skin cells

GENETIC MARKING OF SKIN CELLS

The ability to use gene transfer to introduce a permanent tag into a specific population of skin cells has found extensive use in both fundamental studies of skin function and as preliminary fact-finding in preparation for gene therapy applications. With respect to tracking skin cells as they proliferate and differentiate within skin, the *E. coli* β-galactosidase *lacZ* gene has been used extensively as a reporter.

The use of reporter gene transfer to monitor the fate of *in vitro* cultured keratinocytes returned to an *in vitro* or *in vivo* wound has been studied by many groups (Garlick and Taichman, 1992; Garlick and Taichman, 1993; Setoguchi *et al.*, 1994; Stockschlader *et al.*, 1994; Vogt *et al.*, 1994; Jensen *et al.*, 1994; Ng *et al.*, 1997; Mackenzie, 1997; Kolodka *et al.*, 1998). This has been seen as particularly important since the keratinocyte 'stem cell' has yet to be isolated and its *in vivo* position has not been convincingly mapped. Stem cells are slow cycling cells that give rise to all the cells of the epidermis. They show long-term persistence and therefore might be expected to be required before long-term expression of transduced genes could be achieved (De Luca and Pellegrini, 1997). Clonal proliferation units; columns of marked cells have been identified in mice grafted with cultured murine keratinocytes transduced *in vitro* with *lacZ* retroviral vectors (Mackenzie, 1997). Very similar pictures were identified in human keratinocytes labelled in a comparable fashion and grafted to nude mice (Kolodka *et al.*, 1998). In contrast, work along the same lines failed to find retention of significant expression longer than about four weeks in the form of clonal proliferation units, despite a higher initial transduction frequency (Choate and Khavari, 1997). Interestingly, the appearance of clonal proliferation units within murine keratinocytes on mice (Mackenzie, 1997) and human keratinocytes on mice were nearly equivalent (Kolodka *et al.*, 1998). In pigs retroviral marking and transfer of cultured autologous porcine keratinocytes to regenerating wounds turned up a rare example of a column of *lacZ* marked cells. The marked population stretched the length of what appeared to be two adjacent rete ridges from the basal tip to the epidermal surface (Ng *et al.*, 1997). Some researchers have claimed this image supports the notion that epidermal stem cells lie not at the bottom of rete ridges, but at the tips of dermal papillae (Iizuka and Ishida-Yamamoto, 1997; Iizuka *et al.*, 1996).

SYSTEMIC GENE THERAPY THROUGH SKIN CELL MODIFICATION

The ease of skin culture and grafting of autologous keratinocytes and fibroblasts has fostered a significant number of groups undertaking genetic modification of skin to achieve the secretion of a desirable protein into the systemic circulation (Fenjves *et al.*, 1994; Petersen *et al.*, 1995; Krueger *et al.*, 1994). Depending on the levels of gene product synthesised and secreted, it has been estimated that for some inherited disorders as little of 2% of body surface area would need to be grafted with genetically modified skin. Early experimental approaches focused on the delivery of human growth hormone as a model system (Morgan *et al.*, 1987; Jensen *et al.*, 1994). In addition to the secretion of pharmaco-active proteins, there has also been significant interest in the provision of a metabolic sink to remove unwanted substrates from the circulation (Fenjves *et al.*, 1997; Sullivan *et al.*, 1997).

Treatment of haemophilia

Due to the risk of HIV infection from blood products, the potential to treat patients deficient in either clotting factors VIII and IX has become of particular importance. Early reports described the transduction and high levels of expression of human factor IX in rabbit fibroblasts by retroviral vectors (Lu *et al.*, 1993) and in cultured human primary skin fibroblasts from a haemophilia B patient (Zhou *et al.*, 1993). Other groups have used the keratinocyte as a target. Studies have shown that keratinocytes can secrete factor IX in a fully active form, even though these cells do not normally synthesise it (Gerrard *et al.*, 1996). The length of expression of secreted factor IX *in vivo* is of manifest importance. In some studies detectable levels have remained for only six weeks even though the graft is still present (Fenjves *et al.*, 1996; Page and Brownlee, 1997). Later work, however, has reported human factor IX still present in the blood of mice grafted over a year previously (White *et al.*, 1998).

Treatment of hyperlipoproteinemia

A second area of major research interest for delivery of pharmacoactive proteins to the systemic circulation by transduced keratinocytes is in the treatment of familial type III hyperlipoproteinemia by delivery of apolipoprotein E. ApoE is a plasma protein that serves as a ligand for low density lipoprotein receptors and is involved in transport of cholesterol and other lipids around the body. Hyperlipoproteinemia is caused by a mutant form of ApoE, and is characterised by elevated cholesterol levels and accelerated coronary artery disease. ApoE is normally secreted by keratinocytes, and has been shown to reach the systemic circulation (Barra *et al.*, 1994). Endogenous ApoE is secreted only from basal keratinocytes, but transduced epidermal cells grafted to a murine model indicated secretion from both basal and suprabasal cells indicating that more ApoE per unit area of basal skin could be produced from transduced cells than from normal skin (Fenjves *et al.*, 1994).

Treatment of inherited disease through provision of an epidermal metabolic sink

Two inherited human diseases have received attention with respect to the potential for gene modification to enable keratinocytes to express enzymes that will clear the systemic circulation of a build up of toxic metabolites. Adenosine deaminase (ADA) deficiency causes an accumulation of adenosine and deoxyadenosine that leads to immunodefficiency. ADA gene transfer into keratinocytes established that there could be sufficient enzymatic activity to detoxify the circulation from a modest sized graft; 140 cm^2, if sufficient blood flow to the skin could be achieved (Fenjves *et al.*,

1997). Accumulation of ornithine is found in patients with the progressive blindness disease gyrate atrophy. Studies are underway to see whether autologous keratinocyte grafts modified with retroviral gene transfer of cDNAs encoding ornithine-delta-aminotransferase (OAT) could provide a sufficient metabolic sink to reduce the levels of circulating ornithine (Jensen *et al.*, 1997; Sullivan *et al.*, 1997).

MODELING SKIN DISEASE THROUGH GENETIC MODIFICATION

In common with most other areas of developmental and molecular medicine, the generation of transgenic mice is playing a major role in understanding skin disease. Transgenic technology involves introducing a transgene construct into an early embryo or a totipotent tissue culture cell. The animals develop so that a large proportion of, or indeed the entire cell population of the animal expresses the transgene (Rothnagel *et al.*, 1993; Sellheyer, 1995). By introducing mutated transgenes, the role that a gene plays in both development or disease can be studied, and hopefully lead to better treatment in a clinical setting.

Standard transgenic techniques can be refined to allow expression of the transduced gene in a specific subpopulation of cells by using a tissue specific promoter. In the case of skin, there are numerous such promoters and even cell type-specific promoters. Usefully, there are also promoters which allow expression specifically in basal keratinocytes, such as the keratin 5 (Byrne and Fuchs, 1993) and 14 gene promoters, and in suprabasal cells, such as the keratin 1 and 10 and loricrin (Di Sepio *et al.*, 1995) gene promoters. For example, the keratin 14 gene promoter has been used to express the human growth hormone gene in adult mouse skin. Interestingly, skin grafts from these animals still expressed the protein on non-transgenic host mice (Wang *et al.*, 1997). The skin blistering disease epidermolytic hyperkeratosis has been induced in a transgenic mouse model by expression of phorbol ester-inducible regulatory sequences in a subpopulation of keratinocytes. Application of a phorbol ester produces induction of a mutant keratin product and the onset of disease (Takahashi and Coulombe, 1996). The ability to induce model diseases at will in this way should allow a better understanding of their aetiology and consequently their treatment with the gene modification techniques described in this review (see *Figure 3*).

Caution must be exercised however when interpreting data from transgenic animal models of human disease. In particular, it should be borne in mind that mouse skin is not itself a good model of human skin, being thinner and hairier. For example, in attempts to reproduce the phenotype of psoriasis, an inflammatory disease of skin associated with scaling and hyperproliferative keratinocytes, numerous transgenic mice have been made which express growth factor genes in basal keratinocytes (Vassar and Fuchs, 1991; Turksen *et al.*, 1992) or suprabasal cells (Carroll *et al.*, 1997). The transgenic animals all reproduced some aspects of the psoriatic phenotype but none reproduced all of them. A closer fit to the psoriatic phenotype was obtained when transgenic animals were generated which showed aberrant integrin expression (Carroll *et al.*, 1995). However, expression of transgenes with no obvious connection to psoriasis can also generate a psoriatic phenotype (Wilson *et al.*, 1990) and it may be that a generalised inflammatory response of the mouse to transgene expression in the skin is being misinterpreted as a psoriatic phenotype in some cases.

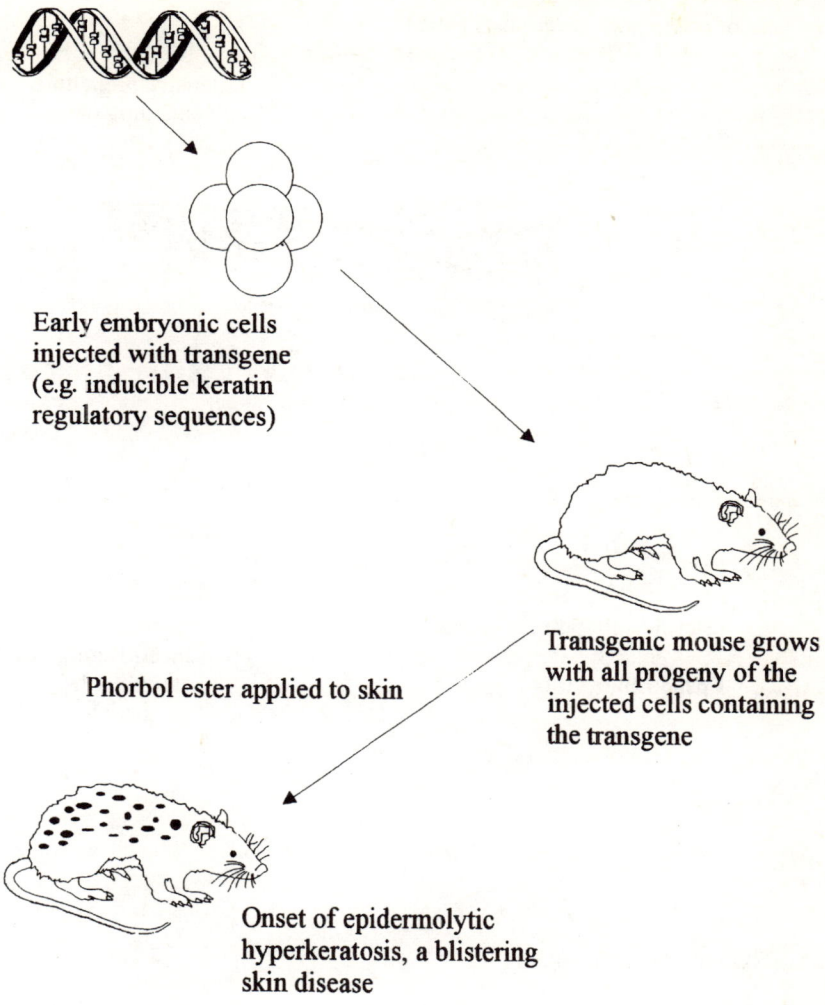

Early embryonic cells
injected with transgene
(e.g. inducible keratin
regulatory sequences)

Transgenic mouse grows
with all progeny of the
injected cells containing
the transgene

Phorbol ester applied to skin

Onset of epidermolytic
hyperkeratosis, a blistering
skin disease

Figure 3. The development of transgenic technology. A gene of interest is introduced into mouse embryos. If the transgene is controlled by an inducible tissue specific promoter, the effect of the transgene can be studied specifically in the skin (Takahashi and Coulombe, 1996).

GENE THERAPY OF SKIN DISEASE INCLUDING CANCER

Genetic modification for inherited skin disease

Perhaps the most obvious use for genetically modified keratinocytes is in the treatment of inherited skin conditions. The skin of X-linked ichthyosis sufferers is characterised by an extensive keratinisation followed by rapid and premature shedding of most suprabasal cell layers (Scheimberg *et al.*, 1996). This is brought about by a deficiency in steroid sulphatase, a recessive single gene defect ideal for correction by gene therapy. The introduction of a normal steroid sulphatase cDNA into keratinocytes by Epstein Barr virus has led to active protein being produced and a slowing down of the maturation of cells from ichthyotic skin *in vitro* (Jensen *et al.*, 1993) (see *Figure 4*).

A. Skin of an X-linked ichythosis patient

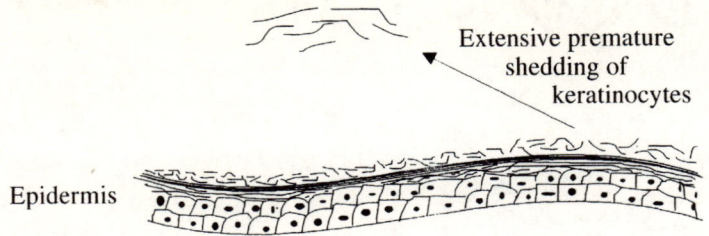

B. Gene therapy

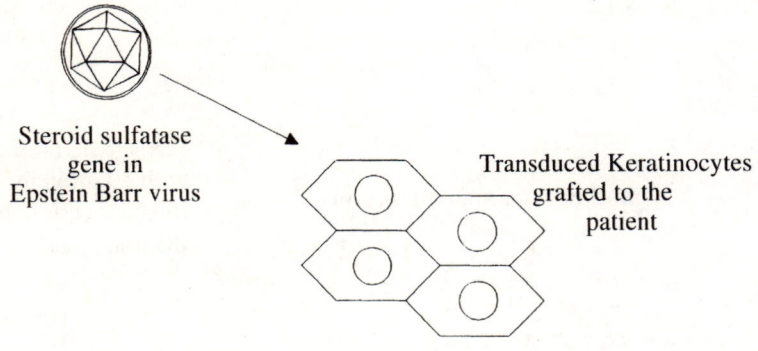

C. Skin of a treated patient

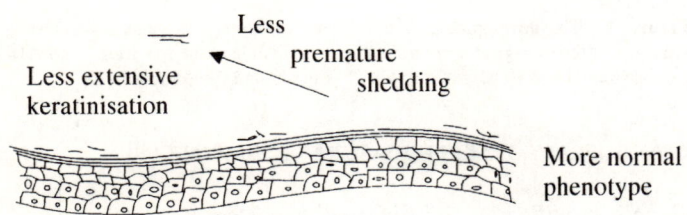

Figure 4.   The treatment of X-linked ichthyosis by genetic modification of skin cells. Cells isolated from the patient are transduced by an Epstein Barr virus vector containing the steroid sulphatase gene. The aim will be to graft these *ex vivo* modified cells back to the patient or to look at direct *in vivo* transfer of the gene to the patients skin (Jensen *et al.,* 1993).

Recent reports have described the use of retroviral gene transfer of an intact steroid sulphatase cDNA into keratinocytes from a patient with X-linked ichthyosis that have been grafted on to a nude mouse. The skin has shown a substantially normal phenotype (Freiberg *et al.*, 1997).

Xeroderma pigmentosum is also an autosomal recessive genetic disorder that could be corrected by gene therapy. This is particularly relevant since there is currently no treatment for the disease (Takayama *et al.*, 1995). By transducing xeroderma pigmentosum fibroblasts with a functional cDNA copy of the defective XPD (ERCC2) gene using retroviruses, cells have shown increased survival and a normal level of DNA repair synthesis *in vitro* (Carreau *et al.*, 1995; Quilliet *et al.*, 1996).

There are a large number of very distressing blistering skin diseases caused by mutations in keratin or basement membrane proteins (Compton, 1994; Bruckner-Tuderman, 1994). Some are lethal in early childhood. A recent report has shown that retroviral transduction of a laminin B3 cDNA can restore the adhesive and organisational properties of this essential basement membrane protein to offer a potential route to treatment of junctional epidermolysis bullosa (Dellambra *et al.*, 1998).

Genetic modification of skin cells for cancer

One of the biggest areas of gene therapy research is in the treatment of cancer. The p53 tumour suppressor gene has been the focus of much attention due to its widespread

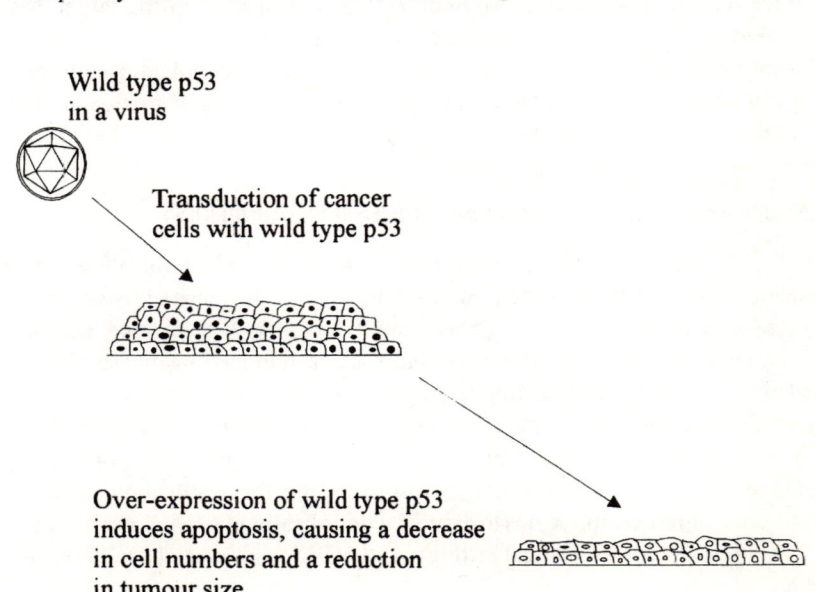

Figure 5.   The potential for treatment of skin cancer by genetic modification. A large number of cancers are caused by mutation or inactivation of both p53 alleles. Transduction of tumour cells by a virus containing wild type p53 restores the normal apoptotic pathway and results in tumour cell death (Liu *et al.*, 1995).

inactivation in tumours. p53 is believed to play a key role in apoptosis as mutation or inactivation of the p53 gene leads to uncontrolled cell proliferation (Wang and Wang, 1996). Frequent clones of p53 mutated keratinocytes have been found in normal skin arising from the dermo-epidermal junction and from hair follicles, and the incidence of mutated clones is increased with exposure to sunlight (Jonason *et al.*, 1996). In one approach researchers have tried to induce apoptosis by adenoviral gene transfer of wild type p53 into squamous cell carcinomas, and shown that tumour growth has been suppressed both *in vitro* and *in vivo* (Liu *et al.*, 1995) (see *Figure 5*).

It is thought that many types of cancer could be treated by modification of the patient's immune system. For example, cytokines such as the interleukins may be delivered directly to the tumour to enable targeted destruction by the host immune system. The production of human IL-2 adjacent to the tumour site by cells has been shown in various murine models to promote a strong immune response leading to tumour growth inhibition or rejection (Quintin-Colonna *et al.*, 1996). In the case of skin, dermal fibroblasts have been isolated and transduced to express IL-4 by a retroviral vector then administered in a vaccine containing irradiated autologous tumour cells. IL-4 production was followed *in vitro* for up to three weeks showing that dermal fibroblasts are suitable for therapeutic delivery via genetic modification (Elder *et al.*, 1996). Again, the short-term expression of the transduced gene gives cause for concern, but such an approach might be suitable for administration after tumour excision to treat any residual tumour cells. Similar results have been shown with IL-6 administered by gene gun transfer into murine skin (Sun *et al.*, 1995) and IL-12 into epidermal cells overlaying an intradermal tumour (Rakhmilevich *et al.*, 1996). An alternative method of treating cancer is to deliver a gene to the cells that will produce a toxic gene product when an activator is applied to the target area. The advantage of this system is that even if non-tumourigenic cells receive the gene, they will not necessarily be destroyed because the activator is applied only to the target area. Head and neck squamous cell carcinomas have been treated in this way (O'Malley *et al.*, 1995; O'Malley *et al.*, 1996).

GENE THERAPY FOR WOUND HEALING AND TISSUE ENGINEERING

The field of cutaneous wound repair and research into wound healing has rapidly expanded over recent years and now sits within the new science of tissue engineering (Nerem and Sambanis, 1995). There are three phases to wound healing – an inflammatory phase, a proliferative phase and a remodelling phase (Kirsner and Eaglstein, 1993). Wound healing strategies have been targeted at all three of these stages. Current treatments vary widely depending upon the type and size of the wound, but the application of cultured keratinocytes is becoming more and more widespread, although not yet in common usage. Research into the application of genetically modified cultured keratinocytes to express wound healing factors is already underway and could be tested in a clinical setting shortly (Feliciani *et al.*, 1996; Svensjo *et al.*, 1998).

Several examples of genetically modified keratinocytes secreting a product that acts on other skin cells have been described in nude mouse models. Eming *et al.* (1995) demonstrated that keratinocytes secreting PDGF-A stimulated the production of vascular and connective tissues. Modification of the autocrine control of keratinocyte proliferation through targeted expression of human IGF-I resulted in keratinocytes that were no longer dependent on exogenously added IGF-I, and when grafted to athymic mice did not show significantly altered epidermal differentiation (Eming *et al.*, 1996).

Conclusion

There are many avenues of active research into the use of genetically modified skin cells. No doubt, many more will become evident as results in this area accumulate. Certainly, more flexibility is needed to produce better and safer gene vectors with more cell-specific gene promoters to produce higher efficiencies of cell transduction and prolonged gene expression. The use of *ex vivo* gene transfer strategies will be intimately linked with progress in the field of tissue engineering so that the genetically modified cells have a prolonged residence in the appropriate tissue. Clinical gene transfer to skin in human subjects cannot be far off. This is a research field that is still in its infancy.

References

ABE, S., FUJITA, T. AND TAKAMI, Y. (1995). Disappearance of Langerhans cells and melanocytes after cryopreservation of skin. *British Journal of Plastic Surgery* **48**, 405–409.

AICHER, A., WESTERMANN, J., CAYEUX, S., WILLIMSKY, G., DAEMEN, K., BLANKENSTEIN, T., UCKERT, W., DORKEN, B. AND PEZZUTTO, A. (1997). Successful retroviral mediated transduction of a reporter gene in human dendritic cells: feasibility of therapy with gene-modified antigen presenting cells. *Experimental Hematology* **25**, 39–44.

ALEXANDER, M.Y. AND AKHURST, R.J. (1995). Liposome-mediated gene transfer and expression via the skin. *Human Molecular Genetics* **4**, 2279–2285.

ANSEL, J.C., TIESMAN, J.P., OLERUD, J.E., KRUEGER, J.G., KRANE, J.F., TARA, D.C., SHIPLEY, G.D., GILBERTSON, D., USUI, M.L. AND HART, C.E. (1993). Human keratinocytes are a major source of cutaneous platelet-derived growth factor. *Journal of Clinical Investigation* **92**, 671–678.

ARASE, S., SADAMOTO, Y., KATOH, S., URANO, Y. AND TAKEDA, K. (1990). Co-culture of human hair follicles and dermal papillae in a collagen matrix. *Journal of Dermatology* **17**, 667–676.

ARASE, S., SHIKIJI, T., UCHIDA, N., KATOH, S., FUJIE, T. AND URANO, Y. (1994). Experimental approaches for the reconstitution of hair *in vitro*. *Skin Pharmacology* **7**, 12–15.

ARTHUR, J.F., BUTTERFIELD, L.H., ROTH, M.D., BUI, L.A., KIERTSCHER, S.M., LAU, R., DUBINETT, S., GLASPY, J., MCBRIDE, W.H. AND ECONOMOU, J.S. (1997). A comparison of gene transfer methods in human dendritic cells. *Cancer Gene Therapy* **4**, 17–25.

AUBOCK, J., IRSCHICK, E., ROMANI, N., KOMPATSCHER, P., HOPFL, R., HEROLD, M., BAUER, M., HUBER, C. AND FRITSCH, P. (1988). Rejection, after a slightly prolonged survival time, of Langerhans cell-free allogeneic cultured epidermis used for wound coverage in humans. *Transplantation* **45**, 730–737.

BADIAVAS, E., MEHTA, P.P. AND FALANGA, V. (1996). Retrovirally mediated gene transfer in a skin equivalent model of chronic wounds. *Journal of Dermatological Science* **13**, 56–62.

BAER, R.P., WHITEHALL, T.E., SARKER, R., SARKAR, M., MESSINA, L.M. AND STANLEY, J.C. (1996). Retroviral mediated transduction of endothelial cells with the Lac Z gene impairs cellular proliferation *in vitro* and graft endothelialisation *in vivo*. *Journal of Vascular Surgery* **24**, 892–899.

BARRA, R.M., FENJVES, E.S. AND TAICHMAN, L.B. (1994). Secretion of apolipoprotein E by basal cells in cultures of epidermal keratinocytes. *Journal of Investigative Dermatology* **102**, 61–66.

BARRANDON, Y. (1989). Restriction of growth potential in paraclones of human keratinocytes by a viral oncogene. *Proceedings of the National Academy of Sciences, USA* **86**, 4102–4106.

BARRANDON, Y. AND GREEN, H. (1987). 3 clonal types of keratinocytes with different capacities for growth. *Proceedings of the National Academy of Sciences, USA* **84**, 2302–2306.

BARRECA, A., DE LUCA, M., DEL MONTE, P., BONDANZA, S., DAMONTE, G., CARIOLA, G., GIORDANO, G., CANCEDDA, R. AND MINUTO, F. (1992). *In vitro* paracrine regulation of human keratinocyte growth by fibroblast-derived insulin-like growth factors. *Journal of Cellular Physiology* **151**, 262–268.

BEVAN, S., WOODWARD, B., NG, R.L.H., GREEN, C. AND MARTIN, R. (1997). Retroviral gene transfer into porcine keratinocytes following improved methods of cultivation. *Burns* **23**, 525–532.

BOYCE, S.T. AND HAM, R.H. (1983). Calcium regulated differentiation of normal human epidermal keratinocytes in chemically defined clonal culture and serum free culture. *Journal of Investigative Dermatology* **81**, 533–540.

BOYCE, S.T., MEDRANO, E.E., ABDEL MALEK, Z., SUPP, A.P., DODICK, J.M., NORDLUND, J.J. AND WARDEN, G.D. (1993). Pigmentation and inhibition of wound contraction by cultured skin substitutes with adult melanocytes after transplantation to athymic mice. *Journal of Investigative Dermatology* **100**, 360–365.

BRUCKNER-TUDERMAN, L. (1994). Epidermolysis bullosa: Pathogenetic pathways from mutations to symptoms. *Annals of Medicine* **26**, 165–171.

BURNETT, T.S. AND GALLIMORE, P.H. (1983). Establishment of a human keratinocyte cell line carrying complete human papillomavirus type I genomes: lack of vegiative viral DNA synthesis upon keratinisation. *Journal of General Virology* **64**, 1509–1520.

BYRNE, C. AND FUCHS, E. (1993). Probing keratinocyte and differentiation specificity of the human K5 promoter *in vitro* and in transgenic mice. *Molecular & Cellular Biology* **13**, 3176–3190.

CARREAU, M., QUILLIET, X., EVENO, E., SALVETTI, A., DANOS, O., HEARD, J.M., MEZZINA, M. AND SARASIN, A. (1995). Functional retroviral vector for gene therapy of xeroderma pigmentosum group D patients. *Human Gene Therapy* **6**, 1307–1315.

CARROLL, J.M., CROMPTON, T., SEERY, J.P. AND WATT, F.M. (1997). Transgenic mice expressing IFN-gamma in the epidermis have eczema, hair hypopigmentation, and hair loss. *Journal of Investigative Dermatology* **108**, 412–422.

CARROLL, J.M., FENJVES, E.S., GARLICK, J.A. AND TAICHMAN, L.B. (1993). Keratinocytes as a target for gene therapy. In *Molecular Biology of the skin: The Keratinocyte*. Eds. M. Darmon and M. Blumenberg, pp 269–284. San Diego: Academic Press.

CARROLL, J.M., ROMERO, M.R. AND WATT, F.M. (1995). Suprabasal integrin expression in the epidermis of transgenic mice results in developmental defects and a phenotype resembling psoriasis. *Cell* **83**, 957–968.

CHEN, B.F., CHANG, W.C., CHEN, S.T., CHEN, D.S. AND HWANG, L.H. (1995). Long term expression of the biologically active growth hormone in genetically modified fibroblasts after implantation into a hypophysectomized rat. *Human Gene Therapy* **6**, 917–926.

CHENG, L., ZIEGELHOFFER, P.R. AND YANG, N.S. (1993). *In vivo* promoter activity and transgene expression in mammalian somatic tissues evaluated by using particle bombardment. *Proceedings of the National Academy of Sciences, USA.* **90**, 4455–4459.

CHOATE, K.A. AND KHAVARI, P.A. (1997). Sustainability of keratinocyte gene transfer and cell survival *in vivo*. *Human Gene Therapy* **8**, 895–901.

CIERNIK, I.F., KRAYENBUHL, B.H. AND CARBONE, D.P. (1996). Puncture-mediated gene transfer to the skin. *Human Gene Therapy* **7**, 893–899.

COMPTON, C.C. (1994). Keratinocyte grafting: animal models. In *The Keratinocyte Handbook*. Eds. I. Leigh, B. Lane, and F. Watt, pp 513–526. Cambridge: Cambridge University Press.

COMPTON, J.G. (1994). Epidermal disease: faulty keratin filaments take their toll. *Nature: Genetics* **6**, 6–7.

CONDON, C., WATKINS, S.C., CELLUZZI, C.M., THOMPSON, K. AND FALO, L.D. (1996). DNA based immunisation by *in vivo* transfection of dendritic cells. *Nature: Medicine* **2**, 1122–1128.

COTTEN, M. AND WAGNER, E. (1993). Non-viral approaches to gene therapy. *Current Opinion in Biotechnology* **4**, 705–710.

DAI, Y.F., QUI, X.F., XUE, J.L. AND LIU, Z.D. (1992). High efficient transfer and expression of human clotting factor IX cDNA in cultured human primary skin fibroblasts from hemophilia B patient by retroviral vectors. *Sci China B.* **35**, 183–193.

DE LUCA, M., BONDANZA, S., DI MARCO, E., MARCHISIO, P.C., D'ANNA, F., FRANZI, A.T. AND CANCEDDA, R. (1994). Keratinocyte-melanocyte interactions in *in vitro* reconstituted normal human epidermis. In *The Keratinocyte Handbook.* Eds. I.M. Leigh, E.B. Lane and F.M. Watt, pp 95–108, Cambridge: Cambridge University Press.

DE LUCA, M. AND PELLEGRINI, G. (1997). The importance of epidermal stem cells in keratinocyte-mediated gene therapy. *Gene Therapy* **4**, 381–383.

DELLAMBRA, E., VAILLY, J., PELLEGRINI, G., BONDANZA, S., GOLISANO, O., MACCHIA, C., ZAMBRUNO, G., MENEGUZZI, G. AND DE LUCA, M. (1998). Corrective transduction of human epidermal stem cells in laminin-5-dependent junctional epidermolysis bullosa. *Human Gene Therapy* **9**, 1359–1370.

DI SEPIO, D. JONES, A., LONGLEY, M.A., BUNDMAN, D., ROTHNAGEL, J.A. AND ROOP, D.R. (1995). The proximal promoter of the mouse loricrin gene contains a functional AP-1 element and directs keratinocyte-specific but not differentiation-specific expression. *Journal of Biological Chemistry* **270**, 10792–10799.

DODDS, W.J. (1982). The pig model for biomedical research. *Federation Proceedings* **41**(2), 247–256.

ELDER, E.M., LOTZE, M.T. AND WHITESIDE, T.L. (1996). Succesful culture and selection of cytokine gene modified human dermal fibroblasts for the biologic therapy of patients with cancer. *Human Gene Therapy* **7**, 479–487.

EMING, S.A., LEE, J., SNOW, R.G., TOMPKINS, R.G., YARMUSH, M.L. AND MORGAN, J.R. (1995). Genetically modified human epidermis overexpressing PDGF-A directs the development of a cellular and vascular connective tissue stroma when transplanted to athymic mice – implications for the use of genetically modified keratinocytes to modulate dermal regeneration. *Journal of Investigative Dermatology* **105**, 756–763.

EMING, S.A., SNOW, R.G., YARMUSH, M.L. AND MORGAN, J.R. (1996). Targeted expression of insulin-like growth factor to human keratinocytes: modification of the autocrine control of keratinocyte proliferation. *Journal of Investigative Dermatology* **107**, 113–120.

FELICIANI, C., GUPTA, A.K. AND SAUDER, D.N. (1996). Keratinocytes and cytokine/growth factors. *Critical Reviews in Oral Biology and Medicine* **7**, 300–318.

FENJVES, E.S., SCHWARTZ, P.M., BLAESE, R.M. AND TAICHMAN, L.B. (1997). Keratinocyte gene therapy for adenosine deaminase deficiency: a model approach for inherited metabolic disorders. *Human Gene Therapy* **8**, 911–917.

FENJVES, E.S., SMITH, J., ZARADIC, S. AND TAICHMAN, L.B. (1994). Systemic delivery of secreted protein by grafts of epidermal keratinocytes: Prospects for keratinocyte gene therapy. *Human Gene Therapy* **5**, 1241–1248.

FENJVES, E.S., YAO, S.N., KURACHI, K. AND TAICHMAN, L.B. (1996). Loss of expression of a retrovirus transduced gene in human keratinocytes. *Journal of Investigative Dermatology* **106**, 576–578.

FERRY, N., DUPLESSIS, O., HOUSSIN, D., DANOS, O. AND HEARD, J.M. (1991). Retroviral-mediated gene transfer into hepatocytes *in vivo. Proceedings of the National Academy of Sciences, USA* **88**, 8377–8381.

FLOWERS, M.E., STOCKSCHLAEDER, M.A., SCHUENING, F.G., NIEDERWIESER, D., HACKMAN, R., MILLER, A.D. AND STORB, R. (1990). Long term transplantation of canine keratinocytes made resistant to G418 through retrovirus mediated gene transfer. *Proceedings of the National Academy of Sciences, USA* **87**, 2349–2353.

FREIBERG, R.A., CHOATE, K.A., DENG, H., ALPERIN, E.S., SHAPIRO, L.J. AND KHAVARI, P.A. (1997). A model of corrective gene transfer in X-linked ichthyosis. *Human Molecular Genetics* **6**, 927–933.

FURTH, P.A., SHAMAY, A. AND HENNIGHAUSEN, L. (1995). Gene transfer into mammalian cells by jet injection. *Hybridoma* **14**, 149–152.

FUSENIG, N.E. (1994). Epithelial-mesenchymal interactions regulate keratinocyte growth and differentiation *in vitro*. In *The Keratinocyte Handbook*. Eds. I. Leigh, B. Lane and F. Watt, pp 71–94. Cambridge: Cambridge University Press.

GARLICK, J.A., KATZ, A.B., FENJVES, E.S. AND TAICHMAN, L.B. (1991). Retrovirus-mediated transduction of cultured epidermal keratinocytes. *Journal of Investigative Dermatology* **97**, 824–829.

GARLICK, J.A. AND TAICHMAN, L.B. (1992). A model to study the fate of genetically-marked keratinocytes in culture. *Journal of Dermatology* **19**, 797–801.

GARLICK, J.A. AND TAICHMAN, L.B. (1993). The fate of genetically marked human oral keratinocytes *in vitro*. *Archives of Oral Biology* **38**, 903–910.

GERRARD, A.J., AUSTEN, D.E. AND BROWNLEE, G.G. (1996). Recombinant factor IX secreted by transduced human keratinocytes is biologically active. *British Journal of Haematology* **95**, 561–563.

GLOVER, M.T., BODMER, J., BODMER, W., KENNEDY, L.J., BROWN, J., NAVARRETE, C., KWAN, J.T. AND LEIGH, I.M. (1993). HLA antigen frequencies in renal transplant recipients and non-immunosuppressed patients with non-melanoma skin cancer. *European Journal of Cancer* **29A**, 520–524.

GORMAN, C. (1985). High efficiency gene transfer into mammalian cells. In *DNA cloning II*. Ed. D.M. Glover, pp 143–190. Oxford: IRL Press.

GREEN, H., KEHINDE, O. AND THOMAS, J. (1979). Growth of cultured human epidermal cells into multiple epithelia suitable for grafting. *Proceedings of the National Academy of Sciences, USA* **76**, 5665–5668.

GREENHALGH, D.A., ROTHNAGEL, J.A. AND ROOP, D.R. (1994). Epidermis: An attractive target tissue for gene therapy. *Journal of Investigative Dermatology* **103 Suppl.**, 63S–69S.

HANSBROUGH, J.F., BOYCE, S.T., COOPER, M.L. AND FOREMAN, T.J. (1989). Burn wound closure with cultured autologous keratinocytes and fibroblasts attached to a collagen-glycosaminoglycan substrate. *Jama* **262**, 2125–2130.

HARMON, C.S. AND NEVINS, T.D. (1994). Hair fibre production by human hair follicles in whole organ culture. *British Journal of Dermatology* **130**, 415–423.

HOEBEN, R.C., FALLAUX, F.J., VAN TILBURG, N.H., CRAMER, S.J., VAN ORMONDT, H., BRIET, E. AND VAN DER EB, A.J. (1993). Toward gene therapy for hemophilia A: long term persistance of factor VIII-secreting fibroblasts after transplantation into immunodeficient mice. *Human Gene Therapy* **4**, 179–186.

HOLBROOK, K.A. (1994). Ultrastructure of the epidermis. In *The Keratinocyte Handbook*. Eds. I.M. Leigh, E.B. Lane and F.M. Watt, pp 3–33. Cambridge: Cambridge University Press.

IHARA, S., WATANABE, M., NAGAO, E. AND SHIOYA, N. (1991). Formation of hair follicles from a single cell suspension of embryonic rat skin by a two step procedure *in vitro*. *Cell and Tissue Research* **266**, 65–73.

IIZUKA, H. AND ISHIDA-YAMAMOTO, A. (1997). Another support for the location of epidermal stem cells residing adjacent to the tips of dermal papillae in the interfollicular epidermis. *Journal of Investigative Dermatology* **109**, 697.

IIZUKA, H., ISHIDA-YAMAMOTO, A. AND HONDA, H. (1996). Epidermal remodelling in psoriasis. *British Journal of Dermatology* **135**, 433–438.

JENSEN, T.G., JENSEN, U.B., JENSEN, P.K., IBSEN, H.H., BRANDRUP, F., BALLABIO, A. AND BOLUND, L. (1993). Correction of steroid sulfatase deficiency by gene transfer into basal cells of tissue-cultured epidermis from patients with recessive X-linked ichthyosis. *Experimental Cell Research* **209**, 392–397.

JENSEN, T.G., SULLIVAN, D.M., MORGAN, R.A., TAICHMAN, L.B., NUSSENBLATT, R.B., BLAESE, R.M. AND CSAKY, K.G. (1997). Retrovirus-mediated gene transfer of ornithine-delta-aminotransferase into keratinocytes from gyrate atrophy patients. *Human Gene Therapy* **8**, 2125–2132.

JENSEN, U.B., JENSEN, T.G., JENSEN, P.K.A., RYGAARD, J., HANSEN, B.S., FOGH, J., KOLVRAA, S. AND BOLUND, L. (1994). Gene transfer into cultured human epidermis and its transplantation onto immunodeficient mice: An experimental model for somatic gene therapy. *Journal of Investigative Dermatology* **103**, 391–394.

JIANG, C.K., CONNOLLY, D. AND BLUMENBERG, M. (1991). Comparison of methods for transfection of human epidermal keratinocytes. *Journal of Investigative Dermatology* **97**, 969–973.

JONASON, A.S., KUNALA, S., PRICE, G.J., RESTIFO, R.J., SPINELLI, H.M., PERSING, J.A., LEFFELL, D.J., TARONE, R.E. AND BRASH, D.E. (1996). Frequent clones of p53-mutated keratinocytes in normal human skin. *Proceedings of the National Academy of Sciences, USA* **93**, 14025–14029.

KANGESU, T., NAVSARIA, H.A., MANEK, S., SHUREY, C.B., JONES, C.R., FRYER, P.R., LEIGH, I.M. AND GREEN, C.J. (1993). A porcine model using skin graft chambers for studies on cultured keratinocytes. *British Journal of Plastic Surgery* **46,** 393–400.

KATZ, A.B. AND TAICHMAN, L.B. (1994). Epidermis as a secretory tissue: An *in vitro* tissue model to study keratinocyte secretion. *Journal of Investigative Dermatology* **102**, 55–60.

KIRSNER, R.S. AND EAGLSTEIN, W.H. (1993). The wound healing process. *Dermatologic Clinics* **11**, 629–640.

KOLODKA, T.M., GARLICK, J.A. AND TAICHMAN, B.L. (1998). Evidence for keratinocyte stem cells *in vitro*: long term engraftment and persistence of transgene expression from retrovirus-transduced keratinocytes. *Proceedings of the National Academy of Sciences, USA* **95**, 4356–4361.

KOZARSKY, K.F. AND WILSON, J.M. (1993). Gene therapy: Adenovirus vectors. *Current Opinion in Genetic Development* **3**, 499–503.

KRUEGER, G.G., MORGAN, J.R., JORGENSON, C.M., SCHMIDT, L., LI, H.L., KWAN, M.K., BOYCE, S.T., WILEY, H.S., KAPLAN, J. AND PETERSEN, M.J. (1994). Genetically modified skin to treat disease: potential and limitations. *Journal of Investigative Dermatology* **103**, 76S–84S.

LANE, E.B., WILSON, C.A., HUGHES, B.R. AND LEIGH, I.M. (1991). Stem cells in hair follicles. Cytoskeletal studies. *Annals of the New York Academy of Sciences* **642**, 197–213.

LANGDON, J., WILLIAMS, D.M., NAVSARIA, H. AND LEIGH, I.M. (1991). Autologous keratinocyte grafting: a new technique for intra-oral reconstruction. *British Dental Journal* **171**, 87–90.

LEIGH, I.M. AND WATT, F.M. (1994). Keratinocyte Culture Systems. In *The Keratinocyte Handbook*. Eds. I.M. Leigh, E.B. Lane and F.M. Watt, pp 43–52. Cambridge: Cambridge University Press.

LI, A., SIMMONS, P.J. AND KAUR, P. (1998). Identification and isolation of candidate human keratinocyte stem cells based on cell surface phenotype. *Proceedings of the National Academy of Sciences, USA* **95**, 3902–3907.

LI, L.N. AND HOFFMAN, R.M. (1995). The feasibility of targeted selective gene therapy of the hair follicle. *Nature: Medicine* **1**, 705–706.

LIU, T.J., EL-NAGGAR, A.K., McDONNELL, T.J., STECK, K.D., WANG, M., TAYLOR, D.L. AND CLAYMAN, G.L. (1995). Apoptosis induction mediated by wild type p53 adenoviral gene transfer in squamous cell carcinoma of the head and neck. *Cancer Research* **55**, 3117–3122.

LU, B., SCOTT, G. AND GOLDSMITH, L.A. (1996). A model for keratinocyte gene therapy: preclinical and therapeutic considerations. *Proceedings of the American Association of Physicians* **108,** 165–172.

LU, D.R., ZHOU, J.M., ZHENG, B., QUI, X.F., XUE, J.L., WANG, J.M., MENG, P.L., HAN, F.L., MING, B.H. AND WANG, X.P. (1993). Stage I clinical trial of gene therapy for hemophilia B. *Science In China – Series B, Chemistry, Life Sciences And Earth Sciences* **36**, 1342–1351.

MACKENZIE, I.C. (1997). Retroviral transduction of murine epidermal stem cells demonstrates clonal units of epidermal structure. *Journal of Investigative Dermatology* **109**, 377–383.

MARINKOVICH, M.P., KEENE, D.R., RIMBERG, C.S. AND BURGESON, R.E. (1993). Cellular origin of the dermal-epidermal basement membrane. *Developmental Dynamics* **197**, 255–267.

MATHOR, M.B., FERRARI, G., DELLAMBRA, E., CILLI, M., MAVILIO, F., CANCEDDA, R. AND DE LUCA, M. (1996). Clonal analysis of stably transduced human epidermal stem cells in culture. *Proceedings of the National Academy of Sciences, USA* **93**, 10371–10376.

MORGAN, J.R., BARRANDON, Y., GREEN, H. AND MULLIGAN, R. (1987). Expression of an exogenous growth hormone gene by transplantable human epidermal cells. *Science* **237**, 1476–1479.

NABEL, G.J., NABEL, E.G., YANG, Z., FOX, B.A., PLAUTZ, G.E., GAO, X., HUANG, L., SHU, S., GORDON, D. AND CHANG, A.E. (1993). Direct gene transfer with DNA-liposome complexes in melanoma: Expression, biologic activity, and lack of toxicity in humans. *Proceedings of the National Academy of Sciences, USA* **90**, 11307–11311.

NAVSARIA, H.A., MYERS, S.R., LEIGH, I.M. AND MCKAY, I.A. (1995). Culturing skin *in vitro* for wound therapy. *Trends in Biotechnology* **13**, 91–100.

NEAD, M.A. AND MCCANCE, D.J. (1995). Poly-L-ornithine-mediated transfection of human keratinocytes. *Journal of Investigative Dermatology* **105**, 668–671.

NEREM, R.M. AND SAMBANIS, A. (1995). Tissue engineering: from biology to biological substitutes. *Tissue Engineering* **1**, 3–13.

NG, R.L., WOODWARD, B., BEVAN, S., GREEN, C.J. AND MARTIN, R. (1997). Retroviral marking identifies grafted autologous keratinocytes in porcine wounds receiving cultured epithelium *1*. *Journal of Investigative Dermatology* **108**, 457–462.

O'MALLEY, B.W., CHEN, S.H., SCHWARTZ, M.R. AND WOO, S.L. (1995). Adenovirus-mediated gene therapy for human head and neck squamous cell cancer in a nude mouse model. *Cancer Research* **55**, 1080–1085.

O'MALLEY, B.W., COPE, K.A., CHEN, S.H., LI, D., SCHWARTA, M.R. AND WOO, S.L. (1996). Combination gene therapy for oral cancer in a murine model. *Cancer Research* **56**, 1737–1741.

PAGE.S.M AND BROWNLEE, G.G. (1997). An ex vivo keratinocyte model for gene therapy of hemophilia B. *Journal of Investigative Dermatology* **109**, 139–145.

PETERSEN, M.J., KAPLAN, J., JORGENSEN, C.M., SCHMIDT, L.A., LI, L., MORGAN, J.R., KWAN, M.K. AND KRUEGER, G.G. (1995). Sustained production of human transferrin by transduced fibroblasts implanted into athymic mice: a model for somatic gene therapy. *Journal of Investigative Dermatology* **104**, 171–176.

QUILLIET, X., CHEVALLIER-LAGENTE, O., EVENO, E., STOJKOVIC, T., DESTÉE, A., SARASIN, A. AND MEZZINA, M. (1996). Long-term complementation of DNA repair deficient human primary fibroblasts by retroviral transduction of the XPD gene. *Mutation Research DNA Repair* **364**, 161–169.

QUINTIN-COLONNA, F., DEVAUCHELLE, P., FRADELIZI, D., MOUROT, B., FAURE, T., KOURILSKY, P., ROTH, C. AND MEHTALI, M. (1996). Gene therapy of spontaneous canine melanoma and feline fibrosarcoma by intratumoural administration of histocompatible cells expressing human interleukin 2. *Gene Therapy* **3**, 1104–1112.

RAKHMILEVICH, A.L., TURNER, J., FORD, M.J., MCCABE, D., SUN, W.H., SONDEL, P.M., GROTA, K. AND YANG, N.S. (1996). Gene gun mediated skin transfection with IL-12 gene results in regression of established primary and metastatic murine tumours. *Proceedings of the National Academy of Sciences, USA* **93**, 6291–6296.

RAMBUKKANA, A., BOS, J.D., IRIK, D., MENKO, W.J., KAPSENBERG, M.L. AND DAS, P.K. (1995). *In situ* behaviour of human Langerhans cells in skin organ culture. *Laboratory Investigation* **73**, 521–531.

RAMESH, N., LAU, S., PALMER, T.D., STORB, R. AND OSBORNE, W.R. (1993). High-level human adenosine deaminase expression in dog skin fibroblasts is not sustained following transplantation. *Human Gene Therapy* **4**, 3–7.

REYNOLDS, A.J. AND JAHODA, C.A. (1994a). Hair follicle reconstruction *in vitro*. *Journal of Dermatological Science* **7 Supplement**, S84–S97.

REYNOLDS, A.J. AND JAHODA, C.A.B. (1994b). Hair follicle stem cells: Characteristics and possible significance. *Skin Pharmacology* **7**, 16–19.

RHEINWALD J G, G. (1977). Epidermal growth factor and the multiplication of cultured human keratinocytes. *Nature* **265**, 421–424.

RHEINWALD, J.G. (1989). Human epidermal keratinocyte cell culture and xenograft systems: applications in the detection of potential chemical carcinogens and the study of epidermal transformation. [Review]. *Progress in Clinical & Biological Research* **298**, 113–125.

RHEINWALD, J.G. AND GREEN, H. (1975). Serial cultivation of strains of human epidermal keratinocytes: the formation of keratinising colonies from single cells. *Cell* **6**, 331–344.

ROTHNAGEL, J.A., GREENHALGH, D.A., XIAO-JING, W., SELLHEYER, K., BICKENBACH, J.R.,

DOMINEY, A.M. AND ROOP, D.R. (1993). Transgenic models of skin diseases. *Archives of Dermatology* **129**, 1430–1436.

SACKMAN, J.E., CEZEAUX, J.L., REDDICK, T.T., FREEMAN, M.B., STEVENS, S.L. AND GOLD-MAN, M.H. (1996). Evaluation of the effect of retroviral gene transduction on vascular endothelial cell adhesion. *Tissue Engineering* **2**, 223–234.

SANDERS, D.A., PHILPOTT, M.P., NICOLLE, F.V. AND KEALEY, T. (1994). The isolation and maintenance of the human pilosebaceous unit. *British Journal of Dermatology* **131**, 166–176.

SCHEIMBERG, I., HARPER, J.I., MALONE, M. AND LAKE, B.D. (1996). Inherited ichthyoses: a review of the histology of the skin. *Pediatric Pathology and Laboratory Medicine* **16**, 359–378.

SELLHEYER, K. (1995). Transgenic mice as models for skin disease. *Hautarzt.* **46**, 755–761.

SETOGUCHI, Y., JAFFE, H.A., DANEL, C. AND CRYSTAL, R.G. (1994). *Ex vivo* and *in vivo* gene transfer to the skin using replication-deficient recombinant adenovirus vectors. *Journal of Investigative Dermatology* **102**, 415–421.

SHIH, C., SHILO, B.Z., GOLDFARB, M.P., DANNENBERG, A. AND WEINBERG, R.A. (1979). Passage of phenotypes of chemically transformed cells via transfection of DNA and chromatin. *Proceedings of the National Academy of Sciences, USA* **76**, 5714–5718.

STOCKSCHLADER, M.A., SCHUENING, F.G., GRAHAM, T.C. AND STORB, R. (1994). Transplantation of retrovirus transduced canine keratinocytes expressing the beta galactosidase gene. *Gene Therapy* **1**, 317–322.

SULLIVAN, D.M., JENSEN, T.G., TAICHMAN, L.B. AND CSAKY, K.G. (1997). Ornithine-delta-aminotransferase expression and ornithine metabolism in cultured epidermal keratinocytes: toward metabolic sink therapy for gyrate atrophy. *Gene Therapy* **4**, 1036–1044.

SUN, W.H., BURKHOLDER, J.K., SUN, J., CULP, J., TURNER, J., LU, X.G., PUGH, T.D., ERSHLER, W.B. AND YANG, N.S. (1995). *In vivo* cytokine gene transfer by gene gun reduces tumour growth in mice. *Proceedings of the National Academy of Sciences, USA* **92**, 2889–2893.

SVENSJO, T., YAO, F., POMAHAC, B. AND ERIKSSON, E. (1998). Gene therapy applications of growth factors. In *Growth factors and receptors: A practical approach*. Eds. I. McKay and K.D. Brown, pp 227–251. Oxford: Oxford University Press.

TAKAHASHI, K. AND COULOMBE, P.A. (1996). A transgenic mouse model with an inducible skin blistering disease phenotype. *Proceedings of the National Academy of Sciences, USA* **93**, 14776–14781.

TAKAYAMA, K., SALAZAR, E.P., LEHMANN, A., STEFANINI, M., THOMPSON, L.H. AND WEBER, C.A. (1995). Defects in DNA repair and transcription gene ERCC2 in the cancer prone disorder xeroderma pigmentosum group D. *Cancer Research* **55**, 5656–5663.

TRAPNELL, B.C. (1993). Adenoviral vectors for gene transfer. *Adv. Drug Deliv. Rev.* **12**, 185–199.

TSUNODA, R., BOSSELOIR, A., ONOZAKI, K., HEINEN, E., MIYAKE, K., OKAMURA, H., SUZUKI, K., FUJITA, T., SIMAR, L.J. AND SUGAI, N. (1997). Human follicular dendritic cells *in vitro* and follicular dendritic cell like cells. *Cell and Tissue Research* **288**, 389.

TURKSEN, K., KUPPER, T., DEGENSTEIN, L., WILLIAMS, J. AND FUCHS, E. (1992). Interleukin 6: insights to its function in skin by overexpression in transgenic mice. *Proceedings of the National Academy of Sciences, USA* **89**, 5068–5072.

VASSAR, R. AND FUCHS, E. (1991). Transgenic mice provide new insights into the role of TGF-alpha during epidermal development and differentiation. *Genes and Development* **5**, 714–727.

VOGT, P.M., THOMPSON, S., ANDREE, C., LIU, P., BREUING, K., HATZIS, D., BROWN, H., MULLIGAN, R.C. AND ERIKSSON, E. (1994). Genetically modified keratinocytes transplanted to wounds reconstitute the epidermis. *Proceedings of the National Academy of Sciences, USA* **91**, 9307–9311.

WAKELING, W.F., SOUBERBIELLE, B.E. AND BENNETT, D.C. (1995). Detection of a human DNA sequence correlated with melanocyte like differentiation and tumour supression after transfection into murine melanoma cells. *Melanoma Research* **5**, 27–40.

WANG, T.H. AND WANG, H.S. (1996). p53, apoptosis and human cancers. *Journal of the Formosan Medical Association* **95**, 509–522.

WANG, X., ZINKEL, S., POLONSKY, K. AND FUCHS, E. (1997). Transgenic studies with a keratin promoter-driven growth hormone transgene: prospects for gene therapy. *Proceedings of the National Academy of Sciences, USA* **94**, 219–226.

WATT, F.M. (1998). Epidermal stem cells in culture. *J. Cell Sci. Suppl.* **10**, 85–94.

WHITE, S.J., PAGE, S.M, MARGARITIS, P. AND BROWNLEE, G.G. (1998). Long-term expression of human clotting factor IX from retrovirally transduced primary human keratinocytes *in vivo. Human Gene Therapy* **9**, 1187–1195.

WILSON, J.B., WEINBERG, W., JOHNSON, R., YUSPA, R. AND LEVINE, A.J. (1990). Expression of the BNLF-1 oncogene of Epstein–Barr virus in the skin of transgenic mice induces hyperplasia and aberrant expression of keratin 6. *Cell* **61**, 1315–1327.

XIAO, W. AND BRANDSMA, J.L. (1996). High efficiency, long-term clinical expression of cottontail rabbit papillomavirus (CRPV) DNA in rabbit skin following particle-mediated DNA transfer. *Nucleic Acids Research* **24**, 2620–2622.

ZHANG, L., LI, L., HOFFMANN, G.A. AND HOFFMAN, R.M. (1996). Depth-targeted efficient gene delivery and expression in the skin by pulsed electric fields: an approach to gene therapy of skin aging and other diseases. *Biochemitsry and Biophysics Research Communications* **220**, 633–636.

ZHOU, J.M., QUI, X.F., LU, D.R., LU, J.Y. AND XUE, J.L. (1993). Long term expression of human factor IX cDNA in rabbits. *Science In China – Series B, Chemistry, Life Sciences And Earth Sciences* **36**, 1333–1341.

10

The Functions of 4-α-glucanotransferases and their use for the Production of Cyclic Glucans

TAKESHI TAKAHA[1] AND STEVEN M. SMITH[2*]

[1]*Biochemical Research Laboratories, Ezaki Glico Co Ltd., Utajima, Nishiyodogawaku, Osaka 555, Japan, and [2]Institute of Cell and Molecular Biology, University of Edinburgh, Mayfield Road, Edinburgh, EH9 3JH, UK.*

Introduction

α-1,4-D-glucans occur widely in nature in the form of glycogen and starch. Bacteria, lower eukaryotes and animals accumulate glycogen, whereas starch is characteristic of higher plants. While these two glucans are chemically the same, they differ in structure, particularly with respect to the number and organisation of the α-1,6-linked branches which they contain, and they differ in their molecular masses. Starch is comprised of two components: amylose which contains very few α-1,6-linked branches, and amylopectin which contains such branches at regular intervals (Hizukuri, 1986, 1996). Glycogen is typically more highly branched than amylopectin (Gunja-Smith *et al.*, 1970; Manners, 1991; Sandhya Rani *et al.*, 1992). These polymers serve as carbon stores which accumulate when glucose is plentiful and are utilised by the organism when needed. Since starch and glycogen are chemically the same, their metabolism, even in diverse organisms, is remarkably similar. This means that information obtained from studies of α-1,4-glucan metabolism in one organism can be of great value in trying to understand such metabolism in another. It also means that genes encoding enzymes of α-1,4-glucan metabolism can be transferred between very different organisms to modify glucan metabolism *in vivo*, or can be used in starch or glycogen processing. From an applied viewpoint, starch is of greatest importance because about 30 million tonnes is produced annually from plants, for food and

Abbreviations: 4αGTase, 4-α-glucanotransferase or 1,4-α-D-glucan:1,4-α-D-glucan, 4-α-D-glucanotransferase; AP, amylopectin; CA, cycloamylose; CA6, CA7, CA17 *etc.*, cycloamylose or cyclodextrins with 6, 7, 17 *etc.* glucose residues; CCD, cyclic cluster dextrin; CD, cyclodextrin; CDase, cyclodextrinase; CGTase, cyclodextrin glucanotransferase; D-enzyme, disproportionating enzyme; DP, degree of polymerisation; G1P, glucose-1-phosphate; G, G2, G3 *etc.*, glucose, maltose, maltotriose, *etc.*; GDE, glycogen debranching enzyme; P_i, inorganic phosphate; R, reducing end of 1,4-α-D-glucan.

*To whom correspondence may be addressed.

industrial uses (Lillford and Morrison, 1997). There is therefore great interest in trying to improve or modify starch production in plants, and in using enzymes from diverse organisms for starch processing.

Of all the enzymes involved in α-1,4-glucan metabolism, one group whose functions are more varied and less well understood than others, are the 4-α-glucanotransferases (1,4-α-D-glucan:1,4-α-D-glucan, 4-α-D-glucanotransferase; here abbreviated to 4αGTase). These enzymes break an α-1,4-link and transfer the resulting glucan moiety to an acceptor molecule through creation of a new α-1,4-link. Studies of the functions of these enzymes are particularly important not only for the understanding of their roles *in vivo*, but also because it has been discovered that they are able to synthesise cyclic glucans *in vitro*, and therefore have great potential for use in starch processing. The purposes of this review are:

- to draw together the available information concerning the functions of these enzymes in diverse organisms
- to compare the structures, relatedness and action patterns of these enzymes
- to introduce a unifying nomenclature
- to consider how they may be employed in starch processing
- to explain the structures and potential applications of the cyclic glucans which they can produce

The action of 4αGTases

The action of 4αGTases can be summarised by the following equation:

$$(\alpha\text{-}1,4\text{-glucan})_m + (\alpha\text{-}1,4\text{-glucan})_n \leftrightarrow (\alpha\text{-}1,4\text{-glucan})_{m\text{-}x} + (\alpha\text{-}1,4\text{-glucan})_{n+x}$$

This action is the inter-molecular glucan transfer reaction, it is readily reversible, and is often called the 'disproportionating reaction'. Such enzymes can potentially also catalyse an intra-molecular glucan transfer reaction, within a single linear glucan molecule, to create a cyclic glucan product, as follows:

$$(\alpha\text{-}1,4\text{-glucan})_m \leftrightarrow \text{cyclic}(\alpha\text{-}1,4\text{-glucan})_x + (\alpha\text{-}1,4\text{-glucan})_{m\text{-}x}$$

This reaction is also reversible, and is often referred to as the 'coupling reaction'

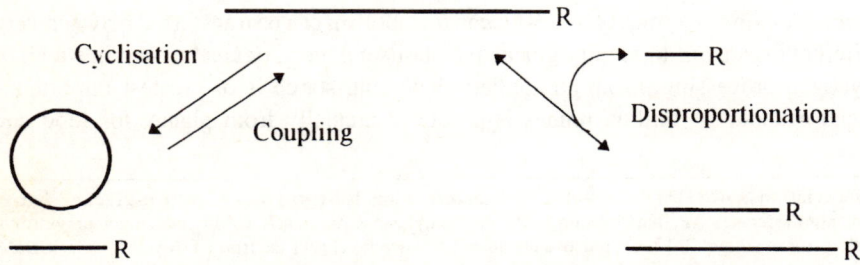

Figure 1. Reactions catalysed by 1,4-α-D-glucanotransferases (4αGTase) on 1,4-α-D-glucans. Cyclisation is achieved by 4αGTase attack a linear glucan (centre top) and transfer of the newly-formed reducing end (R) to the non-reducing end of itself, to form a cyclic molecule and a linear fragment. The reverse reaction is sometimes referred to as the Coupling Reaction. Disproportionation involves transfer from the linear donor glucan to an acceptor, which may be a glucan or glucose, to form two linear products of different size. The specificities of different types of 4αGTase are described in the text.

when the donor molecule is a cyclic glucan. These reactions are summarised diagrammatically in *Figure 1*. Since these reactions are reversible, the products of reaction of 4αGTases with glucan substrates are potentially varied and heterogeneous, particularly when complex glucan substrates (including amylopectin and glycogen) are also considered. However, the products are determined by the specificity of the different enzymes, the substrates employed, and the conditions of the reaction.

Groups within the 4αGTase family

There are three types of 4αGTase which can catalyse the glucan transfer reactions described above, and which have been extensively described and studied. These are:

1. Cyclodextrin glucanotransferase (CGTase) [EC 2.4.1.19]
2. Disproportionating enzyme (D-enzyme) or amylomaltase [EC 2.4.1.25]
3. Glycogen debranching enzyme (GDE) (amylo-1,6-glucosidase/4-α-glucanotransferase) [EC 3.2.1.33 + EC 2.4.1.25]

Unfortunately, the common names of these enzymes are either not informative or are misleading. Sometimes different names are used for the same enzyme, leading to potential confusion. It has also become apparent that there are fewer distinctions between the reaction specificities of these three enzyme types than previously thought. Furthermore, it has recently become clear that there are other 4αGTases which do not apparently belong to these three groups, based on their primary amino acid sequences and reaction specificities. Given the growing importance of these enzymes, it is necessary that the different types are carefully described for the benefit of future research. We propose that the three 4αGTases above should be referred to as Types I, II and III, respectively. In addition, there is a related but significantly different enzyme found in *Thermotoga maritima*, and another type from several thermophilic bacteria which has a completely unrelated primary structure. We propose that these should be Types IV and V, respectively. The charateristics of these five types will now be discussed.

Type I 4αGTase (CGTases)

DISTRIBUTION

This group of enzymes is well known for the synthesis of cyclodextrins (CDs), and most of the work on 4αGTase has been focused on this enzyme group. The enzyme has been reported in increasing numbers of bacteria and at least 20 genes for this enzyme have already been isolated (and more enzymes have been identified biochemically). The enzyme seems to be widespread in *Bacillus* species, since most reports are of CGTase genes from the Bacillaceae. Exceptions are two from *Klebsiella* (Binder *et al.*, 1986; Fiedler *et al.*, 1996) and two from *Themoanaerobacterium* (Bahl *et al.*, 1991; Jorgensen *et al.*, 1997). Recent results from genome sequencing projects also support this view, since no CGTase has been identified in the 15 completed genomes (including 10 Eubacteria, 4 Archaea and 1 Eukaryote). Given the large amount of work on this group of enzymes, the historical background of CDs and CGTases has already been comprehensively reviewed (French, 1957; Kobayashi, 1996) so will not be covered here. Instead, we will briefly summarise some of their properties and discuss some of the most recent findings.

ENZYME ACTION

CGTase converts starch into a mixture of cyclic α-1,4-glucans (cyclodextrins, CD) with degree of polymerisation (DP) of 6, 7 or 8 (α-, β-, γ-CD, respectively), and residual dextrin. This is achieved by the intra-molecular glucan transfer (cyclisation) reaction. Since α-, β-, and γ-CDs have central cavities with distinct dimensions and different specificity for guest molecules (see below), the production of specific types of CD is desirable. The composition of CD produced is primarily determined by the type of enzyme employed, but can be influenced by additions (e.g. complexant or ethanol) to the reaction mixture (Rendleman, 1993; Mori *et al.*, 1995). The yield of CD depends on the enzyme employed, concentration of starch, and degree of hydrolysis of the starch substrate (Kitahata, 1995). Given the importance of CDs, product specificity (proportions of α-, β-, and γ-CDs) is one of the most important properties of this enzyme group. CGTases are often classified according to the major CD which they produce (for reviews see Schmid, 1989; Starnes, 1990; Kitahata, 1995; Kobayashi, 1996). The proportions of specific CDs produced can be changed by single amino acid substitutions in CGTase (Nakamura *et al.,* 1993; Nakamura *et al.*, 1994; Penninga *et al.*, 1995; Wind *et al.*, 1998).

Apart from the cyclisation reaction, Type I 4αGTases also catalyse disproportionation and coupling reactions as described in *Figure 1*. The final products from added substrate thus depend on the equilibrium of the three transferase reactions. These activities have been reviewed by Kitahata (1995). The smallest acceptor, smallest donor, and smallest glucan unit transferred were not determined in most cases, but glucose generally seems to be the smallest acceptor, and a glucose unit is the smallest transferred unit (*Table 1*). The smallest donor molecule might be different for each CGTase, since the smallest donor molecule was reported to be maltotetraose for the enzyme from *B. subtilis* No. 313 (Kato and Horikoshi, 1986), maltotriose for *B. macerans* and *B. ohbensis* enzymes, and maltose for the *B. megaterium* enzyme (Kitahata, 1995). CGTases also have hydrolytic activity, where the degree of hydrolytic activity relative to transferase activity, is variable between different CGTases (Kitahata and Okada, 1982).

Studies of crystal structures have been carried out with five different CGTases (or mutant forms, in some cases), which include enzymes from alkalophilic *Bacillus* sp.1011 (Harata *et al.*, 1996), *Thermoanaerobacterium thermosulfurigenes* EM1 (Knegtel *et al.*, 1996; Wind *et al.*, 1998), *Bacillus circulans* strain 251 (Lawson *et al.*, 1994; Knegtel *et al.*, 1995; Strokopytov *et al.*, 1995, Strokopytov *et al.*, 1996), *B. stearothermophilus* (Kubota *et al.*, 1991, 1994) and *Bacillus circulans* strain 8 (Klein and Schultz, 1991; Klein *et al.*, 1992; Schmidt *et al.*, 1998). CGTases are structually homologous to α-amylases, whose crystal structures have also been determined (Matsuura *et al.*, 1984; Boel *et al.*, 1990; Qian *et al.,* 1993; Kadziola *et al.*, 1994; Machius *et al.*, 1995; Brayer *et al.,* 1995; Ramasubbu *et al.*, 1996; Morishita *et al.*, 1997). Both types of enzyme belong to the α-amylase super-family, all members of which contain a characteristic (β/α)8-barrel domain and four highly conserved regions (Svensson, 1994) (see *Figure 2*). Three acidic residues (Asp229, Glu257 and

Table 1. Characteristics of the activities of each type of 4αGTase

4aGTase	Strain	Smallest donor	Smallest acceptor	Smallest transferred unit	Disproportionated products	Cyclisation reaction	Smallest CA (DP)	References
Type I	*B.subtilis* 313	G4	nt	nt	G2,G3,Gn	+	8	Kato and Horikoshi (1986)
	B.ohbensis	G3	G	G	G,G2,G3,Gn	+	7	Kitahata (1995)
	B.macerans	G3	G	G	G2,G3,Gn	+	6	Kitahata (1995)
	B.megaterium	G2	G	G	G,G2,G3,Gn	+	6	Kitahata (1995)
Type II	Potato	G3	G	G2	G, G3,Gn	+	17	Takaha *et al.* (1993, 1996)
	Barley	G3	G	G2	G, G3,Gn	nt		Yoshio *et al.* (1986)
	Sweet potato	nt	nt	G2	G, G3,Gn	nt		Suganuma *et al.* (1991)
	C.butyricum	G3	G	G	G,G2,G3,Gn	nt		Goda *et al.* (1997)
	E.coli ML308	G3	G	G	G,G2,G3,Gn	nt		Palmer *et al.* (1976)
	E.coli ATCC3806	G2	G	G	G,G2,G3,Gn	nt		Kitahata *et al.* (1989a)
	E.coli K12	G2	G	G	G,G2,G3,Gn	+	17	Unpublished
	T.aquaticus	G2	G	G	G,G2,G3,Gn	+	22	Unpublished
Type III	Yeast	G4	G3	G2	G2,G3,Gn	nt		Tabata and Ide (1988)
Type IV	*T.maritima*	G4	G2	G2	G2,G3,Gn	nt		Liebl *et al.* (1992)
Type V	*T.litoralis*	G2	G	G	G,G2,G3,Gn	+	nt	Jeon *et al.* (1997)
	*Pyrococcus*KOD1	G2	G	G	G,G2,G3,Gn	nt		Tachibana *et al.* (1987)
Others	*S.mitis*	G2	G	G	G,G2,G3,Gn	nt		Walker (1966)
	S.bovis	G3	G	G	G,G2,G3,Gn	nt		Walker (1965)
	B.subtilis	G4	G4	G2	G2,G3,Gn	nt		Pazur and Okada (1968)
	S.mutans	G2	G	G	G,G2,G3,Gn	nt		Medda and Smith (1984)

nt = not tested. G, G1, G2, Gn = glucose, maltose, maltotriose, *etc.* Organism names are given in the text.

Species	Region 1		Region 2		Region 3		Region 4	
Type I 4αGTase (CGTase)								
Bacillus circulans 251	NIKVIIDFAPNH	129	DGIRMDAVKH	224	FTFGEWFL	253	QVTFIDNHD	320
Bacillus sp. 1011	NIKVIIDFAPNH	129	DGIRVDAVKH	224	FTFGEWFL	253	QVTFIDNHD	320
Bacillus circulans **8**	GIKIVIIDFAPNH	129	DGIRVDAVKH	224	FTFGEWFL	253	QVTFIDNHD	320
Thermoanaerobacterium thermosulfurigenes	NIKVIIDFAPNH	130	DGIRLDAVKH	225	FTFGEWFL	254	MVTFIDNHD	321
Bacillus stearothermophilus	GIKVIIDFAPNH	125	DGIRMDAVKH	220	FTFGEWFL	249	QVTFIDNHD	316
Klebsiella pneumoniae	NMKLVLDYAPNH	124	DAIRIDAIKH	118	FFFGEWFG	253	QVVFMDNHD	325
Type II 4αGTase (Amylomaltase / D-enzyme)								
Potato	VGYHSADVWANK	298	DEFRIDHFRG	368	NIIAEDLG	416	QVVYTGTHD	465
Aquifex aeolicus	PSYSSADVWTNP	211	DFLRLDHFRG	281	PFIAEDLG	329	NVYYTSTHD	377
Synechocystis sp.	VAHDSADVWANP	222	DIVRIDHFRG	293	PIVAEDLG	342	AVVYTGTHD	383
Thermus aquaticus	VAEDSAEVWAHP	218	HLVRIDHFRG	288	PVLAEDLG	336	VVVYTGTHD	387
Streptococcus pneumoniae	VAEDSSDMWANP	220	DIVRIDHFRG	290	NIIAEDLG	338	SVMYTGTHD	387
Clostridium butyricum	IAQDSSDVWSNP	209	DILIKIDHFRG	280	EIIAEDLG	328	CVAYTGTHD	377
Borrelia burgdorferi	IAYDSADVWAYQ	232	DIIKIDHFRG	302	KIWYEDFQ	350	CIVYTGSGD	399
Chlamydia psittaci	ISKDSCDVWYYR	245	SLYRLDHIVG	309	LPIGEDLG	359	SVTSLSTHD	409
Haemophilus influenzae	SSRGSADVWSDP	391	GVLRIDHVMG	455	LLIGEDLG	504	AYATIGTHD	551
Escherichia coli	VGTGGAETWCDR	379	GALRIDHVMS	443	MVIGEDLG	492	SMAVAATHD	540
Micobacterium tuberculosis	VHPNGADAWALQ	407	GAVRIDHIIG	471	VVVGEDLG	520	CLSSVTTHD	572
Type III 4αGTase (Glycogen debranching enzyme)								
Human muscle	NVICITDVVYNH	192	QGVRLDNCHS	504	YVVAELFT	534	ALFMDITHD	602
Rabbit muscle	NVLCITDVVYNH	232	QGVRLDNCHS	544	YVVAELFT	574	ALFMDITHD	642
Caenorhabditis elegans	NILTVQDVVWNH	354	HGLRIDNAHG	670	YVFAELFT	700	GLFLDQSHD	768
Saccharomyces cerevisiae	NMLSLTDIVFNH	218	DGFRIDNCHS	530	YVVAELFS	560	ALFMDCTHD	662
Type IV 4αGTase								
Thermotoga maritima	GIKVVLDLPINH	83	DGFRFDAAKH	190	IFLAEIWA	212	PVNFTSNHD	270
α-Amylase								
Aspergillus oryzae	GMYLMVDVVANH b	111	DGLRIDTVKH c bb	200	YCIGEVLD cbbb	226	LGTFVENHD bc	289

Figure 2. Conserved regions in the 4αGTase family of enzymes and identification of four distinct types. The four conserved regions compared are those identified in the α-amylase super-family of enzymes. The same regions from one α-amylase (*Aspergillus oryzae*) are shown for comparison. Numbers denote amino acid positions in each polypeptide. Four types of related 4αGTase are distinguished, and a fifth unrelated type is shown in *Figure 4*. Proposed catalytic (c) and binding (b) sites, in *A. oryzae* α-amylase are indicated (Matsura *et al.*, 1984).

Asp328 – numbers from *B. circulans* strain 251) in conserved regions 2, 3 and 4, respectively, are present in all known α-amylase family enzymes (Svensson, 1994) and are believed to participate in catalysis. Their roles have been studied by mutagenesis in conjunction with crystallographic studies (Klein *et al.*, 1992; Nakamura *et al.*, 1992; Strokopytov *et al.*, 1995; Knegtel *et al.*, 1995) and Glu257 was identified to be the acid/base catalyst and Asp229 the catalytic nucleophile. Identification of Asp229 as a catalytic nucleophile and evidence for a covalent intermediate in CGTase was recently obtained (Mosi *et al.*, 1997).

PHYSIOLOGICAL ROLE

Utilisation of exogenous glucans (starch or glycogen) as a substrate for growth is achieved in some bacteria by secretion of amylases and CGTase. The CGTase can degrade the starch through cyclisation activity, creating CDs, but can also potentially then catalyse coupling reactions to linearise the CDs, and disproportionation reactions to change maltooligosaccharide chain lengths. Until recently, it was unknown if CDs could be imported into the cell, but studies of mutants of *Klebsiella oxytoca* suggest

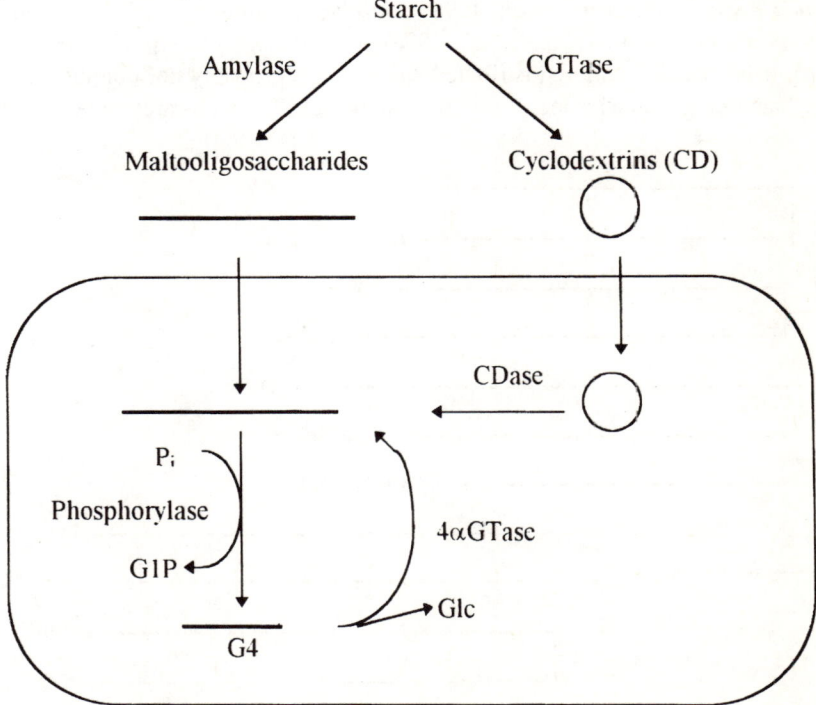

Figure 3. Roles of 4αGTases in glucan utilisation by bacteria. Breakdown of starch or other glucan is initiated by extracellular hydrolases such as amylase, or by CGTase, depending upon the species of bacterium. Maltooligosaccharides or cyclodextrins (CD) are imported by specific transporter proteins. Maltooligosaccharides are broken down by glucan phosphorylase, producing glucose-1-phosphate (G1P). Phosphorylase cannot break down molecules beyond maltotetraose (G4), so 4αGTase (amylomaltase or D-enzyme) catalyses a disproportionation reaction to create larger maltooligosaccharides and release glucose (Glc). In those bacteria which import CD, it is linearised by cyclodextrinase (CDase) and maltooligosaccharides metabolised by phopsphorylase and 4αGtase.

the existence of a specific CD uptake system (Fiedler *et al.*, 1996; Pajatsch *et al.*, 1998). Imported CDs are apparently then hydrolysed by a cytoplasmic cyclodextrinase (Feederle *et al.*, 1996) and further metabolised by a phosphorylase-dependent pathway of a type found in many bacteria (*Figure 3*). This phosphorylase-dependent pathway makes use of a Type II 4αGTase (see below). Presumably, the evolution of such a CD synthesis and uptake mechanism provides a competitive advantage over other bacteria which cannot import CDs.

Type II 4αGTase (D-enzyme and amylomaltase)

DISTRIBUTION AND RELATEDNESS

Amylomaltase was first found in *Escherichia coli* as a maltose-inducible enzyme (Monod and Torriani, 1948). The amylomaltase gene has been cloned from *E. coli* (Pugsley and Dubrevil, 1988), *Streptococcus pneumoniae* (Lacks *et al.*, 1982), *Clostridium butyricum* (Goda *et al.*, 1997), *Chlamidia* (Hsia *et al.*, 1997) and *Thermus aquaticus* (Terada *et al.*, unpublished), but further homologous genes have been identified in the genomes of *Haemophilus influenzae* (Fleischmann *et al.*, 1995), *Aquifex aeolicus* (Deckert *et al.*, 1998), *Borelia* sp. (Fraser *et al*, 1997), *Mycobacterium tuberoculosis* (Cole *et al.*, 1998) and *Synechosystis* sp. (Kaneko *et al.*, 1996). It is clearly a widely-distributed enzyme, but probably not ubiquitous since some bacterial genome projects have failed to reveal an obvious amylomaltase gene.

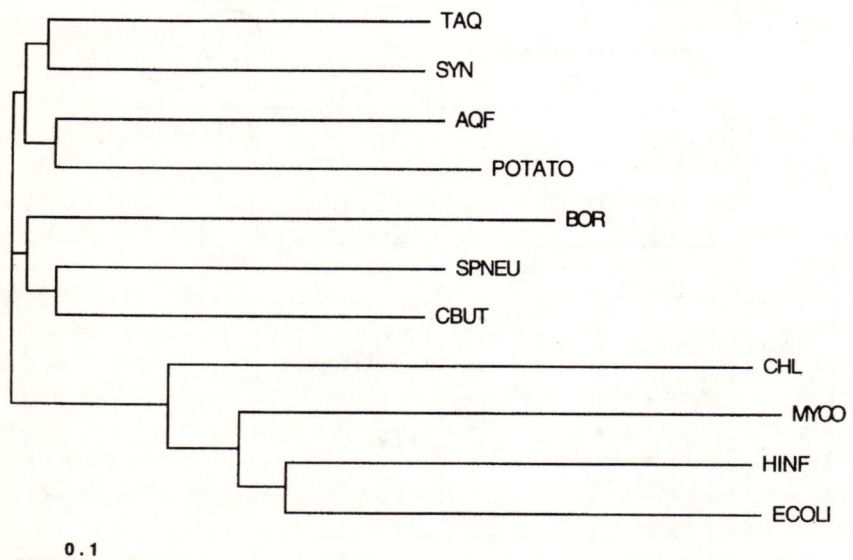

Figure 4. Phylogenetic relationships of Type II 4αGTases. Comparisons of amino acid sequences (all deduced from nucleic acid sequences) for 11 enzymes using the 'Phylip' programme indicate that they fall into at least two distinct groups. At this time it is not possible to distinguish the activities of these groups because enzyme activities have not been fully characterised in most cases. TAQ, *Thermus aquaticus*; SYN, *Synechosystis* sp.; AQF, *Aquifex aeolicus*; POTATO, *Solanum tuberosum*; BOR, *Borelia* sp.; SPNEU, *Strepotococcus pneumoniae*; CBUT, *Clostridium butyricum*; CHL, *Chlamidia* sp.; MYCO, *Mycobacterium tuberculosis*; HINF, *Haemophilus influenzae*; ECOLI, *Escherichia coli*.

A similar 4αGTase is also present in plants and is called disproportionating enzyme (D-enzyme). D-enzyme was first found in potato tubers (Peat *et al.*, 1956) but has since been found in carrot roots (Manners and Rowe, 1969), tomato fruits (Manners and Rowe, 1969), germinated barley seeds (Yoshio *et al.*, 1986), sweet potato tubers (Suganuma *et al.*, 1991), spinach leaves (Okita *et al.*, 1979), pea leaves (Kakefuda *et al.*, 1986) and *Arabidopsis* leaves (Lin and Preiss, 1988). A cDNA for D-enzyme was isolated from RNA of potato tubers (Takaha *et al.*, 1993), where the similarity of plant D-enzyme and bacterial amylomaltase was first confirmed at the molecular level. The phylogenetic tree of nearly all available (eleven) amylomaltase and D-enzyme amino acid sequences shows that these enzymes may be further divided into two subgroups in which sequence homolgy is very high within each subgroup, but less between subgroups (*Figure 4*). The first subgroup comprises enzymes from *E. coli*, *Mycobacterium*, *Chlamidia* and *H. influenzae*, and the second comprises the other amylomaltases and potato D-enzyme. It is too premature to formally propose two subgroups since more sequence analysis may provide further information, and the reaction specificity of only a few of these enzymes is currently available.

It has been suggested that D-enzyme (amylomaltase) is a member of the α-amylase super-family of enzymes (Heinrich *et al.*, 1994; Svensson, 1994). Alignments of these enzymes clearly indicates the presence of the four conserved regions characteristic of the α-amylase super-family of enzymes (*Figure 2*). We also note that the motifs RIDH(FRG), and EDLG in regions 2 and 3, are conserved within this enzyme group (Type II 4αGTase). However, not all of the 7 'invariant' amino acid residues found in these four conserved regions (Svensson, 1994) are found in Type II 4αGTases. In particular, the His residue in region 1 is completely replaced.

ENZYME ACTION

Plant D-enzyme has been partially purified from potato tubers (Jones and Whelan, 1969), germinating barley seeds (Yoshio *et al.*, 1986), *Arabidopsis* leaves (Lin and Preiss, 1988) and sweet potato tubers (Suganuma *et al.*, 1991), and its activity on maltooligosaccharides was investigated. These studies demonstrated that D-enzyme catalyses a disproportionation reaction on maltooligosaccharides in which a glucan moiety is transferred from one α-1,4-glucan molecule to another, or to glucose. The results of these studies indicated that D-enzymes of plants have common reaction characteristics. The smallest donor molecule is maltotriose, the smallest acceptor molecule is glucose, the smallest transferred glucan is a maltose unit, and glucosyl transfer never occurs (*Table 1*). It should also be noted that maltose is not produced during D-enzyme reaction on any substrate, because there are two 'forbidden linkages' in maltooligosaccharides larger than maltotetraose: the non-reducing end linkage and the bond penultimate to the reducing end (Jones and Whelan, 1969).

A limited number of bacterial amylomaltases have been subjected to biochemical analysis of their action. Analysis of amylomaltases from *E. coli* (Palmer *et al.*, 1976; Kitahata *et al.*, 1989a; Kitahata *et al.*, 1989b) and *C. butyricum* (Goda *et al.*, 1997) suggests that they catalyse a similar disproprtionation reaction to plant D-enzymes, but there are some differences. The major difference is that the bacterial amylomaltases catalyse a glucosyl transfer (*Table 1*). Additionally, maltose can act as a donor molecule for the enzyme from *E. coli* strain ATCC380 (Kitahata, 1989a), although the rate with maltose is much lower than that for maltotriose (our unpublished work). However, the work of Palmer *et al*. (1976) with *E. coli* strain ML308 and that of Goda *et al*. (1997) with *C. butyricum* indicates that maltose is not a donor.

PHYSIOLOGICAL ROLE

In *E. coli*, amylomaltase is part of a maltooligosaccharide transport and utilisation system, which includes maltodextrin phosphorylase and maltose transport proteins (Schwartz, 1987). The role of amylomaltase is apparently to convert short maltooligosaccharides into longer chains upon which glucan phosphorylase can act (*Figure 3*). This phosphorylase, like that in plants, degrades maltooligosaccharides to maltotetraose, but no further. The genes for amylomaltase and glucan phosphory-lase constitute the *malPQ* operon. A similar operon structure was also found in *S. pneumoniae* (Lacks *et al*., 1982), *K. pneumoniae* (Bloch *et al*., 1986) and *C. butyricum* (Goda *et al.,* 1997), so the function of these amylomaltases is expected to be the same as the *E. coli* enzyme. On the other hand, the genes for amylomaltase found in the genomes of *H. influenzae* (Fleischmann *et al*., 1995) and *A. aeolicus* (Deckert *et al*., 1998), are part of the glycogen operon, which include genes of glycogen synthesis and degradation. Furthermore, these organisms do not have the genes homologous to *E. coli malE, malF, malG* which are involved in the transport of maltooligosaccharides into the cytoplasm. All these observations suggest that in *H. influenzae* and *A. aeolicus*, amylomaltase may not be involved in exogenous maltooligosaccharide utilisation, but is involved in glycogen metabolism. Thus, the physiological role of amylomaltase may be different in each organism, but these differences do not correlate with the enzyme subgroups identified above (*Figure 4*).

The physiological role of D-enzyme in plants is not clear, but it is a plastidic enzyme and is assumed to be involved in starch metabolism. It is not required for the accumulation of starch or creation of starch structure, so it is likely to be involved in starch turnover (Takaha *et al*., 1998a). The long-held view is that D-enzyme converts small maltooligosaccharides into larger ones which serve as substrates for breakdown by starch phosphorylase, analogous to the maltooligosaccharide utilisation system of *E. coli* (Schwartz, 1987). This view is consistent with the observed preference of D-enzyme for maltooligossaccharide substrates and the observed effects of antisense inhibition in transgenic potato (Takaha *et al*., 1998a). However, other possible functions should not be excluded, since D-enzyme can also act upon high molecular weight starch molecules (see below).

Type III 4αGTase (GDE)

DISTRIBUTION AND STRUCTURE

This is a bifunctional enzyme with amylo-1,6-glucosidase and 4αGTase activities. It appears to be present only in those Eukaryotes which synthesise glycogen (therefore excluding higher plants). The enzyme has been isolated, and primary sequence determined, from human muscle (Yang *et al*., 1992) and rabbit muscle (Liu *et al*., 1993), and gene sequences found in the yeast *Saccharomyces cereviseae* (Bussey *et al*., 1997) and in *Caenorhabditis elegans* (Wilson *et al*., 1998). Alignment of amino acid sequences shows that they are similar enzymes and have the four conserved regions characteristic of the α-amylase super-family (Svensson, 1994). The three invariant acidic residues present in regions 2, 3 and 4 of all known α-amylase super-family enzymes (Svensson, 1994) are also found in Type III 4αGTases, and therefore are expected to participate in catalysis (*Figure 2*). Recently, Asp549 of the rabbit enzyme was identified as the catalytic nucleophile (Braun *et al.*, 1996) in support of this expectation.

PHYSIOLOGICAL ROLE AND ENZYME ACTION

Glycogen phosphorylase is the major enzyme responsible for attacking glycogen during its breakdown in Eukaryotes. This enzyme can sequentially remove glucose units (as glucose-1-phosphate) from the non-reducing ends of branches, but stops when branch lengths are reduced to four residues. At this point, the 4αGTase activity of GDE, catalyses maltotriosyl transfer from the shortened branch to another non-reducing end on the glycogen molecule (thereby creating a new substrate for phosphorylase) which then allows removal of the remaining α-1,6-linked glucosyl residue by the glucosidase activity of GDE. In this way, glycogen is completely debranched and degraded to glucose-1-phosphate (for review see Nelson *et al.*, 1979).

It has been suggested that the enzyme catalyses the two different reactions by means of two different active sites (Nelson *et al.*, 1979). This has more recently been supported by two new pieces of evidence. Firstly, the transferase reaction occurs with net retention of anomeric configuration, but the glucosidase reaction occurs with net inversion of anomeric configuration (Liu *et al.*, 1991). This indicates that the two reactions operate by means of different catalytic mechanisms. Secondly, transferase activity can be irreversibly inactivated by a water-soluble carbodiimide in the presence of amines, without affecting glucosidase activity (Liu *et al.*, 1991).

Yeast and mammalian (rabbit) enzymes catalyse similar reactions (Tabata and Hizukuri, 1992; Tabata *et al.*, 1995) but may have differences in substrate preference (Lee and Carter, 1973). In particular, both enzymes will act on phosphorylase-limit dextrin, but while native glycogen and amylopectin are good substrates for the yeast enzyme, they are not for the rabbit enzyme. The disproportionation (transferase) reaction on linear maltooligosaccharides (G4 to G7) has been extensively studied with the yeast enzyme (Tabata and Ide, 1988). The results show that the smallest donor is maltotetraose, the smallest acceptor is maltotriose, and smallest unit transferred is the maltosyl unit. The rate is higher with a longer substrate, where maltotriosyl units are preferentially transferred. Maltose is the smallest product and glucose is not produced. Glucose may serve as an acceptor for the rabbit enzyme (Brown and Illingworth, 1962). There is no information available on the biochemical properties of the *C. elegans* enzyme.

Type IV 4αGTase

The fourth group includes the 4αGTase found in the hyper-thermophilic bacterium *Thermotoga maritima* MSB8 (Liebl *et al.*, 1992). The enzyme catalyses the disproportionation of maltooligosaccharides and gives products including maltose, but no glucose. Analysis with maltooligosaccharides concluded that the smallest donor is maltotetraose, the smallest acceptor is maltose, and the smallest glucan unit to be transferred by this enzyme is a maltosyl unit (Liebl *et al.*, 1992). Such characteristics of this enzyme are most similar to the transferase action of Type III 4αGTase, but different to any of the other 4αGTases described above (*Table 1*). The enzyme is also reported to act on high molecular weight amylose, soluble starch and amylopectin, and to decrease the iodine colour without increasing reducing power (Liebl *et al.*, 1992). The gene for the enzyme was cloned (Liebl *et al.*, 1992) and sequenced, but does not show overall similarity to bacterial amylomaltases and potato D-enzyme (Heinrich *et al.*, 1994), but shows the highest similarity to α-amylase (*AmyC*) of *Dictyoglomus thermophilum* (Horinouchi *et al.*, 1988), and α-amylase of *Bacillus megaterium* (Metz *et al.*, 1988). From these results, the authors concluded that this 4αGTase is the first member of a new 4αGTase, which is more closely related to α-amylase, α-glucosidase and CGTase than to other 4αGTases (Heinrich *et al.*, 1994).

Type V 4αGTases

Type V 4αGTases were recently found independently in the hyper-thermophilic Archaea, *Thermococcus litoralis* (Jeon *et al.*, 1997) and in *Pyrococcus* sp. KOD1 (Tachibana *et al.*, 1997). These enzymes have similar properties. They both catalyse the disproportionation of maltooligosaccharides (G2 and larger) and produce a series of maltooligosaccharides, including maltose and glucose. These enzymes also transfer glucan chains from starch (soluble starch, amylose) to the acceptor glucose, and produce a series of maltooligosaccharides including maltose. All this information indicates that the smallest donor for these two enzymes is maltose, the smallest acceptor is glucose, and the smallest glucan unit to be transferred is the glucosyl unit (*Table 1*). Therefore, these enzymes are more similar to the amylomaltase of *E. coli* than to the plant D-enzyme. The genes for these 4αGTases were cloned and sequenced (Jeon *et al.*, 1997; Tachibana *et al.*, 1997). Surprisingly, they do not show overall similarity to the 4αGTases described above, but show significant similarity to α-amylase of the hyper-thermophilic Archaeon *Pyrococcus furiosus* (Laderman *et al.*, 1993a, 1993b) and α-amylase (*AmyA*) of the hyper-thermophilic bacterium *Dictyoglomus thermophilum* (Fukusumi *et al.*, 1988). Although these latter two enzymes were reported as α-amylases, they are likely to be 4αGTases. In particular, for the *P. furiosus* enzyme, the presence of disproportionation activity on malto-oligosaccharides was clearly documented (Laderman *et al.*, 1993a) and both enzymes are able to decrease the ability of starch to form a blue complex with iodine, as found for 4αGTases of *T. litoralis* and *Pyrococcus*. sp. KOD1 (Laderman *et al.*, 1993a; Fukusumi *et al.*, 1988).

The alignment of the four enzymes described above, together with the homologous sequence recently identified in the genome of *Pyrococcus horikoshii* (Kawasaki *et al.*, 1998), is shown in *Figure 5*. These five enzymes are highly homologous, but lack the

Figure 5.   Comparisons of deduced polypeptide sequences from Type V 4αGTases. The complete amino acid sequences are shown for five enzymes. Identical amino acids are denoted by asterisks and conserved amino acids by dots. There appears to be no similarity with the sequences of the other four types of 4αGTase shown in *Figure 2*. (TLITGT: *Thermococcus litoralis*; PKODGT: *Pyrococcus* sp. KOD1; PFURIGT: *Pyrococcus furiosus*; PHORIGT: *Pyrococcus horikoshii*; DTHERGT: *Dictyoglomus thermophilum*).

```
TLITGT    MERINFIFGIHNHQPLGNFGWVFEEAYNRSYRPFMEILEEFPEMKVNVHFSGPLLEWIEE  60
PKODGT    MEMVNFIFGIHNHQPLGNFGWVMESAYERSYRPFMETLEEYPNMKVAVHYSGPLLEWIRD  60
PFURIGT   GDKINFIFGIHNHQPLGNFGWVFEEAYEKCYWPFLETLEEYPNMKVAIHTSGPLIEWLQD  60
PHORIGT   MPRINFIFGVHNHQPLGNFEWIIKRAYEKAYRPFLETLEEYPNMKVAVHISGVLVEWLER  60
DTHERGT   TKSIYFSLGIHNHQPVGNFDFVIERAYEMSYKPLINFFFKHPDFPINVHFSGFLLLWLEK  60
          :  * :*:*****:*** :::: **: .* *::: : :.*:: : :* ** *: *:.

TLITGT    NKPDYLDLLRSLIKRGQLEIVVAGFYEPVLAAIPKEDRLVQIEMLKDYAR-KLGYDAKGV  119
PKODGT    NKPEHLDLLRSLVKRGQLEIVVAGFYEPVLASIPKEDRIVQIEKLKEFAR-NLGYEARGV  119
PFURIGT   NRPEYIDLLRSLVKRGQVEIVVAGFYEPVLASIPKEDRIEQIRLMKEWAK-SIGFDARGV  119
PHORIGT   NRPEYIDLLKSLIKKGQVELVVAGFYEPILVAIPEEDRVEQIKLSKGWAR-KMGYEARGL  119
DTHERGT   NHPEYFEKLKIMAERGQIEFVSGGFYEPILPIIPDKDKVQQIKKLNKYIYDKFGQTPKGM  120
          *:*::::* *: : ::**:*:* .*****:* **.:*:: **.  : :   .:*  .:*:

TLITGT    WLTERVWQPELVKSLREAGIEYVVVDDYHFMSAGLSKEELFWPYYTEDGGEVITVFPIDE  179
PKODGT    WLTERVWQPELVKSLRAAGIDYIVVDDYHFMSAGLSKDELFWPYYTEDGGEVITVFPIDE  179
PFURIGT   WLTERVWQPELVKTLKESGIDYVIVDDYHFMSAGLSKEELYWPYYTEDGGEVIAVFPIDE  179
PHORIGT   WLTERVWEPELVKTLREAGIEYVILDDYHFMSAGLSKEELFWPYYTENGGEAIVVFPIDE  179
DTHERGT   WLAERVWEPHLVKYIAEAGIEYVVVDDAHFFSVGLKKEEDLFGYYLMEEQGYKLAVFPISM  180
          **:****:*.*** :  :**:**:.**.:**.:::*:  :**. : *  .:.****.

TLITGT    KLRYLIPFRPVKKTIEYLESLTSDDPSKVAVFHDDGEKFGVWPGTYEWVYEKGWLREFFD  239
PKODGT    KLRYLIPFRPVDKTLEYLHSLDDGDESKVAVFHDDGEKFGVWPGTYEWVYEKGWLREFFD  239
PFURIGT   KLRYLIPFRPVDKVLEYLHSLIDGDESKVAVFHDDGEKFGIWPGTYEWVYERGWLKEFFD  239
PHORIGT   KLRYLIPFRPVNETLEYLHSLADEDESKVAVFHDDGEKFGAWPGTHELVYERGWLKEFFD  239
DTHERGT   KLRYLIPFADPEETITYLDKFASEDKSIALLFDDGEKFGLWPDTYRTVYEEGWLETFVS  240
          *******     .:.: **..:  * **:*:.******* **.*:. ***.***. *..

TLITGT    AITSNEKIN--LMTYSEYLSKFTPRGLVYLPIASYFEMSEWSLPAKQAKLFVEFVEQLKE  297
PKODGT    RVSSDERIN--LMLYSEYLQRFPRGLVYLPIASYFEMSEWSLPARQAKLFVEFVEVNELKV  297
PFURIGT   RISSDEKIN--LMLYTEYLEKYKPRGLVYLPIASYFEMSEWSLPAKQARLFVEFVNELKV  297
PHORIGT   RISSDDKIN--LMLYSEYLSKFRPKGLVYLPIASYFEMSEWSLPARQAKLFFEFIKKLKE  297
DTHERGT   KIKENFLLVTPVNLYT-YMQRVKPKGRIYLPTASYREMMEWVLFPEAQKELEELVEKLKT  299
          :..:.   * *:.: *:* :*** *** **.*** :* ** . : :   *:::*

TLITGT    EGKFEKYRVFVRGGIWKNFFFKYPESNFMHKRMLMVSKAVRDNP------EARKYILKAQ  351
PKODGT    ENKFDRYRVFVRGGIWKNFFFKYPESNYMHKRMLMVSKAVRNNP------EAREFILRAQ  351
PFURIGT   KGIFEKYRVFVRGGIWKNFFYKYPESNYMHKRMLMVSKLVRNNP------EARKYLLRAQ  351
PHORIGT   LNLFEKYRIFVRGGIWKNFLYKYPEGNYMHKRMLMLSKLLRNNP------TARIFVLRAQ  351
DTHERGT   ENLWDKFSPYVKGGFWRNFLAKYDESNHMQKKMLYVWKKVQDSPNEEVKEKAMEEVFQGQ  359
          .  .:::: :*:**:*:**: **  :**:**:** :  ::: *    *   *  . *

TLITGT    CNDAYWHGVFGGIYLPHLRRTVWENIIKAQRYLKPENK---ILDVDFDGRAEIMVENDGF  408
PKODGT    CNDAYWHGVFGGVYLPHLRRAVWENIIKAQSHVKTGNF---VRDIDFDGRDEVFIENENF  408
PFURIGT   CNDAYWHGLFGGVYLPHLRRAIWNNLIKANSYVSLGKV---IRDIDYDGFEEVLIENENF  408
PHORIGT   CNDAYWHGIFGGIYLPHLRRAVWRNLIKAHSYLEPENR---VFDLDFDGGEEIMLENENF  408
DTHERGT   ANDAYWHGIFGGLYLPHLRTAIYEHLIKAENYLENSEIRFNIFDFDCDGNDEIIVESPFF  419
          .:*******:***:****:*  ::.::***  ::.  :     *.* ** *:::*. *

TLITGT    IATIKPHYGGSIFELSSKRKAVNYNDVLPRRWEHYHEVPEATKPEKESEEGIASIHELGK  468
PKODGT    YAVFKPAYGGALFELSSKRKAVNYNDVLARRWEHYHEVPEAATPE-EGGEGVASIHELGK  467
PFURIGT   YAVFKPSYGGSLVEFSSKNRLVNYVDVLARRWEHYHGYVES------QFDGVASIHELEK  462
PHORIGT   ILVVKPHYGGAIFEMSSKKKYVNYLDVVARRWEHYHSL--------------------K  447
DTHERGT   NLYLSPNHGGSVLEWDFKTKAFNLTNVLTRRKEAYHSKLSYVTS---EAQG-KSIHERWT  475
          .* :**::.*  . *  .* :*:.** * **

TLITGT    QIPEEIRRELAYDWQLRAILQDHFIKPEETLDN-YRLVKYHELGDFVNQPYEYEMI--EN  525
PKODGT    QIPDEIRRELAYDSHLRAILQDHFLEPETTLDE-YRLSRYIELGDFLTGAYNFSLI--EN  524
PFURIGT   KIPDEIRKEVAYDKYRRFMLQDHVVPLGTTLED-FMFSRQQEIGEFPRVPYSYELL--DG  519
PHORIGT   DIPEGMKRELSYDKWPRGMLQDHFLLPTEVLDN-YMLSKYRELGDFLMSSYHYQIE--DK  504
DTHERGT   AKEEGLENILFYDNHRRVSFTEKIFESEPVLEDLWKDSSRLEVDSFYEN-YDYEINKDEN  534
          : :..  **  **    :** *:: * .  :: :*:.**  *  .

TLITGT    GVKLWREGGVYAEEEKIPARVEKKIELTED----GFIAKYRVLLEKPYKALFGVEINLAVH  581
PKODGT    GITLERDG---SVAKRPARVEKSVRLTED----GFIVDYTVRSDA-RALFGVELNLAVH  575
PFURIGT   GIRLKREH-------LGIEVEKTVKLVND----GFEVEYIVNNKTGNPVLFAVELNVAVQ  568
PHORIGT   -LRLWRSG---KVKGISVEVEKVLRLNKD----GFTTEYRIVSKEELGLMFGVEINLAVQ  556
DTHERGT   KIRVLFSG-----VFRGFELCKSYILYKDKSFVDVVYEIKNVSETPISLNFGWEINLNFL  589
          :  :      .   . :     :  :**: :: *. *.:*:*

TLITGT    S-----------VMEKPEE-FEAKEFEVNDPYG-IGKVRIE--LDKAAKVWKFPIKTLSQ  626
PKODGT    S-----------VMEEPAE-FEAKEFEVNDPYG-IGKVEIE--LDRRAKVWKPYIKTLSQ  620
PFURIGT   S-----------IMESPGV-LRGKEIVVDDKYA-VGKFALK--FEDEMEVWKFPIKTLSQ  613
PHORIGT   G-----------TVEYPAE-FMSKEIEVKDIF---GKVKIE--SEKEAKIWKFPIKTLSQ  599
DTHERGT   APNHPDYYFLIGDQKYPLSSFGIEKVNNWKIFSGIG-IELECVLDVEASLYRYPIETVSL  648
          .       :* : ::. .: *   .::  :   .::::*::*:*

TLITGT    SEAGWDFIQQGVS----YTMLFPIEKELEFTVRFREL--  659
PKODGT    SESGWDFIQQGVS----YTVLFPVEGELRFRLRFREL--  653
PFURIGT   SESGWDLIQQGVS----YIVPIRLEDKIRFKLKFEEASG  648
PHORIGT   SESGWDFVQQGVS----YTFLYPIEKMLNIKLKFKESM-  633
DTHERGT   SEEGFERVYQGSALIHFYKVDLPVGSTWRTTIRFWVK--  685
          ** *:: : **  :    *   .  :    . ::*
```

four highly conserved sequences found in enzymes of the α-amylase super-family, so form a new group of 4αGTase. Further investigations of these enzymes in terms of their structure and mechanism of action will be very interesting, since they are the first example of 4αGTases which might work by a mechanism different to the one commonly employed by α-amylase-family enzymes. Furthermore, the enzymes belonging to this new group are expected to have industrial applications, since they have optimum temperatures for activity of 90 to 100°C (Laderman *et al.*, 1993a; Jeon *et al.*, 1997; Tachibana *et al.*, 1997; Fukusumi *et al.*, 1988).

Other 4αGTases

Several additional glucanotransferase activities have been identified biochemicaly (in some cases as amylomaltase) in several bacterial species, including *Streptococcus bovis* (Walker, 1965), *Streptococcus mitis* (Walker, 1966), *Bacillus subtilis* (Pazur and Okada, 1968), *Pseudomonas stutzeri* (Schmidt and John, 1979) and *Streptococcus mutans* (Medda and Smith, 1984). Since the primary sequences for these enzymes are not available, we do not know to which group each enzyme belongs. The properties of these enzymes are summarised in *Table 1*.

Synthesis of cycloglucans by 4αGTases

DISCOVERY AND SYNTHESIS OF CYCLOAMYLOSE

Studies aimed at determining whether potato D-enzyme can act upon high molecular weight glucans, led to the discovery that this enzyme can catalyse an intra-molecular transglycosylation reaction on synthetic amylose, to produce cycloamylose (CA) of high DP (Takaha *et al.*, 1996). In the early stages of the reaction with amylose AS-320, CAs with several hundred glucoses were produced, and as the reaction progressed, the mean product size tended towards a DP of about 90, but the yield of CA reached more than 95% (Takaha *et al.*, 1996). D-enzyme apparently breaks an internal α-1,4-bond within the amylose molecule, then transfers the newly-formed reducing end to the non-reducing end of itself, to create the CA molecule. This reaction is readily reversible, because the large CAs produced initially, are converted to smaller CAs as the reaction proceeds (*Figure 6*). The high yield of CA indicates that hydrolytic activity is very low. The smallest product is CA with 17 glucoses (CA17). CAs have properties which suggest that they may be of applied value (see below).

This observation led to examination of the possibility that bacterial Type II enzymes may also catalyse the formation of CA from amylose. 4αGTases from *E. coli* and *Thermus aquaticus* have subsequently been shown to produce CA (Terada *et al.*, unpublished). Similar to potato D-enzyme action, the initial products are larger than the final products. The smallest product of the *E. coli* enzyme is also CA17, but that from *T. aquaticus* is CA22 (Terada *et al.*, unpublished). The yield of CA from the *E. coli* enzyme is less than that of the potato enzyme, which is attributable to its higher hydrolytic activity, as revealed by the increase in reducing power during the reaction (*Figure 7*).

Subsequently, it has been demonstrated that Type V 4αGTases can produce

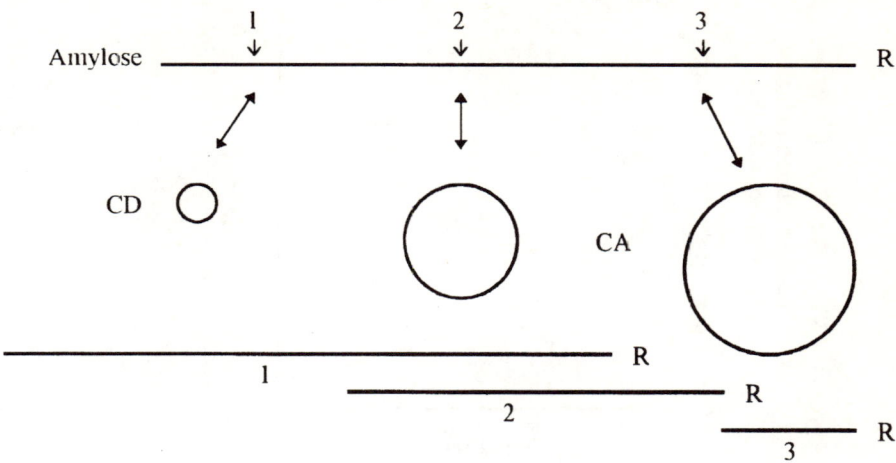

Figure 6.   Cycloglucan synthesis from high molecular weight amylose. 4αGTase is able to attack within the linear amylose molecule (where R is the reducing end) and to create a cyclic product by intramolecular transglycosylation. Smaller linear products are also created in these reactions. In those cases studied, the first cyclic products of such a reaction are cycloamyloses (CA) with a high degree of polymerisation (>50). As the reaction proceeds, smaller cycloglucans are produced through further coupling and cyclisation reactions. The end-products of CGTase (Type I 4αGTase) activity are cyclodextrins (CD) whereas those of Type II 4αGTases are cycloamyloses (CA).

cycloglucans. The enzymes from *T. litoralis* and *Pyrococcus* sp. KOD1 were incubated with amylose and soluble starch, respectively, without an additional acceptor molecule. In both cases the ability to form a blue complex with iodine decreased, but there was little increase in reducing power. The products from the reaction using the *T. litoralis* enzyme contained cyclic α-1,4-glucans indicating that the enzyme catalyses the cyclisation of amylose (Jeon *et al.*, 1997). The formation of such cyclic glucans by Type IV enzymes has not been demonstrated, but is suggested by the observation that such enzymes can reduce the ability of amylose to form a blue complex with iodine, without increasing reducing power (Liebl *et al.*, 1992). It is not known if Type III 4αGTases can produce cyclic glucans from amylose.

These discoveries led to a re-examination of the action of Type I 4αGTases on amylose, and the discovery that this enzyme from two *Bacillus* species also preferentially produced large CAs early in the reaction with amylose (Terada *et al.*, 1997). As the reaction proceeded, the characteristic α-, β- and γ-cyclodextrins (CA6, 7 and 8) were produced. This was a remarkable finding since CAs with 6, 7 or 8 glucoses had been thought to be the only products of CGTase activity since their discovery in 1903 (Szejtli, 1988). This observation therefore demands a re-evaluation of the mode of action of Type I 4αGTases on amylose.

The discovery of CA synthesis with a range of 4αGTases raises important questions about substrate specificity and conformation. It appears that each of these enzymes can accommodate the amylose molecule in its active site and can catalyse the cyclisation reaction. Presumably, the properties of the linear amylose dictate that the initial cleavage and glucanotransferase reaction will produce a CA molecule of high DP. Subsequently, such large CA molecules must be accommodated again by the enzyme

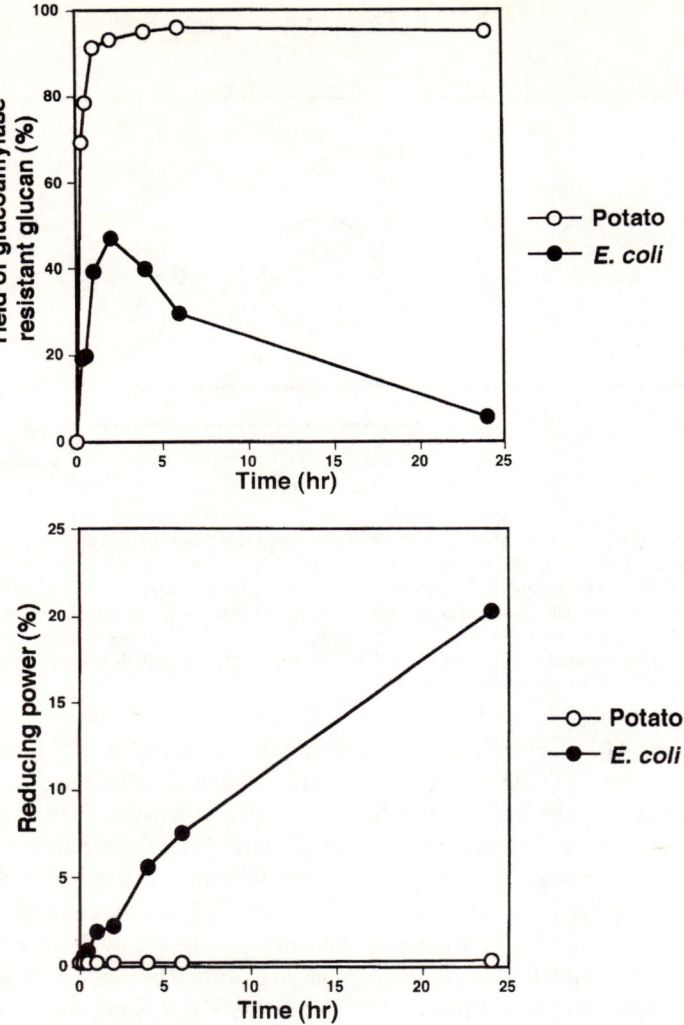

Figure 7.   Cycloamylose production from linear amylose and hydrolytic activity of Type II 4αGTases of potato and *E. coli*. Synthetic amylose (AS-320) was incubated with potato and *E. coli* enzymes. The yield of glucoamylose-resistant products (cycloglucans) was determined (upper panel) and the change in reducing power assayed (lower panel) during the reaction (Takaha *et al.*, 1996).

and a further glucanotransferase reaction takes place to create a smaller CA. Somehow we have to be able to explain how each enzyme determines the final product size. In the extreme cases, potato D-enzyme preferentially produces CA90 while CGTases produces CA7 and CA8. Significant progress has been made with the determination of some CA structures (see below) but we are still remarkably ignorant of substrate conformations and the structures of enzyme-substrate complexes.

ACTION OF 4αGTASES ON AMYLOPECTIN *IN VITRO*

Following the discovery that potato D-enzyme can catalyse a cyclisation reaction on

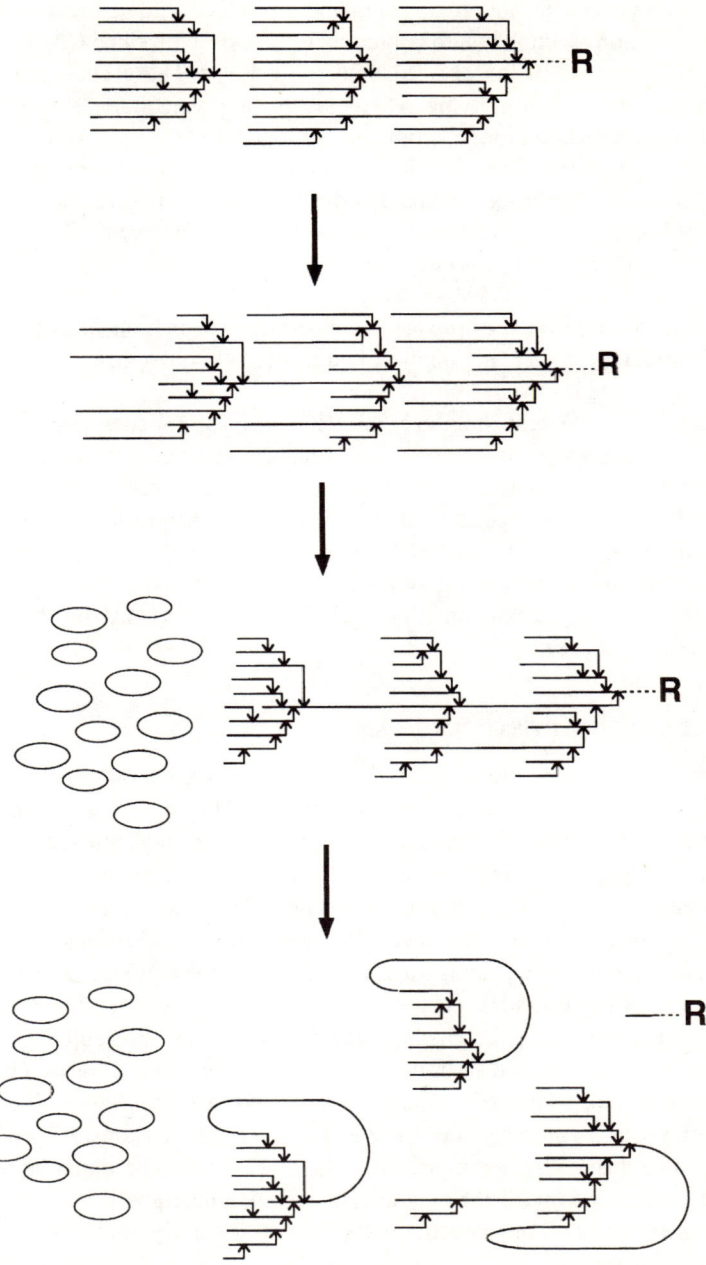

Figure 8.   Action of potato Type II 4αGTase on amylopectin. 4αGTase first attacks outer chains and creates longer branches through disproportionation activity. Next, cycloglucans are produced from the lengthened side chains. Cycloamylose accounts for most of these products but some molecules can also contain α-1,6-links (not shown). 4αGTase next attacks between cluster units and catalyses another intramolecular transglycosylation reaction to create cyclic cluster dextrins.

amylose, its action on amylopectin (AP) was investigated (Takaha *et al.*, 1998b). It was found that D-enzyme can degrade AP to products of smaller size, which are largely glucoamylase-resistant. Analysis of these products showed that D-enzyme catalyses cyclisation reactions on the outer chains of AP to produce CA molecules, and other cycloglucans which also contain α-1,6-links. These cycloglucans are larger than the outer chains of the AP molecule, implying that D-enzyme first carried out disproportionation reactions to produce longer side chains, before catalysing cyclisation reactions. The way in which such structures are produced is shown in *Figure 8*. One result of this activity is that the AP side chains become shortened. Subsequently, D-enzyme attacks glucan chains between AP cluster units and catalyses intra-cluster cyclisation reactions which separate cluster units from each other (Takaha *et al.*, 1998b). The product of this reaction is called cyclic cluster dextrin (CCD). Type I 4αGTases can catalyse formation of cyclodextrins and CCD from AP (Terada *et al.*, unpublished results) but other 4αGTases have not apparently been tested.

It is also of note that branching enzyme (EC 2.4.1.18; 1,4-α-D-glucan:1,4-α-D-glucan 6-α-D-[1,4-glucano]-transferase) can catalyse cyclisation reactions on AP to produce cyclic cluster units. In this case, the enzyme does not necessarily attack outer chains, but cleaves the glucan chain between clusters and then forms an intra-cluster α-1,6-link, producing a long-chain CCD (Takata *et al.*, 1996, 1997). CCDs are of potential applied value. They are highly soluble in water and have a low propensity for retrogradation. Starch pastes containing CCD have low viscosity and high transparency (Takata *et al.*, 1997).

STRUCTURES AND PROPERTIES OF CYCLOAMYLOSES

Cyclodextrins are annular molecules of 6, 7 or 8 glucoses, the internal cavity of which is hydrophobic, while the outer surface is hydrophilic. They are therefore soluble in aqueous solutions, but can accommodate hydrophobic guest molecules in the central cavity with high specificity (Szejtli, 1988). Such inclusion complexes can change the solubility, reactivity or stability of guest molecules. They have been the subject of much research over many decades, and have found many applications in the food, chemical and pharmaceutical industries. For example, they provide the basis for formulations used in pesticides, cosmetics and drug delivery, and can change the flavour or odour of compounds. The high specificity of guest molecule inclusion is reflected in their ability to discriminate isomers of the same compound. Some chemically-modified cyclodextrins also possess catalytic activity (Szejtli, 1988). The central cavities of α-, β- and γ-cyclodextrins have different dimensions, allowing them to accommodate different guest molecules. The discovery of the much larger CAs immediately suggested that if they could also accommodate guest molecules, they would be of different sizes or properties to those of cyclodextrins and could therefore have new applications.

CA molecules of defined size were purified by high performance liquid chromatography using an ODS (C18) column, for structural analysis by NMR and X-ray crystallography (Jacob *et al.*, 1998; Saenger *et al.*, 1998). The first aim was to determine how the structure of the CA molecule changes when the number of glucoses is increased progressively to more than 8 (γ-cyclodextrin). As the macrocycle increases

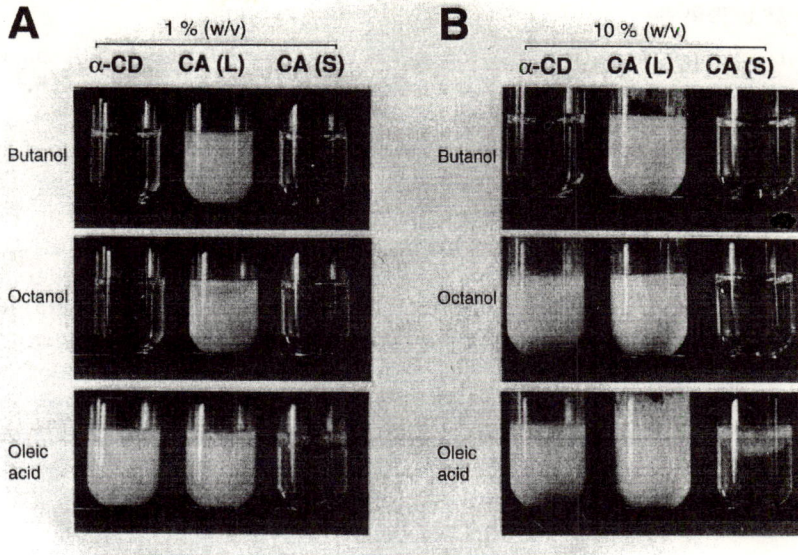

Figure 9. Formation of inclusion complexes between cycloamyloses and guest molecules. Aqueous solutions (1% and 10% [w/v]) of α-CD (CA6) and CA were prepared. In one case the CA was comprised of a mixture of large molecules (L) with DP greater than 50. In the other case the CA was a mixture of smaller molecules (S) with DP between 17 and 50. To 2 ml of these solutions was added about 0.2 ml of butanol, octanol or oleic acid, then mixed. A precipitate indicates that the alcohol or fatty acid has formed an inclusion complex with the cycloglucan.

from 8 to 9 glucoses, the molecule apparently experiences steric strain which results in distortion of the previously annular molecule into one which is boat-shaped. Molecules with 10 and 14 glucoses could be crystallised and their structures were determined. Remarkably, when the molecule increases in size beyond 9 glucoses, the steric strain is relieved by 180° rotations about two diammetrically-opposed glucosidic bonds to create a novel hydrogen bonding pattern. Such rotations create two 'band-flip' motifs in the molecule, which is now described as 'butterfly-shaped'. The CA14 molecule also contains the 'band-flip' motifs and adopts a similar conformation. The central cavities of these CA molecules are likely to be too small to accommodate guest molecules. CA26 was also crystallised and its structure solved. This molecule is comprised of two antiparallel single helices of 12 glucoses each, connected by the 'band-flip' motifs at each end. The helices which have 6 glucoses per turn, contain central cavities with similar diameter to that of α-cyclodextrin, but are much longer (Saenger *et al.*, 1998). This suggests that such CAs could potentially accommodate guest molecules. This possibility has been tested by mixing CA with alcohols and fatty acids (*Figure 9*). It is clear that butanol, octanol and oleic acid can form inclusion complexes with a CA mixture containing molecules with DP greater than 50. The high solubility, non-reducing character and ability to form inclusion complexes suggest that CA has great potential for future applications.

Future prospects

4αGTases are important in α-1,4-glucan metabolism in organisms from all kingdoms. Furthermore, they are of applied value for the production of cycloglucans and for starch processing. There are several aspects which require more research in order to better understand their functions, and to exploit these enzymes. Only the structures of Type I enzymes have been determined at atomic resolution. Structural determination of other 4αGTases will be required in order to understand the reaction specificities of these enzymes, and to explain how they produce distinct cycloglucan products. Rational design of enzymes to produce defined products can then be initiated. A particular aim for starch processing will be to obtain thermostable enzymes with low hydrolytic activity and high reaction specificity. This aim may be achieved both by protein engineering, and by searching natural sources for new enzymes. The evolutionary relationships of different 4αGTases are of interest, but the origin of Type V enzymes is particularly fascinating, and studies of their properties should provide valuable new information. Systems for the preparation of cycloamyloses and cyclodextrins are required so that their properties and potential applications can be effectively investigated.

References

BAHL, H., BURCHHARDT, G., SPREINAT, A., HAECKEL, K., WIENECKE, A., SCHMIDT, B. AND ANTRANIKIAN, G. (1991). *Applied Environmental Microbiology* **57**, 1554–1559.

BINDER, F., HUBER, O. AND BOCK, A. (1986). *Gene* **47**, 269–277.

BLOCH, M.A. AND RAIBAUD, O. (1986). *Journal of Bacteriology* **168**, 1220–1227.

BOEL, E., BRANDY, L., BRZOZOWSKI, A.M., DEREWENDA, Z., DODSON, G.G., JENSEN, V.J., PETERSEN, S.B., SWIFT, H., THIM, L. AND WOLDIKE, H.F. (1990). *Biochemistry* **29**, 6244–6249.

BRAUN, C., LINDHORST, T., MADSEN, N.B. AND WITHERS, S. (1996). *Biochemistry* **35**, 5458–5463.

BRAYER, G.D., LUO, Y. AND WITHERS, S.G. (1995). *Protein Science* **4**, 1730–1742.

BROWN, D.H. AND ILLINGWORTH, B. (1962). *Proceedings of the National Academy of Sciences USA.* **48**, 1783–1787.

BUSSEY, H., STORMS, R.K., AHMED, A., ALBERMANN, K., ALLEN, E., ANSORGE, W., ARAUJO, R., APARICIO, A., BARRELL, B., BADCOCK, K., BENES,V., BOTSTEIN, D., BOWMAN,S., BRUCKNER, M., CARPENTER, J., CHERRY, J.M., CHUNG, E., CHURCHER, C., COSTER, F., DAVIS, K., DAVIS, R.W., DIETRICH, F.S., DELIUS, H., DiPAOLO, T., DUBOIS, E., DUSTERHOFT, A., DUNCAN, M., FLOETH, M., FORTIN, N., FRIESEN, J.D., FRITZ, C., GOFFEAU, A., HALL, J., HEBLING, U., HEUMANN, K., HILBERT, H., HILLIER, L., HUNICKE-SMITH, S., HYMAN, R., JOHNSTON, M., KALMAN, S., KLEINE, K., KOMP, C., KURDI, O., LASHKARI, D., LEW, H., LIN, A., LIN, D., LOUIS, E.J., MARATHE, R., MESSENGUY, F., MEWES, H.W., MIRTIPATI, S., MOESTL, D., MULLER-AUER, S., NAMATH, A., NENTWICH, U., OEFNER, P., PEARSON, D., PETEL, F.X., POHL, T.M., PURNELLE, D., SCHAFER, M., SCHARFE, M., SCHERENS, B., SCHRAMM, S., SCHROEDER, M., SDICU, A.M., TETTELIN, H., URRESTARAZU, L.A., USHINSKY, S., VIERENDEELS, F., VISSERS, S., VOSS, H., WALSH, S.V., WAMBUTT, R., WANG, Y., WEDLER, E., WEDLER, H., WINNETT, E., ZHONG, W.W., ZOLLNER, A., VO, D.H. AND HANI, J. (1997). *Nature* **387,** (6632 Suppl), 103–105.

COLE, S.T., BROSCH, R., PARKHILL, J., GARNIER, T., CHURCHER, C., HARRIS, D., GORDON, S.V., EIGLMEIER, K., GAS, S., BARRY III, C.E., TEKAIA, F., BADCOCK, K., BASHAM, D., BROWN, D., CHILLINGWORTH, T., CONNOR, R., DAVIES, R., DEVLIN, K., FELTWELL, T., GENTLES, S., HAMLIN, N., HOLROYD, S., HORNSBY, T., JAGELS, K., KROGH, A., McLEAN, J., MOULE, S., MURPHY, L., OLIVER, S., OSBORNE, J., QUAIL, M.A., RAJANDREAM, M.A., ROGERS, J., RUTTER, S., SEEGER, K., SKELTON, S., SQUARES, S., SQARES, R., SULSTON, J.E., TAYLOR, K., WHITEHEAD, S. AND BARRELL, B.G. (1998). *Nature* **393,** 537–544.

DECKERT, G., WARREN, P.V., GAASTERLAND, T., YOUNG, W.G., LENOX, A.L., GRAHAM, D.E., OVERBEEK, R., SNEAD, M.A., KELLER, M., AUJAY, M., HUBER, R., FELDMAN, R.A., SHORT, J.M., OLSON, G.J. AND SWANSON, R.V. (1998). *Nature* **392**, 353–358.

FEEDERLE, R., PAJATSCH, M., KREMMER, E. AND BOCK. A. (1996). *Archives of Microbiology* **165**, 206–212.

FIEDLER, G., PAJATSCH, M. AND BOCK, A. (1996). *Journal of Molecular Biology* **256**, 279–291.

FLEISCHMANN, R.D., ADAMS, M.D., WHITE, O., CLAYTON, R.A., KIRKNESS, E.F., KERLAVAGE, A.R., BULT, C.J., TOMB, J-F., DOUGHERTY, B.A., MERRICK, J.M., MCKENNEY, K., SUTTON, G., FITZHUGH, W., FIELDS, C.A., GOCAYNE, J.D., SCOTT, J.D., SHIRLEY, R., LIU, L-I., GLODEK, A., KELLEY, J.M., WEIDMAN, J.F., PHILLIPS, C.A., SPRIGGS, T., HEDBLOM, E., COTTON, M.D., UTTERBACK, T.R., HANNA, M.C., NGUYEN, D.T., SAUDEK, D.M., BRANDON, R.C., FINE, L.D., FRITCHMAN, J.L., FUHRMANN, J.L., GEOGHAGEN, N.S.M., GNEHM, C.L., MCDONALD, L.A., SMALL, K.V., FRASER, C.M., SMITH, H.O. AND VENTER, J.C (1995). *Science* **269**, 496–512.

FRASER, C.M., CASJENS, S., HUANG, W.M., SUTTON, G.G., CLAYTON, R.A., LATHIGRA, R., WHITE, O., KETCHUM, K.A., DODSON, R., HICKEY, E.K., GWINN, M., DOUGHERTY, B., TOMB, J.-F., FLEISCHMANN, R.D., RICHARDSON, D., PETERSON, J., KERLAVAGE, A.R., QUACKENBUSH, J., SALZBERG, S., HANSON, M., VAN-VUGT, R., PALMER, N., ADAMS, M.D., GOCAYNE, J.D., WEIDMAN, J., UTTERBACK, T., WATTHEY, L., MCDONALD, L., ARTIACH, P., BOWMAN, C., GARLAND, S., FUJII, C., COTTON, M.D., HORST, K., ROBERTS, K., HATCH, B., SMITH, H.O. AND VENTER, J.C. (1997). *Nature* **390**, 580–586.

FRENCH, D. (1957). *Advances in Carbohydate Chemistry* **12**, 189–260.

FUKUSUMI, S., KAMIZONO, A., HORINOUCHI, S. AND BEPPU, T (1988). *European Journal of Biochemistry* **174**, 15–21.

GODA, S.K., EISSA, O., AKHTAR, M. AND MINTON, N.P. (1997). *Microbiology* **143**, 3287–3294.

GUNJA-SMITH, Z., MARSHALL, J.J., MERCIER, C., SMITH, E.E. AND WHELAN, W.J. (1970). *FEBS Letters* **12**, 101–104.

HARATA, K., HAGA, K., NAKAMURA, A., AOYAGI, M. AND YAMANE, K. (1996). *Acta Cryst.* **D52**, 1136–1145.

HEINRICH, P., HUBER, W. AND LIEBL,W. (1994). *Syst. Applied. Microbiology* **17**, 297–305.

HIZUKURI, S. (1986). *Carbohydrate Research* **147**, 342–347.

HIZUKURI, S. (1996). Starch: Analytical aspects. In *Carbohydrates in Food*. Ed. A. Eliasson, pp 347–429. New York: Marcel Dekker, Incorporated.

HORINOUCHI, S., FUKUSUMI, S., OHSHIMA, T. AND BEPPU, T. (1988). *European Journal of Biochemistry* **176**, 243–253.

HSIA, R.C., PANNEKOEK, Y., INGEROWSKI, E. AND BAVOIL, P.M. (1997). *Molecular Microbiology* **25**, 351–359.

JACOB, J., GESSLER, K., HOFFMANN, D., SANBE, H., KOIZUMI, K., SMITH, S.M., TAKAHA, T. AND SAENGER, W. (1998). *Angewandte Chemie International Edition* **37**, 606–609.

JEON, B.S., TAGUCHI, H., SAKAI, H., OHSHIMA, T., WAKAGI, T. AND MATSUZAWA, H. (1997). *European Journal of Biochemistry* **248**, 171–178.

JONES, G. AND WHELAN, W. J. (1969). *Carbohydate Research* **9**, 483–490.

JORGENSEN, S.T., TANGNEY, M., STARNES, R.L., AMEMIYA, K. AND JORGENSEN, P.L. (1997). *Biotechnology Letters* **19**, 1027–1031.

KADZIOLA, A., ABE, J., SVENSSON, B. AND HASER, R. (1994). *Journal of Molecular Biology* **239**, 104–121.

KAKEFUDA, G., DUKE, S.H. AND HOSTAK, M.S. (1986). *Planta* **168**, 175–182.

KANEKO, T., SATO, S., KOTANI, H., TANAKA, A., ASAMIZU, E., NAKAMURA, Y., MIYAJIMA, N., HIROSAWA, M., SUGIURA, M., SASAMOTO, S., KIMURA, T., HOSOUCHI, T., MATSUNO, A., MURAKI, A., NAKAZAKI, N., NARUO, K., OKUMURA, S., SHIMPO, S., TAKEUCHI, C., WADA, T., WATANABE, A., YAMADA, M., YASUDA, M. AND TABATA, S. (1996). *DNA Research* **3**, 109–136.

KATO, T. AND HORIKOSHI, K. (1986). *Journal of the Japanese Society of Starch Science* **33**, 137–143.

KITAHATA, S. (1995). In *Enzyme Chemistry and Molecular Biology of Amylases and Related Enzymes*. Ed. The Amylase Research Society of Japan, pp 6–17. Boca Raton (Florida): CRC Press.

KITAHATA, S. AND OKADA, S. (1982). *Journal of the Japanese Society of Starch Science* **29**, 13–18.

KITAHATA, S., MURAKAMI, H. AND OKADA S. (1989a). *Agricultural Biological Chemistry* **53**, 2653–2659.

KITAHATA, S., MURAKAMI, H., SONE, Y. AND MISAKI, A. (1989b). *Agricultural Biological Chemistry* **53**, 2661–2666.

KLEIN, C. AND SCHULZ, G.E. (1991). *Journal of Molecular Biology* **217**, 737–750.

KLEIN, C., HOLLENDER, J., BENDER, H. AND SCHULZ, G.E. (1992). *Biochemistry* **31**, 8740–8746.

KNEGTEL, R.M., STROKOPYTOV, B., PENNINGA, D., FABER, O.G., ROZEBOOM, H.J., KALK, K.H., DIJKHUIZEN, L. AND DIJKSTRA, B.W. (1995). *Journal of Biological Chemistry* **270**, 29256–29264.

KNEGTEL, R.M., WIND, R.D., ROZEBOOM, H.J., KALK, K.H., BUITELAAR, R.M., DIJKHUIZEN, L. AND DIJKSTRA, B.W. (1996). *Journal of Molecular Biology* **256**, 611–622.

KOBAYASHI, S. (1996). In *Enzymes for Carbohydrate Engineering*. Eds. K.H. Park, J.F. Robyt, and Y.-D. Choi, pp 23–41. Amsterdam: Elsevier Science.

KUBOTA, M., MATSUURA, Y., SAKAI, S. AND KATSUBE, Y. (1991). *Denpun Kagaku* **38**, 141–146.

KUBOTA, M., MATSUURA, Y., SAKAI, S. AND KATSUBE, Y. (1994). *Oyo Toshitu Kagaku* **41**, 245–253.

LACKS, S.A., DUNN, J.J. AND GREENBERG, B. (1982). *Cell* **31**, 327–336.

LADERMAN, K.A., DAVIS, B.R., KRUTZSCH, H.C., LEWIS, M.S., GRIKO, Y.V., PRIVALOV, P.L. AND ANFINSEN, C.B. (1993a). *Journal of Biological Chemistry* **268**, 24394–24401.

LADERMAN, K.A., ASADA, K., UEMORI, T., MUKAI, H., TAGUCHI, Y., KATO, I. AND ANFINSEN, C.B. (1993b). *Journal of Biological Chemistry* **268**, 24402–24407.

LAWSON, C.L., VAN MONTFORT, R., STROKOPYTOV, B., ROZEBOOM, H.J., KALK, K.H., DE VRIES, G.E., PENNINGA, D., DIJKHUIZEN, L. AND DIJKSTRA, B.W. (1994). *Journal of Molecular Biology* **236**, 590–600.

LEE, E.Y.C. AND CARTER, J.H. (1973). *Archives of Biochemistry and Biophysics* **154**, 636–641.

LIEBL, W., FEIL, R., GABELSBERGER, J., KELLERMANN, J. AND SCHLEIFER, K.H. (1992). *European Journal of Biochemistry* **207**, 81–88.

LILLFORD, P.J. AND MORRISON, A. (1997). In *Starch: structure and functionality*. Eds. P.J. Frazier, P. Richmond, and A.M. Donald, pp 1–8. Cambridge: The Royal Society of Chemistry.

LIN, T. P. AND PREISS, J. (1988). *Plant Physiology* **86**, 260–265.

LIU, W., MADSEN, N.B., BRAUN, C. AND WITHERS, S. (1991). *Biochemistry* **30**, 1419–1424.

LIU, W., DE CASTRO, M.L., TAKRAMA, J., BILOUS, P.T., VINAYAGAMOORTHY, T., MADSEN, N.B. AND BLEACKLEY, R.C. (1993). *Archives of Biochemistry and Biophysics* **306**, 232–239.

MACHIUS, M., WIEGAND, G. AND HUBER, R. (1995). *Journal of Molecular Biology* **246**, 545–559.

MANNERS, D.J. (1991). *Carbohydrate Polymers* **16**, 37–82.

MANNERS, D.J. AND ROWE, K. L. (1969). *Carbohydrate Research* **9**, 441–450.

MATSUURA, Y., KUSUNIKI, M., HARADA, W. AND KAKUDO, M. (1984). *Journal of Biochemistry* (Tokyo) **95**, 697–702.

MEDDA, S. AND SMITH, E.E. (1984). *Analytical Biochemistry* **138**, 354–359.

METZ, R.J., ALLEN, L.N., CAO, T.M. AND ZEMAN, N.W. (1988). *Nucleic Acids Research* **16**, 5203.

MONOD, J. AND TORRIANI, A.M. (1948). *Compt. Rend.* **227**, 240.

MORI, S., GOTO, M., MASE, T., MATSUURA, A., OYA, T. AND KITAHATA, S (1995). *Bioscience, Biotechnology and Biochemistry* **59**, 1012–1015.

MORISHITA, Y., HASEGAWA, K., MATSUURA, Y., KATSUBE, Y., KUBOTA, M. AND SAKAI, S. (1997). *Journal of Molecular Biology* **267**, 661–672.

MOSI, R., HE, S., UITDEHAAG, J., DIJKSTRA, B.W. AND WITHERS, S.G. (1997). *Biochemistry* **36**, 9927–9934.

NAKAMURA, A., HAGA, K., OGAWA, S., KUWANO, S., KIMURA, K. AND YAMANE, K. (1992). *FEBS Letters* **296**, 37–40.

NAKAMURA, A., HAGA, K. AND YAMANE, K. (1993). *Biochemistry* **32**, 6624–6631.

NAKAMURA, A., HAGA, K. AND YAMANE, K. (1994). *Biochemistry* **33**, 9929–9936.

NELSON, T.E., WHITE, R.C. AND GILLARD, B.K. (1979). *American Chemical Society Symposium Series* **88**, 131–162.

OKITA, T.W., GREENBERG, E., KUHN, D.N. AND PREISS, J. (1979). *Plant Physiology* **64**, 187–192.

PAJATSCH, M., GERHART, M., PEIST, R., HORLACHER, R., BOOS, W. AND BOCK, A. (1998). *Journal of Bacteriology* **180**, 2630–2635.

PALMER, T.N., RYMAN, B.E. AND WHELAN, W.J. (1976). *European Journal of Biochemistry* **69**, 105–115.

PAZUR, J.H. AND OKADA, S. (1968). *Journal of Biological Chemistry* **243**, 4732–4738.

PEAT, S., WHELAN, W.J AND REES, W.R. (1956). *Journal of the Chemical Society*, 44–53.

PENNINGA, D., STROKOPYTOV, B., ROZEBOOM, H.J., LAWSON, C.L., DIJKSTRA, B.W., BERGSMA, J. AND DIJKHUIZEN, L. (1995). *Biochemistry* **34**, 3368–3376.

PUGSLEY, A.P. AND DUBREVIL, C. (1988). *Molecular Microbiology* **2**, 473–479.

QIAN, M., HASER, R. AND PAYAN, F. (1993). *Journal of Molecular Biology* **231**, 785–799.

RAMASUBBU, N., PALOTH, V., LUO, Y., BRAYER, G.D. AND LEVINE, M.J. (1996). *Acta Crystallogr.* **D52**, 435–446.

RENDLEMAN, JUNIOR, J.A (1993). *Carbohydrate Research* **247**, 223–237.

SAENGER, W., JACOB, J., GESSLER, K., STEINER, T., HOFFMANN, D., SANBE, H., KOIZUMI, K., SMITH, S.M. AND TAKAHA, T. (1998). *Chemical Reviews* **98**, 1787–1802.

SANDHYA RANI, M.R., SHIBANUMA, K. AND HIZUKURI, S. (1992). *Carbohydrate Research* **227**, 183–194.

SCHMID, G. (1989). *Trends in Biotechnology* **7**, 244–248.

SCHMIDT, J. AND JOHN. M. (1979). *Biochimica Biophysica Acta* **566**, 100–114.

SCHMIDT, A.K., COTTAZ, S., DRIGUEZ, H. AND SCHULZ, G.E. (1998). *Biochemistry* **37**, 5909–5915.

SCHWARTZ, M. (1987). In *Escherichia coli and Salmonella typhimurium: Cellular and Molecular Biology*. Ed. F.C. Neidhardt, pp 1482–1502. Washington DC: American Society for Microbiology.

STARNES, R.L. (1990). *Cereal Food World* **35**, 1091–1099.

STROKOPYTOV, B., PENNINGA, D., ROZEBOOM, H.J., KALK, K.H., DIJKHUIZEN, L. AND DIJKSTRA, B.W. (1995). *Biochemistry* **34**, 2234–2240.

STROKOPYTOV, B., KNEGTEL, R.M., PENNINGA, D., ROZEBOOM, H.J., KALK, K.H., DIJKHUIZEN, L. AND DIJKSTRA, B.W. (1996). *Biochemistry* **35**, 4241–4249.

SUGANUMA, T., SETOGUCHI, S., FUJIMOTO, S. AND NAGAHAMA, T. (1991). *Carbohydrate Research* **212**, 201–212.

SVENSSON, B. (1994). *Plant Molecular Biology* **25**, 141–157.

SZEJTLI, J. (1988). *Cyclodextrin Technology*. The Netherlands: Kluwer Academic Publishers, Dordrecht.

TABATA, S. AND HIZUKURI, S. (1992). *European Journal of Biochemistry* **206**, 345–348.

TABATA, S. AND IDE, T. (1988). *Carbohydrate Research* **176**, 245–251.

TABATA, S., NAKAYAMA, A. AND HIZUKURI, S. (1995). *Oyo Toshitu Kagaku* **42**, 193–202.

TACHIBANA, Y., FUJIWARA, S., TAKAGI, M. AND IMANAKA, T (1997). *Journal of Fermentation Bioengineering* **83**, 540–548.

TAKAHA, T., YANASE, M., OKADA, S. AND SMITH, S.M. (1993). *Journal of Biological Chemistry* **268**, 1391–1396.

TAKAHA, T., YANASE, M., TAKATA, H., OKADA, S. AND SMITH, S.M. (1996). *Journal of Biological Chemistry* **271**, 2902–2908.

TAKAHA, T., CRITCHLEY, J., OKADA, S. AND SMITH S.M. (1998a). *Planta* **205**, 445–451.

TAKAHA, T., YANASE, M., TAKATA, H., OKADA, S. AND SMITH, S.M. (1998b). *Biochemical and Biophysical Research Communications* **247**, 493–497.

TAKATA, H., TAKAHA, T., OKADA, S., HIZUKURI, S., TAKAGI, M. AND IMANAKA, T. (1996). *Carbohydrate Research* **295**, 91–101.

TAKATA, H., TAKAHA, T., NAKAMURA, H., FUJII, K., OKADA, S., TAKAGI, M. AND IMANAKA, T. (1997). *Journal of Fermentation Bioengineering* **84**, 119–123.

TERADA, Y., YANASE, M., TAKATA, H., TAKAHA, T. AND OKADA, S. (1997). *Journal of Biological Chemistry* **272**, 15729–15733.

WALKER, G.J. (1965). *Biochemical Journal* **94**, 299–308.

WALKER, G.J. (1966). *Biochemical Journal* **101**, 861–872.

WILSON, R., AINSCOUGH, R., ANDERSON, K., BAYNES, C., BERKS, M., BONFIELD, J., BURTON, J., CONNELL, M., COPSEY, T., COOPER, J., COULSON, A., CRAXTON, M., DEAR, S., DU, Z., DURBIN, R., FAVELLO, A., FULTON, L., GARDNER, A., GREEN, P., HAWKINS, T., HILLIER, L., JIER, M., JOHNSTON, L., JONES, M., KERSHAW, J., KIRSTEN, J., LAISTER, N., LATREILLE, P., LIGHTNING, J., LLOYD, C., MCMURRAY, A., MORTIMORE, B., O'CALLAGHAN, M., PARSONS, J., PERCY, C., RIFKEN, L., ROOPRA, A., SAUNDERS, D., SHOWNKEEN, R., SMALDON, N., SMITH, A., SONNHAMMER, E., STADEN, R., SULSTON, J., THIERRY-MIEG, J., THOMAS, K., VAUDIN, M., VAUGHAN, K., WATERSTON, R., WATSON, A., WEINSTOCK, L., WILKINSON-SPROAT, J. AND WOHLDMAN, P. (1998). *Nature* **368**, 32–38.

WIND, R.D., BUITELAAR, R.M. AND DIJKHUIZEN, L. (1998). *European Journal of Biochemistry* **253**, 598–605.

YANG, B.Z., DING, J.H., ENGHILD, J.J., BAO, Y. AND CHEN, Y.T. (1992). *Journal of Biological Chemistry* **267**, 9294–9299.

YOSHIO, N., MAEDA, I., TANIGUCHI, H. AND NAKAMURA, M. (1986). *Journal of the Japanese Society of Starch Science* **33**, 244–252.

11
Thermomechanical Properties of Amorphous Saccharides: Their Role in Enhancing Pharmaceutical Product Stability

FELIX FRANKS

BioUpdate Foundation, 7 Wootton Way, Cambridge, CB3 9LX, UK

Introduction

The thermochemical and thermomechanical properties of, and slow relaxation processes within amorphous carbohydrate matrices have grown into topics of interest and considerable research activity. Following the pioneering studies by Levine and Slade (1993) the significance of glass transitions of anhydrous carbohydrates became to be recognised by the food processing industry, especially in the areas of intermediate and low moisture product development. More recently, interest in the formation and properties of amorphous carbohydrates has also spread to the pharmaceutical process industry, where such compounds find extensive use as stabilisers and processing aids in various types of dry dosage formulations (Ahlneck and Zografi, 1990).

It has thus become clear that metastable and thermodynamically unstable, supersaturated states are of great practical importance, especially for mixtures in which eutectic phase separation does not occur spontaneously within observable periods. The combination of conventional phase coexistence curves with glass transition/composition profiles and, possibly, crystal nucleation information, in single representations, has given rise to the description 'state diagram' (Franks, 1982). The state diagram thus aims to incorporate a time dimension into the conventional phase diagram, in the sense that both vitrification and nucleation phenomena are kinetic rate processes, functionally quite unrelated to equilibrium phase transitions.

Of particular practical interest are mixtures in which one or more components might be able to crystallise spontaneously in real time, either wholly or partially, whereas other, coexisting solute species form supersaturated solutions, leading eventually to vitreous states (solid solutions). Of such systems, the ternary mixtures water/sucrose/glycine and water/sucrose/NaCl have received most attention (Suzuki and Franks, 1993; Shalaev et al., 1996), because they serve as models for the processing and drying of therapeutic preparations designed for injection and infusion, such as blood

Biotechnology and Genetic Engineering Reviews – Vol. 16, April 1999
0264–8725/99/16/281–292 $20.00 + $0.00 © Intercept Ltd, P.O. Box 716, Andover, Hampshire SP10 1YG, UK

coagulating factors and peptide hormones. Even the solid/liquid equilibrium phase diagrams are of a complex nature, displaying not only the three anhydrous crystalline phases, but also several crystal hydrates, with multiple peritectic points, as well as hydrated or anhydrous stoichiometric compounds, *e.g.* between sucrose and NaCl. The state diagrams contain, in addition, ternary glass transition/composition surfaces, bounded by the glass transition temperatures of the three pure components.

It is the purpose of this short review to highlight some salient physical and chemical features of concentrated mixtures containing lower oligosaccharides, with emphasis on their nonequilibrium properties.

The glass transition

A common feature of sugars and other polyhydroxy compounds (PHC) is their reluctance to crystallise from aqueous solutions during drying by freezing or evaporation. Eutectic phase separation within the period of observation is therefore rare. Instead, supersaturated solutions are formed which ultimately undergo a glass transition at a characteristic temperature T_g.

Figure 1 illustrates a typical state diagram for an aqueous carbohydrate solution (Franks, 1994). The systems of most practical interest cover the stippled area that lies beyond the domain bounded by the liquidus and solidus curves, specifically between the saturation solubility, denoted by S, and the glass transition profile. Thus, an unsaturated, dilute solution A can be dried by freezing to the (notional) eutectic T_e and beyond, or by evaporation to S and beyond. The system has to traverse the region of supersaturation and instability in order to reach the vitreous state of 'kinetic stability' at B. All indications are that physical and chemical processes can occur in the region of instability (and even in the vitreous state) that are quite unlike those conventionally described in text books on phase equilibria and chemical kinetics or in the literature devoted to 'solutions'.

As a generic group, PHCs fulfil the requirements of slow crystallisation and 'interesting' solution rheology and chemistry. As water is removed from their dilute

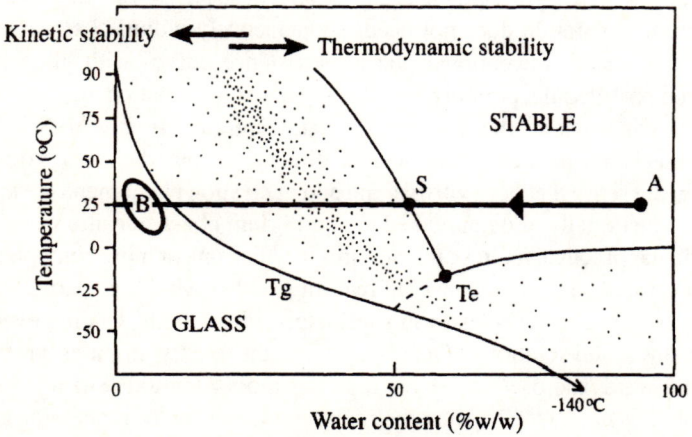

Figure 1.   Regions of stability and instability (stippled) traversed during the drying of a dilute solution (A) to an amorphous glassy solid (B). The density of stippling corresponds to the degree of instability of the supersaturated solution. Reproduced, with permission, from Franks (1994).

Table 1. Glass temperatures of dry anhydrous
PHCs; data from various sources

	$T_g/°C$
Glycerol	−93
Ribosde	−10
Xylose	−10
Sorbitol	−3
Mannitol	crystallises
Fructose	13
Glucose	39
Sucrose	70
Maltose.H_2O	70
Lactose	crystallises
Trehalose	106
Raffinose	109
Maltotriose	95
Stachyose	132
Destran[a]	84

[a] as supplied, without additional drying

solutions, either by freezing or by evaporation at ambient or elevated temperature, the viscosity of the supersaturated residue increases to the point, usually 10^{12}–10^{14} Pa s, where the mixture exhibits vitrification; it may then still contain up to 50% w/w of water, depending on the temperature at which the drying is carried out and the chemical nature of the solute(s).

The glass transition is usually determined by DSC; heating scans display a discontinuity in the heat flow (specific heat) at T_g. Angell (1995) has, however, pointed out that this heat flow discontinuity is not universally observed. This led him to differentiate between so-called strong and fragile fluids, depending on the intermolecular forces, where only members of the latter group, which contains hydrogen-bonded fluids, are expected to display DSC signals at T_g.

There remain large gaps in our understanding of the phenomenon of the glass transition, especially at the molecular level. Indeed, there does not yet exist a generally accepted theory for the origin of the glass transition. Some of the hypotheses that have been advanced from time to time cannot be tested experimentally, nor do they provide a measure of predictability. They are thus of very limited value to the technologist.

As a first approximation, T_g values of PHCs follow their molecular weights, as illustrated in *Table 1*. There are, however, some apparent anomalies. Mannitol, sorbitol, fructose, glucose and galactose are all monosaccharide hexoses displaying a considerable range of T_g values. Presumably, the molecular flexibility plays a role in determining the glass transition, although any relationship between molecular structure and glass 'structure' and T_g is still quite obscure.

Several empirical and semi-empirical relationships exist for the prediction of glass temperatures of single substances and of mixtures. For the group of PHCs it is found that the ratio $T_g/T_m \approx 0.7$ is a good predictor, where T_m is the melting point. For

Note added in proof: In a recent study of the behaviour shown by trehalose dihydrate during heating and drying, Sussich *et al.* (1998) found that the dried, amorphous sugar can be 'cold-crystallised' at 110°C to yield a new polymorph with a melting point in the neighbourhood of 215°C. This finding may have implications in the use of sugars as pharmaceutical excipients and makes a re-examination of the processes which accompany sugar dehydration and annealing highly desirable.

mixtures, T_g can be estimated in terms of the T_g values of the individual components and the mixture composition (Gordon and Taylor, 1952) with an acceptable degree of accuracy. Thus for a binary mixture,

$$T_g = \frac{wT_2 + k(1 - w)T_1}{w + k(1 - w)}$$

where w is the weight (or mol) fraction of solute (PHC) and T_1 and T_2 are the glass transition temperatures of components 1 and 2 (solute), respectively. The constant k is a fitting parameter which does, however, possess some physical significance. In practice, component 1 is usually residual water, where it acts as a ubiquitous plasticiser, i.e. it depresses the glass temperature of the mixture.

Experimental techniques

Of the various physical techniques by means of which slow processes in solids can be probed, thermoanalytical methods take pride of place. In particular, differential scanning calorimetry (DSC) has established itself as the method of choice by most workers. The study of vitrification, nucleation and crystallisation phenomena by temperature scanning methods (e.g. DSC) does however introduce complexities, because the measurements are often affected by the thermal history of the sample under study, so that the experimental procedures may need to rely for their reliability on well controlled annealing protocols and corrections for artefacts due to scanning rate, or change in sample configuration (collapse, powder coalescence) during the course of the measurement (Shalaev and Franks, 1995b). Care must be taken to ensure that the scanning rate does not exceed the rate of the process under observation.

Unfortunately, much of the published literature on aqueous solutions of PHCs is limited to the physical changes taking place in only one of the components, namely water, which may well be the major component of a mixture under study, but its behaviour also happens to be the least interesting aspect of the physical behaviour of such complex mixtures. This is particularly true for studies of physical and chemical changes during freeze-drying, because the ice formed during the initial freezing treatment is subsequently sublimed and plays no further part in any thermally induced changes of the 'product' phase. The advent of modulated DSC (MDSC) has facilitated the interpretation and deconvolution of superimposed reversible and/or irreversible processes, taking place at different rates, in complex mixtures (Izzard et al., 1996). It is now possible to deconvolute complex cooling and heating scans and to obtain information on super-cooling, crystal nucleation/growth during cooling and/or heating, eutectic crystallisation/ melting behaviour and glass transitions, and thus to map ternary state diagrams.

The structures of crystalline solids are usually probed by X-ray diffraction techniques, and this has also been true for the elucidation of PHC crystal structures which tend to be of a complex nature. PHC molecules in the crystal are linked by hydrogen-bonds into infinite, three-dimensional, intermolecular networks, akin to the well-known structure of ice Ih.

For studies of related amorphous PHC states, X-ray methods have been of very limited use, although the rheological behaviour (high viscosity) of these substances in the fused state suggests that they are still extensively hydrogen bonded. Equally, the structural features of PHCs in solutions of hydrogen-bonding solvents remain to be

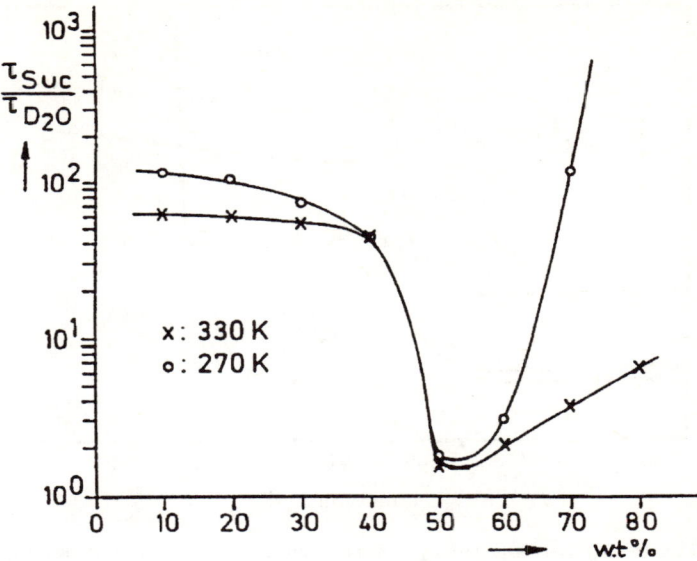

Figure 2. Rotational mobility of sucrose τ_{suc} relative to that of water τ_{D2O} as a function of the sucrose concentration. Redrawn, with changes, from Girlich (1991).

elucidated. In principle, neutron scattering, coupled with isotopic substitution, would appear to be a powerful method, but major experimental and data processing problems have so far prevented their exploitation. Recently, the first published neutron scattering study of glucose in the crystalline, fused and glassy state has highlighted the similarity between the close-range order in the crystal and the amorphous forms (Tromp *et al.*, 1997). Thus, the main differences relate to the heterogeneity in the hydrogen bonding details, i.e. bent hydrogen bonds in the amorphous states, but few changes in the number of hydrogen bonds per glucose molecule. Similarly, vitrification from the fused liquid state produces few *structural* changes in the intermolecular ensemble.

The conformations of PHCs in the presence of solvents capable of interacting by hydrogen bonding also requires further study. The permissibility of extrapolating from crystal structures to *structure* in solution was treated by Jeffrey (1973) who concluded that the PHC conformation observed in the crystal forms a valid starting point for calculations or simulations or as a close approximation to one or more possible rotamers which may exist in solution.

This thesis was subsequently tested on mannitol and sorbitol in aqueous and nonaqueous solvents, by a combination of ^{1}H n.m.r. and Molecular Dynamics simulation methods (Franks *et al.*, 1991). It was found that the nature of the solvent affected several of the torsional bond angles, and that some of the bond angles were identical to those found in the crystal, whereas others deviated significantly from the crystal geometry, giving the PHC molecule a distinctly different time-averaged conformation in solution. This may account for the observation that most PHCs do not crystallise easily from a saturated aqueous solution in real time.

The diffusional dynamics in aqueous PHC solutions have been probed by n.m.r. relaxation measurements (Girlich, 1991). *Figure 2* illustrates some of the surprising results for sucrose and water in their mixtures: the rotational motions of the two

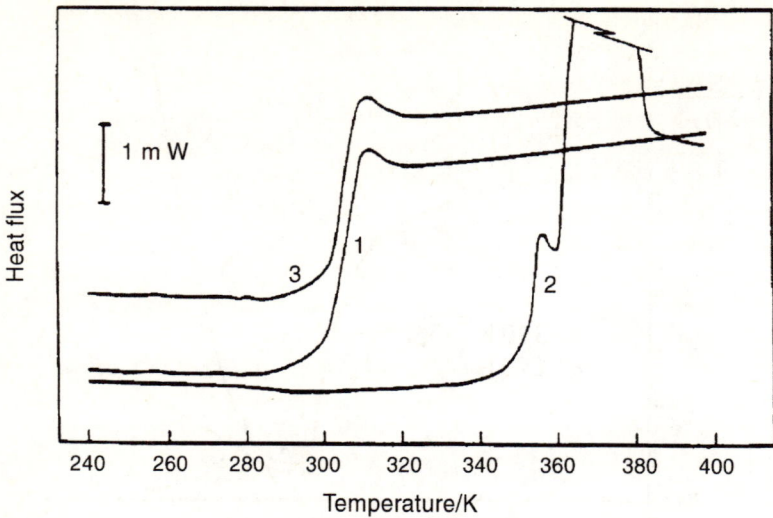

Figure 3.   DSC heating scans of previously cooled amorphous trehalose, containing 8% residual water; 1: first heating scan; 2: heating scan after exposure at 82°C over-night; 3: scanned immediately after recording scan 2 and cooling to –33°C. Reproduced, with permission, from Aldous *et al.* (1995).

molecular species are uncoupled in dilute solution, as would be expected. With increasing concentration, the motions become strongly coupled. Finally, at high sucrose concentrations, as the glass transition is approached, the motions once again become decoupled. Even in the glass, water motions, down to –150°C, are character-istic of those found in liquid water, although the amorphous sugar matrix has all the properties normally associated with a solid. This remarkably high water mobility may be able to explain the influence of residual water in glasses on the chemical stability of substances encapsulated in excipient glasses.

The amorphisation and recrystallisation of PHCs

For PHCs to function as lyoprotectants in labile therapeutic preparations, the formu-lator should be aware of their physical properties, in particular of any changes that might occur in their physical state, either during processing or during long-term storage. If it is desired to produce an amorphous preparation, then inadvertent crystallisation is likely to cause severe deterioration in the biological activity of the labile product. On the other hand, the crystallisation of a hydrated PHC from the solid solution may be beneficial, because it removes water which might have migrated into the product to be stabilised.

Since amorphous states are thermodynamically unstable, the probability of crystal-lisation in real time of PHCs from supersaturated solution, from the melt, or from the amorphous solid cannot be discounted, provided that the appropriate physical conditions exist. In principle, this is equally true for the crystallisation of stoichiometric hydrates, given the required mol ratio sugar:water. Knowledge of the rates of such processes is then required for any predictions of long-term stability.

One may speculate that, above T_g, such rates are proportional to $(T - T_g)$. In preliminary studies of physical transformations in dried trehalose and raffinose containing low amounts of residual water, Aldous *et al.* (1995) observed amorphisation and recrystallisation processes. This is illustrated for trehalose in *Figure 3*. Scan 1 shows the expected glass transition of trehalose, containing 8% water. The prolonged heat treatment at a temperature just below the melting point of crystalline trehalose $2H_2O$ is sufficient to promote partial crystallisation of the dihydrate, with the amorphous residue displaying a much increased T_g (scan 2). Once melted, the hydrate does not recrystallise during the time it takes to cool the melt and to rescan the temperature range to the original T_g of the preparation.

Where carbohydrates are used as stabilising excipients, say for proteins, their devitrification above T_g generally leads to rapid bioinactivation. This process has been described for a freeze-dried preparation of calcitonin gene-related protein (2%), stabilised with lactose (95%) and containing 3% residual moisture. When exposed to a temperature above T_g (40°C), the anhydrous sugar crystallised irruptively, leaving a residual amorphous phase, now consisting of 40% protein and 60% 'residual' moisture with a subzero glass temperature (Franks, 1992). This type of sugar devitrification is therefore highly damaging and must be avoided by storage well below T_g of the preparation. On the other hand, the crystallisation of a sugar hydrate provides additional desiccation, by removing water from the amorphous phase, thereby increasing T_g and the storage stability. The degree of such desiccation depends on the mol ratio sugar:water and would therefore be expected to increase as the number of mols of water per mol of sugar increases, i.e. in the order

$$(\alpha,\alpha\text{-trehalose, melibiose, melezitose}).2H_2O < \text{mannotriose}.3H_2O <$$

$$(\beta,\beta\text{-trehalose, stachyose}) \, 4H_2O < \text{raffinose}.5H_2O$$

It thus appears that an exceptional dry state stabilising potential of some sugars depends on 1) their ability to crystallise from a highly supersaturated solution in the form of a hydrate and 2) a rate of crystallisation that is sufficiently high at room temperature to prevent the degradation of a labile bioproduct being co-dried in the sugar solution. If the above hypothesis is correct, then *iso*-trehalose (β,β-trehalose) should be greatly superior, since on an equal weight basis, it can remove twice the amount of water from the dried preparation. A study of the glass forming and crystallisation potential of *iso*-trehalose revealed, unexpectedly, that this sugar crystallises much more rapidly from a freezing aqueous solution than does the α,α-isomer (Roberts and Franks, 1996).

The trisaccharide raffinose pentahydrate provides a particularly interesting, although not unique, example of amorphous/crystalline transitions. Saleki-Gerhardt *et al.* (1995), who reported on the vacuum dehydration of the sugar, found that progressive reduction in the mol ratio raffinose:water led to a progressive amorphisation of the crystalline hydrate. On the other hand, when the anhydrous, amorphous sugar was exposed to water vapour, it progressively recrystallised, until X-ray diffraction measurements indicated a 100% conversion to the crystal. However, the water uptake at this stage corresponded to the formation of a tetrahydrate, rather than the pentahydrate.

In a more recent study, Kajiwara and Franks (1997) re-examined the phase behaviour of raffinose-water systems. A close study of their X-ray diffraction data at different degrees of drying, coupled with a detailed DSC study, suggested that a crystalline raffinose tetrahydrate does indeed exist, although the shifts in the atomic oxygen positions from those in the pentahydrate are of a minor nature. Distinct

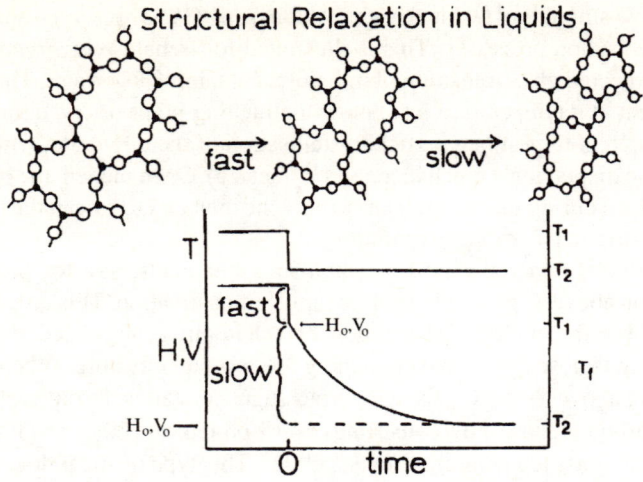

Figure 4. Schematic plot of enthalpy and volume against time during isothermal structural relaxation, following a step change in temperature. Adapted from Moynihan (1995).

melting points and enthalpies of fusion could, however, be identified. The phase diagram, although incomplete, also indicated a eutectic between the tetrahydrate and an even lower, as yet undefined, hydrate. As has also been reported for trehalose (Ding *et al.,* 1996), the (hypothetical) anhydrous crystalline state could not be produced in real time, despite various attempts at annealing and seeding. The rehydration and recrystallisation kinetics of the amorphised sugar show a complex dependence on the relative humidity which remains to be elucidated more thoroughly.

Dynamics and reactivity below the glass transition

Initial studies of PHC/water systems, used to stabilise labile biological substances and food products, e.g. enzymes, drugs, starch-based products, had led most investigators to the assumption that the glass transition provides a borderline between fairly rapid deterioration and complete stability (e.g. Slade and Levine, 1988; Green and Angell, 1989; Franks, 1990). It has become clear, however, that physical and chemical changes can, and do, take place within vitreous matrices, although not always at easily measurable rates.

An amorphous solid is characterised by a lack of a long-range, periodic order. It is an undercooled liquid which is thermodynamically unstable with respect to the crystal form. The energy barrier to viscous flow is high enough to prevent it from reverting to the stable, crystalline state within the normal time scale of observation. In a similar manner, other kinetic rate processes, e.g. chemical reactions, are severely inhibited in the glassy state. This makes the glass a valuable stabilising medium for labile materials.

For processes that are slow, relative to the time of observation, we need to consider the time evolution of equilibrium properties (H, V, structure, etc.). At equilibrium, the average structure is constant with time, but subject to fluctuations (dynamic equilibrium). Where fluctuations involve breaking and remaking of bonds and spatial translation of molecules or groups of molecules, relaxation rates may become low (relative to the time of observation). A liquid can then be 'frozen' on the experimental time scale and take on the thermal and mechanical properties of a solid (on an appropriate time scale). All thermodynamic properties will then become (partially) time-dependent, i.e. they will exhibit an immediate response to a perturbation, followed by a slow approach to equilibrium (Moynihan, 1995). This is shown schematically in *Figure 4*. A step change in T or P in a melt produces crystal-like instant response (e.g. due to bond vibrations), followed by 'slow' structural relaxation (bond break/remake, hydrogen bonds in the case of carbohydrates?) until a new equilibrium is reached. The time decay can be expressed by a 'stretched exponential' and is characterised by a structural relaxation time τ.

Because of their complex hydrogen bonding topologies, saccharides exhibit such bond exchange properties to a high degree (e.g. syneresis and ageing of aqueous gels). Above T_g, the temperature dependence of τ is best expressed by the Vogel-Tammann-Fulcher (VTF) equation

$$\tau = A \exp[B/(T \ C)].$$

where B and C are expressed as temperatures. The physical nature of C, as regards its description of PHC systems, is still under discussion. It used to be equated to T_g, but recent reports cast doubt on this interpretation. Below T_g, $\tau(T)$ exhibits Arrhenius kinetics, with high activation energies. For instance, the relaxation time for glycerol at its T_g (185K) = 33 min; at 170K, i.e. $(T_g - T)$ = 15K, the relaxation time is 4.5 days.

Temperature of zero mobility

The fact that neither physical nor chemical processes are completely inhibited in amorphous solids below T_g has given rise to a re-evaluation of dynamics in glassy matrices. Thus, below T_g, molecular relaxation times are too long for equilibrium to be established within an experimental timescale; this is related to a reduction in the number of accessible configurations (configurational entropy). The probability of a transition, W(T), is given by

$$W(T) = A \exp (-z \ \Delta U/kT)$$

where z is the number of molecules in a given domain, and ΔU is the potential energy barrier opposing the rearrangement. The critical size of the domain (z^*) is related to the *configurational entropy* $S_c = Ns_c^*/S_c$. The critical configurational entropy s_c^* cannot decrease below k ln 2 (i.e. must have at least two configurations).

The average probability of a transition is given by $A.\exp(-const/TS_c)$. If the constant is *not* equal to zero, then we cannot attain the situation where $S_c = 0$ during a cooling process of finite time. Kauzmann (1948) first suggested that a temperature of 'zero mobility' T_o could be defined, such that, as $S_c \to 0$, so $T \to T_o$. Therefore $T_g \neq T_o$, but $T_g > T_o$ and the divergence increases with the magnitude of the constant, i.e. with ΔU. Thus, real glasses have a residual configurational entropy that increases with the difficulty of structural rearrangements. S_c can be obtained experimentally from calorimetric measurements. By putting ΔC = constant and $S_c(T_o)$ = 0, then by integration, $S_c(T_g) = \Delta C \ln (T_g/T_o)$. It is found that for many 'real' materials, (e.g. oxides, silicates), $T_g/T_o = 1.29 \pm 11\%$.

From a semi-empirical treatment of relaxation processes it can be inferred that W(T) is proportional to τ^{-1}, and thus to diffusion and viscous flow.

If $[(T - T_o)/T_o] \ll 1$, then it can be shown that $W(T) = A.\exp[-B/(T - T_o)]$. This is the VTF equation (see above) which adequately describes diffusion and viscous flow, although not in the neighbourhood of T_g.

For multicomponent systems the situation is more complex, because the entropy of mixing must be included, and $\Delta S_{mix} > 0$. Complexities also arise with multicomponent systems made up of molecules of different sizes (and shapes?), e.g. sucrose + water, where decoupling of translational/ rotational motions are observed (Girlich, 1991). Experimental data are scarce. It is however firmly established that water remains mobile within a sucrose glass matrix. This finding may have important implications for chemical stability.

The fragility concept

Examination of reduced viscosity/temperature plots reveals a variety of behaviours. Some materials (SiO_2) display Arrhenius behaviour, i.e. ln (η/η_g) varies in a linear manner with T_g/T. For other materials the viscosity falls off much more steeply than is predicted by the Arrhenius equation.

Angell first advanced the 'fragility' concept: in so-called strong fluids, the structure (mainly covalent bonded) is maintained above T_g, whereas in fragile fluids, the structure is rapidly disrupted above T_g (labile hydrogen bonds?). According to this classification, PHCs and their (solid) aqueous solutions are fragile materials.

A fragility parameter δ is defined by

$$\delta = (\Delta U.s*/k)/(T_g.\Delta C_{g \to l}), \text{ where } T_g/T_o = f(\delta)$$

The VTF equation can then be expressed in terms of the relaxation time, referred to T_o:

$$\tau = \tau_o \exp [\delta T_o/(T - T_o)]$$

Assuming that 17 orders of magnitude separate the relaxation times at T_g and at some common high temperature limit (the constant B in the VTF equation), then a linear relationship is obtained between T_g/T_o and δ:

$$T_g/T_o = 1 + 0.025\delta$$

Solving for δ, using some of the above equations and data from 'real' materials,

$$\delta = \frac{665}{(\Delta H*/2.3RT_g) - 17}$$

Fragile liquids display VTF behaviour: δ is small, narrow glass transitions are observed, detectable by DSC. Strong liquids, on the other hand, display Arrhenius behaviour, with large δ values and broad glass transitions which are difficult to detect by DSC.

Implications for long-term physical/chemical stability

Physical stability is clearly related to viscosity, and hence, to structural relaxation. *Chemical* stability may be subject to additional factors, e.g. possibility of reactions triggered by intramolecular rearrangements. Parameters which describe stability include $\tau(T)$, T_o, δ, and $\eta(T)$. They can be obtained from scanning and isothermal calorimetry.

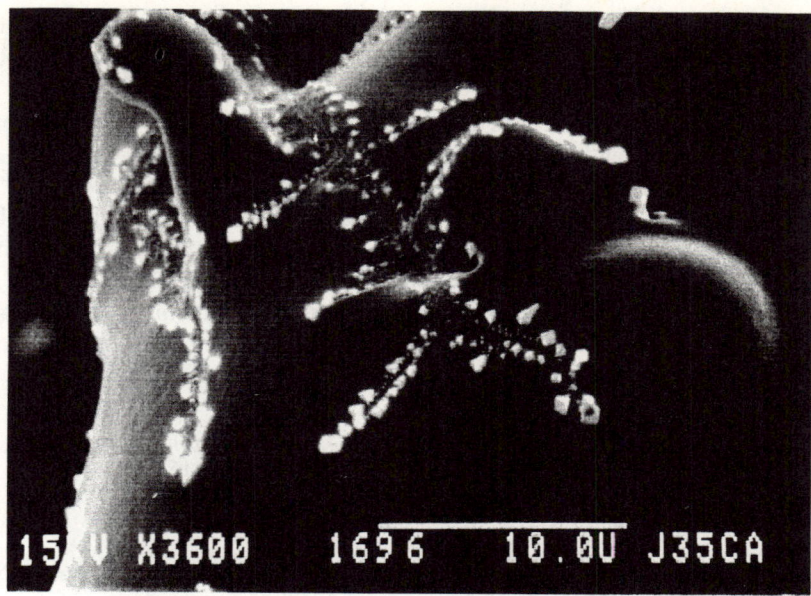

Figure 5. Scanning electron micrograph of freeze-dried sucrose/NaCl solution (mass ratio 5:1), containing 3% residual water and stored for several days at *ca.* 5 degrees below its T_g. The scale bar represents 10 μm.

The relationship between physical and chemical stability (if any) is not yet clear. According to Pikal (personal communication), the chemical stability of 'dry' human growth hormone, formulated with stachyose or trehalose, and measured as aggregation or chemical degradation at 40 and 50°C, is related to $(T-T_o)$, rather than $(T-T_g)$. Another recent report describes the kinetics of chemical bond cleavage reactions in materials held in a variety of amorphous PHC matrices, above and below T_g (Streefland *et al.*, 1998). It was found that the reactions were severely retarded, but not inhibited below T_g, and also that the PHCs with the highest glass temperatures (dextran) were less effective in retarding the reaction than simple disaccharides (sucrose) with lower T_g values.

The effect of slow relaxations on a physical transformation (phase separation) is graphically demonstrated in *Figure 5* which shows an electron micrograph, taken of a freeze-dried ternary solution (water/NaCl/sucrose) after several days storage just below its glass transition (Van den Berg *et al.*, 1993). Concurrent DSC had shown that during the initial freezing, NaCl did not crystallise, i.e. no eutectic phase separation was observed. After the sublimation of ice, the mixture was therefore completely amorphous. The micrograph reveals the slow crystallisation of well-formed cubic NaCl crystals at stress cracks in the amorphous matrix. At the time when the photograph was taken, the crystals had achieved approx. 0.1 μm dimensions. The growth rate would be expected to follow VTF kinetics.

Conclusions

Largely because of its involvement in governing the shelf life of labile pharmaceutical and food products, the thermomechanical behaviour of water-soluble amorphous solids at a fundamental level is now receiving renewed interest. Basically, these materials resemble 'real' materials, such as polymers, oxides and silicates, in their mechanical properties, but they differ from these substances in their structural composition which relies on orientation-specific, hydrogen-bonded networks, rather than on covalent bonds.

It is becoming apparent that most theoretical approaches which have been developed for 'real' materials can also be applied to amorphous PHC-based materials, but experimental data are as yet scarce.

The relationships between thermomechanical attributes and chemical stability remain to be elucidated, as does also the role of residual water in preparations which are usually prepared by drying a dilute aqueous solution. The significance of the temperature of zero mobility as governing long-term stability remains to be demonstrated.

References

AHLNECK, C. AND ZOGRAFI, G. (1990). *International Journal of Pharmaceutics* **62**, 87.

ALDOUS, B.J., AUFFRET, A.D. AND FRANKS, F. (1995). *Cryo-Letters* **16**, 181.

ANGELL, C.A. (1995). *Science* **267**, 1924.

DING, S.-P., FAN, J.L., GREEN, L., LU, Q., SANCHEZ, E. AND ANGELL, C.A. (1996). *Journal of Thermal Analysis* **47**, 1391.

FRANKS, F. (1982). In *Water – A Comprehensive Treatise,* Vol. 7. Ed. F. Franks, Chapter 3. New York: Plenum Press.

FRANKS, F. (1990). *Cryo-Letters* **11**, 93.

FRANKS, F. (1992). *Japanese Journal of Freezing and Drying* **38**, 5.

FRANKS, F. (1994). *Bio/Technology* **12**, 253.

FRANKS, F., DADOK, J., YING, S., KAY, R.L. AND GRIGERA, J.R. (1991). *Journal of the Chemical Society, Faraday Transactions* **81**, 579.

GIRLICH, D. (1991). *Ph.D. Thesis*, University of Regensburg.

GORDON, M. AND TAYLOR, D.S. (1952). *Journal of Applied Chemistry* **2**, 493.

GREEN, J.L. AND ANGELL, C.A. (1989). *Journal of Phyical Chemistry* **93**, 2880.

IZZARD, M.J., ABLETT, S., LILLFORD, P.J., HILL, V.L. AND GROVES, I.F. (1996). *Journal of Thermal Analysis* **47**, 1407–1418.

JEFFREY, G.A. (1973). *Advances in Chemistry, Series Number* **117**, 177.

KAJIWARA, K. AND FRANKS, F. (1997). *Journal of the Chemical Society, Faraday Transactions* **93**, 1779.

MOYNIHAN, C.T. (1995). *Reviews in Mineralogy* **32**, 1.

ROBERTS, C.R. AND FRANKS, F. (1996). *Journal of the Chemical Society, Faraday Transactions* **92**, 1337.

SALEKI-GERHARDT, A., STOWELL, J.G., BYRN, S.R. AND ZOGRAFI, G. (1995). *Journal of Pharmaceutical Science* **84**, 318.

SHALAEV, E.YU. AND FRANKS, F. (1995a). *Thermochimica Acta* **255**, 49.

SHALAEV, E.YU. AND FRANKS, F. (1995b). *Journal of the Chemical Society, Faraday Transactions* **91**, 1511.

SHALAEV, E. YU., FRANKS, F. AND ECHLIN, P. (1996). *Journal of Physical Chemistry* **100**, 1144.

SLADE, L. AND LEVINE, H. (1988). *Pure and Applied Chemistry* **60**, 1841.

SLADE, L. AND LEVINE, H. (1993). In *The Glassy State in Foods*. Eds. J.M.V. Blanshard and P.J. Lillford, pp 35–102. Nottingham: Nottingham University Press.

STREEFLAND, L., AUFFRET, A.D. AND FRANKS, F. (1998). *Pharmaceutical Research* **15**, 843.

SUSSICH, F., URBANI, R., PRINCIVALLE, F. AND CESARO, A. (1998). *Journal of the American Chemical Society* **120**, 7893.

SUZUKI, T. AND FRANKS, F. (1993). *Journal of the Chemistry Society, Faraday Transactions* **89**, 3283.

TROMP, R.H., PARKER, R. AND RING, S.G. (1997). *Journal of Chemical Physics* **107**, 6038.

VAN DEN BERG, C., FRANKS, F. AND ECHLIN, P. (1993). In *The Glassy State in Foods*. Eds. J.M.V. Blanshard and P.J. Lillford, p 249. Nottingham: Nottingham University Press, Nottingham.

12
Transgenic Tomato Technology: Enzymic Modification of Pectin Pastes

GREGORY A. TUCKER*, HOWARD SIMONS AND NEIL ERRINGTON

University of Nottingham, School of Biological Sciences, Sutton Bonington, Leics., LE12 5RD, UK

Introduction

Polysaccharides have been manipulated chemically, or by the use of added enzymes, for many years in order to improve their functionality for various industrial processes (Tucker and Woods, 1995). However, this chemical modification is often difficult to control. Genetic engineering presents the prospect to manipulate polysaccharide structure actually within the organism of interest. It also allows the prospect for these modifications to be more tightly controlled. This modification cannot take place directly but can via the manipulation of enzymes involved in either synthesis or degradation of the polysaccharides. Manipulation of synthesis is currently difficult given the relative scarcity of biochemical, and in particular genetic, information on the enzyme systems involved. However, advances in this area are being made and no doubt manipulation of polysaccharides by the modification of genes encoding biosynthetic enzymes will occur in the near future. In contrast, the genetic modification of polysaccharide degrading enzymes has been achieved and has already resulted in commercial products.

Cell wall polysaccharides are often major determinants in the rheological properties of plant foods and their processed products. This is particularly the case for pectin in determining the texture of intact fruit and the viscosity of pastes made from fruit. The structure of these plant cell wall polymers, in processed products, can be modified by the application of exogenous enzymes. Alternatively, structural alterations can be brought about by the modification of endogenous enzyme activities. In many instances the action of endogenous cell wall hydrolytic enzymes may be detrimental for food quality and in these cases heating is often used to inactivate these enzymes. A particular instance is the use of a '*hot-break*' process for the inactivation of pectolytic enzymes during the production of tomato pastes.

The plant cell wall is a complex structure (Carpita and Gibeaut, 1993) and it is beyond the scope of this paper to describe this structure. Basically, the plant cell wall

*To whom correspondence may be addressed.

Biotechnology and Genetic Engineering Reviews – Vol. 16, April 1999
0264–8725/99/16/293–308 $20.00 + $0.00 © Intercept Ltd, P.O. Box 716, Andover, Hampshire SP10 1YG, UK

is composed of a network of cellulose microfibrils. These fibrils consist of chains of linear β(1–4)-linked glucans held together by intra-polymer hydrogen bonds. The fibrils within any one plane of the wall are often orientated parallel to each other: a typical cell wall being composed of several such planes with adjacent planes of fibrils set at an angle to each other to give a laminated structure to the wall. The typical cell wall also contains hemicellulose polymers. In dicotyledonous plants, such as the tomato, the major hemicellulose is usually xyloglucan. This is composed of a linear β(1–4)-linked glucan backbone substituted with glucose, galactose and some fucose residues. The hemicellulose can hydrogen bond to the surface of the cellulose microfibrils but is also large enough to actually span the gap between two adjacent fibrils. In this way the cellulose and hemicellulose act to form a kind of interconnected framework. This is then embedded in a gel-like matrix composed of pectins and proteins. The major pectins are galacturonan polymers containing a backbone of β(1–4)-linked galacturonic acid residues. These can have either the free carboxylic acid group at the C6 carbon or alternatively this carbon can be methyl esterified. The galacturonic acid backbone can be interrupted by rhamnose residues. In the case of the pectic polymer-rhamnogalacturonan I – these rhamnose residues occur alternate with galacturonic acid residues in runs of up to several hundred residues. The rhamnose residues can also be linked to neutral sugar side chains of galactose and/or arabinose residues.

Fruit softening is often accompanied by degradation of the pectin components of the cell wall. These changes include a decline in the degree of esterification, a loss of neutral sugars such as galactose and arabinose and an increased solubility and depolymerisation (Tucker, 1993). These changes in pectin structure are brought about by the action of cell wall associated enzymes during ripening. In particular polygalacturonase (E.C 3.2.1.15), which has the capacity to depolymerise the polyuronide chains in pectin, pectinesterase (E.C 1.1.1.11), which can de-esterify the galacturonic acid residues and β-galactosidase thought to be important during the loss of neutral sugars. Pectinesterase and β-galactosidase are associated with most fruit. Polygalacturonase is found in a large number of, but not all, fruit. All three enzymes often undergo significant changes in expression during ripening. The expression and molecular biology of these enzymes have been most extensively studied in tomato fruit. Tomato fruit during ripening show all the typical changes in pectin structure outlined above.

This review describes work largely carried out at Nottingham University to reduce the expression of these cell wall degrading enzymes in tomato fruit. This has been achieved in each case by the application of gene silencing techniques. It is again beyond the scope of this review to consider the mechanism of action of gene silencing. Instead, the reader is directed to a recent conference proceedings for further details (Lycett *et al.*, 1996). Silencing can in effect be brought about by two mechanisms: *antisense* or *co-suppression*. Antisense involves the creation of an artificial gene in which the coding region for all, or part, of an endogenous gene is placed in a reverse (antisense) orientation under the control of a suitable promoter. This antisense gene is then transformed into the plant. In the cell the antisense gene directs the transcription of an antisense RNA molecule. This, by definition, is complimentary to the mRNA product of the normal endogenous gene. The theory is that these two complimentary RNA molecules base pair within the nucleus thus preventing the translation of the mRNA and formation of the target enzyme. A similar gene silencing also occurs if the transgene, instead of containing coding sequences in an antisense orientation, has the sequence in the normal orientation. In this instance, both the transgene and the target gene are silenced, resulting in a phenomenon known as co-suppression. The mechanism(s) for co-suppression have not been fully elucidated.

These techniques have been used at Nottingham, and elsewhere, to reduce the expression of polygalacturonase, pectinesterase and β-galactosidase in tomato fruit. The effects on the cell wall metabolism, softening and processing of the fruit have been examined and attempts made to predict the structure/function relationships for pectin in this biological system.

Enzymes involved in fruit ripening and their down regulation

A wide range of enzymes with the potential to degrade cell walls have been found associated with ripening fruit (*Table 1*). In many cases these have been purified, characterised and corresponding DNA sequences identified. Work at Nottingham has concentrated on the pecteolytic enzymes polygalacturonase, pectinesterase and β-galactosidase.

POLYGALACTURONASE

Polygalacturonase (PG) activity is almost non-detectable in green tomato fruit (Hobson, 1964). However, the activity increases dramatically during ripening. This PG activity has been resolved into at least two isoforms (PG1 and PG2) using ion-exchange chromatography (Pressey and Avants, 1973; Tucker *et al.*, 1980). During ripening PG1 is the first detectable isoform, however, PG2 activity rapidly becomes the dominant isoform in ripe fruit (Tucker *et al.*, 1980). Both PG1 and PG2 have been purified and characterised (Moshrefi and Luh, 1984; Tucker *et al.*, 1980). The PG2 isoform has been shown to be composed of a single polypeptide chain with a molecular weight of around 43 kD, and this has been fully sequenced (Sheehy *et al.*, 1987). In contrast, purified PG1 appears to be composed of two heterologous polypeptide chains with molecular weights of around 43 kD and 38 kD (Moshrefi and Luh, 1984). The 43 kD polypeptide cross reacts with antibodies raised against purified PG2 and tryptic digestion of the 43 kD polypeptide from both PG1 and PG2 result in the same digest pattern of peptide fragments (Tucker *et al.*, 1980). These results suggest that the 43 kD polypeptide in each isoform is identical and presumably represents the catalytic subunit. Several groups have isolated clones corresponding to

Table 1.   List of cell wall degrading enzymes commonly associated with ripening fruit

Arabinosidase
Cellulase
α-Galactosidase
β-Galactosidase
Mannosidase
Polygalacturonase
Pectinesterase
Pectate lyase
Xylosidase

Table 2. Enzyme activities in normal and various genetically modified lines of tomato fruit

Line	Polygalacturonase	Pectinesterase	b-Galactanase
Normal	550	82	13
PG antisense	4	67	ND
PE2 antisense	670	9	ND
PGPE chimeric antisense	7	5	ND
β-Galactanase co-suppressed	ND	ND	11

the 43 kD catalytic subunit of PG from fruit cDNA libraries (Grierson *et al.*, 1986; Sheehy *et al.*, 1987). Expression of the mRNA corresponding to these clones occurs in a fruit and ripening specific manner. There appears to be only a single gene encoding the fruit specific PG protein (Bird *et al.*, 1988). This finding suggests that the isoforms of PG in ripe tomato fruit are the product of a single gene and arise simply from differential post translational modification of a common polypeptide.

Several groups have down regulated fruit PG expression using either antisense RNA technology or co-suppression. (Sheehy *et al.*, 1988; Smith *et al.*, 1988). In both instances PG expression in transgenic fruit has been reduced to less than 0.5% of the activity occurring in normal fruit. Levels of activity in the transgenics produced at Nottingham are shown in *Table 2*. As expected both the isoforms of PG were effected by this transformation since only a single gene is present.

Recently, expression of the β-subunit has been down-regulated in transgenic tomatoes using antisense RNA technology (Watson *et al.*, 1994). This resulted in a reduction in the level of extractable PG1 isoform. In these transgenic fruit polyuronide solubilisation and depolymerisation were both significantly greater than in corresponding normal fruit. This suggests that whilst the β-subunit is not required for PG action it may have a possible regulatory role in limiting pectin degradation. The β-subunit has been immunolocalised to the cell wall in green fruit and as such a role in targeting the PG within the wall has been postulated (Pogson *et al.*, 1991). This targeting may represent the mechanism by which the β-subunit could limit pectin degradation.

PECTINESTERASE

Pectinesterase (PE) activity occurs in tomato fruit throughout both development and ripening (Hobson, 1963). Like PG, this activity can be resolved into several isoforms. Pressey and Avants (1972) showed that PE isoforms could be separated by ion-exchange chromatography and that the profile was different in different tomato cultivars. Tucker *et al.* (1982) demonstrated that the PE from *Ailsa craig* tomatoes could be separated into at least two isoforms, which they called PE1 and PE2. More recently, Warrilow *et al.* (1994) have separated three PE isoforms from tomato fruit which they termed PEA, PEB and PEC. Heparin affinity chromatography can be used to separate three PE isoforms from tomato fruit (*Figure 1*). This is a profile in which the fractions have been assayed using a kinetic programme to give a rate assay which shows the relative activities of the three isoforms. For the purpose of this paper these have been termed PE1, PE2 and PE3. The relative activities of these three isoforms in tomato fruit are shown in *Figure 2*. It can be seen that PE1 and PE2 both have peaks of activity at around the mature green-breaker stage of development. In contrast, PE3

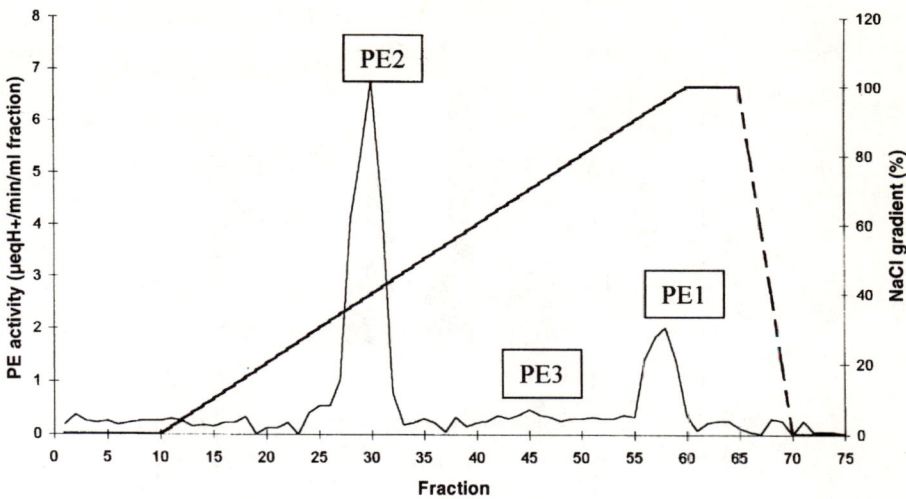

Figure 1. Heparin affinity column profile of pectinesterase isoforms from normal tomato fruit. This is a rate assay to show the relative proportions of activity in each isoform.

activity remains relatively constant throughout both development and ripening. The PE2 isoform represents the major form of the enzyme at all stages of development.

The PE2 isoform has been purified and sequenced (Markovic and Journval, 1986). Using this sequence data Ray *et al.* (1988) succeeded in isolating a clone from a tomato fruit cDNA library.

Characterisation of two further clones with homology to PE2 was later described by Hall *et al.*, 1994. It thus appears that PE2 may be encoded for by a small multi gene family. Two groups have carried out analysis of genomic clones (Harriman *et al.*, 1991; Hall *et al.*, 1994) in both cases they have shown that there are 3 PE2 like genes organised in tandem within the genome. There is some evidence that PE2 in tomato fruit may exist in multiple forms. Gaffe *et al.* (1994) identified five PE isoforms in tomato fruit using isoelectric focusing. Three of these isoforms cross reacted with an antibody raised against an equivalent of PE2 and all three isoforms disappeared in transgenic fruit produced by antisense RNA technology targeted against the PE2. It is possible that these isoforms arise from post translational modification of a single PE2 gene product, as for PG. It is also possible that these three PE2 sub forms may arise from the transcription and translation of two or more of the established PE2 multi gene family.

Several groups have down regulated the expression of the PE2 gene(s) using antisense RNA technology (Tieman *et al.*, 1992; Hall *et al.*, 1993). In all cases, total PE activity in the fruit was reduced to around 10% of normal. The levels of enzyme activities in the plants produced at Nottingham are shown in *Table 2*. The isoform levels during development for these antisense PE2 fruit is shown in *Figure 3*. In comparison to the normal levels shown in *Figure 2*, it can be seen that whilst the activity of PE2 has been reduced to almost zero, that of PE1 and PE3 is unaffected. This result would suggest that the genes for both PE1 and PE3 have insufficient sequence homology to the PE2 antisense gene to result in silencing. This in turn would suggest that these then represent completely different genes. Putative PE1 (Warrilow *et al.*,

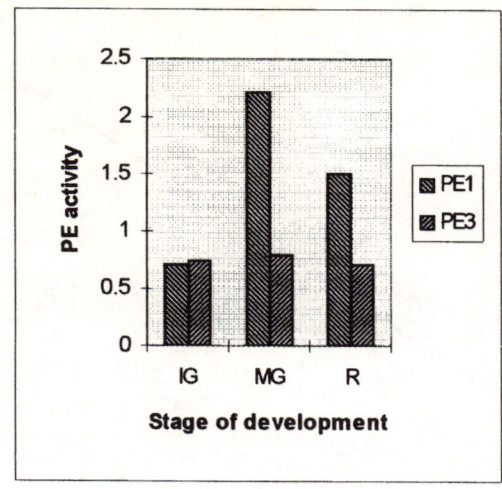

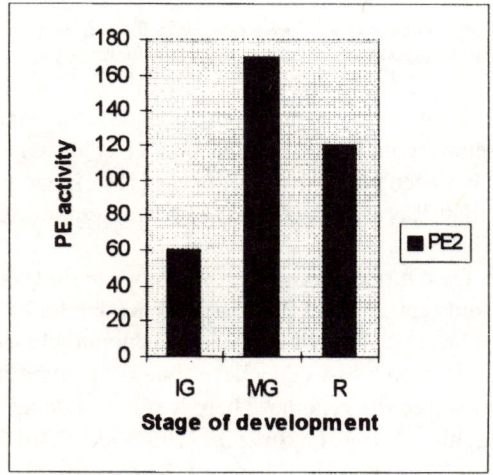

Figure 2.   Pectinesterase activity in normal tomato fruit. Enzyme activity is expressed as (ueq\min\gfwt).

1994) and PE3 (Tucker and Zhang, unpublished data) isoforms have both been purified to homogeneity and their N-terminal amino acid sequences have been determined. These sequences, along with that for the PE2 isoform, are shown in *Figure 4*. It can be seen that all three sequences are very different, again confirming the possibility that the three isoforms are encoded for by separate genes or multi gene families. The identification and characterisation of these genes for PE1 and PE3 remains to be carried out. The study of these other two PE isoforms is particularly important since these appear to be the isoforms which are prevalent in vegetative tissue (Warrilow *et al.*, 1994). Similarly, antisense PE2 plants, whilst showing a marked decline in fruit PE levels, show no effect whatsoever on levels of PE activity in leaves or roots. The vegetative forms of PE appear, at least on the basis of isoelectric focusing and ion-exchange chromatography, to be similar to the PE1 and PE3 isoforms present in the fruit (Tucker, unpublished). This has led Gaffe *et al.* (1994) to describe the

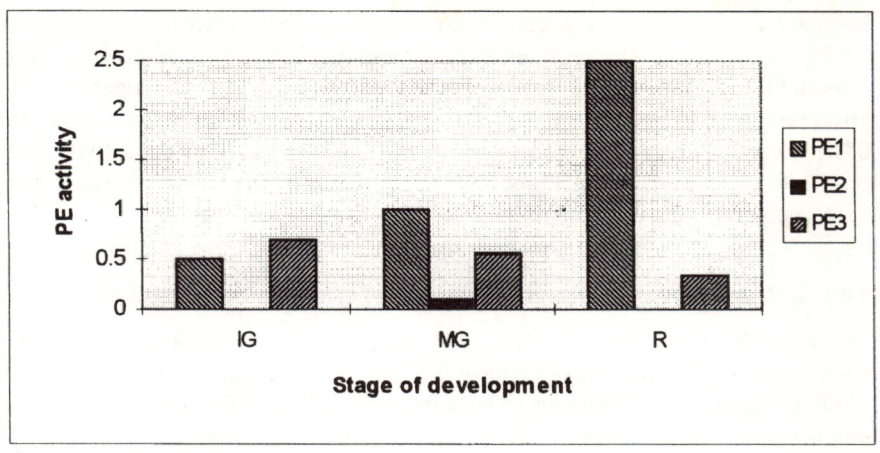

Figure 3.   Pectinesterase activity in transgenic tomato fruit. Activity is expressed as (ueq\min\gfwt).

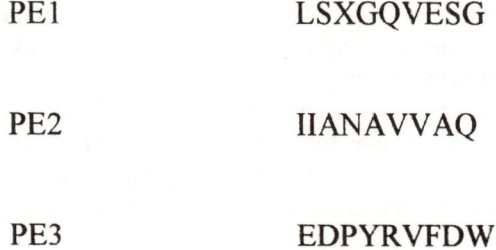

Figure 4.   N-terminal amino acid sequences of putative pectinesterase isoforms from tomato fruit.

presence of two groups of isoforms in plant tissue. Group I are fruit specific isoforms, all of which appear to be related to PE2. Group II are isoforms present in both fruit and vegetative tissue and these presumably correspond to PE1 and PE3.

β-GALACTOSIDASE

In tomato three β–galactosidase isoforms have been separated by a combination of ion-exchange and gel permeation chromatography (Pressey, 1983). However, only one of these three isoforms was found to be capable of degrading a β(1–4)-galactan isolated from tomato cell walls (Pressey and Himmelsbach, 1984; Carey *et al.*, 1995). This isoform was thus recognised as an exo- acting β(1–4)-galactanase. Whilst levels of β-galactosidase activity remain fairly constant during tomato fruit ripening, the β(1–4)-galactanase activity increases dramatically (Carey *et al.*, 1995). This increase does not occur in the tomato ripening inhibitor (*rin*) mutant (Carey *et al.*, 1995).

The β-galactosidase isoform exhibiting galactanase activity has been purified and the sequence data obtained used to isolate related cDNA clones (Carey *et al.*, 1995). Several cDNA clones with homology to the β-galactosidase isoform have been identified (Tucker, unpublished; Gross, personal communication) suggesting that, like PE2, this β-galactosidase isoform may be encoded by a multi gene family. One of the related cDNA clones has been used to down regulate the corresponding enzyme using co-suppression. The resultant enzyme activities are shown in *Table 2*. From northern blot analysis it has been shown that the expression of the endogenous gene corresponding to the cDNA used for the co-suppression has been reduced to around 10% of normal (Carey, 1996). However, the resultant loss of galactanase activity in the transgenics was only marginal, suggesting that there may be other isoform(s) remaining which account for the majority of the activity in fruit.

SIMULTANEOUS SILENCING OF POLYGALACTURONASE AND PECTINESTERASE

Since PG will only cleave the polyuronide chain between two adjacent de-esterified galacturonic acid residues, this enzyme is thought to act synergistically with PE. Thus, the simultaneous down regulation of both genes for PG and PE could be beneficial in reducing pectin degradation in fruit. This could, and indeed, has been achieved by crossing lines of transgenic plants showing either reduced PG or PE activities as described above. However, this process is both time consuming and labour intensive in the selection required. A much simpler method is the use of chimeric constructs as described by Seymour *et al.* (1993). In this method, a 244 bp fragment of PG coding sequence was ligated to the complete 1320 bp coding fragment of PE2 to give a chimeric construct. This construct was then placed under the control of a constitutive promoter and then used to transform tomato plants. The resultant transgenic fruit had both enzyme activities reduced, the levels of activity are shown in *Table 2* alongside those of the individual transgenic lines. It can be seen that the chimeric construct has resulted in a down regulation of the two endogenous genes to a similar extent as in the two individual lines.

This phenomenon has been shown in several independently transformed tomato lines and in each case both target genes are silenced co-ordinately (Jones *et al.,* 1996). In addition, it has been shown that the structure of the chimeric construct can be very flexible. Thus, the chimeric gene will work if both sequences are in the sense orientation (co-supression) or in the antisense orientation (Jones *et al.*, 1998). Indeed, the chimeric gene will still function to silence both target genes if it contains one of the sequences in a sense and the other in an antisense orientation both within the same construct (Jones *et al.,* 1998).

The reduction in PG in these plants is similar to that in the individual lines in that both isoforms of the enzyme are reduced in activity. However, the effect on PE activity is different to that in the individual line. The PE isoform profile from the ripe chimeric transgenic fruit is shown in *Figure 5*. For comparison, a normal profile is also shown in this figure. In this instance, an end point assay has been used to emphasis the presence or absence of particular isoforms. The size of the peaks in this case is not proportional to the actual relative activities of the three isoforms. It is apparent from *Figure 5* that both PE2 and PE3 appear to be markedly reduced in activity whilst PE1 remains relatively active. In comparison, in the individual transgenic line targeted at PE2 (activities shown in *Figure 3*) only PE2 has been down regulated. This difference may be accounted for by the fact that the entire coding sequence for PE2 was used in the generation of the chimeric transgenic plants whereas a smaller fragment was used to produce the antisense PE2 lines. It is possible that the extra coding sequence employed in the chimeric construct contains sufficient homology to a sequence in the PE3 gene to bring about silencing.

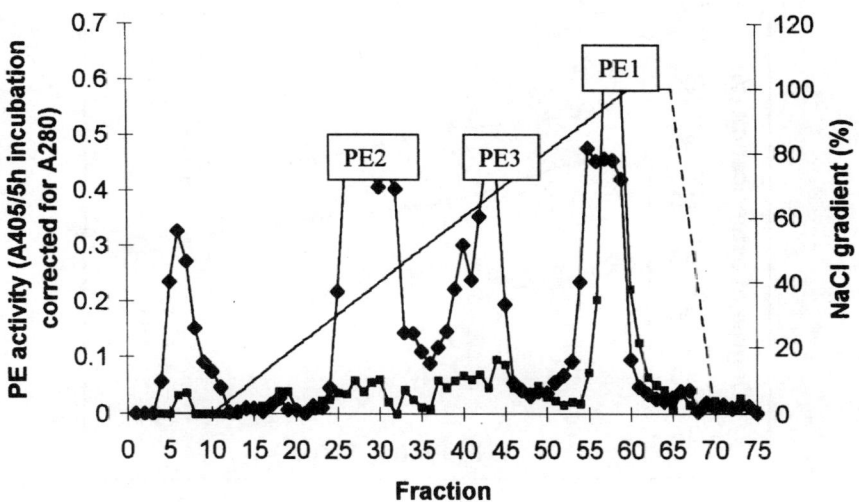

Figure 5.   Heparin affinity column profile of pectinesterase isoforms from normal (filled diamonds) or PGPE (filled squares) tomato fruit. This is the result of an end point assay to emphasis the presence or absence of isoforms. The size of the peaks is not proportional to their relative activities.

Effects of gene silencing on cell wall degradation

The effect of silencing PG activity on cell wall degradation has been described in a previous volume (Tucker, 1990). The effects on pectin solubility, de-esterification and depolymerisation are summarised in *Table 3*.

It can be seen that whilst the level of de-esterification has been unaffected the degree of depolymerisation of the polyuronide has been markedly reduced. This observation was as expected given the mechanism of action of PG. Perhaps an unexpected observation was that the level of chelator solube polyuronide was not effected by the down regulation of PG. This does not mean that the solubility of the pectin has not been effected, however. These experiments used a strong chelating agent to solubilize the pectin and this may have masked more subtle solubility effects. Indeed, Carrington *et al.* (1993) have shown that the level of water soluble pectin in these transgenics is reduced compared to normal.

The effect of reduced PE2 on cell wall metabolism is shown in *Table 3*. In this case, as expected, there was a reduction, although not a complete cessation, of de-esterification with no effect on either depolymerisation or solubility of the polyuronide. This effect on de-esterification has been studied in more detail and the degree of esterification of the pectin in normal and transgenic fruit determined throughout

Table 3.   Effect of down regulation of either PG or PE activity on wall meatabolism in intact tomato fruit

Line	Degree of esterification	Solubility (%)	Average mol.wt (KD)
Normal (green)	75	36	294
Normal (ripe)	56	59	111
PGas (ripe)	53	52	248
PE2as (ripe)	67	66	85

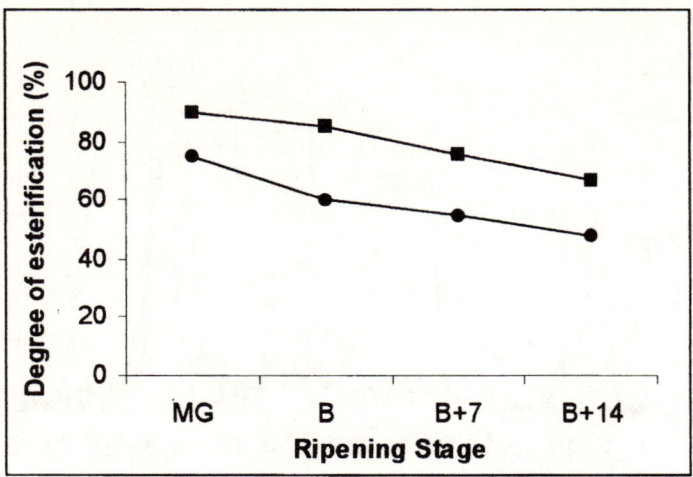

Figure 6.   Changes in degree of esterification of pectin during ripening of normal (filled circles) or PE2 as (filled squares) tomato fruit. MG = Mature green, B = Breaker, B+7 = Breaker plus 7 days, B+14 = Breaker plus 14 days.

development and ripening. The results are shown in *Figure 6*. It can be seen that the antisense PE2 plants have a higher degree of esterification of their pectin at all stages of development. However, the de-esterification accompanying fruit ripening appears to proceed in the transgenic to the same extent as in normal fruit.

No differences have yet been detected between normal fruit and those from transgenics down regulated for β-galactosidase activity. In this case no change in the galactose content of the wall fractions could be seen (Seymour, personal communication). The precise effect on cell wall metabolism of the double down regulation of PG and PE remains to be determined, although preliminary experiments would suggest that depolymerisation of the pectin is reduced, solubility in green fruit enhanced but total solubility in ripe fruit unaffected (Simons, 1998).

Effect of gene silencing on fruit softening and processing

EFFECT ON INTACT FRUIT

The effect of down regulation of PG has also been described in a previous volume (Tucker, 1990). The effect on fruit softening, as determined by compression (Langley *et al.,* 1994) or stress relaxation experiments (Errington *et al.,* 1997), is marginal. The transgenic fruit may be slightly firmer but this is not really significant. The genetic modification does, however, seem to influence the cracking and resistance to transport damage of the fruit. Normal fruit harvested ripe and then transported suffer extensive (up to 50% of the fruit) cracking. Ripe fruit with reduced PG activity, however, only exhibit about 10% losses. This has lead to the introduction in the USA of the *'Flavr Savr'* tomato. This is a commercialised antisense PG tomato which can be harvested ripe, and hence has more flavour.

The down regulation of PE2 in the PE antisense fruit has little or no effect on softening. Similarily, no effect on softening was seen with the simultaneous down regulation of both PG and PE with the chimeric construct. The down regulation of β-galactosidase again had little effect on the compressibility of the fruit but cracking during transport was, as with the reduced PG fruit, markedly reduced. In this case control ripe fruit exhibited 40% cracking compared to zero cracking in the transgenics (Tucker *et al.*, unpublished).

EFFECT ON PROCESSING

A large proportion of the annual tomato crop is processed into products such as paste and ketchup. A major quality attribute in this case is viscosity. A standard industrial measure of viscosity involves the use of a Bostwick tray. This is a perspex box about 1 metre long with a removable trap door sealing off a small reservoir at one end of the tank. A tomato paste is poured into the reservoir, the trap door removed and the distance that the paste flows down the tank in 30 seconds is then recorded. Pastes with highest viscosity thus move the shortest distance. Bostwick values have been obtained for tomato pastes made from fruit from normal, antisense PG (PGas), antisense PE2 (PE2as) and lines in which a chimeric construct has been used to down regulate both PG and PE (PGPEas). These pastes were made by homogenising fruit directly, i.e with no pretreatment, and as such are so called 'cold break' pastes. These values are shown in *Table 4*.

It can be seen that reduction in PE alone (PE2as) has had no effect on the viscosity of the resultant paste. Reduction in PG activity (PGas) results in a clear increase in viscosity, the Bostwick value being almost half that of the control normal paste. Interestingly, the fruit in which both PG and PE have been reduced (PGPEas) show no increase in paste viscosity. It would thus seem that the reduction in PE in these fruit negates the advantage of reduced PG activity. *Table 4* also shows the Bostwick values from these same fruit lines but after the fruit had been subjected to heating prior to being converted to paste. These are so called 'hot break pastes'. The heating has the effect of denaturing the PG and PE enzymes and results in total loss of activity in all fruit lines. In this case it can be seen that the Bostwick values for all lines are about equal. This result shows that in the PGas line the reduced Bostwick value is the result of reduced pectin degradation within the paste and not due to any intrinsic differences in the pectin within the fruit prior to its being converted to paste. This result is supported by the finding that pectin degradation in the fruit of even normal fruit is strictly controlled. It is not until fruit are homogenised (as in paste production) that PG and PE are released and allowed to fully degrade the pectin (Seymour *et al.*, 1987).

The results shown above for the Bostwick tests would indicate that whilst PG activity in tomato homogenates would seem to be disadvantageous for high

Table 4.   Viscosity of pastes from normal and genetically modified tomato fruit as measured by the Bostwick method

Line	Bostwick value (cold break)	Bostwick value (hot break)
Normal	235	202
PGas	78	196
PE2as	232	172
PGPE chimeric	220	180
β-galactanase	153	ND

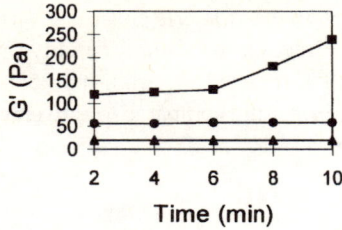

Figure 7.   Effect of time on the viscosity of pastes from normal (filled circles), PG (filled squares) or PE2 (filled triangles) tomato fruit.

viscosity the presence of PE, at least at high levels, would seem to be an advantage. Also, since the enzyme activity appears to be most important in the paste, post homogenisation, rather than in the intact fruit, it was decided to examine the effect of time on paste viscosity. This was achieved by preparing tomato paste, placing a sample in a Bohlin rheometer and recording G' with time. The results are shown in *Figure 7*. It can be seen that the viscosity of paste from normal fruit does not change with time. This is also true of paste made from the PE2as line. However, the viscosity of pastes made from the PGas fruit showed a steady increase with time. The suggestion is that it is the action of PE in the PGas paste that is resulting in the increased viscosity with time.

The availability of these tomato fruit lines showing variable paste viscosity and presumably variable pectin structure in the pastes allowed a study of the possible structure/function relationships of the pectin to be undertaken. It was thought likely that the viscosity of the paste, as determined by the Bostwick measurement, was dependent on the viscosity of the serum component of the paste. To test this, pastes from normal and transgenic fruit were centrifuged to separate the serum from the solid fraction and then the serum viscosity determined using an Ostwald viscometer. *Table 5* shows the viscosity of the serum for all the cold and hot break pastes. This is shown alongside the corresponding Bostwick values.

It is apparent that there is, in fact, no correlation whatsoever between the Bostwick value for the whole paste and the corresponding serum viscosity. The composition of these sera have been determined (Errington *et al*., 1998) but this does not help to explain any effect of serum on total paste viscosity.

Table 6 shows the solid volume fraction (the percentage of solids in the paste) again for normal and transgenic pastes. In this case a weak correlation may be seen with the Bostwick values. It is likely that it is the nature of the solid, or colloidal, material which is important, thus solids from cold break pastes were examined by confocal

Table 5.   Comparison of paste and serum viscosities in normal and genetically modified tomato lines

Line	Bostwick value	Relative serum viscosity
Normal (hot break)	202	14.1
Normal (cold break)	235	1.5
PG as (hot break)	196	31.5
PG as (cold break)	78	1.3
PE2 as (hot break)	172	17.5
PE2 as (cold break)	232	5.6

Table 6. Comparison of paste viscosity and volume fraction of solids in normal and genetically modified tomato lines

Line	Bostwick value	Volume fraction of solids (%)
Normal (hot break)	202	24
Normal (cold break)	235	29
PGas (hot break)	196	24
PGas (cold break)	78	35
PE2as (hot break)	172	37
PE2as (cold break)	232	23

microscopy (Errington *et al.,* 1998). Paste from normal fruit showed massive cell disruption with only wall fragments visible under the microscope. This is consistent with the combined action of PG and PE totally degrading the pectin, weakening the cell wall and thus allowing cells to fragment. In contrast, pastes from the cold break PGas paste showed mainly intact cell walls, although whether this represented intact cells was not determined. The paste from the PE2as fruit also showed mainly intact cell walls yet in this case the paste viscosity is not as high as for the PGas paste. These results do not demonstrate unequivocally what factors within a paste determine its final viscosity. However, a theory can be proposed. In order to have a high viscosity the cell walls must remain relatively intact. The prevention of PG activity, either by down regulation of the enzyme or by reduction in PE activity preventing the genera- tion of de-esterified target sites for PG on the pectin, can both bring this about. However, the presence of intact walls is not sufficient in itself to give high viscosity. The continued action of PE in the paste, as in the PGas line, presumably results in an increased negative charge on the cell walls. This could in turn increase the interactions between either the individual cells or the cells and serum components and result in an increased viscosity.

Cold break pastes made the tomato fruit transformed with an antisense β-galacto- sidase gene also show a marked increase in their viscosity as determined by the Bostwick method (*Table 4*). This is despite the fact that total galactanase activity has hardly been reduced at all and that no detectable change in cell wall metabolism has been detected in these fruit. It remains to be seen whether these pastes contain intact cell walls.

Conclusion

Genetic modification of plants provides a valuable tool for both fundamental studies and commercial applications. The use of gene silencing techniques, as described in this paper, can provide valuable information concerning the role of specific enzymes in metabolism. It has in particular been used, as shown here, to investigate the role of polysaccharide hydrolases in the metabolism of cell walls during fruit ripening. This technique can also be used to probe structure-function relationships by allowing precise and controllable modifications to these polysaccharides and then assessing the effect on physiological factors such as fruit softening or commercially important factors such as paste viscosity.

The genetic modification of plants holds potential for a wide range of commercial applications. Its use to modify the structure of plant derived polysaccharides as described in this paper being only one such application. This technology is still, relatively speaking, in its infancy. It is, however, a technology that is advancing with astonishing speed. It has nonetheless already been employed to modify several polysaccharide structures and in some cases has already lead to successful commercial applications. Currently, such modifications have been carried out on a small number of polymers in a limited range of species and employing the modification of only a minority of the enzymes involved in polysaccharide synthesis or degradation. The potential for using this technology to study the roles of other enzymes in other plant systems is thus enormous. Similarly, the potential for further commercial applications must also be immense.

Acknowledgements

Much of the work presented in this paper was carried out with the support of BBSRC grants to both HS and NE. The authors would also like to acknowledge the contributions of Graham Seymour, Annette Carey, Jian Liang Zhang and Brett Burridge.

References

BIRD, C.R., SMITH, C.J.S., RAY, J.A., MOREAU, P., BEVAN, M.W., BIRD, A.S., HUGHES, S., MORRIS, P.C., GRIERSON, D. AND SCHUCH, W. (1988). The tomato polygalacturonase gene and ripening specific expression in transgenic plants. *Plant Molecular Biology* **11**, 651–662.

CARPITA, N.C. AND GIBEAUT, D.M. (1993). Structural models of primary cell walls in flowering plants: consistency of molecular structure with the physical properties of the walls during growth. *Plant Journal* **3**, 1–30.

CAREY, A. (1996). Purification and characterisation of tomato exo (1–4)-β-D-galactanase. *Ph.D Thesis*. Nottinghan: Nottingham University.

CAREY, A., HOLT, K., PICARD, S., WILDE, R., TUCKER, G.A., BIRD, C.R., SCHUCH, W. AND SEYMOUR, G.B. (1995). Tomato exo-(1–4)-β-D-galactanase: isolation and changes during ripening in normal and mutant tomato fruit and characterisation of a related cDNA clone. *Plant Physiology* **108**, 1099–1107.

CARRINGTON, C.M.S., GREVE, L.C. AND LABOVITCH, J.M. (1993). Cell-wall metabolism in ripening fruit. 6: effect of the antisense polygalacturonase gene on cell-wall changes accompanying ripening in transgenic tomatoes. *Plant Physiology* **103**, 429–434.

ERRINGTON, N., MITCHELL, J.R. AND TUCKER, G.A. (1997). Changes in the force relaxation and compression responses of tomatoes during ripening. The effect of continual testing and polygalacturonase activity. *Postharvest Physiology and Biotechnology* **11**, 141–147.

ERRINGTON, N., TUCKER, G. AND MITCHELL, J. (1998). Effect of genetic Down-regulation of Polygalacturonase and Pectinesterase Activity on Rheology and Composition of Tomato Juice. *Journal of the Science of Food and Agriculture* **76**, 515–519.

GAFFE, J., TIEMAN, D.M. AND HANDA, A.V. (1994). Pectinmethylesterase isoforms in tomato (*Lycopersicon esculentum*) tissues. Effects of expression of a pectinmethylesterase antisense gene. *Plant Physiology* **105**, 199–203.

GRIERSON, D., TUCKER, G.A., KEEN, J., RAY, J., BIRD, C.R. AND SCHUCH, W. (1986). Sequencing and identification of a cDNA clone for tomato polygalacturonase. *Nucleic Acid Research* **14**, 8595–8603.

HALL, L.H., TUCKER, G.A., SMITH, C.J., WATSON, C.F., SEYMOUR, G.B., BUNDICK, Y., BONIWELL, J.M., FLETCHER, J.D., RAY, J.A., SCHUCH, W., BIRD, C.R. AND GRIERSON, D. (1993). Antisense inhibition of pectin esterase gene expression in transgenic tomatoes. *The Plant Journal* **3**, 121–129.

HALL, L.N., BIRD, C.R., PICTON, S.P., TUCKER, G.A., SEYMOUR, G.B. AND GRIERSON, D. (1994). Molecular characterisation of cDNA clones representing pectin esterase isozymes from tomato. *Plant Molecular Biology* **25**, 313–318.

HARRIMAN, R.W.,TIEMAN, D.M. AND HANDA, A.K. (1991). Molecular cloning of tomato pectinmethylesterase gene and its expression in Rutgers, ripening inhibitor, non-ripening and never-ripe tomato fruit. *Plant Physiology* **97**, 80–87.

HOBSON, G.E. (1963). Pectinesterase in normal and abnormal tomato fruit. *Biochemical Journal* **86**, 358–365.

HOBSON, G.E. (1964). Polygalacturonase in normal and abnormal tomato fruit. *Biochemical Journal* **92**, 324–332.

JONES, C., SEYMOUR, G., BIRD, C. AND TUCKER, G.A. (1996). Down regulation of two non-homologous endogenous genes with a single chimeric sense gene construct. In *Mechanisms and applications of gene silencing*. Eds. G.W. Lycett, D. Grierson and G.A. Tucker, pp 85–96. Nottingham: Nottingham University Press.

JONES, C.G, SCOTHERN, G.P, LYCETT, G.W. AND TUCKER, G.A. (1998). The effect of chimeric transgene architecture on co-ordinated gene silencing. *Planta* **204**, 499–505.

LANGLEY, K.R., MARTIN, A., STENNING, R., MURRAY, A.J., HOBSON, G.E., SCHUCH, W. AND BIRD, C. (1994). Mechanical and optical assessment of the ripening of tomato fruit with reduced polygalacturonase activity. *Journal of the Science of Food and Agriculture* **66**, 547–554.

LYCETT, G.W., GRIERSON, D. AND TUCKER, G.A.(Eds.) (1996). *Mechanisms and applications of gene silencing*. Nottingham: Nottingham University Press.

MARKOVIC, O. AND JORNVAL, L.H. (1986). Pectinesterase: The primary structure of the tomato enzyme. *European Journal of Biochemistry* **158**, 455–462.

MOSHREFI, M. AND LUH, B.S. (1984). Purification and characterisation of two tomato polyga-lacturonase isoenzymes. *Journal of Food Biochemistry* **8**, 39–54.

POGSON, B.J., BRADY, C.J. AND ORR, G.R. (1991). On the occurrence and structure of subunits of endo-polygalacturonase isoforms in mature-green and ripening tomato fruit. *Australian Journal of Plant Physiology* **18**, 65–79.

PRESSEY, R. (1983). β-Galactosidases in ripening tomatoes. *Plant Physiology* **71**, 132–135.

PRESSEY, R. AND AVANTS, J.K. (1972). Multiple forms of pectinesterases in tomatoes. *Phytochemistry* **11**, 3139–3142.

PRESSEY, R. AND AVANTS, J.K. (1973). Two forms of polygalacturonase in tomatoes. *Biochimica et Biophysica Acta* **309**, 363–369.

PRESSEY, R. AND HIMMELSBACH, D.S (1984). ^{13}C- N.m.r spectrum of a D-galactose-rich polysaccharide from tomato fruit. *Carbohydrate Research* **127**, 356–359.

RAY, J.,KNAPP, J.,GRIERSON, D., BIRD, C.AND SCHUCH, W. (1988). Identification and sequence determination of a cDNA clone for tomato pectinesterase. *European Journal of Biochemistry* **174**, 119–124.

SEYMOUR, G.B., LASSLETT, Y. AND TUCKER, G.A. (1987). Differential effects of pectolytic enzymes on tomato *in vivo* and *in vitro*. *Phytochemistry* **26**, 3137-3139.

SEYMOUR, G.B., FRAY, R.F., HILL, P. AND TUCKER, G.A. (1993). Down regulation of two non-homologous endogenous genes with a single chimeric sense gene construct. *Plant Molecular Biology* **23**, 1–9.

SHEEHY, R.E., PEARSON, J., BRADY, C.J. AND HIATT, W.R.(1987). Molecular characterisation of tomato fruit polygalacturonase. *Molecular and General Genetics* **208**, 30-36.

SHEEHY, R.E., KRAMER, M. AND HIATT, W.R. (1988). Reduction of polygalacturonase activity in tomato fruit by antisense RNA. *Proceedings of the National Academy of Sciences, USA* **85**, 8805-8809.

SIMONS, H. (1998). Use of multi-gene down regulation to study cell wall degradation during fruit ripening. *Ph.D Thesis*. Nottingham: University of Nottingham.

SMITH, C.J.S., WATSON, C.F., RAY, J., BIRD, C.J., MORRIS, P.C., SCHUCH, W. AND GRIERSON, D. (1988). Antisense RNA inhibition of polygalacturonase gene expression in transgenic tomatoes. *Nature* **334**, 724–726.

TIEMAN, D.M., HARRIMAN, R.W., RAMAMOHAN, G. AND HANDA, A.K. (1992). An antisense pectin methylesterase gene alters pectin chemistry and soluble solids in tomato fruit. *The Plant Cell* **4**, 667–679.

TUCKER, G.A. (1990). Genetic manipulation of fruit ripening. *Biotechnology and Genetic Engineering Reviews* **8**, 133–159.

TUCKER, G.A. (1993). Fruit ripening. In *The Biochemistry of Fruit Ripening*. Eds. G.B Seymour, J.E. Taylor and G.A. Tucker, pp 1–51. London: Chapman and Hall.

TUCKER, G.A. AND WOODS, L. (Eds.) (1995). *Enzymes in Food Processing* 2nd Edition. London: Chapman and Hall.

TUCKER, G.A., ROBERTSON, N.G. AND GRIERSON, D. (1980). Changes in polygalacturonase isoenzymes during the ripening of normal and mutant tomato fruit. *European Journal of Biochemistry* **112**, 119-124.

TUCKER, G.A., ROBERTSON, N.G. AND GRIERSON, D. (1982). Purification and changes in activities of tomato pectinesterase isoenzymes. *Journal of the Science of Food and Agriculture* **33**, 396–400.

WARRILOW, A.G.S., TURNER, R.J. AND JONES, G.M. (1994). A novel form of pectinesterase in tomato. *Phytochemistry* **35**, 863–868.

WATSON, C.F., ZHENG, L.S. AND DELLAPENNA, D. (1994). Reduction of polygalacturonase beta-subunit expression affects pectin solubilisation and degradation during fruit ripening. *Plant Cell* **6**, 1623–1643.

13
Pullulan from Agro-industrial Wastes

CLEANTHES ISRAILIDES[1]*, ALAN SMITH[2], BERNARD SCANLON[2] AND
CHRISTIAN BARNETT[2]

[1]*Institute of Technology of Agricultural Products, National Agricultural Research
Foundation, 1, S. Venizelou St., Lycovrissi 141 23, Athens, Greece and
[2]Department of Life Sciences, The Nottingham Trent University, Clifton Lane,
Clifton, Nottingham, NG11 8NS, UK*

Introduction

Extracellular polysaccharide production is encountered in many microorganisms.
Many strains of bacteria, yeasts and fungi have been selected and are used commercially because they have been found to produce enough extracellular polysaccharides
in broth culture to be of economic interest. There are two different types of extracellular polysaccharides, homopolysaccharides such as dextrans and levans which
contain one type of monomeric unit, and heteropolysaccharides which contain two or
more different monomeric units such as hyaluronic acid.

Pullulan is a homopolysaccharide of industrial interest and economic importance
produced by the yeast-like fungus *Aureobasidium pullulans*. It was first reported by
Bernier (1958) and its structure was elaborated by Bender *et al.* (1959). It is a glucan
composed mainly of maltotriose units linked by α-1,6-glycosidic linkages (Le Duy *et
al.*, 1983; Shin *et al.*, 1989). The presence of maltotetraosyl units has also been
reported (Catley and Whelan, 1970; *Figure 1*). The molecular weight varies greatly
from 50,000 up to 40×10^6 in different strains (Fujii and Shinohara, 1987; Pollock *et
al.*, 1992).

Pullulan has many industrial applications (Yuen, 1974; Hayashibara bulletin, 1990;
Krochta and De Mulder-Johnston, 1997). It can be used as a partial replacement for
starch ingredients as a low calorie viscosity enhancer, binder and quality improver of
foods. It finds uses as an oxygen-impermeable food packaging material preventing
oxidation and retaining the aroma, flavour and freshness of food and prolonging shelf
life.

Pullulan dissolves readily in cold water and forms a viscous solution, but does not
gel. By drying in hot air, the solution forms transparent water soluble films which are
odourless, tasteless and oxygen impermeable. The strength of pullulan films compares

---

*To whom correspondence may be addressed.

Biotechnology and Genetic Engineering Reviews – Vol. 16, April 1999
0264–8725/99/16/309–324 $20.00 + $0.00 © Intercept Ltd, P.O. Box 716, Andover, Hampshire SP10 1YG, UK

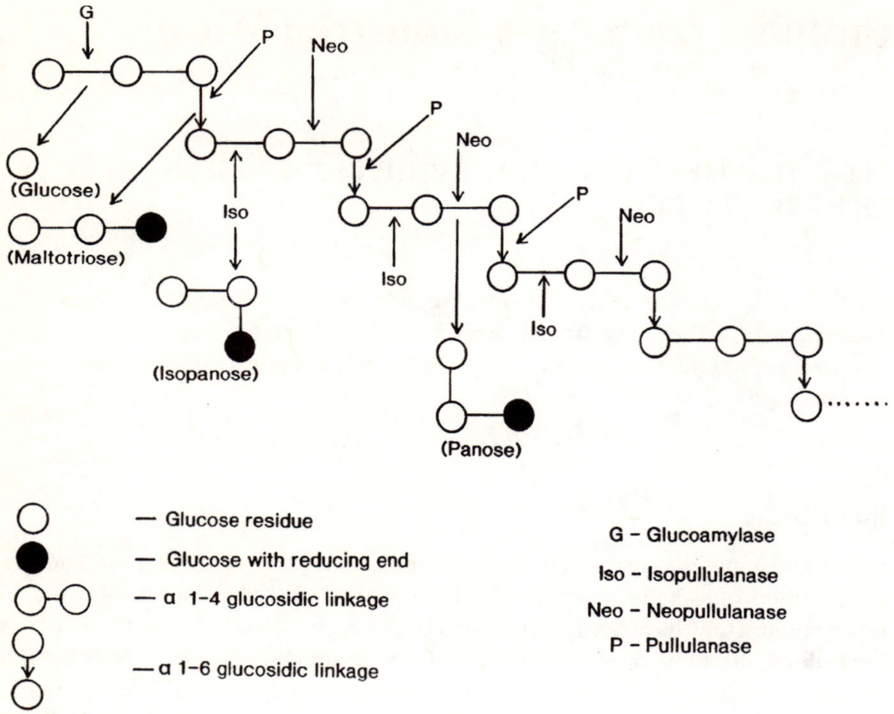

Figure 1.   Structure of pullulan and sites of enzymatic hydrolysis.

with polystyrene, but with the advantage that they are edible and biodegradable (Kaplan *et al.*, 1993). They are also heat sealable, oil resistant and printable. Due to these properties, pullulan can help solve the recycling problem by replacing the aluminium foil layer in paperboard drinks cartons. Additionally, pullulan-film used for packaging meat or vegetables can be readily dissolved on addition of hot water. Another similar application is a water-soluble sugar packet for coffee and tea bags. In the pharmaceutical industry, pullulan is used as a bulking agent for tablets and as an encapsulating agent for soft and hard capsules. The solubility of capsules can be retarded by chemical modification of pullulan so as to release their content in different gastro-intestinal locations.

Pullulan can also be used as an adhesive, as a gelling agent, moisture retainer and stabilizer in cosmetics, as a strong film former in paints, as a coating material to prevent glass bottles shattering and as a contruction material. It can also be formed into a fibre. Pullulan has some uses in the fertilizer industry. It (or its derivatives) is added in amounts up to 50% (w/w) to the fertilizers to provide a higher water solubility (Matsunaga *et al.*, 1976).

The market price of food and pharmaceutical grade pullulan is about 50 Pounds Sterling per kg. The current world supply (circa 10,000 tonnes pa) is produced almost exclusively by the Japanese company Hayashibara Co. Ltd., Okayama, Japan.

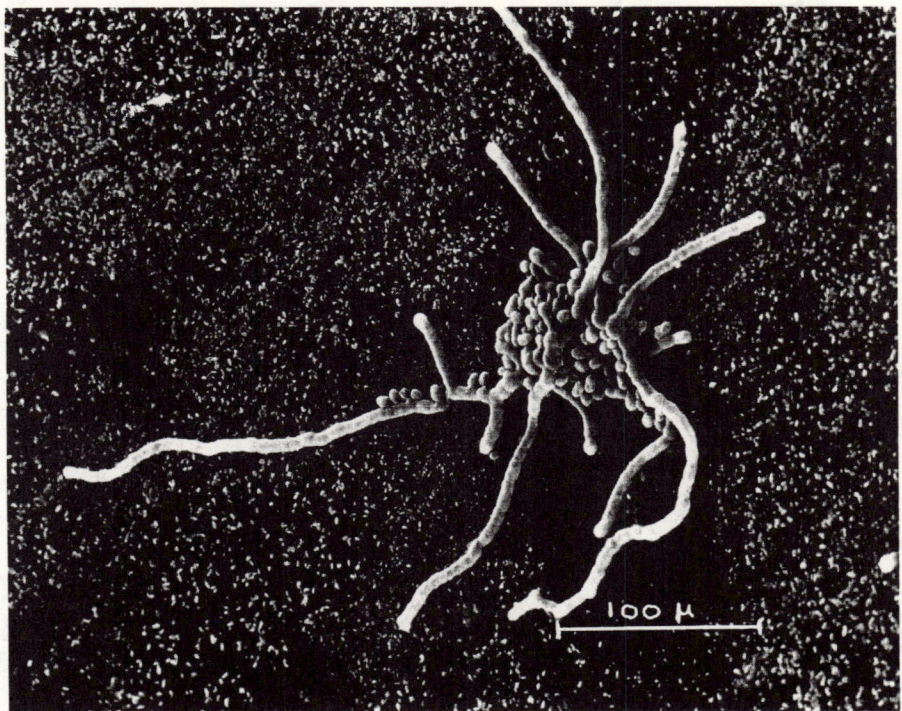

Figure 2. *Aureobasidium pullulans.* Scanning electron micrograph showing chlamydospores originating from the hyphal wall.

Pullulan Production by *Aureobasidium pullulans*

The production of pullulan depends on the fermentation parameters, the morphological state and the fungal strains. *A. pullulans* (de Bary) has been placed in the *Deuteromycota* (Fungi Imperfecti) as its sexual stage has not been established (Cooke, 1959; Hudson, 1965). It is considered to be a 'black yeast' (Hermanides-Nijhof, 1997; de Hoog and McGinnis, 1987). The existence of a perfect stage of *A. pullulans* has been a matter of controversy, and the organism is placed by some in the *Ascomycetes* and by others in the *Basidiomycetes.*

It is polymorphic in that it generally appears as a unicellular yeast-like organism in liquid media and as a mycelium on solid media. The *Aureobasidiaceae* are characterised by the presence of thick-walled spores (chlamydospores or radulospores) produced on minute spicules projecting from the hyphal wall. (*Figures 2* and *3*). Furthermore, there is more than one spore type, namely blastospores (conidia), and resting forms, swollen cells and chlamydospores some of which occur in pairs (*Figure 4*). A pH range of 3.6–5.2 and a low nitrogen to carbon ratio facilitates the conversion of blastospores to swollen cells and chlamydospores (Kockova-Kratochilova *et al.*, 1980).

Although there is no general agreement as to the role that the various morphological stages play in pullulan production, there seems to be a consensus that the main

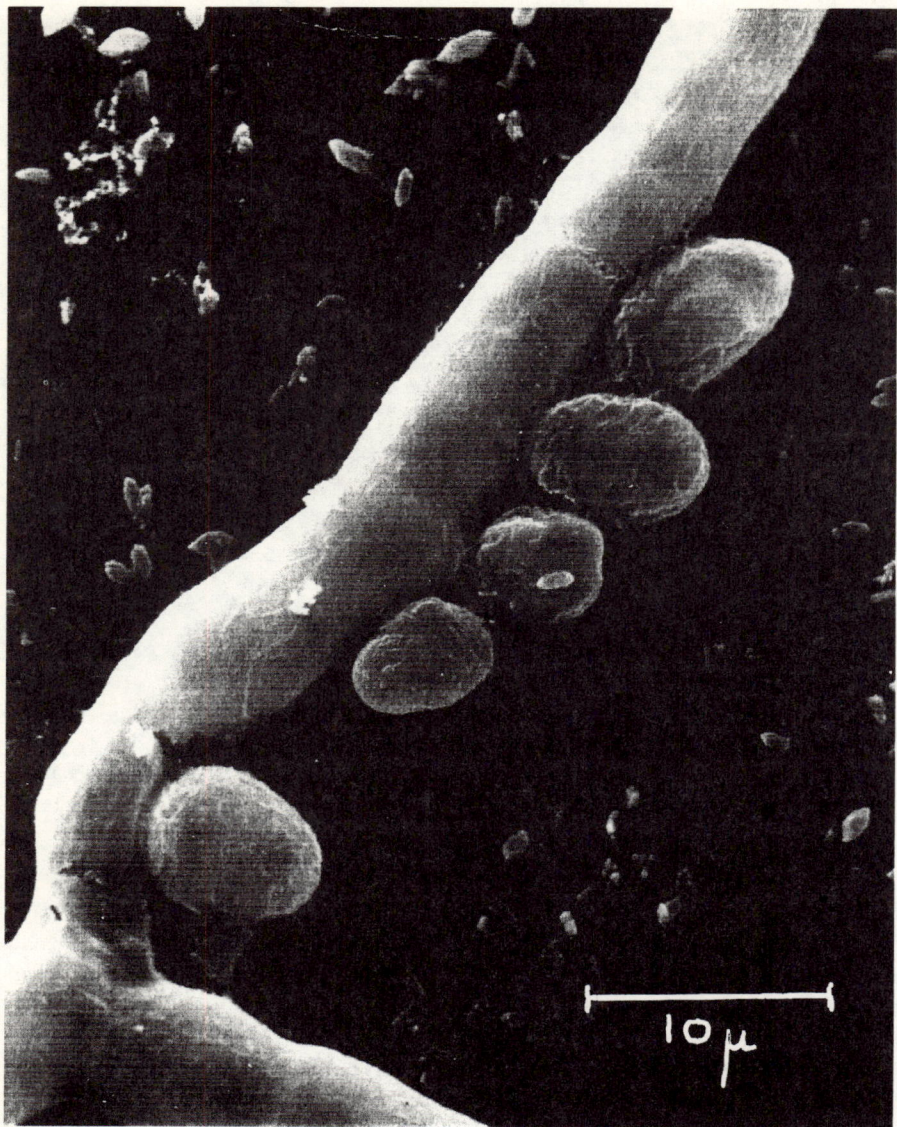

Figure 3.   *Aureobasidium pullulans.* Hypha with chlamydospores (close up).

morphological types responsible for the elaboration of pullulan are the chlamydospore and swollen cell (Simon *et al.*, 1993; Simon *et al.*, 1995). The factors affecting shifting from one morphological type to another are not quite clear but there are reports that it can be related to inoculum size; a low inoculum yielding exclusively mycelium and a high inoculum giving rise to spore types (Catley, 1980; Ramos and Garcia-Acha, 1975).

A. pullulans produces a black pigment, melanin, with a characteristic absorption peak at between 400–600 nm. It is a secondary metabolite produced at the end of exponential growth.

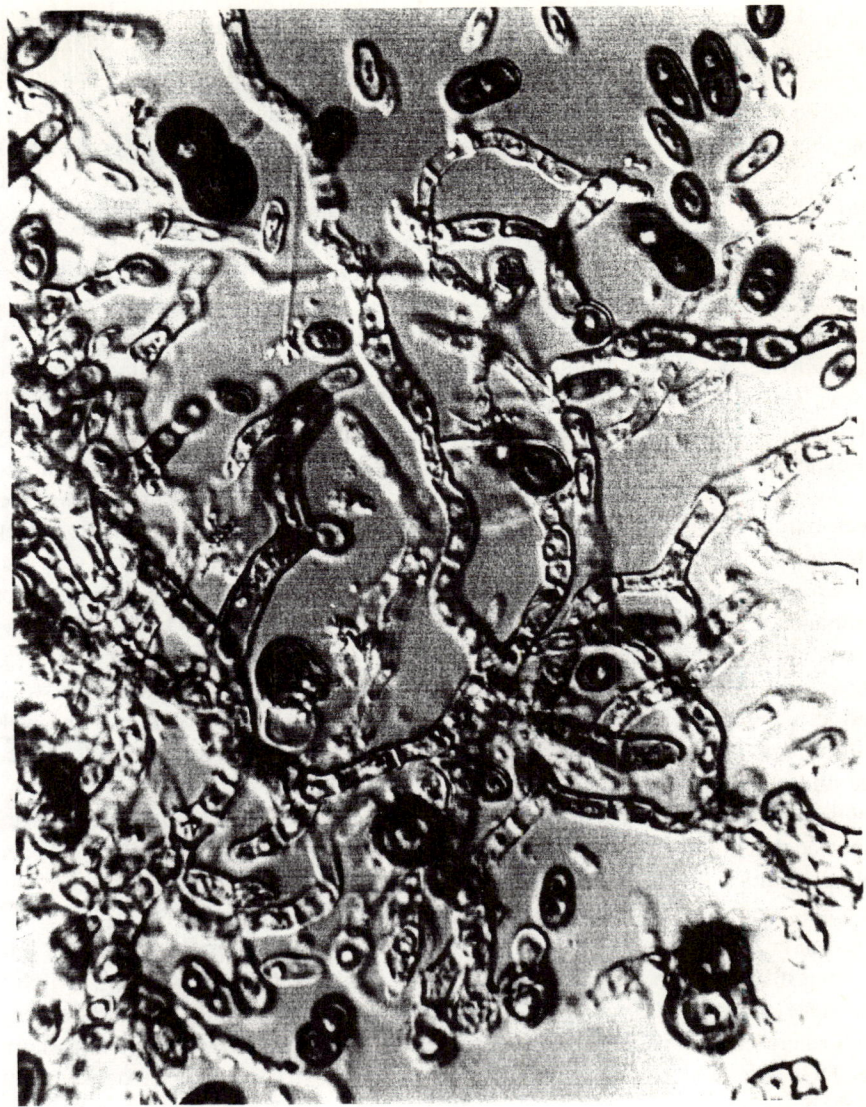

Figure 4. *Aureobasidium pullulans,* under Nomarski microscopy (x 1600), showing hyphae, blastospores, swollen cells and chlamydospores mostly in pairs.

The pigment is released into the growth medium where it covalently binds to pullulan, which is undesirable industrially. The role of melanin is not clear, but it may have a protective function against UV radiation, heavy metal toxicity, free radical formation and enzymatic attack. Melanisation is also thought to increase the susceptibility of plants to attachment and penetration by fungi. Fungal melanin differs from animal melanins in that it is not synthesised from tyrosine but from acetate via the polyketide biosynthetic pathway (Siehr, 1981). This pathway may be blocked by several antifungal compounds.

The formation of melanin and the morphology of the organism depends strongly on cultural conditions used (medium composition, temperature, light, etc.) (Lingapa *et al.*, 1963; Wickerham and Kurtzman, 1975). Current studies also indicate that the production of melanin in the fermentation broth is related to dissolved oxygen tension. It has been observed that in stirred tank reactors where the dissolved oxygen is limited, pigment develops very late in fermentation and sometimes not at all (Rho *et al.*, 1988). The production of pigment-free pullulan has also been reported by maintaining the *A. pullulans* cells in the yeast form in a two stage fermentation (Shabtai and Mukmenev, 1995).

A. pullulans utilises a variety of monosaccharides as carbon sources (e.g. xylose, glucose, fructose). Among disaccharides, sucrose and maltose were better substrates than trehalose, while lactose gave poor growth (Cooke and Matsura, 1963). *A. pullulans* cannot utilize cellulose. Of the polysaccharides on which it can grow, it grows best on dextrins, followed by starch, pectin and inulin. The organism is capable of utilizing cutin, grows excellently on tannic acid and has enzymatic activity (polyphenol oxidase) against lignin (Cooke, 1958). Furthermore, it has the ability to produce a range of important enzymes, including glucoamylase, sucrase, xylanase, pectolytic enzymes, amylases, urease, DNase and RNase (Deshpande *et al.*, 1992). This makes it a serious threat to paints and plastic surfaces. *A. pullulans* was also shown to be a potential microorganism for the production of single cell protein for animal feed, being classified as risk Group I by the World Health Organisation. The palatability (acceptability) as well as digestibility and protein content of straw was enhanced by acid hydrolysis, followed by ammonification of the acid treated straw and fermenting it with *A. pullulans* (Israilides *et al.*, 1981). *A. pullulans* grew well with asparagine, $(NH_4)_2 SO_4$, NH_4NO_3, KNO_3, $NaNO_2$ or urea as a nitrogen source (Israilides, 1979). Pullulan production, however, occurs under nitrogen limiting conditions, and high levels of nitrogen favour the production of biomass. It has been reported that pullulan synthesis proceeded only when all the available NH_4^+ had been utilised (Catley, 1971), while the presence of NH_4^+ inhibited polysaccharide production independently of pH (Seviour and Kristiansen, 1983).

It can grow over a wide range of pH from 1.9 to 10.1 and temperatures from 14° to 37°C (Han and Anderson, 1975). Furthermore, supplementation of K_2HPO_4, NH_4NO_3 and yeast extract to a certain level had a favourable effect in pullulan yields (Israilides, 1998). Fermentation parameters which influence pullulan production, (e.g. temperature, pH, DO_2 agitation, carbon and nitrogen source, etc.) have been studied extensively by many investigators (Catley, 1971; Imshenetsky *et al.*, 1981; Mc Neil and Kristiansen, 1987; 1990; Madi *et al.*, 1996; Wecker and Onken, 1991; West and Reed-Hamer, 1991; Reeslev *et al.*, 1991; Auer and Seviour, 1990; Gibbs and Seviour, 1996; Lacroix *et al.*, 1985).

In general, pullulan is best produced under nitrogen limiting conditions and at a pH value between 5.0 and 6.5; below pH 3 pullulan production is inhibited (Catley, 1971). Most researchers agree that pullulan synthesis is favoured under aerobic conditions and high oxygen transfer rate (Rho *et al.*, 1988; McNeil and Kristiansen, 1987; Moscovici *et al.*, 1996). It has been shown (Rho *et al.*, 1988) that when aeration was stopped, pullulan production also ceased, but synthesis continued on the resumption of aeration. In the same paper it was reported that levels of 10% dissolved oxygen tension (DOT) gave good yields of pullulan and that with further increases in DOT there was a decrease in pullulan production. There are other reports suggesting that beyond certain levels of DOT pullulan synthesis may be reduced, although this may be strain specific (Imshenetsky *et al.*, 1981; Wecker and Onken, 1991).

Pullulan assays

Pullulan is usually precipitated from the cell-free fermentation medium by the addition of two volumes of ethanol or propanol. The precipitated material is refererred to as the agglutinating substance. Different laboratories have used a variety of analytical methods to assess the purity of pullulan produced in this way. Simply weighing the agglutinating substance or measuring total sugars (Dubois *et al.*, 1956) is very unreliable. Hydrolysing the agglutinating substance with pullulanase and determining the maltotriose produced by chromatography, radiometry or other method (Catley, 1971; Finkelman and Vardanis, 1982) is a far better way, but sometimes can be very time consuming. Good resolution and sensitivity can be achieved using high performance anion exchange chromatography (HPAEC) quantified by a pulsed amperometric detector (PAD) (Israilides *et al.*, 1998) or by capillary electrophoresis, a method which was shown to be very powerful in carbohydrate separations (Hertmentin *et al.*, 1994; Barnett *et al.*, 1998).

Detection of oligosaccharides like maltotriose eluted from HPLC columns is the biggest challenge and weakest link in the analysis of oligosaccharides. Non-specific detectors such as the refractive index (R.I.) are used routinely, but their lack of sensitivity and restriction to isocratic elution severely limits the development of improved techniques. Low wavelength UV detection (<210 nm) has been shown to have comparable sensitivity to HPAEC-PAD whilst allowing the use of limited gradient elution, but requires ultra-pure solvents. The use of HPAEC-PAD offsets the above drawbacks.

A coupled enzyme assay method (CEA) was developed for the estimation of pullulan in ethanol agglutinating substances (Israilides *et al.*, 1994a). The assay involves the concerted action of pullulanase, amyloglucosidase and glucose oxidase. The combined action of these enzymes allows pullulan to be measured spectrophotometrically. The rate of colour development was correlated to the concentration of pullulan. *Figure 5* shows the Lineweaver-Burk kinetic plot which depicts the correlation of the rate of such colour development with the concentration of pullulan. Similar correlations were achieved with other kinetic plots.

The CEA can differentiate between pullulan and highly branched polyglucans by repeating the assay omitting the pullulanase. The reaction rate is now dependent on the number of non-reducing ends, which in turn largely depends on the degree of branching. Thus, linear polysaccharides such as pullulan give very low rates, while highly branched polysaccharides will give high rates. High values (>10) for the ratio of the rate of coupled enzyme reaction to amyloglucosidase reaction rate (A/B) indicate a linear polyglucan with predominately α-1,4- and α-1,6-glycosidic bonds. The reproducibility of the CEA was good with a standard error of ±5%, based on reference pullulan from Sigma.

Simon *et al.* (1993) used the number of residues in repeating units (ratio of glucose produced by total hydrolysis to the reducing sugar produced by hydrolysis with pullulanase) as a means of differentiating pullulan from other polysacccharides produced in the broth. Under various cultural conditions, the number of residues in repeating units ranged from 3.2 to 4.2 for pullulan.

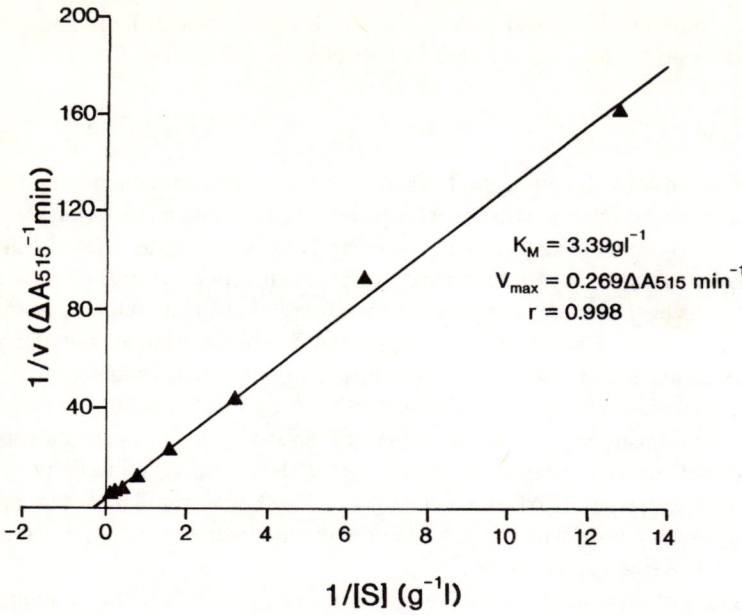

Figure 5.   Lineweaver-Burk kinetic plot in pullulan hydrolysis with the coupled enzyme assay.

A. *pullulans* elaborates polysaccharides other than pullulan. One of these polysaccharides isolated by Hamada and Tsujisaka (1983), was a β-1,3 polyglucan with single glucose residues attached β-1,6 every third or fourth residue along the backbone of the polysaccharide.

Molecular weight determination

The molecular weight of pullulan is important in terms of its industrial applications. The molecular weight of pullulan produced by *A. pullulans* is dependent on pH (Lee and Yoo, 1993), phosphate content of the medium (Hayashibara, 1990), impeller speed (McNeil and Kristiansen, 1987) and the carbon source and age of the culture (Israilides *et al.*, 1994b). Furthermore, there are occasional maltotetraose units incorporated randomly within pullulan. These represent sites particularly susceptible to endoamylase catalysed hydrolysis. Hence, there is a relationship between the fine structure of pullulan and the molecular weight (Catley, 1970; Carolan *et al.*, 1983).

When different substrates from various agro-industrial wastes were used, the pullulan produced differed in molecular weight, purity and other physicochemical properties (Israilides *et al.*, 1994b). For example, grape skin pulp, hot water extracts and hydrolysed starch waste proved to be good substrates for pullulan production. Molasses and olive oil wastes produced small amounts of agglutinating substances, heterogeneous in composition and with very small amounts of pullulan.

Biosynthesis of pullulan

Significant progress has been made in recent years in elucidating the pathways of polysaccharide biosynthesis. Similar progress has not yet been made in determining the pathway for pullulan biosynthesis in *Aureobasidium pullulans*. Taguchi *et al.* (1972) reported the biosynthesis of pullulan from both cell-free preparations and acetone-dried cells. Frozen cells were disrupted by grinding with alumina, the debris removed by centrifugation and the supernatant precipitated with 80% saturated ammonium sulphate. This protein precipitate catalysed pullulan biosynthesis from uridine-5-diphosphoglucose (UDPG) and ATP giving a yield of 54% of pullulan with respect to the UDPG added. In the absence of ATP, no pullulan was formed and neither sucrose nor ADPG could act as substrates. The acetone dried cells when resuspended were incubated with sucrose in a phosphate buffer produced pullulan. When ^{14}C-labelled sucrose was used, ^{14}C activity was found in the lipid fraction (chloroform-methanol extract). An aliquot of this extract was dried under nitrogen, hydrolysed under acid conditions and partitioned between chloroform and an aqueous phase. Most of the radioactivity (85%) was found in the aqueous phase and was shown to be glucose by paper chromatography. The properties of this glycolipid suggested that it was lipid-pyrophosphate-glucose. Under conditions where pullulan synthesis was taking place, the ^{14}C-labelled glycolipid pool was turned over rapidly. The authors suggested that lipid intermediates were involved in pullulan biosynthesis. Catley and McDowell (1982) also showed the incorporation of ^{14}C-labelled glucose into glycolipid, pullulan and other extracellular polysaccharides. The glycolipid extract was separated into three fractions by chromatography on DEAE-cellulose. They chacterised the glyco moiety in the second fraction as a mixture of glucose, isomaltose, panose and isopanose. Since cell wall saccharides do not contain α–(1,4)- or α-(1,6)-D-glucosidic bonds and glycogen biosynthesis does not involve a glycolipid intermediate (Finkelman and Vardanis, 1987), it was proposed that these glycolipids were intermediates in pullulan biosynthesis. The lipid involved was assumed to be a polyprenyl phosphate of the dolichol type partly on the basis of its stability in alkaine solution (Hemming, 1974) and the isolation of polyprenyl-linked saccharides from both *Saccharomyces cerevisiae* (Lehle and Tanner, 1978) and mammalian liver (Behrens and Tabora, 1978).

In a study of the oxygen-dependence of growth and pullulan biosynthesis by *Aureobasidium pullulans,* Rho *et al.* (1988) found that both were absolutely dependent on oxygen. In terms of pullulan biosynthesis, oxygen may be required to recycle the UDPG and ATP that Taguchi *et al.* (1973) showed were required for pullulan biosynthesis in their cell-free preparation. Studies by Rho *et al.* (1988) with *Aureobasidium pullulans* in non-growth media (lacking nitrogen) showed that both biomass and pullulan increased with time (biomass increasing from around 3 to 15 g.l^{-1} over 40 hours and up to 10 g.l^{-1} of pullulan was synthesised over the same time). The increase in biomass was accompanied by an increase in the carbon to nitrogen ratio indicating that the cells were storing some carbonaceous compound(s). Based on an analysis of their data, they proposed that there were two pathways involved in pullulan biosynthesis in nitrogen-free media. Part of the glucose taken up by the cell was converted directly to pullulan, the remaining glucose accumulated within the cell and could be converted to pullulan by a different pathway. Simon *et al.* (1995) showed that glycogen and lipidic granules were concentrated at the plasmalemma in swollen cells actively synthesising pullulan. They proposed that glycogen might act as the source of 1,4-linked glucose units and the lipid granules the source of the lipid for glycolipid synthesis. These glycolipids would then be transported across the plasma membrane and act as precursors for pullulan biosynthesis.

Catly and McDowell (1982) and Barnett (1997) were unable to repeat the work of Taguchi *et al.* (1972). Like Finkelman and Vardanis (1987), we found that the 100,000g pellet from cell-free preparations of *Aureobasidium pullulans* was active in glycogen synthesis but there was no evidence for pullulan synthesis in this membrane-enriched preparation. Cell-free preparations of *Aureobasidium pullulans* can be prepared that are active in pullulan biosynthesis but efforts to purify this activity have so far failed (Barnett, 1997).

Given that the role of lipids in peptidoglycan, glycopeptides and oligosaccharides is well established (Bugg and Brandish, 1994), the genetics and enzymology of polysaccharide biosynthesis have been studied extensively (Glucksman *et al.*, 1993; Keller *et al.*, 1995; Villar-Palasi and Guinovart, 1997) and transformation of *Aureobasidium pullulans* reported (Cullen *et al.*, 1991; Thornwell *et al.*, 1995), our understanding of pullulan biosynthesis should improve significantly in the near future.

Production of pullulan from agro-industrial wastes

Pullulan can be synthesised from a variety of carbohydrate substrates incorporated into either defined (synthetic) or non-defined media. Within the latter are several agro-industrial wastes which have been shown to be suitable for pullulan production (Israilides *et al.*, 1993; Le Duy and Boa, 1982; Le Duy *et al.*, 1983; Shin *et al.*, 1989; Zajic *et al.*, 1979). Utilisation of these substrates would seem to be ecologically sound and economically advantageous as they have low or even negative costs. In this way, regional development seems to be feasible with the creation of small or medium enterprises (SME's), which will give rise to new jobs and protect the environment from pollution.

Among the substrates reported as potentially suitable for pullulan production were hemicelluloses in waste water from the production of viscose fibres (Biely *et al.*, 1978), peat hydrolysate, which is the liquid resulting from the heat treatment of raw peat in an acid solution (Le Duy and Boa, 1983), spent sulphite liquor (Zajic *et al.*, 1979), starch waste and corn glucose syrup (Hayashibara Inc., 1990), olive oil wastes, grape skin pulp extract, molasses (Israilides *et al.*, 1993, 1994b; Bambalov and Jordanov, 1993), carob pod extract (Roukas and Biliaderis, 1995).

Although fermentation parameters for pullulan production have usually been studied with defined substrates such as glucose and sucrose, the results from agro-industrial wastes as substrates have shown that pullulan can be produced in yields similar or better to those obtained with conventional substrates (Israilides *et al.*, 1998). However, it has been shown (Israilides *et al.*, 1994a; 1994b) that the purity of the pullulan in the crude precipitated substance may vary widely with the different carbon substrate and fermentation conditions. Furthermore, pullulan produced in such fermentations is often characterised by heterogeneity of both composition and molecular weight (Israilides *et al.*, 1994b).

Starch waste

Waste potato starch from the manufacture of potato crisps or from other potato processing industries (e.g. frozen pre-fried potatoes) is a fairly homogeneous substrate and relatively free from extraneous material. Certain strains of *A. pullulans* possess starch degrading enzymes (Saha and Bothast, 1993). However, this activity was greater against linear α-1,4-glucans, but very little if any against polysaccharides with α-1,6-linkages. Therefore, in order for starch waste to be considered a good substrate for pullulan fermentation, it must be partially hydrolysed.

Studies using starch waste as a substrate revealed that *A. pullulans* showed preference for oligosaccharides over glucose when it was grown for pullulan production; in fact, the yield of pullulan was dependent on the degree of hydrolysis of starch or dextrose equivalent (DE). This is an indication of total reducing sugar as a percentage of glucose. Unhydrolysed starch has a DE of zero and D-glucose has a DE of 100. The optimum DE was around 55% for pullulan yield, (Barnett *et al.*, 1998). Using hydrolysed starch waste as a substrate, it was also found that the pullulan content of the agglutinating substance increased during the course of fermentation and reached more than 90% (w/w) on the sixth day. This becomes important in determining the optimum time for the production of the highest amount of pullulan.

Grape skin pulp

Grape skin pulp is a by-product of the wine industry and amounts to thousands of tonnes in wine producing countries. Some of this waste material goes for animal feed but most of it is disposed of as waste. Hot water extracts of the pulp can serve as a good substrate for the production of pullulan. Pullulan has been produced in high yield, with a high molecular weight (4.22×10^6) and was pure as assessed by its gel elution profile, glucose content and the number of residues in repeating units (Israilides *et al.*, 1994b). Hence, this is a good substrate for pullulan production.

Olive oil wastes

Olive oil wastes are effluents from the mills producing olive oil. It is considered a major pollutant and causes great problems in olive tree cultivation areas in many Mediterranean countries. Fresh waste is particularly phytotoxic, mainly due to the presence of phenolic compounds (Israilides *et al.*, 1997). At present, there is not an ecological or economic solution for the problem of these wastes. Therefore, many biological and physichochemical methods are used for its treatment, depending on local conditions and various other factors.

Although many microorganisms are growth-inhibited by oil waste because of the presence of phenols, *Aureobasidium pullulans* is able to grow in it to produce pullulan (Iniotakis *et al.*, 1991; Israilides *et al.*, 1993). It has been observed that *A. pullulans*, when grown in the effluent, can reduce the phenolic content by 41% in six days. The pullulan concentration of the ethanol agglutinating substance, however, increased when phenols were removed, indicating that their presence plays an inhibitory role in pullulan production.

The yield of pullulan was very low, 2–3% (w/w) of the agglutinating substance, based on the coupled enzyme and HPAEC (PAD) assays (Israilides *et al.*, 1994b).

Molasses

Fermentation using molasses as a substrate produced mainly low molecular weight

$(0.55 \times 10^6$ Da) material and the amount of pullulan produced was about 5–8% (w/w) of the agglutinating substance (Israilides *et al.*, 1994b). Given the fact that molasses is used widely as a substrate for many industrial fermentations for the production of high value compounds, the potential use of it for the production of pullulan would depend on purification and separation costs, as well as other marketing factors.

Carob pod

Carob tree (*Ceratonia siliqua*, L.) grows naturally on barren soils which are unproductive for most crops. It is found in most warm regions of the Mediterranean area and other countries having a similar climate (Rhodesia, parts of USA, Australia, South America, India and Philippines). The ripe de-seeded carob pod, which contains high levels of tannins, is partially utilised for the production of health confections as a cocoa substitute. It is also used as animal feed, although its high tannin content minimizes the pod's nutritional value. Due to difficulties and high harvesting costs, most carob beans are left unutilised.

The carob pod contains high amounts of water soluble sugars (40–60%) and can serve as a substrate for pullulan production. The yields and fermentation efficiencies reported were 30% and 89% respectively (Roukas and Biliaderis, 1995). However, the pullulan concentration based on the coupled enzyme assay, HPAEC-PAD assay, was low and, in most cases, under 2%. Further, the low A/B ratio in the CEA and the unfavourable number of residues in the repeating units is not in accord with the polysaccharide being pullulan (unpublished data).

Concluding remarks

Pullulan is an edible and biodegradable polysaccharide with numerous commercial applications and relatively high market value. Current research shows that the production of pullulan from agro-industrial wastes is feasible and ecologically sound. Utilization of agro-industrial wastes for the production of an industrially important biopolymer, pullulan, is both challenging and reasonable. Improvements in both yields and purity of pullulan from fermentations from agro-industrial wastes could be achieved by admixture of various wastes in an effort to improve the C/N ratio of the substrate. Improvements in production practices, economies of scale and selection of the proper substrate and strain of *A. pullulans* could all be developed to produce a more favourable product, while protecting the environment from pollution.

References

Auer, D.P.F. and Seviour, R.J. (1990). Influence of varying nitrogen sources on polysaccharide production by *Aureobasidium pullulans* in batch culture. *Applied Microbiology and Biotechnology* **32**, 637.

Bambalov, G. and Jordanov, P. (1993). Production of pullulan polysaccharide from wine-producing wastes. *Scientific Works of the HIFFI-BG* **40**, 229–240.

Barnett, C., Smith, A., Scanlon, B. and Israilides, C. (1998). Pullulan production by *Aureobasidium pullulans* growing on hydrolysed starch. *Carbohydrate Polymers* (in press).

Barnett, C. (1997). *Ph.D. Transfer report*. Nottingham Trent University.

BEHRENS, H. AND TABORA, E. (1978). Dolichol intermediates in the glycosylation of proteins. *Methods in Enzymology* **50**, 402–435.

BENDER, H., LEHMANN, J. AND WALLENFELLS, K. (1959). Pullulan, ein extracellulares Glucan von *Pullularia pullulans*. *Biochimica et Biophysica Acta* **36**, 309–316.

BERNIER, B. (1958). The production of polysaccharides by fungi active in the decomposition of wood forest litter. *Canadian Journal of Microbiology* **4**, 195–204.

BIELY, P., KRATKY, Z., PETRAKOVA, E. AND BAUER, S. (1978). Growth of *Aureobasidium pullulans* on waste water hemicelluloses. *Folia Microbiologia* **24**, 328–333.

BUGG, T.D.H. AND BRANDISH, P.E. (1994). From peptidoglycan to glycoprotein: Common features of lipid-linked oligosaccharide biosynthesis. *FEMS Microbiology Letters* **119**, 263–270.

CAROLAN, G., CATLEY, B.J. AND MCDOUGLAS, F.J. (1983). The location of tetrasaccharide units in pullulan, *Carbohydrate Research* **114**, 237–243.

CATLEY, B.J. (1970). Pullulan, a relationship between molecular weight and fine structure. *FEBS Letters* **10**, 190–193.

CATLEY, B.J. AND WHELAN, W.J. (1970). Observations on the structure of pullulan. *Archives of Biochemistry and Biophysics* **143**, 138–142.

CATLEY, B.J. (1971). The role of pH and nitrogen limitation in the elaboration of the extracellular polysaccharide pullulan by *Aureobasidium pullulans*. *Applied Microbiology* **22(4)**, 650–654.

CATLEY, B.J. (1971). Utilization of carbon sources by *Pullularia pullulans* for the elaboration of extracellular polysaccharides. *Applied Microbiology* **22(4)**, 641–649.

CATLEY, B.J. (1980). Short Communication: The extracellular polysaccharide, pullulan produced by *Aureobasidum pullulans*. A relatioship between elaboration rate and morphology. *Journal of General Microbiology* **120**, 265–268.

CATLEY, B.J. AND MCDOWELL, W. (1982). Lipid-linked saccharides formed during pullulan biosynthesis in *Aureobasidium pullulans*. *Carbohydrate Research* **103**, 65–75.

COOKE, W.B. (1958). The ecology of the fungi. *The Botanical Review* **24**, 341–429.

COOKE, W.B. (1959). An ecological life history of *Aureobasidium pullulans* (de Bary). *Mycopathologia et Mycologia Applicata* **12**, 1–45.

COOKE, W.B. AND MATSUURA, G. (1963). Physiological studies in the black yeasts. *Mycopathologia et Mycologia Applicata* **21**, 3–4.

CULLEN, D., YANG, V., JEFFRIES, T., BOLDUC, J. AND ANDREWS J.H. (1991). Genetic transformation of *Aureobasidium pullulans*. *Journal of Biotechnology* **21**, 283–288.

DE HOOG, G.S. AND MCGINNIS, M.R. (1987). Ascomycetous black yeasts. In *The Expanding Realm of Yeast-like Fungi*. Eds. G.S. de Hoog, M. T. Smith and A.C.M. Weijman. Study Mycol No. 30, pp 187–199.

DESHPANDE, S.M., RALE, V.B. AND LYNCH, J.M. (1992). *Aureobasidium pullulans* in Applied Microbiology: A status report. *Enzyme Microbial Technology* **14**, 514–525.

DUBOIS, M. GILLES, K.A., HAMILTON, J.K., REBERS, P.A. AND SMITH, F. (1956). Colorimetric method for the determination of sugars and related substances. *Analytical Chemistry* **28**, 350–357.

FINKELMAN, M.A.J. AND VARDANIS, A. (1987). Glycogen metabolism in *Aureobasidium pullulans*: a glycogen synthetase with unusual activation properties. *CRC Critical Reviews in Biotechnology* **5(3)**, 185–193.

FUJII, N. AND SHINOHARA. (1987). Polysaccharide products by *Aureobasidium pullulans*. Treatment of various industrial waste waters with agglutinating polysaccharide. *Bulletin of the Faculty of Agriculture, Miyazaki University*, **34(1)**, 221–228.

GIBBS, P.A. AND SEVIOUR, R.J. (1966). Does the agitation rate and/or oxygen saturation influence exopolysaccharide production by *Aureobasidium pullulans* in batch culture? *Applied Microbiology and Biotechnology* **46**, 503–510.

GLUCKSMAN, M.A., REUBER, T.L. AND WALKER, G.C. (1993). Genes needed for the modification, polymerisation, export and processing of succinoglycan by *Rhizobium melitoti*: a model for succinoglycan biosynthesis. *Journal of Bacteriology* **175**, 7045–7055.

HAMADA, N. AND TSUJISAKA, Y. (1983). The structure of the carbohydrate moiety of an acidic polysaccharide produced by Aureobasidium sp. K-1. *Agriculture and Biological Chemistry* **47**, 1167–1172.

HAN, Y.W. AND ANDERSON, A.W. (1973). Semisolid fermentation of ryegrass straw. *Applied Microbiology* **30**, 930–934.

HAYASHIBARA BIOCHEMICAL LABS, INC. (1990). *Pullulan: Features and applications.* Company Bulletin.

HEMMING, F.W. (1974). *MTP International Review of Science. Biochemistry. Series One* **186**, 411–421.

HERMANIDES-NIJHOF, E.J. (1977). *Aureobasidium* and related genera. In *The black yeasts and allied hyphomycetes.* Eds. G.S. de Hoog and E.J. Hermanides-Nijhof. Study of Mycology Number 15, pp 141–181.

HERMENTIN, P., DOENGES, R., WITZEL, R., HOKKE, C.H., VEOGENTART, J.F.G., KAMERNING, J.P., CONDRADT, H.S., NIMTZ, M. AND BRAZEL, D. (1994). Strategy for mapping N-glycans by high performance capillary electrophoresis. *Analytical Biochemistry* **221**, 29–41.

HUDSON, H.J. (1965). An ascomycete with *Aureobasidium pullulans*-type conidia. *Nova Hedwigia* **10**, 319–328.

IMSHENETSKY, A.A., KONDRATYEVA, T.F. AND SMUTKO, A.N. (1981). The effect of acidity of the medium aeration and temperature on pullulan biosynthesis by *Aureobasidium pullulans* polyploid strains. *Microbiologia* **50(3)**, 771–775.

IMSHENETSKY, A.A., KONDRATYEVA, T.F. AND SMUTKO, A.N. (1981). The effect of carbon and nitrogen sources on pullulan biosynthesis by the polyploid strains of *Pullularia pullulans*. *Microbiologia* **50(1)**, 102–105.

INIOTAKIS, N., ISRAILIDES, C., KATSABOXAKIS, K., MICHAILIDES, P., ICONOMOU, D. AND PAPANICOLAOU, D. (1991). Ecological and economical utilisation of waste water from olive oil production with physicochemical and biotechnological methods. In *Treatment and use of sewage sludge and liquid agricultural wastes.* Ed. P.L. Hermite, pp 393–399. London and New York: Elsevier Applied Science Publishers.

ISRAILIDES, C.J. (1979). Microbial enhancement of acid treated rye grass straw for animal feed *Doctoral Thesis.* Univ. Microfilms International 300, North Road. Ann Arbor. Michigan 48100. Ref. No. 7927253, pp 102–107.

ISRAILIDES, C.J., HAN, Y.W. AND ANDERSON, A.W. (1981). Process improving the palatability of straw as animal feed. *U.S. Patent* No.4243686. Jan 1981.

ISRAILIDES, C., SMITH, A. AND BAMBALOV, G. (1993). Production of pullulan from agro-industrial wastes. *Proceedings of the Sixth European Congress on Biotechnology* **2**, 362 Firenze, Italy.

ISRAILIDES, C., BOACKING, M., SMITH, A. AND SCANLON, B. (1994a). A novel rapid coupled enzyme assay for the estimation of pullulan. *Biotechnology and Applied Biochemistry* **19**, 285–291.

ISRAILIDES, C., SCANLON, B., SMITH, A., JUMEL, K. AND HARDING, S.E. (1994b). Characterization of pullulans produced from agro-industrial wastes. *Carbohydrate Polymers* **25**, 203–309.

ISRAILIDES, C.J., VLYSSIDES, A.G. MOURAFERI, V.N. AND KARVOUNI, G. (1977). Olive oil waste water treatment with the use of an electrolysis system. *Bioresource Technology* **61**, 163–170.

ISRAILIDES, C.J., SMITH, A., HARTHILL, J.E., BARNETT, C., BAMBALOV, G. AND SCANLON, B. (1998). Pullulan content of the ethanol precipitate from fermented agro-industrial wastes. *Applied Microbiology and Biotechnology* **49**, 613–617.

KAPLAN, D.L., MAYER, J.M., BALL, D., McCASSIE, J., ALLEN, A.L. AND STENHOUSE, D. (1993). Fundamentals of biodegradable polymers. In *Biodegradable Polymers and Packaging.* Eds. C. Ching, D.Kaplan, and E.Thomas, pp 1–42. Lancaster PA: Technomic Publishing Co Inc.

KELLER, M., ROXLAU, A., WENG, W.W., SCHMIDT, M., QUANDT, J., NIEHSAUS, K., JORDING, D., ARNOLD, W. AND PUHLER, A. (1995). Molecular analysis of the *Rhizobium melitoti mucR* gene regulating the biosynthesis of exopolysaccharides succinoglycan and galactoglycan. *Molecular Plant-Microbe Interactions* **8(2)**, 267–277.

KOCKOVA-KRATOCHILOVA, A., GERNAKOVA, A. AND SLAVIKOVA, E. (1980). Morphological changes during the life cycle of *Aureobasidium pullulans*. (de Bary) Arnaud. *Folia*

Microbiologica **25**, 56–67.

KROCHTA, J.M. AND DE MULDER-JOHNSTON, C.L.C. (1997). Edible and Biodegradable polymer films. *Food Technology* **51(2)**, 61–74.

LACROIX, C., LE DUY, A., NOL, G. AND CHOPLIN, L. (1985). Effect of pH on the batch fermentation of pullulan from sucrose medium. *Biotechnology and Bioengineering* **27**, 202–207.

LE DUY, A. AND BOA, J.M. (1983). Pullulan production from peat hydrolysate. *Canadian Journal of Microbiology* **29**, 143–146.

LE DUY, A., YARMOFF, J.I. AND CHAGRAU, A. (1983). Enhanced production of pullulan from lactose by adaptation and mixed culture techniques. *Biotechnology Letters* **5(1)**, 49–54.

LEE, K. Y. AND YOO, Y.J. (1993). Optimization of pH for high molecular weight pullulan. *Biotechnology Letters* **15**, 1021–1024.

LEHLE, L. AND TANNER, W. (1978). Biosynthesis of large dolichyl diphosphate-linked oligosaccharides in *Saccharomyces cerevisiae*. *Biochim. Biophys. Acta* **539**, 218–229.

LINGAPA, Y., SUSSMAN, A.S. AND BERNSTEIN, I.A. (1963). Effect of light and media upon growth and melanin formation in *Aureobasidium pullulans* (De Bary) ARN. *Mycopathologia et Mycologia Applicata* **20**, 109–128.

MADI, N.S., MCNEIL, B. AND HARVEY, L.M. (1996). Influence of culture pH and aeration on ethanol production and pullulan molecular weight. *Journal of Chemical Technology and Biotechnology* **65**, 343–350.

MATSUNAGA, H., FUJIMURA, S., NAMIOKA, H., TSUJI, K. AND WATANABE, M. (1976). Fertilizer Composition containing pullulan (Sumitomo Chem. Co., Ltd., Hayashibara, Biochem. Lab., Inc.) Ger. Offen 2607347 (CI.CO5 G3/00), 9 Sept. 1976, Japan. Appl. 75/25, 310, 24 Feb. 1975, pp 23.

MCNEIL, B. AND KRISTIANSEN, B. (1987). Influence of impeller speed upon the pullulan fermentation. *Biotechnology Letters* **9(2)**, 101–104.

MCNEIL, B. AND B. KRISTIANSEN. (1990). Temperature effects on polysaccharide formation by *Aureobasidium pullulans* in stirred tanks. *Enzyme Microbial Technology* **12**, 521–526.

MOSCOVICI, M., IONESCU, C., ONISCU, C., FOTEA, O., PROTOPOPESCU, P. AND HANGANU, L.D. (1996). Improved exopolysaccharide production in fed-batch fermentation of *Aureobasidium pullulans* with increased impeller speed. *Biotechnology Letters* **18**, 787–790.

POLLOC, T.J., THORNE, L. AND ARMENTROUT, R.W. (1992). Isolation of new *Aureobasidium* strains that produce high molecular weight pullulan with reduced pigmentation. *Applied and Environmental Microbiology* **58(2)**, 877–883.

RAMOS, S. AND GARCIA-ACHA, I. (1975). A vegetative cycle of *Pullularia pullulans*. *Transactions of the British Mycological Society* **64**, 19–135.

REESLEV, M., NEILSEN, J.C., OLSEN, J., JENSEN, B. AND JACOBSEN, T. (1991). Effect of pH and the initial concentration of yeast extract on regulation of dimorphism and exopolysaccharide formation of *Aureobasidium pullulans* in batch culture. *Mycology Research* **95**, 220.

RHO, D. MULCHANDANI, A., LUONG, J.H.T. AND LE DUY, A. (1988). Oxygen requirement in pullulan fermentation. *Applied Microbiology and Biotechnology* **28**, 361–366.

ROUKAS, T. AND BILIADERIS, C. (1995). Evaluation of carob pod as a substrate for pullulan production by *Aureobasisium pullulans*. *Applied Biochemistry and Biotechnology* **55**, 27–44.

SAHA, B.C. AND BOTHAST, R.J. (1993). Starch conversion by amylases from *Aureobasidium pullulans*. *Journal of Industrial Microbiology* **12**, 413–416.

SEVIOUR, R.J. AND KRISTIANSEN, B. (1983). Effect of ammonium ion concentration on polysaccharide production by *Aureobasidium pullulans*. *European Journal of Applied Microbiology and Biotechnology* **17**, 178.

SHABTAI, Y. AND MUKMENEV, J. (1995). Enhanced production of pigment-free pullulan by a morphogenetically arrested *Aureobasidium pullulans* (ATCC 42033) in a two – stage fermentation with shift from soybean oil to sucrose. *Applied Microbiology and Biotechnology* **43**, 595–603.

SHIN, Y.C., KIM, Y.H, LEE, H.S., CHO, S.J. AND BYUM, S.M. (1989). Production of exopolysaccharide pullulan from inulin by a mixed culture of *Aureobasidium pullulans* and *Kluyveromyces fragilis*. *Biotechnology and Bioengineering* **33**, 129–133.

SIEHR, D.J. (1981). Melanin biosyntheis in *Aureobasidium pullulans*. *Journal of Coatings Technology* **53**, 23–25.

SIMON, L., CAYE-VAUGIEN, C. AND BOUCHONNEAU, M. (1993). Relationship between pullulan production, morphological state and growth conditions in *Aureobasidium pullulans*: new observations. *Journal of General Microbiology* **139**, 979–985.

SIMON, L., BOUCHET, B., CAYE-VAUGIEN, C. AND GALLANT, D.J. (1995). Pullulan elaboration and differentiation of the resting forms in *Aureobasidium pullulans*. *Canadian Journal of Microbiology* **41**, 35–45.

TAGUCHI, R., SAKANO, Y., KIKUCHI, Y., SAKUMA, M. AND KOBAYASHI, T. (1972). Synthesis of pullulan by acetone-dried cells and cell-free enzyme from *Pullularia pullulans* and the precipitation of lipid intermediates. *Agricultural and Biological Chemistry* **37(7)**, 1635–1641.

THORNWELL, S.J., PERRY, R.B. AND SKATRUD, P.L. (1995). Integrative and replicative genetic transformation of *Aureobasidium pullulans*. *Current Genetics* **29**, 66–72.

VILAR-PILASI, C. AND GUINOVART, J.J. (1997). The role of glucose 6-phosphate in the control of glycogen synthetase. *FASEB Journal* **11**, 544–558.

WECKER, A. AND ONKEN, U. (1991). Influence of dissolved oxygen concentration and shear rate on the production of pullulan by *Aureobasidium pullulans*. *Biotechnology Letters* **13(3)**, 155–160.

WEST, T.P. AND REED-HAMER, B. (1993). Effect of pH on pullulan production relative to carbon source and yeast extract composition of growth medium. *Microbios* **75**, 75–82.

YUEN, S.(1974). Pullulan and its applications. *Process Biochemistry* **9**, 7–22.

ZAJIC, J.E., HO, K.K. AND KOSARIC, N. (1979). Growth and pullulan production by *Aureobasidium pullulans* on spent sulphite liquor. *Developments in Industrial Microbiology* **20**, 631–639.

14
Xylans of Industrial and Biomedical Importance

ANNA EBRINGEROVÁ* AND ZDENA HROMÁDKOVÁ

Institute of Chemistry, Slovak Academy of Science, 842 38 Bratislava, Slovak Republic

Introduction

Xylans are the second most abundant biopolymer in the plant kingdom. They are the most common hemicellulose as well as the major non-cellulosic cell wall polysaccharide of angiosperms, grasses and cereals, where they exist in many different compositions and structures (Stephen, 1983). In terrestrial plants, xylans have a variety of side chains attached to the linear β-(1,4)-D-xylopyranan backbone. They include mainly single α-L-arabinofuranosyl and α-D-glucopyranosyl uronic acid (and its 4-O-methyl ether) units. In addition, rhamnose, xylose, galactose, glucose and a variety of di- and trimeric side chains, next to acetyl groups and phenolic acids, like ferulic and coumaric acid, have been identified. Some families of green and red algae utilize xylans in their architecture. These skeletal xylans are homoglycans with β-(1,3)- or mixed β-(1,3;1,4) linkages (Painter, 1983). The extent of our knowledge on the role of xylans in cell walls and their biosynthesis has recently been reviewed by Gregory *et al.* (1998). The authors summarized novel results about the localization of xylans in cell walls and their interactions with other cell wall constituents. They dealt in detail with the influence of xylans on pulping and bleaching in connection with the application of enzymic treatments, the prospects for genetic engineering of lignification in tree species, cloning genes of xylan biosynthesis and xylan manipulation.

In the current trend for a complex and more effective utilization of biomass, increasing attention has been paid during the last few years to the exploitation of xylans as biopolymer resources. Xylans are available in very large amounts in organic wastes from renewable forest, and agricultural residues such as wood meal and shavings, stems, stalks, hulls, cobs, husks, etc. They can be relatively easily extracted from biomass. Nowadays, algal xylans have also been included in biopolymer research. However, the potential of xylans has not yet been completely realized. The great variety of xylan structures within even a single plant (Stephen, 1983; Neto *et al.*, 1997) makes their individual use difficult. An understanding of this diversity and more

*To whom correspondence may be addressed.

Biotechnology and Genetic Engineering Reviews – Vol. 16, April 1999
0264–8725/99/16/325–346 $20.00 + $0.00 © Intercept Ltd, P.O. Box 716, Andover, Hampshire SP10 1YG, UK

detailed knowledge of the molecular structure, physico-chemical and functional properties is necessary if an effective use of this resource is to be even partially achieved. The previous reviews on the isolation, modification, structure and properties of xylans contain data from about 1970 (Dudkin *et al.*, 1991; Stscherbina and Philipp, 1991; Ebringerová, 1992; Visser *et al.*, 1992). Evidently, xylan biopolymers have a very wide variety of direct food and non-food applications. More importantly, the modification or derivatization of these molecules creates novel opportunities to maximally exploit the various valuable properties of xylans for previously unperceived applications. However, to date only a relatively few attempts have been made to commercialize xylans. An exception is the highly branched heteroxylan from corn hulls, a by-product of starch production, which was tried on the market as a new food gum many years ago (Whistler, 1989), without success. This xylan is now being reinvestigated (Hromádková and Ebringerová, 1995; Saulnier *et al.*, 1998).

This present review attempts to describe the recent advances in extraction, modification and characterization of the structure and properties of xylans and xylan derivatives and to give an overview of the perspectives these polymers can offer in various technologies and non-technical applications. Because topics concerning cereal arabinoxylans and rye arabinoxylans, in particular, have been relatively recently considered elsewhere (Izydorczyk and Biliaderis, 1995; Vinkx and Delcour, 1996), the present article will focus on the growing wealth of new or previously unreported data.

Xylan sources and extraction

Corn hulls (Chanliaud *et al.*, 1995) are a conventional source of xylan: however, there are other potential xylan sources available in high amounts such as *sunflower hulls*, a by-product of sunflower-oil production (Bazus *et al.*, 1993), *sweet sorghum stalks* (Billa *et al.*, 1997), and *husks of red gram* (Swamy and Salimath, 1990). Xylans have also been isolated from steamed bamboo grass (Aoyama and Seki, 1994; Aoyama *et al.*, 1995), ramie fibres (Bhaduri *et al.*, 1995), olive pulp (Coimbra *et al.*, 1994), fibres of *Hibiscus cannabinus* (Neto *et al.*, 1996), sisal (Stewart *et al.*, 1997) and flax (Van Hazendonk *et al.*, 1996) and from pressure-refined wheat straw (Sun *et al.*, 1998).

Problems associated with the liberation of the xylan component from wood and annual plant cell walls are still under investigation in connection with the delignification process or bioconversion of lignocellulosic wastes. Moreover, suitable extraction procedures for a potential commercial production of polymeric xylans have to be developed. For the isolation of xylan from hardwoods, a combination of alkaline extraction and steam treatment (Košíková and Ebringerová, 1991; Ishihara *et al.*, 1996) as well as the application of aqueous ammonia (Ebringerova and Hromádková, 1996a) have been proposed. The use of an extruder-type twin-screw reactor makes the extraction more feasible (N'Diaye *et al.*, 1996). The extractibility of xylan from annual plants is easier in comparison to that of wood xylan due to the lower amounts and different structure of lignin. It can be affected by the alkali type and conditions (Lawther *et al.*, 1996) and improved by a multistep mechanical-chemical treatment, important in the case of straw and similar materials (Papatheofanus *et al.*, 1998). The mechanochemical effect of ultrasonication on the cell wall material during alkaline extraction of annual plants was shown to be very effective. Higher yields of xylan can be achieved at lower temperatures and shorter extraction times (Hromádková *et al.*, 1997; 1999). Due to the great variety of xylans and other plant constituents in the case of cereal grains, multistep extraction and purification procedures have been proposed (Izydorczyk and Biliaderis, 1995; Hromádková and Ebringerová, 1995). Barium hydroxide was reported to be an effective tool in the fractional extraction of arabinoxylans from wheat flour (Gruppen *et al.*, 1991), rye grain (Nilsson *et al.*, 1996) as well as sorghum endosperm (Verbruggen *et al.*, 1995). Recently, the isolation and purification of arabinoxylan from cereal brans and flours has been performed in a pilot scale operation and the conditions subsequently optimized (Annison *et al.*, 1992; Chanliaud *et al.*, 1995; Faurot *et al.*, 1995; Bataillon *et al.*, 1998).

The extractibility of xylans is associated with their interactions with the other cell wall constituents. In woody tissues, xylan is usually ester-linked through the glucuronic acid side chains to lignin (Fengel, 1984). Multiple forms of bonding between lignin and arabinoxylan in the cell wall of graminaceous plant tissues has been reported (Wallace *et al.*, 1995). The results from these studies indicate that lignin polymers are attached to arabinosyl and xylosyl residues by both ester and aryl-ether linkages. In another recent study, Ralph *et al.* (1995) have demonstrated the active incorporation of ferulate polysaccharide esters into ryegrass lignin. Ferulic acid is a widespread component of grass and cereal cell walls (Grabber *et al.*, 1995; Saulnier *et al.*, 1995a; Wende and Fry, 1997a,b; Ishi, 1997; Lempereur *et al.*, 1997). Its presence in the arabinoxylan chains provides some potential for the covalent interaction of xylan with other phenolic acid-containing cell wall polymers. An arabinoxylan-protein complex has been isolated from rye bran (Ebringerová *et al.*, 1994a). Linkages to structural cell wall proteins have been claimed to be the cause of insolubility of maize bran heteroxylan (Saulnier *et al.*, 1995b). However, the precise role of the small levels of protein in annual plant xylans and the nature of their interactions are still unclear. Glucuronoxylan-xyloglucan complexes were isolated from olive pulp (Coimbra *et al.*, 1995) and a glucuronoxylan-pectin complex, rich in arabinosyl and rhamnosyl units, from beechwood (Hromádková *et al.*, 1996). The nature of the linkages was not, however, established. Recently, the existence of ether linkage between arabinogalactan type II chains and a β-(1,4)-xylan backbone has been published (Kwan and Morvan, 1998). Also, the demonstration of the presence of xylose in pectin of pea hulls (Renard *et al.*, 1997) has indicated a close association between xylan and pectin polymers in cell walls.

Structural features

The detailed structural characteristics of arabinoxylans present in the main cereals of commercial importance, namely wheat, rye, barley, oat, rice and sorghum have been presented in the review articles of Izydorczyk and Biliaderis (1995) and Vinkx and Delcour (1996). Since the appearance of those articles, more recent studies have been directed to the water-inextractable arabinoxylans (Nilsson *et al.*, 1996; Harkone *et al.*, 1997) which have similar but greater bread-improving properties than their water extractable components (Vinkx and Delcour, 1996). The highly branched 4-O-methylglucuronoxylan, isolated from the seed coat mucilage *of Hyptis suaveolens* has been reported to have, in addition, 2-O-L-fucopyranosyl-D-xylopyranose side chains linked at position O-3 (Aspinall *et al.*, 1991). The water-soluble, neutral arabinoxylan, isolated from the leaves of *Litsea gardneri* has, in contrast to cereal arabinoxylans, 2-linked β-arabinofuranosyl units in terminal and internal positions (Wimalasiri and Kumar, 1995). From *Pasteurella multorida,* an extracellular β-(1,4)-D-xylan has been isolated together with hyaluronic acid (Rosner *et al.*, 1992). The extracellular β-(1,4)-D-xylan present in the cell-suspension of *Silene alba* was shown to carry etherically linked arabinogalactan type II chains of different size (Kwan and Morvan, 1995). Further information on the structural features of mixed-linked xylans isolated from various seaweeds has been obtained by characterization of derived β-1,3-xylooligo-saccharides using matrix-assisted laser-desorption ionization time-of-flight mass spectrometry (Yamagaki *et al.*, 1996) and by ^{13}C NMR spectroscopy studies of the native (Fukishi *et al.*, 1988; Matulewicz *et al.*, 1992; Yamagaki *et al.*, 1997a) and sulfated xylans (Yamagaki *et al.*, 1997b).

The distribution pattern of side chains in heteroxylans is an important feature affecting their solubility, interactions with other polymeric cell wall substances, degradability by enzymes, solution behaviour and other functional properties. It is suggested to be non-random and may reflect, together with the variety of primary structural features, the functional diversity of xylan in plants. Enzymic studies on the distribution of 4-O-methylglucuronic acid residues in glucuronoxylan from sunflower hulls indicated a regular pattern (Bazus *et al.*, 1992). Similarly, a regular distribution pattern was established by a physical method for the 4-O-methylglucuronoxylans of the herbal plants *Althaea officinalis* (Kardošová, 1990) and *Rudbeckia fulgida* (Kardošová *et al.*, 1998). This is in contrast to a rather blockwise distribution suggested for hardwood glucuronoxylans (Kohn *et al.*, 1985). Using xylan-degrading enzymes of known mode of action, structural models describing the substitution pattern of arabinosyl side chains in cereal arabinoxylans were created (Gruppen *et al.*, 1993; Vinkx and Delcour, 1996). Similarly, as wheat arabinoxylans, also those of barley, malt and sorghum showed a non-random distribution pattern. Isolated unsubstituted xylose residues are separated by one or two substituted residues and this pattern is interrupted with longer unsubstituted sequences (Vietor *et al.*, 1994; Cleemput *et al.*, 1995).

Examination of naturally occurring 1,4-linked xylans in plant cell walls and gums have indicated a three-fold, left-handed helical structure (Atkins, 1992). This structure was confirmed by both X-ray diffraction and conformational analysis in the case of the arabinoxylan from rice endosperm flour (Yui *et al.*, 1995). However, such structure does not seem to be a desirable conformation to make a complex firmly associated with cellulose or xyloglucan present in the cell walls, although the existence of such interactions are documented (Attala *et al.*, 1993).

Physicochemical properties

The molecular weights reported for cereal arabinoxylans vary depending on the method of their estimation. This problem was discussed in the mentioned reviews on cereal arabinoxylans (Izydorczyk and Biliaderis, 1995; Vinkx and Delcour, 1996). For water-extractable arabinoxylans, the values of molecular weights obtained by ultracentrifugation are much lower than those obtained by gel filtration methods. Extremely high values (~5000 kDa) were also obtained by light scattering. These results emphasize the difficulties in accurately measuring the molecular weight of asymmetric molecules by the two last mentioned methods. Chain aggregation was suggested to be partially responsible for the large variation in the estimates of molecular weight and also the presence of undissolved microgel particles.

As a result of a rather stiff conformation, arabinoxylans exhibit very high intrinsic viscosity. It is structure-dependent and was reported to be related more strongly to the content of di-substituted xylose units than to the content of monosubstituted units (Izydorczyk and Biliaderis, 1995). The water-soluble arabinoglucuronoxylan from corn cobs (Ebringerová *et al.*, 1992,), having a much lower substituted backbone (DS ~0.25), adopted an extended wormlike conformation what was confirmed also by ultracentrifugation (Dhami *et al.*, 1995). The unexpected higher viscosity of the low-branched arabinoxylan (Ara/Xyl 0.14) in comparison to that of its higher-substituted (Ara/Xyl 0.78) counterpart (Ebringerová and Hromádková, 1992) indicate that the behaviour of arabinoxylans in solution would be influenced not only by the asym-metrical conformation or the DP, but also by the type and arrangement of the substituents along the xylan backbone. Static and dynamic light scattering were used to determine the macromolecular features of corn bran heteroxylans (Chanliaud *et al.*, 1996). After elimination of aggregates by filtration, the weight-average molecular weight values estimated were about 270 and 370 kD. The structural parameters indicate a compact structure of the rigid polymers. The corn bran heteroxylans behave as typical electrolytes and have a rather homogeneous repartition of the charges along the macromolecules (Chanliaud *et al.*, 1997).

The molecular weight distribution of xylans is usually determined by size exclusion chromatography using dextran or pullulan standards for calibration. Most xylans are polydisperse and often have a high molecular weight component (HMC) eluting near to the void volume. In the case of a rye bran arabinoxylan (Ebringerová *et al.*, 1994a), this fraction was shown to be linked to protein. The nature of HMC in beechwood xylan and corn cob arabinoglucuronoxylan may be associated with residual lignin and/or protein, respectively. During degradation of the xylans from corn cobs and corn hulls by ultrasonication, the HMC fraction gradually disappeared and molecular chains of the same size as that of the main component were generated before the mean molecular weight shifted to lower values (Ebringerová *et al.*, 1997; Ebringerová and Hromádková, 1997). The results indicate that this fraction represent rather supramolecular structures than solubilized molecular chains.

The molecular weight determination of lower substituted heteroxylans which are either poorly soluble or even completely insoluble in the commonly used poly-saccharide solvents, is still an unsolved problem. The solution properties of a water-insoluble, low-branched arabinoxylan from rye bran in various solvents have been studied by viscosity and light scattering techniques (Ebringerová *et al.*, 1994b) as a function of time over a period of more than three years. The results suggest that the xylan, and probably other low-branched xylan types, had been isolated either as single strands or at most dimerized strands. These structures have a high tendency to aggregate and form clusters with time. Complexing solvents, used for cellulose dissolution, only dissolve the xylans down to a 6–7 stranded bundle.

The rheological properties of xylans play an important role in many practical applications. The water-insoluble low-substituted 4-O-methylglucuronoxylan, isolated from beech sulphite pulp, forms thixotropic aqueous dispersions of high apparent viscosity at rest which decreased by application of low shear rates (Lenz *et al.*, 1986).

Aqueous dispersions of this xylan type isolated from beechwood exhibit substantial shear thinning, typical of pseudoplastic materials. At higher concentrations and in dependence on the proportion of the water-insoluble fraction, they behave as plastic materials (Hromádková and Ebringerová, 1991; 1993). Whereas, the water-soluble xylans from corn cobs and rye bran show only weak or no thixotropy, their respective water-insoluble counterparts exhibit strong thixotropy and distinct plastic behaviour at lower concentrations (Ebringerová et al., 1992; Ebringerová and Hromádková, 1992). The interactions between the insoluble but swollen particles seem to produce a stronger intrinsic structure than the solubilized xylan chains. As has been established from viscoelastic measurements (Ebringerová et al., 1998a), the beechwood xylan of higher viscosity is able to form gel-like systems, whereas the mechanical spectra of the water-soluble corn cob xylans indicate a 'weak-gel' character and those of both rye bran and corn hull xylans are typical of liquid systems. Rheological studies on wheat arabinoxylans in relation to the structure and molecular size (Izydorczyk and Biliaderis, 1995) have shown that they are shear thinning and exhibit two critical concentrations which correspond to onset of coil overlap among the polymer chains. The existence of three domains provides additional evidence for a rigid, rod-like conformation of arabinoxylans in solution.

Functional properties

Many still unresolved problems in pulping and bleaching of pulps are connected with the xylan component of the plant sources. Knowledge of the distribution of xylan and lignin (Purina et al., 1991), their reactions during pulping (Imai et al., 1997; Buchert et al., 1995) and the molecular weight distribution of xylan/lignin complexes in pulps (Yokota et al., 1995) may help to a better understanding of the xylanase pre-bleaching process. Recently, the importance of xylans in xylanase-based bleaching technologies has been reviewed (Gregory et al., 1998). The brightness reversion of kraft pulps is closely related to the presence of residual lignin and oxidatively modified poly-saccharides (Buchert et al., 1997). Of particular interest are the xylan-derived chromophores affecting the xylanase pre-bleaching process (Wong et al., 1995). The degradative reactions of glucuronoxylan during kraft pulping give rise to novel uronic acid units (Teleman et al., 1996) as well as to formation of hexaneuronic acid groups (Teleman et al., 1995) which contribute to the kappa number of pulps (Li and Gellerstedt, 1997). The beneficial effect of some xylans in papermaking was confirmed in the case of ramie hemicellulose that might be used as a beater additive (Bhaduri et al., 1995).

Xylans contribute to the effects of dietary fibre upon some biochemical and physiological processes in human and animal organisms (Asp et al., 1993; Hromádková and Ebringerová, 1994; Chesson, 1995; Baghurst et al., 1996). The best documented physiological effects of cereals, which represent the most abundant xylan fibre sources, are the faecal bulking effect and the lowering of blood cholesterol and decrease of postprandial glucose and insulin responses. These effects have been connected with the viscous character of the fibre polysaccharides. Water-extractable polysaccharides of cereals were claimed to alleviate alcoholic liver disorder (Aoe et al., 1992). However, only fragmentary knowledge is available on the mode of action of the xylan component of foods. Also, the possible contribution to the observed physiological effects of some of the xylan constituents – such as ferulic acid, which is an effective scavenger of free radicals and potential anti-carcinogen (Garcia-Conessa *et al.*, 1997) – needs to be studied. Algal polysaccharides used as foodstuffs are a new source of dietary fibre. From this point of view, the polysaccharides, including xylans, of *Palmaria palmata* have been characterized by chemical and physicochemical methods, and *in vitro* fermentation tests (Lahaye *et al.*, 1993; Bentoulmu *et al.*, 1997; Bentoulmu and Cherbut, 1997; Rochet and Bernalier, 1997).

A great deal of effort has been made to investigate the role of xylans in bread-making. Recent reviews on cereal arabinoxylans (Izydorczyk and Biliaderis, 1995; Vinkx and Delcour, 1996) have shown that the xylan component of cereal is primarily responsible for the effects on the mechanical properties of dough as well as the texture and other end-product quality characteristics of baked products. Many of these effects have been studied by the addition of pentosan or purified arabinoxylans to wheat flour, such as the increase of water absorption of dough, development of loaf volume, and texture of bread crumbs. Lenz *et al.* (1986) demonstrated the valuable effects of a water-insoluble beechwood xylan, the by-product of viscose production, on dough preparation and properties. In the reviews on arabinoxylans (Izydorczyk and Biliaderis, 1995; Vinkx and Delcour, 1996), attention was paid also to the importance of oxidative gelation of wheat and rye arabinoxylans in relation to bread-making. The gelation results from the cross-linking reactions of the ferulic acid component of arabinoxylans (Wallace and Fry, 1995; Ng *et al.*, 1997; Greenshields and Waldron, 1997). Recently, the effects of endogeneous arabinoxylan hydrolysing enzymes during breadmaking and the changes in molecular weight distribution and solubilization of wheat flour arabinoxylans has been reported (Cleemput *et al.*, 1997). The effect of oxidizing agents, enzymes, ferulic acid and cysteine on rheological properties of the water-soluble xylan-rich polysaccharides of whole grain rye flour was studied in relation to the baking quality of the flour (Girhammar and Nair, 1995).

A further useful functional property of arabinoxylans is their ability to retain gas in dough and protect protein foam against thermal disruption (Izydorczyk and Biliaderis, 1995). These effects were related to the viscosity and film-forming properties of arabinoxylans. However, the contribution of the protein component that is present in most preparations cannot be ruled out.

On a series of structurally different water-soluble heteroxylans, the surface active properties have been investigated using several tests (Ebringerová *et al.*, 1998a). The lowering of the surface tension of water as well as foamability of all tested xylans were low. However, all of them gave stable emulsions of the oil/water type and exhibit remarkable stabilizing effects on protein foam after heating. It should be noted that the xylans contain very small but distinct amounts of protein and/or phenolic substances which may have hydrophobic effects contributing to the observed surface active properties.

The contribution of xylans to the malting and brewing qualities of barley grains has not yet been properly elucidated. Only some evidence exists that technological problems during beer production such as impaired wort-filtration and haze-formation could in fact be associated with arabinoxylans and β-glucans (Izydorczyk and Biliaderis, 1995). Quite recently, the arabinoxylan present in barley and malt cell wall material (Vietor *et al.*, 1992; 1994) as well as in beer (Schwarz and Han, 1995; Han and Schwarz, 1996) have been characterized. As a source of lager-type beers, sorghum flour has been investigated from the viewpoint of changes in content and composition of sorghum sources varying in hardness (Kavitha and Chandrashekar, 1992) and of the presence of non-starch polysaccharides (Verbruggen *et al.*, 1993).

Arabinoxylans in cell wall of cereal grains inhibit the intercellular ice formation, ensuring winter survival of cereals (Kindel *et al.*, 1989). The anomalous behaviour of ice in solutions of ice-binding arabinoxylans has been reported by Williams (1992). Enhancement of viscosity and mechanical interference of the arabinoxylan gel network to the propagation of ice was suggested. These properties might be useful in the production of ice cream and frozen foods. A 'supergel' hemicellulose powder, substantially an arabinoxylan ferulate, produced from corn bran by GB Biotechnology Ltd (Greenshields and Rees, 1992), can be converted to a thermostable, cold-setting gel with peroxidase and hydrogen peroxide. The nature and extent of formed ferulate dehydrodimer cross-links was reported by Ng *et al.* (1997). The products may find applications in food and pharmaceutical industries. Similarly, the arabinoxylanbranan from corn bran, containing phenolic acids, represents a novel polysaccharide material *'Sterigel'*, useful as a wound management aid (Methacanon *et al.*, 1998). For application in other areas of biotechnological importance, a further xylan-based substrate for testing xylanases has been prepared (Chen and Buller, 1995). Thermoplastic xylan-rich polysaccharides have been obtained from *biomass* (Glasser *et al.*, 1996). Xylan can be used also as a filler in polypropylene composites (Amash and Zugenmaier, 1998).

Biologically active xylans

Arabinoglucuronoxylans possessing immunostimulating activities have been isolated from various herbal plants such as *Echinacea purpurea, Acanthopanax senticosus, Eleutherococcus senticosus, Eupatorium perfoliatum, Sabal serrulata, Chamomilla recutita*, and *Arnica montana* (Wagner *et al.*, 1985; Proksch and Wagner, 1987). The xylans from the last three mentioned herbs showed also antiphlogistic effects. From the bark of *Cinnamomum cassia* (Kanari *et al.*, 1989), an arabinoxylan relating to the reticuloendothelial system has been isolated. It comprises highly substituted β-1,4-D-xylan main chains bearing β-L-arabinopyranosyl units as single units as well as in disaccharide side chains. The acidic mucous polysaccharide isolated from the seed of *Plantago asiatica* has a highly branched, partially O-acetylated β-1,4-D-xylan backbone carrying terminal β-D-xylopranosyl units and acidic disaccharides side chains. It showed strong anti-complementary activity (Yamada *et al.*, 1985). Water-soluble, highly-branched arabinoxylans have been isolated from various Litsea species (Herath *et al.*, 1990; Wimalasiri and Kumar, 1995). The aqueous decoction of these plants is used in native medicine in Sri Lanka.

Some of the xylan-rich hemicelluloses isolated from annual plant wastes such as bamboo leaves, corn stalks, wheat straw, etc. (Whistler *et al.*, 1976) and the 4-O-methylglucuronoxylan from Japanese beechwood (Hashi and Takeshita, 1979) have been reported to inhibit the growth rate of sarcoma-180 and other tumours, probably due to the indirect stimulation of the non-specific immunological host defence. Carboxymethylated xylan-rich wood hemicelluloses (Fan and Feng, 1987) have been reported to activate T-lymphocytes and immunocytes and claimed as a new chinese anti-tumour drug. However, no conclusions were drawn about the structural and molecular features essential for the biological activity of the xylans.

A series of endotoxin-free heteroxylans differing in the primary structure and water-solubility, which had been isolated from beechwood meal, corn cobs, rye bran, and corn hulls, were investigated for their mitogenic and comitogenic activities *in vitro* thymocyte tests (Ebringerová *et al.*, 1995a). All the water-insoluble xylans were inactive, and solubilization of the xylans by introduction of carboxymethyl or quaternary ammonium groups had no positive effect on the activity. The highest response in both tests, comparable to that of the commercial immunomodulator, *Zymosan*, was manifested by the water-soluble arabinoglucuronoxylan (ws-AGX) isolated from corn cobs. Disaccharide side chains, comprising 3-linked 2-O-β-D-xylopyranosyl-α-L-arabinofuranosyl unit, are a basic feature of this xylan (Ebringerová *et al.*, 1992). They were estimated by NMR spectroscopy to be in much lower amounts in the less active corn hull xylan (Hromádková and Ebringerová, 1995). With increasing the content of the disaccharide branches of ws (water soluble)-AGX by enzymic treatments (Ebringerová *et al.*, 1995a), the response of the xylan in both mitogenic and comitogenic tests increased significantly. Molecular degradation of ws-AGX by ultrasonication in water and alkali, combined with a decrease in the proportion of the disaccharide side chains, had an adverse effect on the biological activity (Ebringerová *et al.*, 1997). The results indicated that the disaccharide side chains might be important for the expression of the biological activity of ws-AGX in the above mentioned tests. Application of other *in vitro* and *in vivo* tests are needed to gain further knowledge on the biological activity of this xylan-type which is widespread in grass cell walls (Stephen, 1983), and esterified with ferulic acid (Wende and Fry, 1997).

An interesting biological effect has been reported for xylan in plants. The aflatoxin inhibitory activity observed in the developing cotton seed has been suggested to be associated with a seed coat-specific xylan (Mellon *et al.*, 1995). To date, no further information is available on the xylan characteristics. However, the most promising biological effects were discovered in connection with attempts to substitute heparin by xylan sulfates in medical applications. This topic will be dealt with in the section on xylan derivatives which now follows.

Xylan derivatives

As in previous years, attempts have recently been made to modify the xylan component of lignocellulosic materials *in situ*. Alkylation of baggase (Šimkovic *et al.*, 1990), aspenwood meal (Antal *et al*, 1991) and corn cobs (Šimkovic *et al.*, 1992) with 3-chloro-2-hydroxypropyltrimethylammonium chloride (CHMAC) in aqueous alkali yielded water-extractable modified xylan-rich polysaccharides in yields up to 60% of the originally present xylan. The trimethylammonium-2-hydroxypropyl (TMAHP) xylan from aspen wood may be used as a *beater additive*. This substance doubled the beating resistance and increased significantly the tear strength of a bleached spruce organosolv pulp (Antal *et al.*, 1991). Recently, the *in situ* modification of the xylan component of lignocellulosic materials by esterification of sawdust with octanoylchloride was reported to yield esterified hemicelluloses in the liquid fraction (Thiebaud and Borredon, 1998).

A series of structurally different xylans have been modified by quarternary ammonium groups (Ebringerová *et al.*, 1994c). The TMAHP derivatives prepared from beechwood xylan and corn cob xylan, in particular, were shown to be useful in papermaking (Antal *et al.*, 1997). Both derivatives improved the strength properties of bleached hardwood kraft pulp and unbleached thermomechanical spruce pulp and increased the retention of fines. The cationic xylans exhibit antimicrobial activity against some Gram-negative and Gram-positive bacteria (Ebringerová *et al.*, 1995b). The biological activity increases with increasing degree of substitution (DS) and is strongly structure-dependent. Probably, the arrangement of the glycosyl and TMAHP substituents on the xylan chains (Ebringerová and Hromádková, 1996b) play a very important role in the interactions with the polymers of the bacterial cell wall surface. The introduction of TMAHP groups affected also the rheological properties of the xylan chains (Ebringerová *et al.*, 1993). At higher DS (> 0.5), the former shear-thinning xylans became dilatant, probably as a result of strong inter- and intramolecular interactions.

Water-soluble amphiphilic derivatives of beechwood xylan and its sulfoethyl derivative were obtained by introduction of low amounts of long alkyl chains using 1-bromododecane in aprotic solvent (Ebringerová *et al.*, 1998b). The derivatives exhibit excellent emulsifying properties and stabilized protein foam against thermal disruption. Except for the foamability, which was high only in the case of the C_{12}-glucuronoxylan derivative, both emulsifying and foam-stabilizing properties seem not to be significantly influenced by the primary structure of the parent xylan polymers. The modification of beechwood xylan (Ebringerová *et al.*, 1996) as well as other heteroxylans with p-carboxybenzyl bromide in aqueous alkali imparted water-solubility to xylans as well as moderate hydrophobic properties demonstrated by emulsifying and foam-stabilizing activities (Sroková *et al.*, 1997).

The neutral xylan with mixed β-(1,3; 1,4)-linkages, isolated from the red seaweed *Palmaria decipiens*, was oxidised with bromine in aqueous alkali (Jerez *et al.*, 1997). The product having carbonyl groups preferentially on C-2 was coupled with p-chloroaniline in heterogeneous medium to give a water-soluble stable Schiff base. Conjugates with bovine serum albumin were obtained by reductive amination. Periodate oxidation of the seaweed xylan introduced aldehyde functions which after reaction with p-chloroaniline (Barroso *et al.*, 1997) gave ligands for the coordination of Cu (II).

Xylans fully substituted with aromatic carbamate groups, obtained in good yields (Vincendon, 1993), were thermoplastic at high temperatures and decompose above 300°C. Recently, a novel alkylation procedure has been used to prepare thermoplastic 2,3-bis(benzyl ether) xylans in one step with a yield of 80% (Vincendon, 1998). They are soluble in most organic solvents and can be processed at high temperature. A new xylan-based insoluble dye substrate for screening and assay of xylan-degrading enzymes was prepared by crosslinking the Cibachron blue 3GA dyed xylan with 1,4-butanedioldiglycidyl ether (Lee and Lee, 1997).

A water-soluble xylan phosphate monoester has been prepared by phosphorylating the xylan via its thrimethyl silyl derivative (Schnabelrauch *et al.*, 1992). The anti-coagulant action of phosphorylated xylan and other polysaccharides was comparable to that of the sulfated polysaccharides (Dace *et al.*, 1997). Xylan polysulfates can find application in reagents as a non-specific binding blocker in ion-capture binding assay (Adamczyk *et al.*, 1993). A novel drug for prophylaxes and treatment of degenerative articular diseases is based on polysaccharides, including xylans, that have been substituted with non-aromatic long-chain esters and sulfate groups in the form of a physiologically tolerated cation (Raiss and Wiesner, 1992). Texas red-labelled xylan sulfate is useful as a novel fluorescent probe for the location of tumour cells in frozen sections of human colon tissues (Anees, 1996). It could also have potential as a vehicle for the transport of cytotoxic compounds to carcinoma cells of the colon.

Pentosan polysulfate (PPS), usually derived from beechwood glucuronoxylan, has been known as an anticoagulant for nearly thirty years in Europe. However, its range of biological activities is much broader, as documented in the increasing number of papers on this topic. The anticoagulant activity of PPS has been shown to be comparable to that of heparin (Doctor *et al.*, 1991; Kiesel *et al.*, 1991; Kloecking *et al.*, 1992; Hoffmann *et al.*, 1997). A synergistic, anticoagulant action was reported to exist between PPS and a lipoprotein-associated inhibitor (Wun, 1992). In contrast to sodium heparin, sodium PPS has a much higher delay of allergic skin reactions (Koch *et al.*, 1996). The PPS in gel form can be used in treatment of infusion trombophlebitides (Kollar *et al.*, 1994).

PPS antagonises the binding of the basic fibroblast growth factor (bFGF) to cell surface receptors and the evaluation of its anti-tumour activity in animal models and human tumour cell lines has been continued very intensively during the last years. Clinical trials of anti-angiogenic agents (Hawkins, 1995) pointed at PPS as a potential cancer chemotherapeutic agent. Phase-I studies have shown that the coagulation effect of PPS is the dose-limiting toxicity (Swain, *et al.*, 1995) and determined the tolerable duration of the treatment (Lush *et al.*, 1996). Pharmacokinetic analyses indicated marked accumulation of PPS upon chronic administration and thus PPS has been suggested to be more effective as an anti-cancer agent when it is given intermittently and on a weekly schedule (Marshall *et al.*, 1997). It suppresses prostate tumour growth *in vivo* (Pienta *et al.*, 1992; Nguyen *et al.*, 1993). Texas red-labelled PPS is suggested to be a potent inhibitor of colonic carcinoma (Anees, 1996).

PPS has been administrated to patients with aids-related kaposis-sarcoma (Schwartsmann *et al.*, 1996). As an inhibitor of bFGF and due to the lack of significant toxicity, PPS was suggested for further experiments. The antiviral activity of PPS has in fact been documented by several studies (Von Briesen, 1990; Holmes *et al.*, 1991; Schols *et al.*, 1992; Thormar *et al.*, 1995; Este *et al.*, 1996). PPS exhibits anti-metastatic and/or anti-inflammatory activities (Parish and Snowden, 1997). It is very efficient in the treatment of pain, urgency, and frequency associated with interstitial cystitis (Hwang *et al.*, 1997).

As a further effect of PPS, the inhibition of calcium oxalate crystal growth and prevention of aggregation which leads to formation of renal calculi has been reported (Fujisawa *et al.*, 1992; Senthil *et al.*, 1996). PPS decreases the cholesterol and triglyceride levels in the serum of stone forming rats (Shuba *et al.*, 1992) and may be a suitable alternative to heparin when used in conjunction with a triglyceride emulsion for the elevation of plasma free fatty acids before exercise of horses (Orme and Harris, 1997).

PPS was shown to be an anti-arthritic agent for dogs having chronic osteoarthritis (Rogachefsky *et al.*, 1994; Read *et al.*, 1996). Due to the relatively high molecular weight, the ability of PPS to enter connective tissues rich in proteoglycans and interact with the resident cells has been questioned. Laboratory studies (Francis *et al.*, 1993; Ghosh and Hutadilok, 1996) on PPS with a molecular weight of ~5 700 Da) indicated that this drug exhibits multiple actions, including the preservation of articular cartilage proteoglycans in animal models and stimulation of hyaluronan synthesis by synovial fibroblasts *in vitro* and *in vivo*. Those authors suggested that PPS first binds to the cell membrane and is then internalized. The data of recent clinical experiments on patients with osteoarthritis (Anderson *et al.*, 1997) indicate the ability of PPS to selectively recruit lymphocytes into the circulation and modulate the expression of peripheral blood mononuclear cell procoagulant activity. PPS affects the type I collagen synthesis by adult human dermal fibroblast (Ferao and Mason, 1993) and increases, similarly as the chondroprotective drug, *arteparon*, the collagenase activity (Nethery *et al.*, 1992). Despite the extensive study of the biological activities of PPS that has been undertaken over the last few decades, there is still little known about the mechanism of these now well documented effects.

Several patents protect the method of preparation of polysulfates (Wagenknecht *et al.*, 1992) as well as various pharmacological products like polyelectrolyte complexes in microparticulate form (Krone *et al.*, 1991), a pentosan polysulfate 'Elmiron' (Elliot *et al.*, 1997), preparations for inhibition of fibroblast proliferation (Gillespie, 1991), prevention against viral diseases (Diringer *et al.*,1991), irrigation of internal bladder surfaces in mammals (Parsons, 1992), and treatment of degenerative articular ailments (Raiss and Wiesner, 1992).

Concluding remarks

The number of reports that have been covered by this review indicate that, during the last decade, the importance of xylan-type polysaccharides as plant constituents and isolated polymers has significantly increased. Attention has been paid not only to the primary structure and differences in the fine structure of xylans in relation to their functional properties, but also to the physicochemical properties of xylan polymers and their derivatives. The variability in sugar constituents, glycosidic linkages structure of glycosyl side chains offer a number of possibilities for direct site-specific chemical and enzymic modifications. The usefulness of these materials in an industrial and biomedical context is now beyond dispute, and will hopefully stimulate further research into their characteristics.

Acknowledgement

The authors are grateful to the British Council and the Slovak grant agency VEGA (project number 2/4148) for their support.

References

ADAMCZYK, J., BERRY, D.S., FICO, R., JOU, Y.H. AND STROUPE, S.D. (1992). Reagents containing a nonspecific binding blocker in ion-capture binding assay. *International Application* WO 92 21,769 (*Chemical Abstracts* **118**, 143000).

AMASH, A. AND ZUGENMAIER, P. (1998). Study on cellulose and xylan filled polypropylene composites. *Polymer Bulletin* **40**, 251–258.

ANDERSON, J.M., EDELMAN, J. AND GHOSH, P. (1997). Effects of pentosan polysulfate on peripheral blood leukocyte populations and mononuclear cell procoagulant activity in patients with osteoarthritis. *Current Therapeutic Research Clinical and Experimental* **58**, 93–107.

ANEES, M. (1996). Location of tumor cells in colon tissue by texas red labeled pentosan polysulfate, an inhibitor of a cell surface protease. *Journal of Enzyme Inhibition* **10**, 203–208.

ANNISON, G., CHOCT, M. AND CHEETHAM, N.W. (1992). Analysis of wheat arabinoxylans from a large-scale isolation. *Carbohydrate Polymers* **19**, 151–159.

ANTAL, M., EBRINGEROVÁ, A. AND MICKO, M.M. (1991). Kationisierte Hemi-cellulosen aus Espenholzmehl und ihr Einsatz in der Papierherstellung. *Das Papier* **45**, 232–235.

ANTAL, M., EBRINGEROVÁ, A., HROMÁDKOVÁ, Z., PIKULÍK, I.I., LALEG, M. AND MICKO, M.M. (1997). Struktur und papiertechnische Eigenschaften von Aminoalkylxylanen. *Das Papier* **51**, 223–226.

AOE, S., ODA, H. AND TATSUMI, K. (1992). Water soluble polysaccharide extraction from cereals for alleviating alcoholic liver disorder. *Japan Kokai Tokyo Koho* JP 04,360,835. (*Chemical Abstracts* **118**, 132123).

AOYAMA, M. AND SEKI, K. (1994). Chemical characterization of solubilized xylan from steamed bamboo grass. *Holz als Roh-und Werkstoff* **52**, 388–388.

AOYAMA, M., SEKI, K. AND SAITO, N. (1995). Solubilization of bamboo grass xylan by steaming treatment. *Holzforschung* **49**, 193–196.

ASP, N.G., BJORCK, I. AND NYMAN, M. (1993). Physiological effect of cereal dietary fibre. *Carbohydrate Polymers* **21**, 183–187.

ASPINALL, G.O., CAPEK, P., CARPENTER, R.C., GOWDA, D.CH. AND SZAFRANEK, J. (1991). A novel L-fuco-4-O-methyl-D-glucurono-D-xylan from *Hyptis suaveolens*. *Carbohydrate Research* **214**, 107–113.

ATALLA, R.H., HACKNEY, J.M., UHLIN, I. AND THOMPSON, N.S. (1993). Hemicelluloses as structure regulators in the aggregation of native cellulose. *International Journal of Biological Macromolecules* **15**, 109–112.

ATKINS, E.D.T. (1992). Three-dimensional structure, interactions and properties of xylans. In *Xylans and Xylanases*. Eds. J. Visser, G. Beldman, M.A. Kusters van Someren and A.G.J. Voragen, pp 39–50. Amsterdam: Elsevier Science Publishers B.V.

BAGHURST, P.A., BAGHURST, K.I. AND RECORD, S.J. (1996). Dietary Fiber, non-starch polysaccharides and resistant starch – a review. *Food Australia* **48**, S3–S35.

BARROSO, N.P., COSTAMAGNA, J., MATSUHIRO, B. AND VILLAGRAN, M. (1997). The xylan from *Palmaria decipiens*: Chemical modification and formation of Cu(II). *Boletin de la Sociedad Chilena de Quimica* **42**, 301–306.

BATAILLON, M., MATHALY, P., CARDINALI, A.P.N. AND DUCHIRON, F. (1998). Extraction and purification of arabinoxylan from destarched wheat bran in a pilot scale. *Industrial Crops and Products* **8**, 37–43.

BAZUS, A., RIGAL, L., FONTAINE, T., FOURNET, B., GOSSELIN, M. AND DEBEIRE, P. (1992). Enzymic studies of the distribution pattern of 4-O-methylglucuronic acid residues in glucuronoxylan from sunflower hulls. *Bioscience, Biotechnology and Biochemistry* **56**, 508–509.

BAZUS, A., RIGAL, L., GASET, A., FONTAINE, T., WIERUSZESKI, J.-M. AND FOURNET, B. (1993). Isolation and characterisation of hemicelluloses from sunflower hulls. *Carbohydrate Research* **243**, 323–332.

BENTOULMOU, N., MEKKI, N., LAIRON, D. AND CHERBUT, C. (1997). Viscosity of algal fibres determines glycemic and insulemic postprandial response in healthy humans. *Reproduction, Nutrition and Development* **37**, 361.

BENTOULMOU, N. AND CHERBUT, C. (1997). Digestive effects of algal dietary fibres in humans. *Reproduction, Nutrition and Development* **37**, 356–357.

BHADURI, S.K., GHOSH, I.N. AND DEB SARKAR, N.L. (1995). Ramie hemicellulose as a beater additive in papermaking from jute-stick kraft pulp. *Industrial Crops and Products* **4**, 79–84.

BILLA, E., KOULLAS, D.P., MONTIES, B. AND KOUKIOS, E.G. (1997). Structure and composition of sweet sorghum stalk components. *Industrial Crops and Products* **6**, 297–302.

BUCHERT, J., TELEMAN, A., HARJUNPAA, V., TENKANEN, M., VIIKARI, L. AND VUORINEN, T. (1995). Effect of cooking and bleaching on the structure of xylan in conventional pine kraft pulp. *Tappi Journal* **78**, 125–130.

BUCHERT, J., BERGNOR, E., LINDBLAD, G., VIIKARI, L. AND EK, M. (1997). Significance of xylan and glucomannan in the brightness reversion of kraft pulps. *Tappi Journal* **80**, 165–171.

CHANLIAUD, E., SAULNIER, L. AND THIBAULT, J.F. (1995). Alkaline extraction and characterisation of heteroxylans from maize bran. *Journal of Cereal Science* **21**, 195–203.

CHANLIAUD, E., ROGER, P., SAULNIER, L. AND THIBAULT, J.F. (1996). Static and dynamic light scattering studies of heteroxylans from maize bran in aqueous solution. *Carbohydrate Polymers* **31**, 41–46.

CHANLIAUD, E., SAULNIER, L. AND THIBAULT, J.F. (1997). Heteroxylans from maize bran in aqueous solution. Part II: Studies of the polyelectrolyte behaviour. *Carbohydrate Polymers* **32**, 315–320.

CHEN, P. AND BULLER, C.S. (1995). Activity staining of xylanases in polyacrylamide gels containing xylan. *Analytical Biochemistry* **226**, 186–188.

CHESSON, A. (1995). Dietary fiber. In *Food Polysaccharides and Their Applications*. Ed. A.M. Stephen, p 547. New York: Marcel Dekker, Inc.

CLEEMPUT, G., VANOORT, M., HESSING, M., BERGMANS, M.E.F., GRUPPEN, H., GROBET, P.J. AND DELCOUR, J.A. (1995). Variation in the degree of D-xylose substitution in arabinoxylans extracted from a European wheat flour. *Journal of Cereal Science* **22**, 73–84.

CLEEMPUT, G., BOOIJ, C., HESSING, M., GRUPPEN, H. AND DELCOUR, J.A. (1997). Solubilisation and changes in molecular weight distribution of arabinoxylans and protein in wheat flours during bread-making, and the effects of endogenous arabinoxylan hydrolysing enzymes. *Journal of Cereal Science* **26**, 55–66.

COIMBRA, M.A., WALDRON, K.W. AND SELVENDRAN, R.R. (1994). Isolation and characterisation of cell wall polymers from olive pulp (*Olea europaea* L.). *Carbohydrate Research* **252**, 245–262.

COIMBRA, M.A., RIGBY, N.M., SELVENDRAN, R.R. AND WALDRON, K.W. (1995). Investigation of the occurrence of xylan-xyloglucan complexes in the cell-walls of olive pulp (*Olea europaea*). *Carbohydrate Polymers* **27**, 277–284.

DACE, R., MCBRIDE, E., BROOKS, K., GANDER, J., BUSZKO, M. AND DOCTOR, V.M. (1997). Comparison of the anticoagulant action of sulfated and phosphorylated polysaccharides. *Thrombosis Research* **87**, 113–121.

DHAMI, R., HARDING, S.E., ELIZABETH, N.J. AND EBRINGEROVÁ, A. (1995). Hydrodynamic characterisation of the molar mass and gross conformation of corn cob heteroxylan. *Carbohydrate Polymers* **28**, 113–119.

DIRINGER, H., EHLERS, B., SCHRINNER, E. AND WINKLER, I. (1991). Prevention of viral diseases with polysaccharide sulfates. *European Patent Application* EP 464,759 (*Chemical Abstracts* **116**, 143835).

DOCTOR, V.M., LEWIS, D., COLEMAN, M., KEMP, M.T., MARBLEY, E. AND SAUL, V. (1991). Anticoagulant properties of semisynthetic polysaccharide sulfates. *Thrombosis Research* **64**, 413–25.

DUDKIN, M.S., GROMOV, V.S., VEDERNIKOV, N.A., KATKEVITCH, R.G. AND TSCHERNOV, N.K. (1991). Gemicelljulozy. Riga: Zinatne.

EBRINGEROVÁ, A. (1992). Hemicellulosen als biopolymere Rohstoffe. *Das Papier* **46**, 726–734.

EBRINGEROVÁ, A. AND HROMÁDKOVÁ, Z. (1992). Flow properties of rye bran arabinoxylan dispersions. *Food Hydrocolloids* **6**, 437–442.

EBRINGEROVÁ, A., HROMÁDKOVÁ, Z., ALFÖLDI, J. AND BERTH, G. (1992). Structural and solution properties of corn cob heteroxylans. *Carbohydrate Polymers* **19**, 99–105.

EBRINGEROVÁ, A., HROMÁDKOVÁ, Z., KACURÁKOVÁ, M. AND ANTAL, M. (1993). Quaternized D-Xylans: Synthesis, Structure and Properties. *Book of Abstracts on the VIIth European Carbohydrate Symposium*, August 22–27, Cracow, p D003.

EBRINGEROVÁ, A., HROMÁDKOVÁ, Z. AND BERTH, G. (1994a). Structural and molecular properties of water-soluble arabinoxylan-protein complex from rye-bran. *Carbohydrate Research* **264**, 97–109.

EBRINGEROVÁ, A., HROMÁDKOVÁ, Z., BURCHARD, W., DOLEGA, R. AND VORWERG, W. (1994b). Solution properties of water-insoluble rye-bran arabinoxylan. *Carbohydrate Polymers* **24**, 161–169.

EBRINGEROVÁ, A., HROMÁDKOVÁ, Z., KACURÁKOVÁ, M. AND ANTAL, M. (1994c). Quaternized xylans: synthesis and structural characterization. *Carbohydrate Polymers* 24, 301–307.

EBRINGEROVÁ, A., HROMÁDKOVÁ, Z. AND HRÍBALOVÁ, V. (1995a). Structure and mitogenic activities of corn cob heteroxylans. *International Journal of Biological Macromolecules* 17, 327–332.

EBRINGEROVÁ, A., BELICOVÁ, A. AND EBRINGER, L. (1995b). Antimicrobial activity cf quaternized heteroxylans. *Journal of Microbiology and Biotechnology* 10, 640–644.

EBRINGEROVÁ, A. AND HROMÁDKOVÁ, Z. (1996a). Der Einsatz von Ammoniak-lösungen bei der Gewinnung von Hemicellulosen des D-Xylantyps aus Laubholz. *Holz als Roh- und Werkstoff* 54, 127–129.

EBRINGEROVÁ, A. AND HROMÁDKOVÁ, Z. (1996b). Zur Substituentenverteilung in katic-nischen Xylanderivaten. *Angewandte Makromolekulare Chemie* 24, 97–104.

EBRINGEROVÁ, A., NOVOTNÁ, Z., KACURÁKOVÁ, M. AND MACHOVÁ, E. (1996). Chemical modification of beechwood xylan with p-carboxybenzyl bromide. *Journal of Applied Polymer Science* 62, 1043–1047.

EBRINGEROVÁ, A. AND HROMÁDKOVÁ, Z. (1997). The effect of ultrasound on the structure and properties of the water-soluble corn hull heteroxylan. *Ultrasonics Sonochemistry* 4, 305–309.

EBRINGEROVÁ, A., HROMÁDKOVÁ, Z., HRÍBALOVÁ, V. AND MASON, T.J. (1997). Effect cf ultrasound on the immunogenic corn cob xylan. *Ultrasonics Sonochemistry* 4, 311–315.

EBRINGEROVÁ, A., SROKOVÁ, I., TALÁBA, P. AND HROMÁDKOVÁ, Z. (1998A). Novel D-xylan based functional biopolymers. In *Carbohydrate as Organic Raw Materials IV*. Eds. W. Praznik and A. Huber, pp 118–131.Wien, Austria: WUV-Universitätsverlag.

EBRINGEROVÁ, A., SROKOVÁ, TALÁBA, P., KACURÁKOVÁ, M. AND HROMÁDKOVÁ, Z. (1998b). Amphiphilic beechwood glucuronoxylan derivatives. *Journal of Applied Polymer Science* 67, 1523–1530.

ELLIOT, S.J., STRIKER, G.E., JACOT, T.A. AND STRIKER, L. (1997). Elmiron(R) (pentosan polysulfate) modifies extracellular matrix production (ECM) in mouse mesangial cells. *Journal of the American Society of Nephrology* 8, A2388.

ESTE, J.A., DE VREESE, K., WITVROUW, M., SCHMIT, J., VANDAMME, A.M., ANNE, J., DESMYTER, J., HENSON, G.W., BRIDGER, G. AND DECLERCQ, E. (1996). Antiviral activity of the bicyclam derivative JM3100 against drug-resistant strains of human immunodefi-ciency virus types 1. *Antiviral Research* 29, 297–307.

FAN, Y.R. AND FENG, Z.H. (1987). Effect of carboxymethyl-modified hemicellulose on activity of T lymphocytes and amount of immunocytes. *Acta Pharmacologica Sinica* 8, 166–173

FAUROT, A.L., SAULNIER, L., BEROT, S., POPINEAU, Y., THIBAULT, J.F., PETIT, M.D. AND ROUAU, X. (1995). Large scale isolation of water-soluble and water-insoluble pentosans from wheat flour. *Lebensmittel, Wissenschaft und Technologie* 28, 436–441.

FENGEL, D. AND WEGENER, G. (1984). *Wood Chemistry, Ultrastructure, Reactions*, p 167. Berlin: Walter de Gruyter & Co.

FERAO, A.V. AND MASON, R.M. (1993). The effect of heparin on the cell proliferation and type I collagen synthesis by adult human dermal fibroblast. *Biochimica et Biophysica Acta* 1180, 225–230.

FRANCIS, D.J., HUTALIDOK, N., KONGTAWELERT, P. AND GHOSH, P. (1993). Pentosan poly-sulphate and glycosylaminoglycan polysulfaphate stimulate the synthesis of hyaluronan*in vivo*. *Rheumatology International* 13, 61–64.

FUJISAWA, M., ARIMA, S. AND YACHIKU, S. (1992). A study of the inhibitory effect of chondroitin polysulfate on the stone formation of calcium oxalate. *Nippon Ninyokika Gakkaishi Zasshi* 83, 1647–1654.

FUKISHI, Y., OTSURU, O. AND MAEDA, M. (1988). The chemical structure of the D-xylan from the main cell-wall constituents of *Bryopsis maxima*. *Carbohydrate Research* 182, 313–320.

GARCIA-CONESA, M.T., PLUMB, G.W., KROON, P.A., WALLACE, G. AND WILLIAMSON, G. (1997). Antioxidant properties of ferulic acid dimers. *Redox Report* 3, 239–244.

GHOSH, P. AND HUTADILOK, N. (1996). Interactions of pentosan polysulfate with cartilage matrix proteins and synovial fibroblasts derived from patients with osteoarthritis. *Osteoarthritis and Cartilage* 4, 43–53.

GILLESPIE, L. (1991). Xylan sulfates for affecting growth factor function and for inhibiting fibroblast proliferation. *European Patent Application* EP 466,315 *(Chemical Abstracts* **116**, 144830).

GIRHAMMAR, U. AND NAIR, B.M. (1995). Rheological properties of water soluble non-starch polysaccharides from whole grain rye flour. *Hydrocolloids* **9**, 133–140.

GLASSER, W.G., JAIN, R.K. AND SJOSTEDT, M.A. (1996). Thermoplastic pentosan-rich polysaccharides from biomass. *Biotechnology Advances* **14**, 605.

GRABBER, J.H., HATFIELD, R.D., RALPH, J., ZON, J. AND AMRHEIN, N. (1995). Ferulate cross-linking in cell walls isolated from maize cell suspensions. *Phytochemistry* **40**, 1077–1082.

GREENSHIELDS, R.N. AND REES, A.L. (1992). Gel production from plant matter. *UK Patent* 2 261 671.

GREENSHIELDS, R.N., NG, A. AND WALDRON, K.W. (1997). Oxidative cross-linking of corn bran hemicellulose: formation of different ferulic acid dehydrodimers. *Carbohydrate Research* **303**, 459–462.

GREGORY, A.C.E., O'CONNELL, A.P. AND BOLWELL, G.P. (1998). Xylans. *Biotechnology and Genetic Engineering Reviews* **15**, 439–455.

GRUPPEN, H., HAMER, R.J. AND VORAGEN, A.G.J. (1991). Barium hydroxide as a tool to extract pure arabinoxylans from water-insoluble cell wall material of wheat flour. *Journal of Cereal Science* **13**, 275–290.

GRUPPEN, H., KOMERLINK, F.J.M. AND VORAGEN, A.G.J. (1993). Water unextractable cell wall material from wheat flour. 3. A structural model for arabinoxylan. *Journal of Cereal Science* **18**, 111–128.

HAN, J.Y. AND SCHWARZ, P.B. (1996). Arabinoxylan composition in barley, malt, and beer. *Journal of the American Society of Brewing Chemists* **54**, 216–220.

HARKONE, H., PESSA, E., SUORTTI, T. AND POUTANEN, K. (1997). Distribution and some properties of cell wall polysaccharides in rye milling fractions. *Journal of Cereal Science* **26**, 95–105.

HASHI, M. AND TAKESHITA, T. (1979). Antitumor effect of 4-O-methylglucuronoxylan on solid tumor in mice. *Agricultural and Biological Chemistry* **43**, 961–967.

HAWKINS, M.J. (1995). Clinical trials of anti-angiogenic agents. *European Journal of Cancer* **31A**, S14.

HERATH, H.M.T.B., KUMAR, N.S. AND WIMALASIRI, K.M.S. (1990). Structural studies of an arabinoxylan isloated from *Litsea glutinosa* (Lauraceae). *Carbohydrate Research* **198**, 343–351.

HOFFMANN, P., BERNAT, A., DUMAS, A., PETITOU, M., HERAULT, J.P. AND HERBERT, J.M. (1997). The synthetic pentasaccharide SR 90107A/Org 31540 does not release lipase activity into the plasma. *Thrombosis Research* **86**, 325–332.

HOLMES, H.C., MAHMOOD, N., KARPAS, A., PETRIK, J., KEVICHINGTON, D., O'CONNOR, T. AND JEFFRIES, D. (1991). Screening of compound for activity against HIV: a collaborative study. *Antiviral Chemistry* **2**, 287–293.

HROMÁDKOVÁ, Z. AND EBRINGEROVÁ, A. (1991). Rheologische Eigenschaften des Buchenholzxylans. *Das Papier* **45**, 157–162.

HROMÁDKOVÁ, Z. AND EBRINGEROVÁ, A. (1993). Rheologische Eigenschaften des Buchenholzxylans. II. Einfluss der Uronsäureseitenketten. *Das Papier* **56**, 587–593.

HROMÁDKOVÁ, Z. AND EBRINGEROVÁ, A. (1994). Hemicelluloses in nutrition and health. (in Slovak). *Chemical Letters/Chemické Listy* **88**, 591–603.

HROMÁDKOVÁ, Z. AND EBRINGEROVÁ, A. (1995). Isolation and characterization of hemicelluloses from corn hulls. *Chemical Paper* **49**, 97–101.

HROMÁDKOVÁ, Z., EBRINGEROVÁ, A., KACURÁKOVÁ, M. AND ALFÖLDI, J. (1996). Interactions of the beechwood xylan component with other cell wall polymers. *Journal of Wood Chemistry and Technology* **16**, 221–234.

HROMÁDKOVÁ, Z., EBRINGEROVÁ, A. AND MACHOVÁ, E. (1997). Application of ultrasound in isolation of the xylan component of annual plants. *Proceedings of the 8th Bratislava Symposium on Saccharides*, September, 1–5, Smolenice, p 95.

HROMÁDKOVÁ, Z., KOVÁCIKOVÁ, J. AND EBRINGEROVÁ, A. (1999).Study of the classical and ultrasound-assisted extraction of the corn cob xylan. *Industrial Crops and Products* **9**, 101–109.

HWANG, P., AUCLAIR, B., BEECHINOR, D., DIMENT, M. AND EINARSON, T.R. (1997). Efficacy of pentosan polysulfate in the treatment of interstitial cystitis. *Urology* **50**, 39–43.

IMAI, T., YASUDA, S. AND TERASHIMA, N. (1997). Determination of the distribution and reaction of polysaccharides in wood cell walls by the isotope tracer technique. 5. Behavior of xylan during kraft pulping studied by the radiotracer technique. *Mokuzai Gakkaishi* **43**, 241–246.

ISHIHARA, M., NOJIRI, M., HAYASHI, N. AND SHIMIZU, K. (1996). Isolation of xylan from hardwood by alkali extraction and steam treatment. *Mokuzai Gakkaishi* **42**, 1211–1220.

ISHI, T. (1997). Structure and function of ferulated polysaccharides. *Plant Science* **127**, 111–127.

IZYDORCZYK, M.S. AND BILIARDERIS, C.G. (1995). Cereal arabinoxylans: advances in structure and physicochemical properties. *Carbohydrate Polymers* **28**, 33–48.

JEREZ, J.R., MATSUHIRO, B. AND URZUA, C.C. (1997). Chemical modifications of the xylan from *Palmaria decipiens*. *Carbohydrate Polymers* **32**, 155–159.

KANARI, M., TOMODA, M., GONDA, R., SHIMIZU, N., KIMURA, M., KAWAGUCHI, M. AND KAWABE, C. (1989). A reticuloendothelial system-activating arabinoxylan from the bark of *Cinnamomum cassia*. *Chemical and Pharmceutical Bulletin* **37**, 3191–3194.

KARDOŠOVÁ, A., MALOVÍKOVÁ, A., ROSÍK, J. AND CAPEK, P. (1990). Distribution pattern of 4-O-methyl-D-glucuronic acid units in 4-O-methyl-D-glucurono-D-xylan isolated from the leaves of marsh mallow (*Althaea officinalis* L., var. Rhobusta). *Chemical Papers* **44**, 111–117.

KARDOŠOVÁ, A., MATULOVÁ, M. AND MALOVÍKOVÁ, A. (1998). (4-O-methyl-α-D-glucurono)-β-D-xylan from *Rudbeckia fulgida*, var. sullivanti (Boynton et Beadle). *Carbohydrate Research* **308**, 99–105.

KAVITHA, R. AND CHANDRASHEKAR, A. (1992). Content and composition of nonstarch polysaccharides in endosperms of sorghums varying in hardness. *Cereal Chemistry* **69**, 440–446.

KIESEL, J., HARBAUER, G., WENZEL, E., PINDUR, G. AND BOHNERTH, S. (1991). Comparison of the effects of sodium pentosan polysulfate and unfractionated heparin on venous thrombosis: an experimental study in rats. *Thrombosis Research* **64**, 301–308.

KINDEL, P.K., LIAO, S.-Y., LISKE, M.R. AND OLIEN, C.R. (1989). Arabinoxylan from rye and wheat seed that interact with ice. *Carbohydrate Research* **187**, 173–185.

KLOECKING, H.P., DORNHEIM, G. AND SCHULZE-RIEWALD, H. (1992). Affect of sodium pentosan polysulfate on the thrombogenicity of prothrombin complex concentrates. *Thrombosis Research* **67**, 41–48.

KOCH, P., HINDI, S. AND LANDWEHR, D. (1996). Delayed allergic skin reactions due to subcutaneous heparin-calcium, enoxaparin-sodium, pentosan polysulfate and acute skin lesions from systemic sodium-heparin. *Contact Dermatitis* **34**, 156–158.

KOHN, R., HROMÁDKOVÁ, Z., EBRINGEROVÁ, A. AND TOMAN, R. (1985). Distribution pattern of uronic acid units in 4-O-methyl-D-glucurono-D-xylan of beech (*Fagus sylvatica*. L.). *Collection Czechoslovak Chemical Communications* **51**, 2243–2258.

KOLLAR, L., SCHOLZ, M.E. AND ROZSOS, I. (1994). Pentosan polysulfate sodium gel and heparoid gel in the treatment of infusion trombophlebitides: A randomised double-blind study. *Perfusion* **7**, 18–21.

KOŠÍKOVÁ, B. AND EBRINGEROVÁ, A. (1991). Advantage of non-cellulosic polymer extraction from TMP and steamed wood. *Apitta* **45**, 425–430.

KRONE, V., MAGERSTAEDT, M., WALCH, A., GROENER, A. AND HOFFMANN, D. (1991). Pharmacological products containing polyelectrolyte complexes in microparticulate form. *European Patent Application* EP 454,0044 (*Chemical Abstracts* **116**, 67213d).

KWAN, J.S. AND MORVAN, H. (1995). Characterization of extracellular beta-(1,4)-xylan backbone O-substituted by arabinogalactans type-II in a plant-cell suspension. *Carbohydrate Polymers* **26**, 99–107.

LAHAYE, M., MICHEL, C. AND BARRY, N.J. (1993). Chemical, physicochemical and *in vitro* fermentation characteristic of dietary fiber from *Palmaria palmata* (L). *Food Chemistry* **47**, 29–36.

LAWTHER, J.M., SUN, R. AND BANKS, W.B. (1996). Effects of extraction conditions and alkali type on yield and composition of wheat straw hemicellulose. *Journal of Applied Polymer Science* **60**, 1827–1845.

LEE, S.T. AND LEE, J.J. (1997). Insoluble dye substrate for screening and assay of xylan-degrading enzymes. *Journal of Microbiological Methods* **29**, 1–5.

LENZ, J. AND WUTZEL, H. (1984). Effect of the addition of hemicelulose on the flow properties of wheat flour dough. *Rheologica Acta* **23**, 570–572.

LEMOPEREUR, I., ROUAU, X. AND ABECASSIS, J. (1997). Genetic and agronomic variation in arabinoxylan and ferulic acid contents of durum wheat (*Triticum durum*) grain and its milling fractions. *Journal of Cereal Science* **25**, 103–110.

LI, J.B. AND GELLERSTEDT, G. (1997). The contribution to kappa number from hexeneuronic acid groups in pulp xylan. *Carbohydrate Research* **302**, 213–218.

LUSH, R.M., FIGG, W.D., PLUDA, J.M., BITTON, R., HEADLEE, D., KOHLER, D., REED, E., SARTOR, O. AND COOPER, M.R. (1996). A phase I study of pentosan polysulfate sodium in patients with advanced malignancies. *Annals of Oncology* **7**, 39–944.

MARSHALL, J.L., WELLSTEIN, A., RAE, J., DeLAP, R.J., PHIPPS, K., HANFELT, J., YUNMBAM, M.K., SUN, J.X., DUCHIN, K.L. AND HAWKINS, M.J. (1997). A Phase trial I of orally administered pentosan polysulfate in patient with advanced cancer. *Clinical Cancer Research* **3**, 2347–2354.

MATULEWICZ, M.C. AND CEREZO, A.S. (1987). Alkali-soluble polysaccharides from *Chaetangium fastigiatum*: structure of a xylan. *Phytochemistry* **26**, 1033–1035.

MATULEWICZ, M.C., CERESO, A.S. AND JARRET, R.M. (1992). High resolution carbon 13-NMR spectroscopy of mixed linkage xylans. *International Journal of Biological Macromolecules* **14**, 29–32.

MELLON, J.E., COTTY, P.J., GODSHALL, M.A. AND ROBERTS, E. (1995). Demonstration of aflatoxin inhibitory activity in a cotton seed coat xylan. *Applied and Environmental Microbiology* **61**, 4409–4412.

METHACANON, P., KENNEDY, J.F., LLOYD, L.L., PATERSON, M. AND KNILL, C.J. (1997). Chemical characterisation of water soluble arabinoxylanbranan ferulates isolated from maize bran. *Carbohydrate Polymers* **34**, 435–436.

MOHRI, T. AND CHAMBERS, E.L. (1995). Effect of heparin and pentosan polysulfate on the sperm-induced elevation of cytosolic calcium and the activation current. *Journal of General Physiology* **106**, 28.

N'DIAYE, S., RIGAL, L., LAROQUE, P. AND VIDAL, P.E. (1996). Extraction of hemicelluloses from poplar, populus tremuloides, using an extruder-type twin-screw reactor: a feasibility study. *Bioresource Technology* **57**, 61–67.

NETHERY, A., GILES, I., JENKINS, K., JACKSON, CH., BROOKS, P., BURKHARDT, F., GOSH, P., WHITELOCK, J. AND O'GRADY, R.L. (1992). The chondroprotective drugs, arteparon and sodium pentosan polysulfate, increased collagenase activity and inhibit stromelysin *in vitro*. *Biochemical Pharmacology* **44**, 1549–1553.

NETO, C.P., SECA, A., FRADINHO, D., COIMBRA, M.A., DOMINIGUES F., EVTUGUIN D., SILVESTRE, A. AND CAVALEIRO, J.A.S. (1996). Chemical composition and structural features of the macromolecular components of *Hibiscus cannabinus* grown in Portugal. *Industrial Crops and Products* **5**, 189–196.

NETO, C.P., SECA, A., NUNES, A.M., COIMBRA, M.A., DOMINGUES, F., EVTUGUIN, D., SILVESTRE, A. AND CAVALEIRO, J.A.S. (1997). Variations in chemical composition and structure of macromolecular components in different morphological regions and maturity stages of *Arundo donax*. *Industrial Crops and Products* **6**, 51–58.

NG, A., GREENSHIELDS, R.N. AND WALDRON, K.W. (1997). Oxidative cross-linking of corn bran hemicellulose: formation of ferulic acid dehydrodimers. *Carbohydrate Research* **303**, 459–462.

NGUYEN, N.M., LEHR, J.E. AND PIETA, K.J. (1993). Pentosan inhibits angiogenesis *in vitro* and suppresses prostate tumor growth *in vivo*. *Anticancer Research* **13**, 2143–2145.

NILSSON, M., SAULNIER, L., ANDERSSON, R. AND AMAN, P. (1996). Water unextractable polysaccharides from three milling fractions of rye grain. *Carbohydrate Polymers* **30**, 229–237.

ORME, C.E. AND HARRIS, R.C. (1997). A comparison of the lipolytic and anticoagulative properties of heparin and pentosan polysulphate in the thoroughbred horse. *Acta Physiologica Scandinavica* **159**, 179–185.

PAINTER, T.J. (1983). Algal Polysaccharides. In *The Polysaccharides*. Ed. G.O. Aspinal, pp 196–285. Orlando: Academic Press, Inc.

PAPATHEOFANUS, M.G., BILLA, E., KOULLAS, D.P., MONTIES, B. AND KOUKIS, E.G. (1998). Optimising multistep mechanical-chemical fractionation of wheat straw components. *Industrial Crops and Products* **7**, 249–256.

PARISH, C.R. AND SNOWDEN, J.M. (1997). Sulphated polysaccharides having anti-metastatic and/or anti-inflammatory activity. *Biotechnology Advances* **15**, 525.

PARSONS, C.L. (1992). Irrigation of internal bladder surfaces in mammals with sodium pentosan polysulfate. *US Patent* 5,180,715 (*Chemical Abstracts* **118**, 139825).

PIENTA, K.J., MURPHY, B.C., ISAACS, W.B., ISAACS, J.T. AND COFFEY, D.S. (1992). Effect of pentosan, a novel cancer chemotherapeutic agent, on prostate cancer cell growth and mobility. *Prostate* (N.Y.) **20**, 233–241.

PROKSCH, A. AND WAGNER, H. (1987). Structural analysis of a 4-O-methylglucuronoarabino-xylan with immuno-stimulating activity from *Echinacea Purpurea Phytochemistry* **26**, 1989–1993.

PURINA, L., TREIMANIS, A., BRENCE, A., TIMRMAN, G. AND IOZEF, R. (1991). Distribution of lignin and hemicellulose in fiber walls of pine sulfite-sulfate pulps. *Wood Chemistry/ Khimija Drevesiny* (**4**), 37–44.

RALPH, J., GRABBER, J.H. AND HATFIELD, R.D. (1995). Lignin-ferulate cross-link in grasses: active incorporation of ferulate polysaccharide esters into ryegrass lignins. *Carbohydrate Research* **275**, 167–178.

RAISS, R. AND WIESNER, M. (1992). Substituted polysaccharides for treatment of degenerative articular ailments. *PCT International Application* WO 92 13,541 (*Chemical Abstracts* **117**, 245597).

READ, R.A., CULLISHILL, D. AND JONES, M.P. (1996). Systemic use of pentosan polysulfate in the treatment of osteoarthritis. *Journal of Small Animal Practice* **37**, 108–114.

RENARD, C.M.G.C., WEIGHTMAN, R.M. AND THIBAULT, J.-F. (1997). Xylose-rich pectins have been isolated from pea hulls. *International Journal of Biological Macromolecules* **21**, 155–162.

ROCHET, V. AND BERNALIER, A. (1997). Utilization of algal polysaccharides by human colonic bacteria, in anexic culture or in association with hydrogenotrophic microorganisms. *Reproduction, Nutrition and Development* **37**, 221–229.

ROGACHEFSKY, R.A., DEAN, D.D., HOWELL, D.S. AND ALTMAN, R.D. (1994). Treatment of canine osteoarthritis with sodium pentosan polysulfate and insulin-like growth-factor-I. *Annals of the New York Academy of Sciences* **732**, 392–394.

ROSNER, H., GRIMNECKE, H.D., KNIREL, Y.A. AND SHASKOV, A.S. (1992). Hyaluronic acid and a (1-4) beta-D-xylan extracellular polysaccharides of *Pasteurella multorida* (Carter type A) strain 880. *Carbohydrate Research* **223**, 329–333.

SAULNIER, L., VIGOUROUX, J. AND THIBAULT, J.-F. (1995a). Isolation and partial characterization of feruloylated oligosaccharides from maize bran. *Carbohydrate Research* **272**, 241–253.

SAULNIER, L., MAROT, C., CHANLIAUD, E. AND THIBAULT, J.-F. (1995b). Cell wall polysaccharide interactions in maize bran. *Carbohydrate Polymers* **26**, 279–287.

SAULNIER, L., CHANLIAUD, E., THIBAULT, J.-F., DESPRÉ, D. AND MESSAGER, A. (1998). Extraction, structure and functional properties of maize bran heteroxylans. In *Carbohydrates as Organic Raw Materials IV*. Eds. W. Praznik and A. Huber, pp 132–138. Wien: WUV.

SCHOLS, D., PAUWELS, R., WITVROEW, M., DESMYTER, J. AND DELLERQ, E. (1992). Differential activity of polyanionic compound and castanospermine against HIV replication and HIV-induced syncytium formation depending on virus strain and cell type. *Antiviral Chemistry and Chemotherapy* **3**, 23–29.

SCHNABELRAUCH, M., WAGENKNECHT, W., STEIN, A., PHILIPP, B., KLEMM, D. AND NEHLS, I. (1992). Preparation of phoshates of cellulose and related polysaccharides. *German (East) DP* 299,312 (*Chemical Abstracts* **117**, 36095v).

SHUBA, K., SIVAMURUGESAN, A. AND VARABAKSHMI, P. (1992). Changes in serum lipids and lipoproteins in calcium oxalate stone forming rats treated with sodium pentosan polysulfate. *Biochemistry International* **27**, 1011–1018.

SCHWARZ, P.B. AND HAN, J.Y. (1995). Arabinoxylan content of commercial beers. *Journal of the American Society of Brewing Chemists* **53**, 157–159.

SCHWARTSMANN, G., SPRINZ, E., KALAKUN, L., YAMAGUSHI, N., SANDER, E., GRIVICICH, I., KOYA, R. AND MANS, D.R.A. (1996). Phase-II study of pentosan polysulfate (PPS) in patients with aids-related kaposis-sarcoma. *Tumori* **82**, 360–363.

SENTHIL, D., SUBHA, K., SARAVANAN, N. AND VARALAKSHMI, P. (1996). Influence of sodium pentosan polysulfate and certain inhibitors on calcium-oxalate crystal-growth. *Molecular and Cellular Biochemistry* **156**, 31–35.

ŠIMKOVIC, I., MLYNÁR, J. AND ALFÖLDI, J. (1990). New aspects in cationization of lignocellulose materials XI. Modification of bagasse with quarternary ammonium groups. *Holzforschung* **44**, 113–116.

ŠIMKOVIC, I., MLYNÁR, J. AND ALFÖLDI, J. (1992). Modification of corn cob meal with quarternary ammonium groups. *Carbohydrate Polymers* **17**, 285–288.

SROKOVÁ, I., TALÁBA, P., HROMÁDKOVÁ, Z., EBRINGEROVÁ, A. AND ALFÖLDI, J. (1997). Structure and properties of partially hydrophobised D-xylan type polysaccharides. *Proceedings of the 8th Bratislava Symposium on Saccharides,* September 1–5, Smolenice, Slovakia, p 97.

STEPHEN, A.M. (1983). Other plant polysaccharides. In *The Polysaccharides.* Ed. G.O. Aspinall, pp 98–193. Orlando: Academic Press, Inc.

STEWART, D., AZZINI, A., HALL, A.T. AND MORRISON, I.M. (1997). Sisal fibres and their constituent non-cellulosic polymers. *Industrial Crops and Products* **6**, 17–26.

STSCHERBINA, D. AND PHILIPP, B. (1991). Isolation, modification and uses of xylans. Literature review. *Acta Polymerica* **42**, 345–351.

SUN, R., LAWTHER, J.M. AND BANKS, W.B. (1998). Isolation and characterization of hemicellulose B and cellulose from pressure refined wheat straw. *Industrial Crops and Products* **7**, 121–128.

SWAIN, S.M., PARKER, B., WELLSTEIN, A., LIPPMAN, M.E., STEAKLEY, C. AND DELAP, R. (1995). Phase-I trial of pentosan polysulfate. *Investigational New Drugs* **13**, 55–62.

SWAMY, N.R. AND SALIMATH, P.V. (1990). Structural features of acidic xylans isolated from red gram (*Cajanus cajan*) husk. *Carbohydrate Research* **197**, 327–337.

TELEMAN, A., HARJUNPAA, V., TENKANEN, M., BUCHERT, J., HAUSALO, T., DRAKENBERG,T. AND VUORINEN, T. (1995). Characterization of 4-deoxy-beta-1-threo-hex-4-enopyranosyluronic acid attached to xylan in pine kraft pulp and pulping liquor by H-1 and C-13 NMR-spectroscopy. *Carbohydrate Research* **272**, 55–71.

TELEMAN, A., SIIKAAHO, M., SORSA, H., BUCHERT, J., PERTTULA, M. HAUSALO, T. AND TENKANEN, M. (1996). 4-O-Methyl-beta-1-idopyrano-syluronic acid linked to xylan from kraft pulp isolation procedure and characterization by nmr-spectroscopy. *Carbohydrate Research* **293**, 1–13.

THIEBAUD, S. AND BORREDON, M.E. (1998). Analysis of liquid fraction after esterification of sawdust with octanoyl chloride – production of esterified hemicelluloses. *Bioresource Technology* **63**, 139–145.

THORMAR, H., BALZARINI, J., DEBYSER, Z., WITVROUW, M., DESMYTER, J. AND DECLERCQ, E. (1995). Inhibition in visna virus replication and cytopathic effect in sheep choroid plexus cell cultures by selected anti-HIV agents. *Antiviral Research* **27**, 49–57.

VAN HAZENDONK, J.M., REINERINK, E.J.M., DE WAARD, P. AND VAN DAM, J.E.G. (1996). Structural analysis of acetylated hemicellulose polysaccharides from fibre flax *(Linum usitatissimum* L.). *Carbohydrate Research* **291**, 53–62.

VERBRUGGEN, M.A., BELDMAN, G., VORAGEN, A.G.J. AND HOLLEMANS, M. (1993). Waterunextractable cell wall material from Sorghum: Isolation and characterization. *Journal of Cereal Science* **17**, 71–82.

VERBRUGGEN, M.A., BELDMAN, G. AND VORAGEN, A.G.J. (1995). The selective extraction of glucuronoarabinoxylans from sorghum endosperm cell walls using barium and potassium hydroxide solutions. *Journal of Cereal Science* **21**, 271–282.

VIETOR, R.J., ANGELINO, S.A.G.F. AND VORAGEN, A.G.J. (1992). Structural features of arabinoxylans from barley and malt cell wall material. *Journal of Cereal Science* **15**, 213–222.

VIETOR, R.J., KOMERLINK, F.J.M., ANGELINO, S.A.G.F. AND VORAGEN, A.G.J. (1994). Substitution patterns of water-unextractable arabinoxylans from barley and malt. *Carbohydrate Polymers* **24**, 113–118.

VINCEDON, M. (1993). Xylan derivatives: aromatic carbamates. *Macromolecular Chemistry* **194**, 321–28.

VINCENDON, M. (1998). Xylan derivatives: Benzyl ethers, synthesis, and characterization. *Journal of Applied Polymer Science* **67**, 455–460.

VINKX, C.J.A. AND DELCOUR, J.A. (1996). Rye (*Secale cereale* L.) arabinoxylans: a critical review. *Journal of Cereal Science* **24**, 1–14.

VON BRIESEN, H. (1990). Polysulfated polyxylan HOC/Bay-946 inhibits HIV replication on human monocytes/macrophages. *Research in Virology* **141**, 251–257.

VISSER, J., BELDMAN, G., KUSTERS VAN SOMEREN, M.A. AND VORAGEN, A.G.J. (1992). *Xylan and Xylanases*. Amsterdam: Elsevier Science Publishers B.V.

WAGENKNECHT, W., STEIN, A., PHILLIP, B., KLEMM, D., SCHNABELRAUCH, M., DAUTZEN-BERG, H., HOPL, G. AND WALENTA, K. (1992). Preparation of water-soluble acid sulfates of cellulose and related polysaccharides. *German* (East) *DD* 298,643 (*Chemical Abstracts* **117**, 133201).

WAGNER, H., PROKSCH, A., RIESS-MAURER, I., VOLLMAR, A., ODENTHAL, S., STUPPNER, H., JURCIC, K., LE TURDU, M. AND FANG, J.N. (1985). Immunstimuliwerend wirkende Polysaccharide (Heteroglykane) aus höheren Pflanzen. *Arzneimittel-Forschung/Drug Research* **35**, 1069–1075.

WALLACE, G., RUSSEL, W.R., LOMAX, J.A., JARVIS, M.C., LAPIERRE, C. AND CHESSON, A. (1995). Extraction of phenolic-carbohydrate complexes from graminaceous cell walls. *Carbohydrate Research* **272**, 41–53.

WALLACE, G. AND FRY, S.C. (1995). *In vitro* peroxidase-catalysed oxidation of ferulic acid esters. *Phytochemistry* **39**, 1293–1299.

WENDE, G. AND FRY, S.C. (1997a). O-feruloylated, O-acetylated oligosaccharides as side-chains of grass xylans. *Phytochemistry* **44**, 1011–1018.

WENDE, G. AND FRY, S.C. (1997b). 2-O-β-D-Xylopyranosyl-(5-O-feruoyl)-L-arabinose, a widespread component of grass cell walls. *Phytochemistry* **44**, 1019–1030.

WHISTLER, R.L., BUSHWAY, A., SINGH, P.P. AND NAKAHARA, W. (1976). Noncytotoxic, antitumor Polysaccharides. *Advances in Carbohydrate Chemistry and Biochemistry* **32**, 235–275.

WHISTLER, R.L. (1989). Forthcoming opportunities for hemicelluloses. In *Frontiers in Carbohydrate Research – I. Food Applications*. Eds. R.P. Millane, J.N. BeMiller and R. Chandrasekaran, pp 289–296. London: Elsevier Science Publishers Ltd.

WILLIAMS, R.J. (1992). Anomalous behaviour of ice in solutions of ice-binding arabinoxylans. *Thermochimica Acta* **221**, 105–109.

WIMALASIRI, K.M.S. AND KUMAR, N.S. (1995). A water-soluble polysaccharide from the leaves of *Litsea gardneri* (Lauraceae). *Carbohydrate Polymers* **26**, 19–23.

WONG, K.K.Y., CLARKE, P. AND NELSON, S.L. (1995). Possible roles of xylan-derived chromophores in xylanase prebleaching of softwood kraft pulp. *ACS Symposium series* **618**, 352–362.

WUN, T.CH. (1992). Lipoprotein-associated coagulation inhibitor (LACI) is a cofactor for heparin: synergistic, anticoagulant action between LACI, and sulfated polysaccharides. *Blood* **79**, 439–438.

YAMADA, H., NAGAI, T., CYONG, J.-C. AND OTSUKA, Y. (1985). Relationship between chemical structure and anti-complementary activity of plant polysaccharides. *Carbohydrate Research* **144**, 101–110.

YAMAGAKI, T., MAEDA, M., KANAZAWA, K., ISHIZUKA, Y. AND NAKANISHI, H. (1996). Structures of caulerpa cell-wall microfibril xylan with detection of beta-1,3-xylooligosaccharides as revealed by matrix-assisted laser-desorption ionization time-of-flight mass-spectrometry. *Bioscience Biotechnology and Biochemistry* **60**, 1222–1228.

YAMAGAKI, T., MAEDA, M., KANAZAWA, K., ISHIZUKA, Y. AND NAKANISHI, H. (1997a). Structural clarification of Caulerpa cell wall beta-1,3-xylan by NMR spectroscopy. *Bioscience Biotechnology and Biochemistry* **61**, 1077–1080.

YAMAGAKI, T., TSUJI, Y., MAEDA, M. AND NAKANISHI, H. (1997b). NMR spectroscopic analysis of sulfated beta-1,3-xylan and sulfation stereochemistry. *Bioscience Biotechnology and Biochemistry* **61**, 1281–1285.

YOKOTA, S., WONG, K.K.Y., SADDLER, J.N. AND REID, I.D. (1995). Molecular-weight distribution of xylan/lignin mixtures from kraft pulps-toward an understanding of xylanase prebleaching. *Pulp & Paper – Canada* **96**, 39–41.

YUI, T., IMADA, K., SHIBUYA, N. AND OGAWA, K. (1995). Conformation of an arabinoxylan isolated from the rice endosperm cell-wall by x-ray-diffraction and a conformational analysis. *Japan Biotechnology and Biochemistry* **59**, 965–968.

15
Polysaccharide Film Technologies: Interfacial Order and Chain Thermodynamic Rigidity

GEORGES M. PAVLOV* AND ALEXEY E. GRISHCHENKO

Department and Institute of Physics of St Petersburg University, Ulianovskaya str. 1, Petergof, 198904, St Petersburg, Russia

Introduction

The properties of various substances at or near the surface layers differ greatly from those in the bulk of the material. The contribution of surface layers to the overall characteristics of materials increases with decreasing characteristic dimensions. These effects are caused by the self-organization of molecular structures in surface layers. The results of the study of density, mechanical properties, and birefringence indicate that averaged physical characteristics of materials are profoundly affected by sample dimensions (Rostiashvili *et al.*, 1987; Grishchenko, 1996). One of the reasons for these effects is the appearance of orientational order at the interface. This is particularly pronounced in the case of chain-like linear molecules, the parts of which are characterized by high anisotropy. The published data available on these materials to date deal mainly with investigations of induced orientational order that appear under the influence of tensile stress or induced by electomagnetic fields (Hermans, 1949; Paul and Newman, 1978). There are only a relatively small number of papers that deal with the spontaneous organisation of physical properties in the vicinity of interfaces and the dependence of this organisation on the chemical structure of chain molecules (McNally and Sheppard, 1930; Cherkasov *et al.*, 1976; Grishchenko *et al.*, 1983; Grishchenko and Cherkasov, 1997). It should also be pointed out that definitive quantitative criteria for evaluating self-organization at interfaces are generally absent. In the present paper the *'orientational order parameter'* (de Gennes, 1974) is used for the quantitative characterization of chain molecular self-organization at the interface.

The idea that the orientation of polymer chains in surface layers influences the optical properties of films was first suggested by McNally and Sheppard (1930) and has been further developed by Hermans (1949), Stein and Norris (1956) and Stein (1978). Orientational order can be evaluated experimentally by using the method of an

---

*To whom correspondence may be addressed.

Biotechnology and Genetic Engineering Reviews – Vol. 16, April 1999
0264–8725/99/16/347–359 $20.00 + $0.00 © Intercept Ltd, P.O. Box 716, Andover, Hampshire SP10 1YG, UK

inclined polarized beam in combination with sedimentation-diffusion analysis and the method of flow birefringence in solution. The present short review will attempt to establish correlations between the chemical structure and conformational characteristics of polysaccharides on the one hand and the properties of interfaces formed by them on the other, and is based mainly on work from our St. Petersburg laboratory. For this purpose, we include the results for investigations on dextran, pullulan, mannan, methylcellulose, and xanthan.

The inclined polarized beam method

The inclined polarized beam method provides an effective way of enabling us to obtain quantitative information about the structure and orientation of anisotropic elements of polymer chains in surface layers (Cherkasov *et al.*, 1976; Grishchenko, 1996; Grishchenko and Cherkasov, 1997). This method is based on measurement of the birefringence occurring when polarized light passes through a film at an angle to the surface different from the normal angle. The appearance of birefringence in this mode of light transmission through a film has been explained in terms of the orientation of polymer chains (McNally and Sheppard, 1930). An optical method for determining the type of molecular orientation with respect to the surface has been proposed by Cherkasov *et al.* (1976). In this method, birefringence was measured at two fixed incident angles of polarized beam to the film. Cherkasov *et al.* (1976) have modified this method in such a way that it has become possible to obtain more detailed information about orientational order in polymer samples.

An equation has been derived (Cherkasov *et al.*, 1976) for relating the phase difference δ between beams polarized in two perpendicular directions and forming an angle i with the polymer film surface

$$\delta = B(1-\cos 2i) \tag{1}$$

Here B is the surface birefringence given by

$$B = -\pi N_o(\alpha_1-\alpha_2)/n^3\lambda\{(n^2+2)/3\}^2 HS \tag{2}$$

and where N_o is the number of segment per unit volume, n the refractive index of the polymer, λ is the wavelength of the incident light, $(\alpha_1-\alpha_2)$ is the difference between the main polarizabilities of the anisotropic chain element, $S = (3<\cos^2\vartheta>1)/2$ is the factor of orientational order of molecular fragments with respect to the surface, and H is the film thickness.

It follows from equation (2) that surface birefringence B is proportional to film thickness H. This outcome of the theory is in agreement with experimental data for small values of H. However, with increasing H the experimental values of B attain a limiting value and no longer depend on film thickness (Cherkasov *et al.*, 1976; Grishchenko, 1996). This discrepancy between the theory and experiment follows from the assumption underpinning the derivation of equation (2). It was assumed that at low H the orientational order factor does not depend on film thickness and is characterized only by a certain average value of S.

In order to describe adequately the dependence of B on H over a wide range of film thickness, the dependence of the orientational order factor S on thickness should be taken into account.

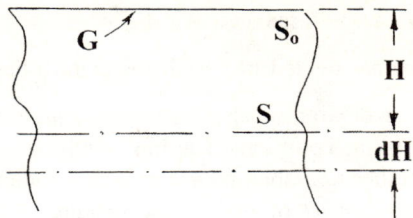

Figure 1.   Scheme for calculating the surface birefringence B versus thickness, H of a film. G: surface of film.

Let us consider the air-polymer interface (*Figure 1*). An infinitely thin layer dH parallel to film surface will be singled out. Let us assume that when the layer depth changes from H to H+dH, the order parameter decreases by dS. This decrease is proportional to S at a distance H from the surface and to layer thickness dH, i.e. we have

$$dS = -kSdH \tag{3}$$

where the minus sign shows that the order parameter *decreases* with the distance from the interface, in other words with increasing H.

Integrating equation (3) from zero

$$\int_0^H S^{-1}\, dS = -k\int_0^H dH$$

and assuming that at H=0, S is equal to S_o, the following equation can easily be obtained:

$$S = S_o e^{-kH} \tag{4}$$

It is clear that at $H = k^{-1}$ the order parameter decreases e times. The distance from the surface H_o at which S decreases e times can be called the effective thickness of the anisotropic surface layer, i.e. we have

$$S = S_o e^{-H/Ho} \tag{5}$$

Equation (5) makes it possible to obtain the dependence of B on H over a wide range of changes in polymer film thicknesses

$$B = -[\pi N_o(\alpha_1 - \alpha_2)/n^3\lambda]((n^2+2)/3)^2 \int_0^H S_o e^{H/Ho}\, dH =$$

$$-[\pi N_o(\alpha_1 - \alpha_2)/n^3\lambda]((n^2+2)/3)^2 S_o H_o (1 - e^{-H/Ho}) = B_o(1 - e^{-H/Ho}) \tag{6}$$

It can be seen that at $H/H_o \ll 1$ the value of B $\sim S_o H$ and at $H \to \infty$ it acquires the maximum value equal to $B = B_o \sim S_o H_o$.

A comparison of the theoretical prediction (6) with experimental data makes it possible to evaluate the effective thickness of an optically anisotropic surface layer, the orientational order and of the parameter of molecular fragments near the interface.

Examples for polysaccharide films

The films that have been investigated are based around the following polysaccharides:

1. *Dextran*: the chains in dextran consist of glucose residues linked mainly by $\alpha(1-6)$ bonds (Yalpani, 1988; Tombs and Harding, 1998).
2. *Pullulan*: this is another microbial polysaccharide produced by *Aureobasidium pullulans*. Its chains consist of maltotriose residues linked by $\alpha(1-6)$ links (Yalpani, 1988).
3. *Mannan* (Elinov and Vitovskaya, 1979): this is an extracellular polysaccharide produced by yeast of the *Rhodotorula rubra* strain. The repeat unit of this mannan contains $\beta(1-3)$ and $\beta(1-4)$ bonds in an equimolar ratio (Elinov and Vitovskaya, 1979).
4. *Methylcellulose*: this is the first member in a series of O-alkyl substituted cellulose ($\beta(1-4)$ glucan). The mean content of OCH_3 groups was 28.4%, which corresponds to the degree of substitution 1.68 (Pavlov *et al.*, 1996).
5. *Xanthan:* this is an anionic polysaccharide produced by *Xanthomonas campestris* bacteria (Yalpani, 1988). A polysaccharide of the Kelko/Keltrol company has been the subject of study with regards to film formation. Its degree of substitution with acetate groups is 0.81 and that with pyruvyl groups is 0.59.

The films investigated had been cast from aqueous solutions at different concentrations on horizontal glass surfaces. The films usually were of greater thickness on the periphery. Homogeneous central part of the films was chosen for investigations. Film thickness was measured with a micrometer. Where necessary, multi-layered 'stacks' consisting of several homogeneous films were used. Film thickness is calculated from the solution concentration C and the area Q according to the equation $H = C \times m/Q\rho$ where m is the mass of solution distributed on the area Q and ρ is the polymer density. The assumption had been made that film thickness for a given film studied was homogeneous througout. It should be noted that the study of birefringence of thick methyl cellulose and xanthan films (H>0.1 mm) was difficult because they were not transparent. This fact probably results from microphase segregation of the films.

Methods for evaluating the equilibrium rigidity of polysaccharide molecules

The equilibrium chain rigidity of all polysaccharides used here was evaluated from data obtained in solution investigations by the methods of molecular hydrodynamics (Tanford, 1961; Tsvetkov *et al.*, 1970; Cantor and Schimmel, 1980). In this case, the main characteristics are the velocity sedimentation coefficient s_0, the translational diffusion coefficient D_0, and intrinsic viscosity $[\eta]$. Useful information about the hydrodynamic study of polysaccharides may be obtained from the articles of Pavlov (1989, 1995, 1997); Pavlov *et al.* (1997); Harding (1992, 1995). To obtain quantitative evaluations of thermodynamic rigidity (Kuhn segment length), fractions (samples) over a wide molecular weight range M have been investigated. In the case of translational friction, the most complete result taking into account the effects of intramolecular draining (percolation) and volume interaction had been obtained on the basis of the Gray, Bloomfield, Hearst theory (1967):

$$[s]P_0N_A = [3/(1-\varepsilon)(3-\varepsilon)]M_L^{(1-\varepsilon)/2} A^{-(1-\varepsilon)/2} M^{(1-\varepsilon)/2} + (M_L P_0/3\pi)$$

$$[\ln(A/d) - (1/3)(A/d)^{-1} - \varphi(\varepsilon)]$$

where $[s] \equiv s_0 \eta_0 (1 - \upsilon \rho_0)^{-1}$, η_0 is the viscosity of solvent and $(1 - \upsilon \rho_0)$ is the buoyancy factor or density increment; $M_L = M_0/\lambda'$ is the mass of the unit length of macro-molecules, M_0 is the molecular weight of the repeating unit, λ' is the projection of the unit in the chain direction; A is the Kuhn segment length, d is the hydrodynamic chain diameter; $P_0 = 5.11$ is the Flory hydrodynamic parameter, N_A is Avogadro's number; ε is the parameter characterizing the volume effects where $<h^2> \sim M^{1+\varepsilon}$ and $<h^2>$ is the mean-square end-to-end distance of the chain and $\varphi(\varepsilon)$ is the function that can be tabulated utilizing the theory of Gray, Bloomfield, Hearst (1967).

Viscometric data is usually interpreted on the basis of an assumption that chain dimensions in phenomena of translational and rotational friction are equivalent. In the analytical form, the relationship between hydrodynamic parameters is expressed by the following equation (Pavlov *et al.*, 1990; 1992)

$$[s]P_0N_A = (M^2\Phi/[\eta])^{1/3}$$

where Φ is the Flory hydrodynamic parameter, and $[\eta]$ is the intrinsic viscosity.

The dermination of segmental optical anisotropy

The intrinsic segmental anisotropy $(\alpha_1 - \alpha_2)$ is a fundamental characteristic necessary for interpreting the results of birefringence in films. The principal method for its determination is flow birefringence (the '*Maxwell effect*') of macromolecules in solutions when the contour length of macromolecules L greatly exceeds the Kuhn segment length, A (so that L/A>15). Under these circumstances, the dependence on molecular parameters is given by the following expression (Tsvetkov and Andreeva, 1981; Tsvetkov, 1989)

$$[n]/[\eta]\{45kTn_0/4\pi(n^2+2)^2\} = \{(\alpha_1 - \alpha_2)$$
$$+ (\Delta n/\Delta c)^2 M_S/2\pi\upsilon N_A + (2.61\Phi(\Delta n/\Delta c)^2/\pi^2 N_A^2)(M/[\eta])\} \tag{7}$$

where $(\alpha_1 - \alpha_2)$ is the intrinsic segmental anisotropy of optical polarizabilities, M_S is the molar weight of the segment, n_0 is the refractive index of the solvent, u is the partial specific volume of the polymer, k is the Boltzmann constant, T is the absolute temperature.

The first term in equation (7) reflects the contribution of intrinsic polarizability anisotropy to the birefringence effect, the second one corresponds to the microform effect, and the third term corresponds to the effect of the form of random coil in solution. If refractive indices of the polymer and the solvent coincide, birefringence is determined only by the difference between main polarizabilities of a statistical segment $(\alpha_1 - \alpha_2)$. If the refractive index increment of the polymer-solvent system $\Delta n/\Delta c$ is non-zero, then to determine the value of $(\alpha_1 - \alpha_2)$, the effects of macro- and microform should be taken into account. As can be seen from equation (7), the extrapolation of the dependence of $[n]/[\eta]$ on $M/[\eta]$ to the zero value of the parameter $M/[\eta]$ makes it possible to take into account the contribution of the macroform effect to the total birefringence value. The values of $\lim[n]/[\eta]$ at $M/[\eta] \to 0$ have in this way been determined for pullulan in water and DMSO.

The second extrapolation of the values of $\lim[n]/[\eta]$ at $M/[\eta]=0$ to the zero value of refractive index increment ($\Delta n/\Delta c \rightarrow 0$) enable us to exclude from consideration the microform effect as well, which follows from equation (7):

$$\lim_{M\rightarrow 0; \; \Delta n/\Delta c \rightarrow 0}[n]/ [\eta]=\{4\pi(n^2+2)^2/45 \, kTn_s \}(\alpha_1-\alpha_2)$$

The above procedure has made it possible to estimate the intrinsic optical segmental anisotropy of pullulan ($\alpha_1-\alpha_2$) to be $+33 \times 10^{-25}$ cm^3 (Pavlov *et al.*, 1998a). The intrinsic optical anisotropy of a dextran segment has also been obtained in an analogous way ($\alpha_1-\alpha_2$)= $+12.5 \times 10^{-25}$ cm^3 (Pavlov *et al.*, 1998b). The results from these studies have in this way shown that the polarizabilities of pullulan and dextran in the main chain direction slightly exceed those in the transverse direction, which is in good agreement with the chemical structure of their monomer units.

Experimental evaluation of orientational order

Orientational order of polysaccharide chain fragments in surface layers has indeed been succesfully probed by the inclined polarized beam method, and a standard optical scheme with a visual system of birefringence recording has been used to obtain this information.

To achieve this, the optical phase difference, or 'optical retardation' δ induced by the polymer film is determined with the aid of a penumbral Brace compensator, $\delta = \delta_o \sin 2\Delta\varphi$ (Tsvetkov *et al.*, 1970). Here $\Delta\varphi = \varphi-\varphi_o$ is the difference between compensator readings when the film forms an angle with the optical axis and when the polarized beam is normal to the film and δ_o is the phase difference induced by the mica plate of the compensator ($\delta_o = 0.076$ radians).

Figure 2 shows as an example the dependence of $\Delta\varphi$ on the incidence angle i of the

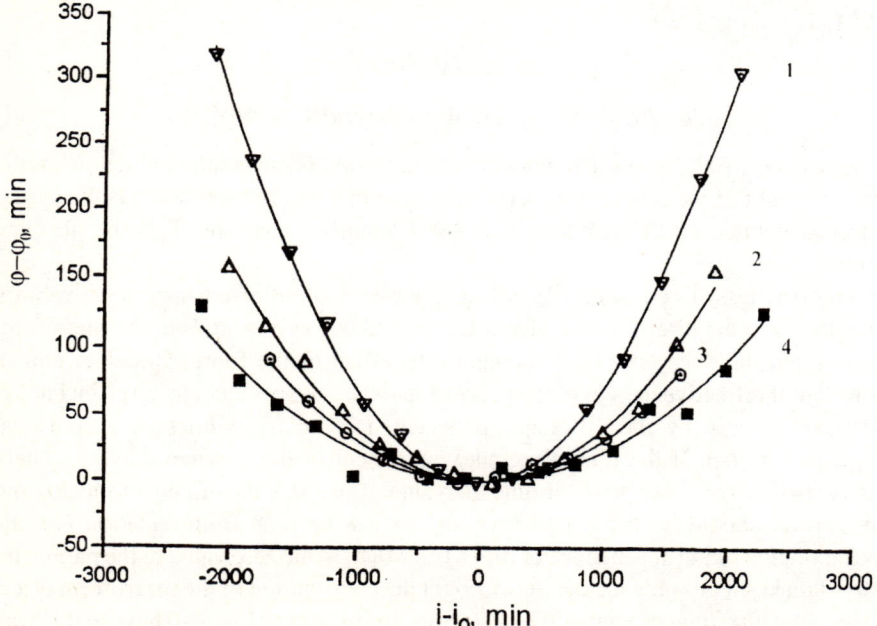

Figure 2.   Plot of $\Delta\varphi$ (see text) *versus* angle i between the beam and normal to the film for dextran films of different thickness H. 1– H=0.47 mm, 2 – 0.23 mm, 3 – 0.13 mm, 4 – 0.075 mm.

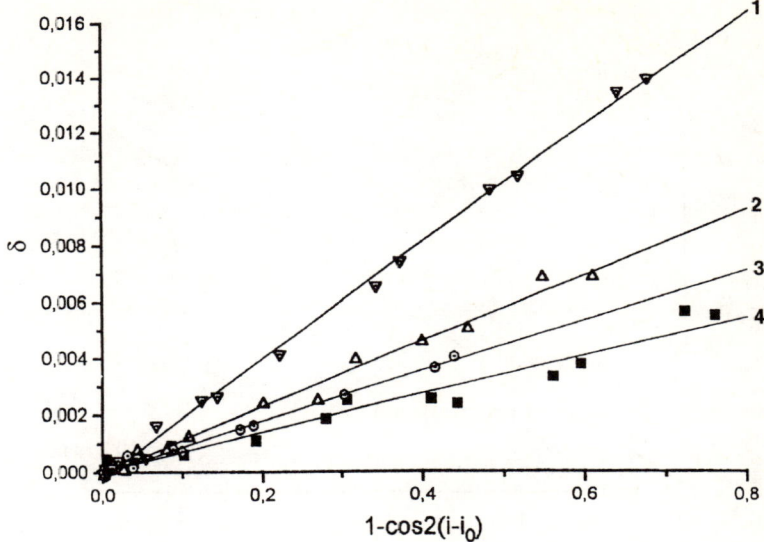

Figure 3. The optical retardation δ vs $(1-\cos 2i)$ for dextran films of different thickness: 1– H=0.47 mm, 2 – 0.23 mm, 3 – 0.13 mm, 4 – 0.075 mm.

polarized beam for dextran films of different thicknesses (Pavlov *et al.*, 1998b). *Figure 3* shows the dependence of δ on $(1-\cos 2i)$ for the same films. According to equation (1), the tangents to the slopes of such plots make it possible for us to evaluate the coefficients of surface birefringence B for all dextran films. Plots similar to those in *Figures 2* and *3* have also been constructed for other polysaccharides, and such plots have enabled the determination of the surface birefringence B for films of different thicknesses for all polysaccharides investigated. The dependences of B on H for these

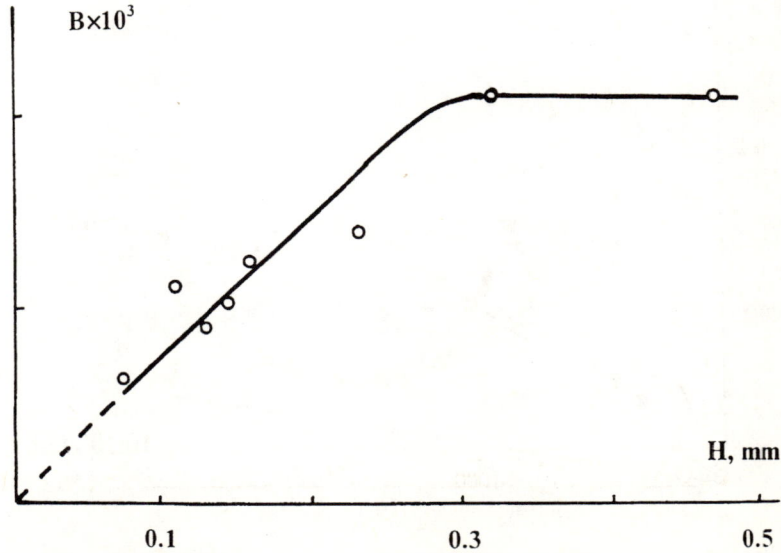

Figure 4. Dependence of surface birefringence B on thickness H of dextran films.

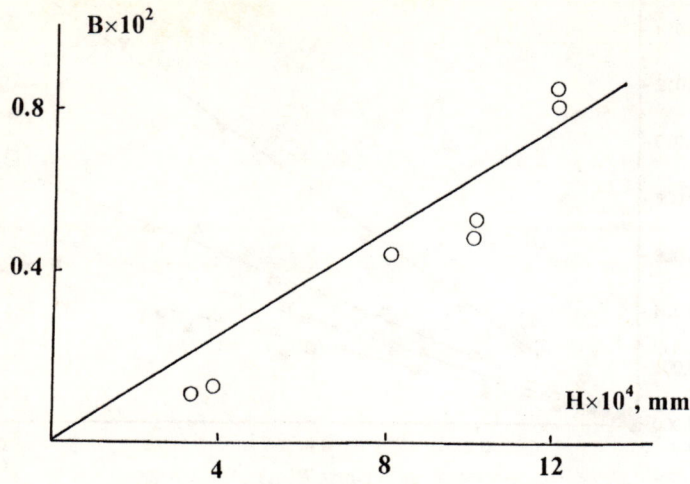

Figure 5.   Dependence of surface birefringence B on thickness H of xanthan films.

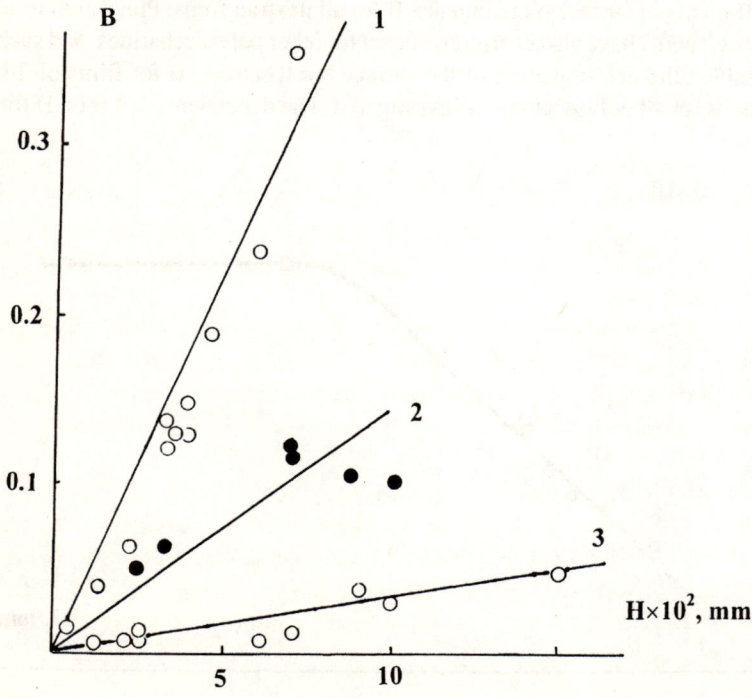

Figure 6.   Dependence of surface birefringence B on thickness H of methylcellulose (1), mannan (2) and pullulan (3) films.

Table 1. The film birefringence, orientational order parameter and equilibrium rigidity of some polysaccharide molecules

Polysaccharide	Structural type	dB/dH × 100 mm^{-1}	S	A × 10^8 cm	Reference
dextran	$\alpha(1-6)$	7.2	−0.007	13	Gekko (1971); Huber (1991)
pullulan	$\alpha(1-6)$, $\alpha(1-6)$	33	−0.012	21	Kawahara (1984); Pavlov (1994)
mannan	$\beta(1-3)$, $\beta(1-4)$	167	−(0.06−0.16)	51	Pavlov (1992)
methyl cellulose	$\beta(1-4)$	463	−(0.17−0.45)	155	Pavlov (1995)
xanthan	double helix	650	−(0.24−0.5)	2400	Sato (1984)

polysaccharides are shown in *Figures 4–6*: all these data are recent work from our St. Petersburg laboratory.

At low values for the ratio $H/H_0 \ll 1$, equation (6) becomes

$$B = -[\pi N_o (\alpha_1 - \alpha_2)/n^3\lambda]((n^2+2)/3)^2 S_o H \qquad (8)$$

It is clear that the slope of the plot of B against H in the region of low H makes it possible to evaluate the orientational order parameter near the interface $S_o = (dB/dH)_{H\to 0}$. Taking into account that $N_o = \rho N_A/M_S$ and using values for the optical anisotropy of unit mass of polysaccharide chains $\beta = (\alpha_1 - \alpha_2)/M_S$, it is easy to estimate S_o.

Table 1 gives the values of orientational order parameter S_o for dextran and pullulan. No reliable data are unfortunately available on intrinsic optical anisotropy of mannan, methyl cellulose, and xanthan. Therefore, the parameter S_o for these polymers has been evaluated by proceeding from the assumption that the specific optical anisotropy values, $\beta = (\alpha_1 - \alpha_2)/M_S$, differ only slightly from those for dextran and pullulan. The S_o values for mannan, methyl cellulose, and xanthan have thus been evaluated in this way (*Table 1*). The analysis of equation for the orientational order factor $S = (3<\cos^2\vartheta>-1)/2$ shows that when the chains are oriented in the plane of film surface, i.e. $125.3° > \vartheta > 54.7°$, then we have $S<0$ ($S=0.5$ at complete planar order). In the case when the chains are largely arranged normally to the surface ($54.7° > \vartheta > -54.7°$), the orientational order parameter $S>0$ and attains a value of unity ($S=+1$) at complete normal order (90°).

Analysis of the experimental data in this way thus shows that fragments of chain molecules of at least all the polysaccharides investigated so far are oriented at the interface mainly *parallel to the film surface*.

As follows from theoretical equation (6), the dependence of B on H makes it possible for us to determine the effective value of the thickness of the optically anisotropic surface layer. This evaluation has thus become possible for films made of dextran for which the dependence of B on H was investigated over a wide range of film thicknesses (Pavlov *et al.*, 1999).

Figure 4 shows that at low H value, B value is proportional to H. However, with increasing film thickness, B attains the limiting value. The experimental dependence of B on H for dextran is best described by the theoretical dependence (equation (6)) when the effective value of surface layer thickness $H_o = 0.14$ mm. The comparison of the value for H_o with segment length A for dextran shows that H_o exceeds A by five order of magnitude.

ORIENTATIONAL ORDER AND THERMODYNAMIC RIGIDITY

Let us consider the problem of the relationship between the capacity of polysaccharide

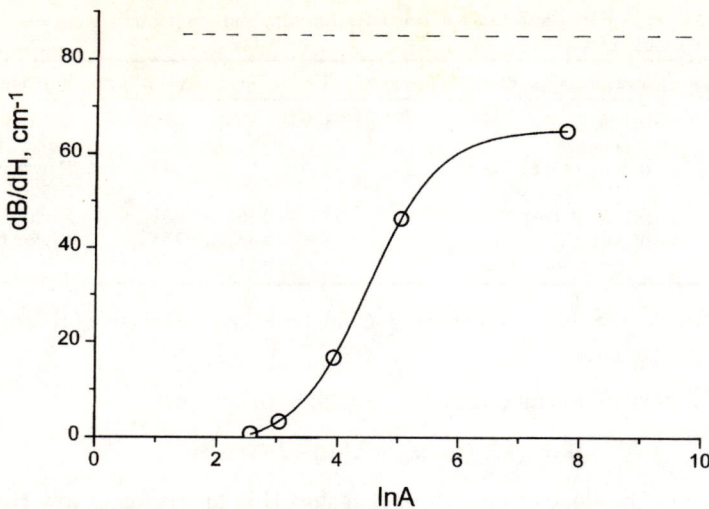

Figure 7.   Plot of (dB/dH) *versus* lnA. A: Kuhn segment length.

molecules to form orientational order in surface layers and the thermodynamic rigidity of molecules.

The surface birefringence from a film reflects the ability of polysaccharide chains to undergo self-organization at the interface or to form molecular orientational order. The value of dB/dH depends on specific optical anisotropy $\beta = (\alpha_1 - \alpha_2)/M_s$ (a value characterizing the structure of the polymer repeat unit; it is effectively a *structural factor* of the surface birefringence) and on the orientational order parameter of molecules in surface layers at the interface S_0 (orietational factor of the surface birefringence).

To a first approximation the value of $\beta = (\alpha_1 - \alpha_2)/M_s$ may be considered to depend slightly on the molecular structure of the polysaccharides investigated. Therefore, the value of dB/dH can serve as a quantitative measure of molecular orientational order of fragments of polysaccharide chains in surface layers.

Figure 7 shows the dependence of dB/dH on the logarithm of length of polysaccharide statistical segments. It is clear that for dextran and pullulan, the molecules of which are characterized by high flexibility, the value of surface birefringence is not high. However, with increasing thermodynamic rigidity of polysaccharides, the value of dB/dH increases abruptly, tending to the limiting value. The experimental values of dB/dH and S_0 (*Table 1*) enable us to evaluate the limiting value of dB/dH for dextran and pullulan proceeding from the assumption that the orientational order parameter of these polysaccharides is characterized by the theoretical maximum value $S_{max} = -0.5$.

$$\lim_{S \to S_{max}}(dB/dH) = (dB/dH)(S_{max}/S_0)$$

The average limiting value lim(dB/dH) calculated for dextran and pullulan was found to be 85cm^{-1}. This limiting value is designated by a broken line in *Figure 7*. It can be seen that the dB/dH values found in this work for five polysaccharides are characterized by the common dependence on lnA and have the same limiting value of dB/dH corresponding to the maximum value of the orientational order parameter of chain fragments with respect to the surface $S_{max} = 0.5$.

Consequently, the results of this work confirm the concept that the tendency to form ordered structures, i.e. the liquid crystalline phase is most pronounced in macromolecular substances which have an elongated shape. All these observations make it possible for us to estimate quantitatively the geometric parameters of molecules, which correspond to the conditions of orientational order formation. As indeed can be seen in *Figure 7*, the formation of orientational order already begins for chain molecules with segment length A=1 nm. The most abrupt change in this parameter takes place at segment length A= 10 nm. The results of this work show unequivocally that *chain thermodynamic rigidity is the determining parameter in the formation of orientational order* and can serve as the principal quantitative criterion for molecular self-organization on the interface.

Final remarks

It is interesting to mention that the relationship between orientational order and thermodynamic flexibility was first pointed out in the interpretation of data on flow birefringence of comb-like polymer solutions (Tsvetkov and Andreeva, 1981). To explain adequately the values of optical anisotropy, it was neccessary to assume that steric interactions of regularly spaced side chains result in an orientational order which corresponds to increasing equilibrium rigidity of these chains.

It should be also noted that quantitative evaluations of equilibrium polysaccharide rigidity that correspond to a transition to a maximum degree of self-organization in surface layers are in agreement with the understanding that the equilibrium rigidity plays an important role in the formation of lyotropic mesophase (Flory, 1956; Khokhlov, 1988).

References

BOGDANOVA, L.M., GRISHCHENKO, A.E., IRZHAK, V.I. AND ROSENBERG, B.A. (1987). Mechanooptical properties and structure of surface layers of epoxide network polymers. *Vysokomol Soedin* **29**, 1588–1592.

CANTOR, C.R. AND SCHIMMEL, P.R. (1980). *Biophysical Chemistry, Part 2*. San Francisco: W.H. Freeman.

CHERKASOV, A.N., VITOVSKAYA, M.G. AND BUSHIN, S.V. (1976). Preferential orientation of macromolecules in the surface layers of polymer blends. *Vysokomol Soedin* **18**, 1628–1635

DE GENNES, P.-J. (1974). *The Physics of Liquid Crystals*. Oxford, UK: Clarendon Press.

ELINOV, N.P. AND VITOVSKAYA, G.A. (1979). Mannan produced by *Rhodotoruoa rubra* strain 14. *Carbohydrate Research* **75**,185–193.

FLORY, P.J. (1956). Phase equilibria in solutions of rod-like particles. *Proceedings of the Royal Society of London* **234A**,73–89.

GEKKO, K. (1971). Physicochemical studies of oligodextran. Intrinsic viscosity molecular weight relations. *Makromol Chem* **148**, 229–238.

GRAY, H.B., BLOOMFIELD, V.A. AND HEARST, J.E. (1967). Sedimentation coefficients of linear and cyclic wormlike coils with excluded-volume effects. *Journal of Chemical Physics* **46**, 1493–1499.

GRISHCHENKO, A.E., RUCHIN, A.E., KOROLIOVA, S.G., SKAZKA, V.S., IRZHAK, V.I., ROSENBERG, B.A. AND ENIKOLOPIAN, N.S. (1983). Study of structure of surface layers of epoxide networks. *Doklady Akademii Nauk SSSR* **269**, 1384–1387.

GRISHCHENKO, A.E. (1996). *Mechanooptics of Polymers*. St. Petersburg: St. Petersburg University.

GRISHCHENKO, A.E. AND CHERKASOV, A.N. (1997). Orientational order in surface layers of polymers. *Advances in Physical Science (Russian)* **167**, 269–285.

HARDING, S.E. (1992). Sedimentation analysis of polysaccharides. In *Analytical Ultracentrifugation in Biochemistry and Polymer Science*. Eds. S.E. Harding, A.J. Rowe and J.C. Horton, pp 495–516. Cambridge: Royal Society of Chemistry.

HARDING, S.E. (1995). On the hydrodynamic analysis of macromolecular conformation. *Biophysical Chemistry* **55**, 69–93.

HERMANS, P.H. (1949). *Physics and Chemistry of Cellulose fibres*. New York: Elsevier.

HUBER, A. (1991). Characterization of branched and linear polysaccharides by size-exclusion chromatography/low-angle laser light scattering. *Journal of Applied Polymer Science Applied Polymer Symposium* **48**, 95–102.

KAWAHARA, K., OHTA, K., MIYAMOTO, H. AND NAKAMURA, S. (1984). Preparation and solution properties of pullulan fractions as standard samples for water-soluble polymers. *Carbohydrate Polymers* **4**, 335–346.

KHOKHLOV, A.R. (1988). Statistical physics of liquid crystal ordering in polymer systems. In *Liquidcrystal Polymers*. Ed. N.A. Plate, pp 98–137. Moscow: Khimia.

MCNALLY, J.G. AND SHEPPARD, S.E. (1930). Double refraction in cellulose acetate and nitrate films. *Journal of Physical Chemistry* **34**, 165–177.

PAVLOV, G.M. (1989). Sedimentation parameter of cellulose, cellulose derivatives and some other polysaccharides. *Khimia Drevesiny (Wood Chem)* **4**, 3–13.

PAVLOV, G.M., PANARIN, E.F., KORNEEVA, E.V., KUROCHKIN, C.V., BAIKOV, V.E. AND USHAKOVA, V.N. (1990). Hydrodynamic properties of poly(1-vinyl-2-pyrrolidone) molecules in dilute solution. *Makromol Chem* **191**, 2889–2899.

PAVLOV, G.M., KORNEEVA, E.V., MICHAILOVA, N.A. AND ANANYEVA, E.P. (1992). Hydrodynamic properties of the fraction of mannan formed by Rhodotorula rubra yeast. *Carbohydr Polymers* **19**, 243–248.

PAVLOV, G.M. (1995). Investigations of cellulose and lignins by molecular hydrodynamic methods. In *Cellulose and cellulose derivatives*. Eds. J.F. Kennedy, G.O. Phillips, P.O. Williams and L. Piculell, pp 541–546. Cambridge: Woodhead Publishers Limited.

PAVLOV, G.M., MICHAILOVA, N.A., TARABUKINA, E.B. AND KORNEEVA, E.V. (1995). Velocity sedimentation of water soluble methyl cellulose. *Progr Colloid Polym Sci* **99**, 109–113.

PAVLOV, G.M., MICHAILOVA, N.A., KORNEEVA, E.V. AND SMIRNOVA, G.N. (1996). Hydrodynamic and molecular characteristics of water-soluble methylcellulose. *Vysokomol Soedin* **38A**, 1582–1586

PAVLOV, G.M. (1997). The concentration dependence of sedimentation for polysaccharides. *European Biophysical Journal* **25**, 385–397.

PAVLOV, G.M., ROWE, A.J. AND HARDING, S.E. (1997). Conformational zoning of large molecules using the analytical ultracentrifuge. *Trends in Analytical Chemistry* **16**, 401–405.

PAVLOV, G.M., YEVLAMPIEVA, N.P. AND KORNEEVA, E.V. (1998), Flow birefringence of pullulan molecules in solution. *Polymer* **39**, 235–240.

PAVLOV, G.M., GRISHCHENKO, A.E., RJUMTSEV, E.I. AND YEVLAMPIEVA, N.P. (1999). Optical properties of dextran in solution and films. *Carbohydrate Polymers* **38**, 267–271.

PAUL, D.R. AND NEWMAN, S. (Eds.) (1978). *Polymer Blends*. Orlando: Academic Press.

SATO, T., NORISUYE, T. AND FUJITA, H. (1984.) Double-stranded helix of xanthan. Dimensional and hydrodynamic properties in 0.1 M aqueous sodium chloride. *Macromolecules* **17**, 2696–2703.

STEIN, R.S. AND NORRIS, F.H. (1956). The X-ray diffraction, birefringence, and infra-red dichroism of stretched polyethylene. *Journal of Polymer Science* **21**, 381–396.

STEIN, R.S. (1978). Optical behavior of polymer blends. In *Polymer Blends*. Eds. D.R. Paul and, S. Newman, pp 393–444. Orlando: Academic Press.

ROSTIASHVILI, V.G., IRZHAK, V.I. AND ROSENBERG, B.A. (1987). Glass-transition in polymers. Leningrad: Chimiya.

TANFORD, C. (1961). *Physical Chemistry of Macromolecules*. New York: J. Wiley and Sons.

TOMBS, M.P. AND HARDING, S.E. (1999). *An Introduction to Polysaccharide Biotechnology*. London: Taylor and Francis.

TSVETKOV, V.N. AND ANDREEVA, L.N. (1981.) Flow and electric Birefringence in rigid-chain polymer solutions. *Advances in Polymer Science* **39**, 96–205.

TSVETKOV, V.N. (1989). *Rigid-chain polymers. Hydrodynamic and Optical Properties in Solution.* New York and London: Consultant Bureau.

TSVETKOV, V.N., ESKIN, V.E. AND FRENKEL, S.YA. (1970). *Structure of Macromolecules in Solutions.* London: Butterworths.

TURKOV, V.K. AND GRISHCHENKO, A.E. (1990). Properties and structure of surface layers of polyesterureaurethanes. *Vysokomol Soedin* **32**,1032–1035.

YALPANI, M. (1988). *Polysaccharides.* Amsterdam: Elsevier.

16

Pectins, Pectinases and Plant-Microbe Interactions

ROLF A. PRADE[1]*, DONGFENG ZHAN[2], PATRICIA AYOUBI[1] AND ANDREW J. MORT[2]

Departments of Microbiology and Molecular Genetics[1] and Biochemistry and Molecular Biology[2], Oklahoma State University, Stillwater, OK 74078, USA

Introduction

Pectins contain two types of covalently linked backbones. Homopolygalacturonic acid (HG) is a linear helix, $\alpha(1{\rightarrow}4)$–linked, that contains a few xylose substitutions in the proximity where the rhamnogalacturonan (RG) is connected. Rhamnogalacturonans consist of linear $\alpha(1{\rightarrow}4)$-linked repeats of the dimer of galacturonic acid linked $\alpha(1{\rightarrow}2)$ to rhamnose (Rha) residues forming $\alpha(1{\rightarrow}2)$-linked dimers. Methyl esterification is found on HG segments. RGs are acetylated and frequently substituted with galactans, arabinans and arabinogalactans linked to Rha residues. Plant cell wall degrading enzymes, including pectinases, are ubiquitous among pathogenic or saprophytic bacteria and fungi. Pectate lyases cleave non-esterified HG-pectate chains by elimination, resulting in unsaturated reaction products. Pectin lyases are able to cleave highly methoxylated HG-pectin and pectin methylesterases remove methyl groups rendering HG-pectin polymers accessible to other depolymerizing enzymes. RG specific enzymes such as rhamnogalacturonase, galactanase, arabinase, arabinosidase and galactosidase have frequently been reported. There are only a few links for which we remain uncertain about the existence of a specific cleaving enzyme. These links are; Ara-Rha, Gal-Rha, Xyl-GalA and unsaturated rhamnose containing products. Microorganisms recognize pectins as complex bonded carbon sources, respond by activating the synthesis and secretion of proteins involved in degradation and metabolize released sugar residues. Pathogenic bacteria and fungi induce pectinases early during infection (penetration). Even though genetically it remains unclear whether these activities are absolutely required for infection, it seems clear that plants respond to invasive pectin degradation by producing polygalacturonase inhibiting proteins (PGIP), indicating a specific biochemical interplay at the pectin degradation level.

Higher plants have developed complex defense systems against potential pathogens. There are numerous structural and chemical features present in plants that discourage

---

*To whom correspondence may be addressed.

Biotechnology and Genetic Engineering Reviews – Vol. 16, April 1999
0264–8725/99/16/361–391 $20.00 + $0.00 © Intercept Ltd, P.O. Box 716, Andover, Hampshire SP10 1YG, UK

discourage visitors from establishing a parasitic interaction. With few exceptions, plant cells are enclosed by multi-layered walls with specialized structures that confer protection against invaders, mechanical strength, shape and size to tissues, organs and entire plants. The primary, secondary and middle lamella are sub-structural levels found in typical plant cell walls. Primary and secondary layers contain variable amounts of cellulose, hemicellulose and pectins. The secondary wall is not always present and is usually involved in providing structural support. Therefore, the cellulose content is increased and pectin content diminished in secondary walls. The middle lamella, also known as the intercellular substance, fills the spaces between primary walls of adjacent cells.

Carbohydrates are the main components of plant cell walls that form the bulk of the supporting structure. As a result, plant cell walls represent the most abundant natural and renewable organic material available on earth. This resource will be used increasingly as a source of energy, as well as raw materials for biotechnological processes.

When microorganisms initiate a colonization cycle on plants, regardless of whether a pathogenic or saprophytic (*i.e.* biomass decay) association, the first complex carbon source they encounter is the pectin present in the middle lamella. Since we know that bacteria and fungi are effective in depolymerizing pectin and other polysaccharides into metabolizable energy sources, the natural interactions that take place at this level are of importance to several aspects of biology.

Complete hydrolysis of pectins involves numerous enzymes – several of them are at least partially dependent on the outcome of one or more prior enzyme/substrate interactions. Thus, natural interactions between pectin degrading enzymes and structural pectins in plants are complex, but specific, and in many instances exhibit 'quasi' redundant functions (e.g. lyases and hydrolases). Moreover, pectin/pectinase interactions between entire plants (tissue) and microorganisms (bacteria or fungi) are highly specific and localized, contingent upon the expression of an infection regulated genetic circuit.

In this discussion we will initially focus on key aspects of pectin structure, then discuss in detail microbial pectin degrading systems, and finally analyze several examples of natural pectin/pectinase interactions. Even though pectins are important industrial food additives and pectinases are at the core of developments in fruit and vegetable preservation technologies, we will not focus on applied aspects of pectins. The discussion presented here is intended to be an updated description of what we know about interactions between plants and microbes that occur in nature. This information should be complementary to several excellent reviews that have recently been published, describing the economics and the utility of pectins (Sakai *et al.*, 1993; Hamilton *et al.*, 1995; Sutherland, 1995; Hugouvieux-Cotte-Pattat *et al.*, 1996; De Lorenzo and Cervone, 1997; Thakur *et al.*, 1997; Hadfield and Bennett, 1998)].

Pectin structure

Pectin means different things to different people. In this discussion, we will describe pectins as those cell wall polysaccharides containing a relatively high proportion of $\alpha(1\rightarrow4)$-linked galacturonic acid (GalA). Others have often used operational definitions of pectin as those polysaccharides extracted from cell walls by chelators or hot water.

The definition is complicated further by having to decide how far, out from the GalA-containing region, in the molecule is still pectin. Within the pectin molecule, extensive side chains can be attached to the GalA-containing backbones, and these might be cross-linked to other polymeric cell wall components, such as proteins (Qi *et al.*, 1995), and hemicellulose (Keegstra *et al.*, 1973; Fu and Mort, 1996). In an attempt to be consistent and perfectly general, we will consider first the two major types of pectin backbone structures, and then describe the many substituents found on them.

PECTIN BACKBONES

There is strong evidence for two types of backbones in pectins. The best known, homogalacturonan (HG), depicted in *Figure 1*, contains long stretches of $\alpha(1{\rightarrow}4)$-linked D-galactopyranosyl uronic acid residues (Thibault *et al.*, 1993). The other, rhamnogalacturonan (RG), depicted in *Figure 2*, contains stretches of the disaccharide α-D-GalA*p* $(1{\rightarrow}2)$ α-L-Rha*p* linked $(1{\rightarrow}4)$ to each other (Lau *et al.*, 1985). Both backbones show a wide variety of decorations on them such as acylations on the 2-, or 3-hydroxyl positions of galacturonic acid residues (GalA), esterifications of GalA acidic groups, and glycosyl side chains on 2- or 3-hydroxyl positions of GalA or 4-hydroxyl position of rhamnose (Rha). Three major questions remain about pectin backbone structures: a) the length and variability of polyGalA stretches, b) the length and variability of polyGalA-Rha stretches, and c) whether or not HG and RG regions are covalently linked together.

The question of how long and uniform HG regions are is difficult to address. Determining the molecular weight of pectin fragments is tricky due to their polyanionic character (Fishman *et al.*, 1984; Mort *et al.*, 1991). In addition, RG regions (as suggested by the presence of rhamnose) are often found in association with HG regions. Thus, it has been difficult to determine whether or not HG regions are interrupted by RG regions or vice versa. Thibault *et al.* (1993) found that prolonged mild acid hydrolysis of de-esterified pectins causes solubilization of most of the

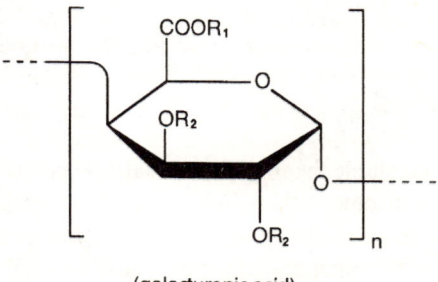

(galacturonic acid)

R_1 = H, Me or non-methyl esters

R_2 = H or acetyl

Figure 1. The Homogalacturonan (HG) Region. Homogalacturonan (HG) regions contain most of the galacturonic acid content found in native pectin molecules. They form long linear helical, often heavily methylesterified (R_1) and rarely acetylated (R_2) structures with essentially no side chains. n, indicates the degree of HG polymerization which can be greater than 100 residues (Chambat and Joseleau, 1980).

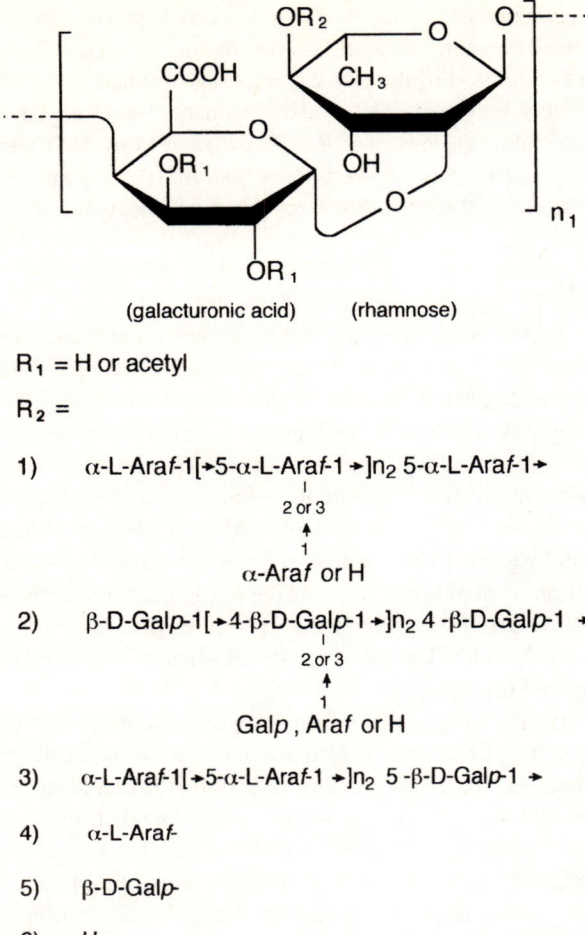

R$_1$ = H or acetyl

R$_2$ =

1)   α-L-Ara*f*-1[→5-α-L-Ara*f*-1 →]n$_2$ 5-α-L-Ara*f*-1→
                                    |
                                  2 or 3
                                    ↑
                                    1
                              α-Ara*f* or H

2)   β-D-Gal*p*-1[→4-β-D-Gal*p*-1 →]n$_2$ 4 -β-D-Gal*p*-1 →
                                    |
                                  2 or 3
                                    ↑
                                    1
                              Gal*p* , Ara*f* or H

3)   α-L-Ara*f*-1[→5-α-L-Ara*f*-1 →]n$_2$ 5 -β-D-Gal*p*-1 →

4)   α-L-Ara*f*-

5)   β-D-Gal*p*-

6)   H

Figure 2. The Rhamnogalacturonan (RG) Region. Rhamnogalacturonan (RG) regions contain galacturonic acid, rhamnose (GalA-Rha) dimer repeats with frequent and variable side chains containing arabinose (Ara) and galactose (Gal). n$_1$, indicates the number of GalA-Rha disaccharide repeat units, reported to be up to at least 30 (Zhan *et al.*, 1998) and as high as 200 (McNeil *et al.*, 1980). n$_2$, indicates the number of sugar residues in the side chains (from 0 to 40).

neutral sugars, leaving insoluble the majority of GalA. Pectic acids containing 23 or more residues are acid insoluble (Hotchkiss and Hicks, 1990). In addition, the rhamnose content also decreases with time in these acid precipitates, indicating gradual destruction of RG containing regions. Thus, it is likely that most of the insoluble material is composed of HG. Furthermore, this insoluble precipitate is soluble after neutralization, and the approximate molecular weight of 20,000 suggests HG fragments containing, on average, 115 GalA residues.

It has been generally accepted that HGs contain periodic interruptions caused by single rhamnose residue insertions which would produce kinks in an otherwise extended helical structure (Rees and Wright, 1971; Powell *et al.*, 1982; Jarvis, 1984). However, we were unable to find the expected rhamnose containing GalA oligomers in enzyme-digested citrus pectin (Zhan *et al.*, 1998). The existence of Rha inserts has not yet been investigated in other pectins.

The helical nature of pectic acid (HG with no esterifications or sugar side chains) has been determined by three-dimensional structural modeling (Rees and Wright, 1971), investigated by x-ray diffraction, and analyzed by various forms of spectroscopy (Rees, 1982). Even though there seems to be agreement on the helical nature of pectic acid chains, there remains the question as to whether there are two or three GalA residues per helix turn. Ca^{2+} ions interact with pectic acid molecules to form strong complexes, if the participating GalA-oligomers are 14 residues or longer. Grant and coworkers (Grant *et al.*, 1973) proposed an 'egg box' type of model in which two helical polyGalA fragments interact with each other via ionic interactions and coordination of Ca^{2+} ions between chains.

The length of RG backbones has also been elusive. In sycamore the length of RG backbones was calculated to be around 200 GalA-Rha repeats, based on gel permeation chromatography elution profiles (McNeil *et al.*, 1980). The above number of GalA-Rha repeats is an approximate prediction, because it was based on dextran standards and side chains had to be artificially discounted. In addition, in most plant cell wall extracts the RG region cannot be separated from HG regions. In apple (Schols *et al.*, 1990), various vegetables (Schols *et al.*, 1994; Schols *et al.*, 1995), watermelon, and cotton (Yu and Mort, 1996), RG appears to be strongly associated with a portion of HG heavily substituted with $\beta(1\rightarrow3)$-linked xylose. The xylose substitution on the HG fragment prevents endopolygalacturonase (EPG) digestion, making it difficult to determine whether a stretch of RG contains HGs at one or both ends, or if the backbone consists of several interspersed RG sections flanked by xylosylated HG segments. In beet pectin, RGs contain at least ten GalA-Rha repeat units (Renard *et al.*, 1995) while citrus pectin has up to at least 30 units (Zhan *et al.*, 1998). RG backbone oligomers of up to these lengths have been detected as the result of controlled (partial) acid hydrolysis that causes selective release of neutral sugar side chains from Rha before affecting the majority of Rha-GalA linkages of the backbone, producing a homologous series of RG oligomers. High performance anion exchange chromatography (HPAEC) or capillary zone electrophoresis (CZE) can further analyze these oligomers. Each oligomer can also be compared to known standards, and in the case of HPAEC, eluted for further compositional and structural analysis.

So far, all of the evidence indicates the presence of covalent linkage between HG and RG regions. All methods of extracting pectins yield preparations containing GalA and Rha along with varying amounts of arabinose (Ara), galactose (Gal), xylose (Xyl), mannose (Man), or glucose (Glc). Usually, there is a great preponderance of GalA, indicating the dominance of HG, but the presence of Rha shows RG is also present. Gel permeation or ion exchange chromatography does not yield fractions containing only GalA if the extraction did not involve HG cleaving enzymes, or if the tissue from which the pectin was extracted was not rich in pectinases.

DECORATIONS ON HG BACKBONES

Methyl esters

Methanol, as the methyl ester of carboxylic acid groups, is a well studied adornment on HGs. The degree of methyl esterification (DM) varies widely from plant to plant, cell to cell, and from one location within a cell wall to another (Liners and Van Cutsen, 1992; McCann *et al.*, 1994; Femenia *et al.*, 1998). The pattern of esterification has been suggested to be random in some cases (de Vries *et al.*, 1983; Mort *et al.*, 1993) and well defined in others (de Vries *et al.*, 1986; Mort *et al.*, 1993). The activity of enzymes that degrade pectin is directly affected by the presence of methyl esters (Chen and Mort, 1996). It is widely believed (convincing evidence is available for only a few cases) that newly synthesized pectin has a high degree of esterification and that pectin methylesterases (PME) produce the lower DM pectins (Goldberg *et al.*, 1986).

Plant PMEs are reported to cause blockwise de-esterification (Taylor, 1982), and PMEs of pathogenic or saprophytic microbes are thought to be random de-esterifiers. Grasdalen and collaborators (Grasdalen *et al.*, 1988) found that the chemical shift of the H-5 and H-1 signals in the proton NMR spectrum of GalA reflects its esterification and esterification of neighboring residues. DeVries and collaborators (de Vries *et al.*, 1986) used the digestion products from pectin lyase to infer patterns of esterification. The lyase was postulated to require three or four adjacent esterified GalA residues for activity. Mort and coworkers (Mort *et al.*, 1993) devised a quantitation method for the various lengths of contiguous non-esterified GalA residues in a pectin fragment. The experiment was based on a combination of reduction of esterified residues to galactose, specific hydrogen fluoride (HF) solvolytic cleavage of galactosyl glycosidic linkages and separation of the resulting GalA containing oligomers. We found that commercial pectins with around 50% esterification have a random pattern, but some HGs extracted from cotton suspension cultures show an almost strictly alternating pattern of esterified and non-esterified residues.

Non-methyl esters on GalA carboxyl groups

There are several reports of esters in pectins involving alcohols other than methanol. In no case however, has the type of alcohol been determined. The presence of these esters was implied indirectly, and in some cases up to 30% of the esters has been presumed to be something other than methyl ester (Kim and Carpita, 1992; McCann *et al.*, 1994; Needs *et al.*, 1998). When Needs and collaborators (Needs *et al.*, 1998) attempted to isolate oligosaccharides containing putative non-methyl esters from carrot roots after driselase (a cocktail of digestive enzymes) digestion, they failed, suggesting that driselase may contain activities which hydrolyze non-methyl esters or that they do not exist. Additional evidence for the presence of a small amount of alcohols other than methanol esterified to pectin was presented by Brown and coworkers (Brown and Fry, 1993) who detected anomalous behavior on thin-layer-chromatography (TLC) of driselase digested pectin fragments. Saponification of the oligomers converted them to $GalA_3$. In our opinion, the presence of non-methyl esters in pectins will not be certain until they have been identified by direct methods.

Finally, it has been suggested multiple times that GalA carboxyls could be esterified to other sugar hydroxyls to form cross-links (Fry, 1986). However, no convincing evidence has been published, although there is an intriguing suggestion that pectin from peas can be cross-linked by a trans-esterification induced by pectin methylesterases (Hou and Chang, 1996).

Acetate esters

Acetate esters on the 2- and 3- hydroxyl position of GalA residues in HGs have been reported frequently on sugar beet pectin (Rombouts and Thibault, 1986; Thibault *et al.*, 1993), but they are only a minor substituent on most HGs. Ishii has isolated small amounts of acetylated GalA trimers from bamboo (Ishii, 1995) and potatoes (Ishii, 1997) and characterized them by extensive NMR and mass spectral analysis.

Xylogalacturonan

Xylose has been recognized as a minor component in most pectin preparations and is found concentrated at particular sites, often referred to as xylogalacturonan, XGA (*Figure 3*). Bouveng (1965) described a xylogalacturonan extracted from pine pollen as being an HG with xylose linked $\beta(1\rightarrow3)$ to approximately one in every two GalA residues. In their generalized pectin structure, Cook and Stoddart (1973) included side chains of up to three sugar residues linked to HG regions via the $\beta(1\rightarrow3)$-xylose residues. These included: $\beta(1\rightarrow2)$-galactosyl-xylose, $\alpha(1\rightarrow2)$-fucosyl-xylose, and both of these oligomers elongated by addition of a $\beta(1\rightarrow3)$-glucuronosyl residue on the fucose or $\beta(1\rightarrow6)$-glucuronosyl residue on the galactose. In recent review articles, the presence of xylogalacturonan regions is mostly ignored (McNeil *et al.*, 1984; Bacic *et al.*, 1988; Carpita and Gibeaut, 1993; Albersheim *et al.*, 1996). However, recent work from Voragen's (Schols and Voragen, 1994; Schols *et al.*, 1995) and Mort's (Yu and Mort, 1996) groups has shown the presence of xylogalacturonans in

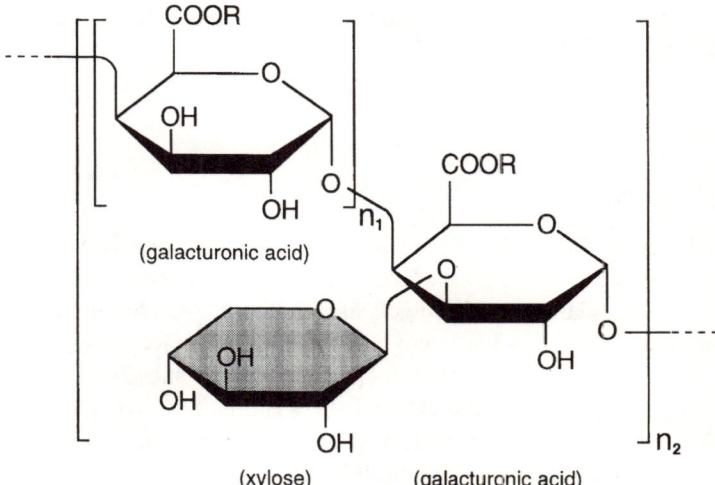

R = H or Me

Figure 3. The Xylogalacturonan (XGA) Region. Xylogalacturonan (XGA) regions are HG segments containing xylose substituents located close to the region where the rhamnogalacturonan (RG) region is believed to interact in native pectin molecules. n_1, indicates the number of non-substituted, interspersed GalA residues in the XGA backbone (average of 1) and n_2, the number of Xyl-GalA-GalA$_{n1}$ repeated units (Yu and Mort, 1996).

many fruit pectins. Kikuchi and coworkers (Kikuchi *et al.*, 1996) have also found substantial amounts of xylogalacturonan in cultured carrot cells.

Rhamnogalacturonan II

In 1978 Darvill and coworkers (Darvill *et al.*, 1978) reported that endopoly-galacturonase digestion of sycamore cells releases a polymeric region of pectin rich in GalA, Rha, and a variety of other sugars. They designated this region rhamnogalacturonan II (RGII) because of its high Rha and GalA content and to distinguish it from the rhamnogalacturonan region which had already been designated rhamnogalacturonan I (RG). *Figure 4* shows a structural description of RGII. Over the last 20 years, members of what now is named the CCRC (Complex Carbohydrate Research Center, Athens GA, USA) have increasingly refined the structure of this region (O'Neill *et al.*, 1996 and references therein). RGIIs have been found in all plants in which its presence was investigated, and it appears that all are almost structurally identical. Moreover, it appears that RGII is the site at which most of the boron found in plants is bound (Matoh *et al.*, 1993). Thus, it is tempting to conclude that RGIIs form cross-links between pectin molecules (Kobayashi *et al.*, 1996; O'Neill *et al.*, 1996) via borate interactions between apiose residues from independent RGII sections.

Apiogalacturonan

In a few plant species such as *Lemna minor* and *Zostera nana* there is a high degree of apiose substituents on the HG backbone, and the apiose occurs as apiobiose linked directly to an as yet unidentified position on GalA residues (Hart and Kindel, 1970a; Hart and Kindel, 1970b).

DECORATIONS ON RG BACKBONES

Acetate and methyl esters

Komalavilas and Mort (1989) first reported acetylation specifically on rhamno-galacturonan after they isolated the GalA-Rha disaccharide repeat unit acetylated at position O-3 of the GalA after HF solvolysis of cotton suspension culture walls. Their conclusion that position O-2 was not acetylated in native RGs was discounted by Lerouge and collaborators (Lerouge *et al.*, 1993) who showed, using ^{1}H NMR spectroscopy, acetylation on both O2- and O3- positions of GalA residues recovered from EPG solubilized sycamore RG fragments. Ishii also found acetate esters on both, 2- and 3-hydroxyl positions in bamboo RGs (Ishii, 1995) and potato (Ishii, 1997).

GalA linked sugar side chains

Most, or all, of the side chains found in RGs are connected to the backbone through Rha residues. There have been reports about a small proportion of the GalA residues in the rhamnogalacturonan region having xylose linked to them (Cheethan *et al.*, 1993; An *et al.*, 1994). However, we strongly suspect that, since in most pectins

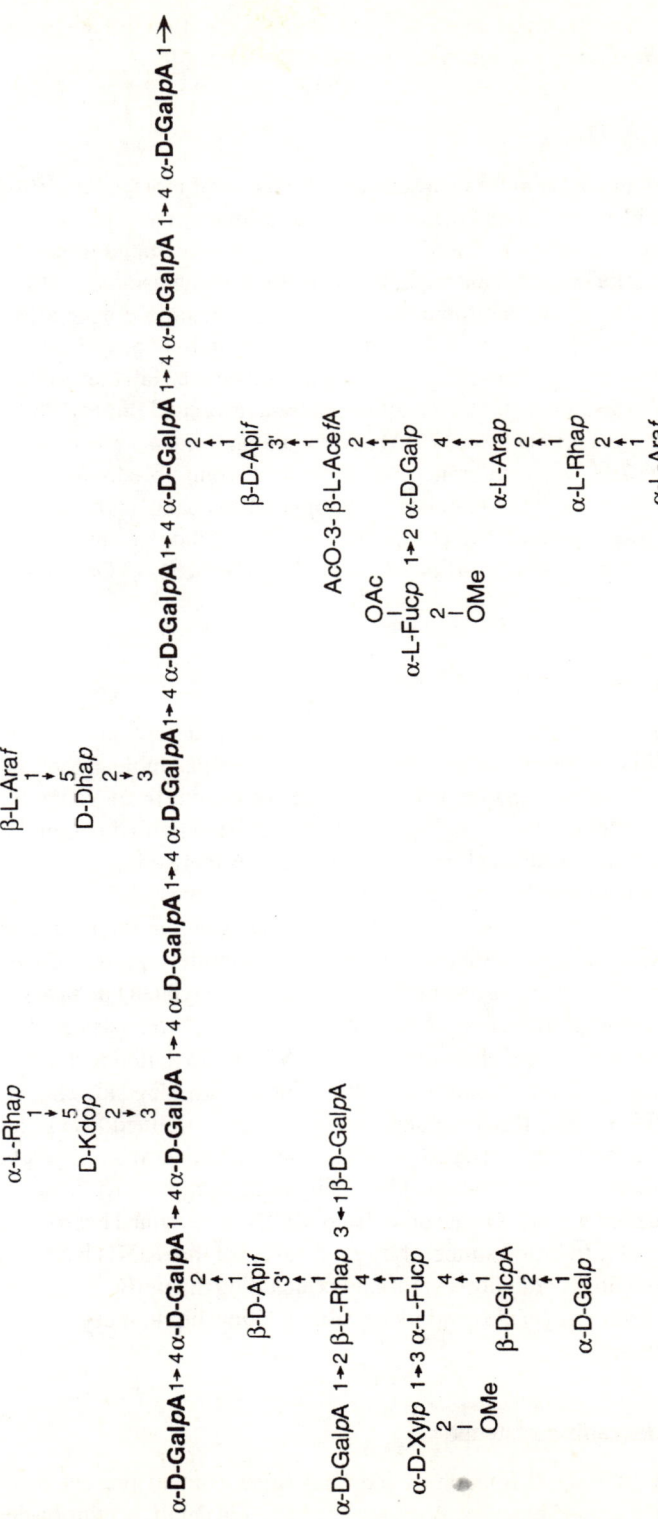

Figure 4. The Rhamnogalacturonan–II (RGII) Region (partial structure). Rhamnogalacturonan–II (RGII) is found as oligomeric regions present in all native pectins, with an almost identical structural configuration. This region contains a GalA backbone with extensive side chains containing rhamnose (Rha), galactose (Gal), xylose (Xyl), glucose (Glc), fucose (Fuc), apiose (Api), arabinose (Ara), 2-keto-3-deoxy-D-manno-octulopyranosylonic acid (Kdo) and 3-deoxy-D-lyxo-2-heptulopyranosylaric acid (Dha) and aceric acid (AceA). RGIIs are resistant to endopolygalacturonase (EPG) digestion. The order and position of side chain attachments on the GalA backbone has not been determined and were arbitrarily assigned. Structural representation redrawn from O'Neill *et al.* (1996).

xylogalacturonans are localized in close proximity to the RG-HG junction, the xylosylated GalA is probably not internal to the RG region itself.

Rha linked sugar side chains

Since the discovery that RGs make up a substantial fraction of most pectins (McNeil *et al.*, 1980), it has been shown that many of the Rha residues have sugar side chains attached to them at position O-4. These side chains are quite variable in length and sugar composition. The predominant sugars are arabinose and galactose, but fucose, glucuronic acid, 4-O-methyl glucuronic acid, xylose, and rhamnose have also been reported. Lau and coworkers (Lau *et al.*, 1987) isolated and characterized a large variety of oligomers linked to rhamnitol after they destroyed the GalA residues in the RG backbone using a dissolving metal reduction reaction (Mort and Bauer, 1982). The arabinose appears most often as $\alpha(1\rightarrow5)$-linked arabinofuranans with $\alpha(1\rightarrow3)$ arabinose and some $\alpha(1\rightarrow2)$ side branches. Galactose is found predominantly as a single galactose residue, or $\beta(1\rightarrow4)$-linked galactopyranose chains with galactose or arabinose side branches through the O-3 positions. An arabinogalactan based on a backbone of $\beta(1\rightarrow3)$-linked Gal with $\beta(1\rightarrow6)$-linked side branches of Gal and/or Ara is also quite common.

Pectin nomenclature

There are quite a few names used to describe pectins and sections of pectins. Protopectin is an old term used to refer to pectin that is in its insoluble native state in cell walls. Protopectinases are enzymes which solubilized pectins from plant material (Sakai *et al.*, 1993). Pectate, or polypectate, is the salt of de-esterified pectin. Pectic acid is the acid form of de-esterified pectin. Within pectins there are usually various regions, some of which have been given very specific names such as rhamno-galacturonan II, and I (McNeil *et al.*, 1984), and some more general, such as hairy and smooth regions (Schols and Voragen, 1996). The smooth regions consist of homogalacturonan (HG) which may be methylesterified or acetylated but have few, if any, side chains. The hairy regions are very complex, consisting of both rhamnogalacturonan and xylogalacturonan regions. Within the rhamnogalacturonan (RG) regions there may also be subdivisions into sections with long side chains and others with short side chains. Rhamnogalacturonan I (RG) is defined as a rhamno-galacturonan with a perfectly repeated disaccharide backbone with side chains attached to O-4 positions of Rha residues. Thus, hairy regions contain RGs, but RG is only a section of the hairy region. On the other hand, RGIIs were named based on their GalA and Rha content. Structural studies, however, revealed that RGIIs have an HG backbone with side chains containing several sugars including rhamnose. Localization of RGIIs in native pectin molecules and its possible relationship to hairy or smooth regions remains unknown.

Microbial pectin degrading systems

Pectins make up a substantial fraction of the total sugar content present in plant materials collectively termed biomass. As described earlier in detail, pectins basically utilize the majority of their sugar residues in the assembly of a linear molecule with a homopolygalacturonic (HG), non-branched methyl-esterified segment, and a rhamno-galacturonan (RG) branched and acetylated fragment. A simplified and schematic representation of a generic pectin molecule is shown in *Figure 5.*

Microorganisms in general recognize pectins, under a variety of physiological circumstances, as a potential, but complex bonded carbon source. They respond positively to the extracellular presence of pectins by activating the synthesis and secretion of a wide range of proteins (all enzymes that degrade a portion of a polysaccharide defined as pectin) involved in modification, degradation, transport, assimilation and metabolism of the released sugar residues. Complete enzymatic degradation by pectinases results primarily in galacturonic acid and rhamnose with production of lesser amounts of galactose, arabinose, methanol, acetate and traces of numerous other sugars (e.g. xylose, fucose, apiose), all of which are assimilated and metabolized by microorganisms, to promote vegetative growth.

The foremost biochemical activity of a microbial pectinase is to cleave glycosidic bonds in native pectin molecules, initially resulting in the production of intermediate sized fragments containing multiple residues followed by the release of individual sugars. In general, enzymes involved in depolymerization are extracellular and the final products are assimilated and metabolized. Typically, monomers and dimers are transported into the cytoplasm, which are then degraded by cytoplasmic enzymes. However, in certain bacteria (e.g. *Erwinia* and *Bacterioides*) specific enzymes are found in association with the outer membrane, periplasm or cytoplasmic membrane. In *Figure 5*, the mode of action and predicted products for most types of microbial pectinases is shown. In addition, an extensive survey of bacterial pectinases, including extracellular and intracellular steps can be found in *Table 1.*

Pectinolytic microorganisms produce an array of pectin degrading enzymes reflecting the complexity of the natural substrate. Pectinases are classified according to their site of cleavage, endo or exo (if they cleave within or at the end of the substrate chain, respectively), preferred substrate (pectin or pectate) and the mode by which they cleave the glycosidic bond (hydrolase or lyase). Pectin hydrolases transfer a single proton from a donor acidic amino acid to the glycosidic oxygen and cleavage involves the net consumption of a water molecule. Eliminases, commonly known as lyases, cleave glycosidic bonds via an elimination mechanism that generates unsaturated oligomeric or monomeric reaction products. The proton, α to the carbonyl group, is abstracted by a basic amino acid and followed by elimination of the sugar, resulting in a 4,5 GalA unsaturation (Linhardt *et al.*, 1986).

Pectin lyases act on highly methyl-esterified pectins and do not require calcium for enzymatic activity. Pectate-lyases are catalytically similar to pectin-lyases, except that they cleave non-esterified GalA residues and require calcium for optimal activity (Pickersgill *et al.*, 1994; Mayans *et al.*, 1997). Until recently, it was not clear whether calcium is part of the enzyme (Rao *et al.*, 1996) or is involved in substrate binding (Crawford and Kolattukudy, 1987). Recent x-ray studies on a pectate lyase from *Bacillus subtilis* complexed with calcium, shows that calcium interacts with an arginine residue that is conserved across all pectin and pectate lyases (Pickersgill *et al.*, 1994).

Pectin- and pectate-lyases are common among fungi and have been found in most species where they have been looked at (Dean and Timberlake, 1989; Gysler *et al.*,

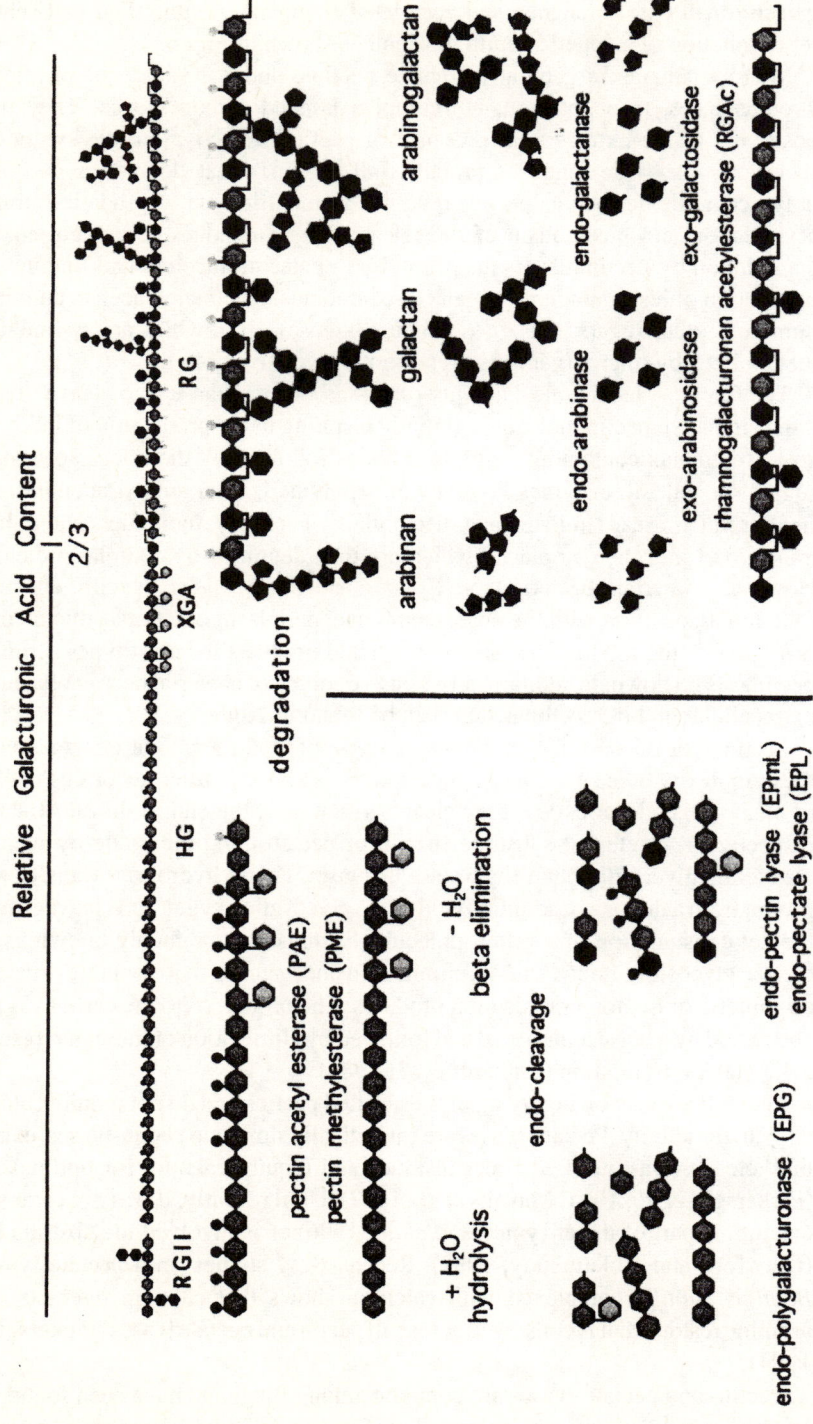

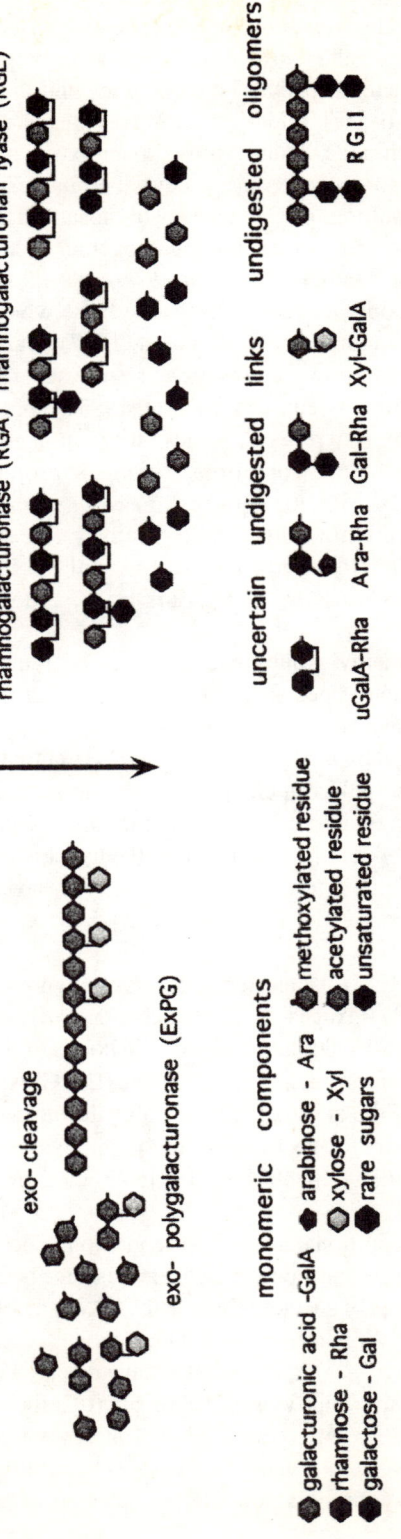

Figure 5. Mode of Action and Predicted Products for Known Pectinases. A schematic of a generic pectin molecule and its complete degradation by extracellular enzymes is shown. Length and distribution of linear and branched portions were based on the relative content of galacturonic acid recovered from backbones containing exclusively polygalacturonic acid (HG, XGA and RGII) and rhamnogalacturonan (RG). Only enzymes directly involved in degradation are listed. For detailed information of related metabolic activities refer to *Table 1*. At the bottom of the figure sugar residues found as constituents in native pectins and fragments that cannot be degraded with known enzymes are shown. HG, homogalacturonan (polygalacturonic acid), RG, rhamnogalacturonan (region 1 or hairy region), RGII, rhamnogalacturonan region 2 as defined by Darvill and collaborators (Darvill *et al.*, 1978) and XGA, xylogalacturonan. Positions of covalent links are drawn as reported for the backbone regions only. Covalent link representations in side chains are approximations.

1990; Polizeli *et al.*, 1991; Gonzalez-Candelas and Kolattukudy, 1992; Bowen *et al.*, 1995; Ho *et al.*, 1995; Kopecny and Hodrova, 1995). In bacteria, pectate lyases represent the largest group of bacterial pectinolytic enzymes and are directly involved in the destruction of plant tissues. Endo-pectate lyases cleave non-esterified pectate chains at internal sites within the chain resulting in a mixture of unsaturated oligomers. Typically, bacterial exo-pectate lyases cleave the non-reducing end of non-esterified pectate chains resulting in the release of unsaturated di- or tri-galacturonide (*Table 1*). Interestingly, many phytopathogenic bacteria produce more than one pectate lyase. For example, *Erwinia chrysanthemi* secretes eight endo-pectate lyase isozymes and only a single exo-pectate lyase which is retained in the periplasm (Shevchik and Hugouvieux-Cotte-Pattat, 1997). Based on amino acid sequence relatedness, bacterial pectate lyases are currently separated into five different classes (Shevchik and Hugouvieux-Cotte-Pattat, 1997). This functionally redundant group of enzymes apparently reflects, in part, differential enzymatic sets recruited under varying physiological conditions – *e.g.* saprophytic versus pathogenic growth (Salmond, 1994). Phytopathogenic *Erwinia* species and *Pseudomonas marginalis* also produce extracellular endo-pectin lyases which, unlike the pectate lyases, are induced by DNA damaging agents and possibly by plant host phytoalexins (Tsuyumu and Chatterjee, 1984; Zink *et al.*, 1985; Sone *et al.*, 1988).

Pectin methylesterases remove methyl groups rendering pectin polymers accessible to the depolymerizing enzymes such as pectate lyases and polygalacturonases. Two different pectin methylesterases have been identified from *E. chrysanthemi*. The first was identified as an extracellular pectin methylesterase (*PmeA*) (Laurent *et al.*, 1993) while the second, *PmeB*, was identified as an outer membrane lipoprotein (Shevchik *et al.*, 1996). *PmeB* has a greater activity on small pectic oligomers which can easily diffuse into the periplasm (Shevchik *et al.*, 1996). Pectin methylesterases have also been found in other phytopathogenic bacteria (Schell *et al.*, 1994), several intestinal bacteria (Cornick *et al.*, 1994; Tierny *et al.*, 1994) and in fungi (Khanh *et al.*, 1991; Wojciechowski and Fall, 1996).

Polygalacturonases (PG), poly (1,4-α-D-galacturonide) glycan hydrolases, catalyze the cleavage of glycosidic α(1→4) linkages on HG backbones. Polygalacturonases are of the endo (EPG) type if they cleave internal glycosidic bonds of the HG chain, the products are oligomeric and monomeric, and are of the exo (ExPG) type when they cleave at the non-reducing end of the substrate chain. Polygalacturonases are only active on non-esterified pectin regions, thus demanding the removal of methylester groups for complete depolymerization (Khanh *et al.*, 1991; Wojciechowski and Fall, 1996).

In the literature there is a disproportional number of reports on endo- versus exo-acting fungal pectinases. For example, endo-polygalacturonases have been cloned and genetically studied in a large number of species (Keon and Waksman, 1990; Bussink *et al.*, 1991; Reymond *et al.*, 1994; Whitehead *et al.*, 1995; Centis *et al.*, 1996; Gao *et al.*, 1996; Guo *et al.*, 1996; Centis *et al.*, 1997), and also have received considerable attention from physiologists (Kester and Visser, 1990; Scott-Craig *et al.*, 1990; Polizeli *et al.*, 1991; Clay *et al.*, 1997; Patino *et al.*, 1997; Takasawa *et al.*, 1997). In contrast, less attention has been given to the genes for exo-polygalacturonases (Scott-Craig *et al.*, 1998), although this activity is consistently detected in physiological studies (Kester and Visser, 1990; Polizeli *et al.*, 1991; Aguilar and Huitron, 1993; Di Pietro and Roncero, 1996; Patino *et al.*, 1997).

In bacteria, endo- and exo-polygalacturonases have been identified as extracellular depolymerizing enzymes of phytopathogenic *Erwinia* (Kotoujansky, 1987; He and Collmer, 1990). In contrast to pectate lyases, polygalacturonases of *Erwinia* do not require calcium but are stimulated by sodium, potassium and ammonium ions (Nasuno and Starr, 1966). Polygalacturonase has also been identified in *Agrobacterium tumefaciens* (Rodrigues-Palenzuela *et al.*, 1991) and several intestinal bacteria (see *Table 1*).

Rhamnogalacturonan acetyl-esterase (Kauppinen *et al.*, 1995), rhamnogalacturonan hydrolase (Mutter *et al.*, 1994) and rhamnogalacturonan lyase (Mutter *et al.*, 1996) deacetylate and cleave the rhamnogalacturonan backbone (RG) of the substituted regions of native pectins, respectively (Schols *et al.*, 1990; Schols and Voragen, 1994). These enzymes have been found frequently in fungi (Schols *et al.*, 1990; Sakamoto *et al.*, 1994; Suykerbuyk *et al.*, 1995; Chen *et al.*, 1997; Nelson *et al.*, 1997). Except for a pectin acetyl esterase (Shevchik and Hugouvieux-Cotte-Pattat, 1997), we are unaware of similar reports in bacteria. Rhamnogalacturonases cleave de-acetylated regions of the RG backbone, producing linear hexamers and tetramers with Gal branches (see *Figure 5*). Additional enzymes that degrade the RG backbone have been found and include: rhamnogalacturonan α-D-galactopyranosyluronohydrolase, which releases GalA residues from the non-reducing end of RGs but not HGs (Mutter *et al.*, 1998a; Mutter *et al.*, 1998b) and rhamnogalacturonan α-L-rhamnopyranohydrolase, which cleaves Rha residues from the non-reducing end of RGs (Mutter *et al.*, 1994).

Enzymes that degrade RG side chains have also been found and include: arabinases, α-L-arabinofuranosidase (v.d. Veen *et al.*, 1991; Ramon *et al.*, 1993; v. d. Veen *et al.*, 1993; Flipphi *et al.*, 1994; Ruijter *et al.*, 1997), endo-(1→5)-α-L-arabinases (Flipphi *et al.*, 1993), and galactanases, exo-β-(1→4)-galactanase (Bonnin *et al.*, 1995; Christgau *et al.*, 1995), endo-β-(1→4)-D-galactanase (Yamaguchi *et al.*, 1995), exo-(1→3)-β-D-galactanase – able to by-pass the branching points of galactan backbones (Pellerin and Brillouet, 1994) and endo-β-(1→5)-galactofuranase (Reyes *et al.*, 1992). Only arabinases have been found in bacteria so far (McKie *et al.*, 1997; Sakamoto *et al.*, 1997).

Enzymes that remove Xyl residues from the XGA region have not yet been reported. However, activities that release GalA residues from the non-reducing end of XGA trimers, have been observed. This enzyme also acts as an exo-galacturonase, releasing GalA monomers from HGs. Thus, it is likely that this enzyme is an exo-polygalacturonase, not inhibited by xylosylated GalA residues. Other links for which a suitable enzyme has not been identified (or tested) include the entire RGII region, the RG portions, Ara-Rha as well as Gal-Rha branching bonds, and Rha containing unsaturated elimination (lyase) products.

In bacteria, the major end products of polygalacturonases and pectate lyases are saturated and unsaturated di-galacturonate, respectively, which enter cells and are further metabolized intracellularly. Extracellular galacturonate enters the cell through active transport in *E. chrysanthemi* or is produced intracellularly by the action of oligo-galacturonate lyase. Oligogalacturonate lyase is an intracellular enzyme in *E. chrysanthemi* which cleaves the di-galacturonates into two different monomers: 5-keto-4-deoxy-uronate (DKI) and galacturonate. The monomers DKI and galacturonate

Table 1. Survey of bacterial pectin degrading enzymes

Enzyme[1]	Organism(s)	Location	Inducer	Substrate(s)	Final Product(s)
Pectate lyases					
endo-pectate lyase (EPL)	Ecar, Ecry	extracellular	KDG	HG	unsaturated oligogalacturonate
	Sbov	extracellular	constitutive	HG	unsaturated tri- and tetra-galacturonate
exo-pectate lyase (ExPL)	Pvir	extracellular	NR	HG	NR
	Lmul, Lpec	extracellular	NR	HG	unsaturated digalacturonate
	Bfib, Bgal, Bpec	extracellular	NR	HG	unsaturated trigalacturonate
	Ecar, Ecry	periplasm	KDG	HG, methyl-pectin	unsaturated digalacturonate
	Bthe	intracellular	HG	HG	unsaturated digalacturonate
pectate lyase (PL)	Xcam	extracellular	HG	HG	NR
	Bsub, Pflo	extracellular	constitutive	HG	NR
	Pmar	extracellular	NR	HG	NR
	Yent, Ypse	periplasm and cytoplasm	NR	HG	NR
	Bfra, Bova, Bsp., Bvul	intracellular	HG	HG	NR
	Koxy, Kpne	intracellular	NR	HG	NR
	Bpol, Cmul, Psyr, Rsol	NR	NR	HG	NR
Pectin lyases					
endo-pectin lyase (EPmL)	Ecar, Ecry, Pmar	extracellular	DNA damage	pectin	NR
pectin lyase (PmL)	Pvir	NR	NR	pectin	NR
Esterases					
pectin methylesterase (PME)	Cthe[2]	extracellular	pectin, pectate and oligogalacturonate	methyl-pectin	HG, methanol
	Ecry, Bfib, Bgal, Bpec	extracellular	NR	methyl-pectin	HG, methanol
	Ecry	outer membrane	NR	methyl-oligogalacturonides	HG, methanol
	Bthe	periplasm and cytoplasm	NR	methyl-pectin	HG, methanol
pectin acetyl esterase (PAE)	Cmul, Lpec, Rsol	NR	NR	methyl-pectin	HG, methanol
	Ecry	NR	NR	demethylated, acetyl-pectin	pectin, acetate
Polygalacturonases					
endo-polygalacturonase (EPG)	Ecar, Pmar, Rsol	extracellular	NR	NR	NR
	Atum	extracellular	constitutive	HG	NR
	Bcep	NR	NR	HG	NR

Enzyme	Organism	Location	Substrate		Product
exo-polygalacturonase (ExPG)	Che², Ecry	extracellular	pectin, pectate and oligogalacturonate	HG, nonreducing end	di- and tri-galacturonate
	Ecry	extracellular	NR	HG, nonreducing end	di- and tri-galacturonate
	Bhe	inner membrane	HG	NR	galacturonate
	Bfib	intracellular	NR	triglacturonate digalacturonate	galacturonate,
polygalacturonase (PG)	Lmul	NR	NR	NR	NR
	Koxy	extracellular	NR	NR	NR
Catabolism of branched pectin					
endo-arabinase	Bsub	extracellular	NR	RGI linked arabinan	mono-, di- and tri-arabinan
endo (exo)-arabinase	Pflo	extracellular	NR	RGI linked arabinan	arabinotriose
Intracellular catabolism of pectin oligomers and monomers					
exo-oligogalacturonate hydrolase	Che	intracellular	pectin, pectate	di- and tri-galacturonate	galacturonate
oligogalacturonate lyase	Ecar, Ecry	intracellular	NR	di- and oligo-galacturonides	galacturonate and DKI
galacturonate isomerase	Che, Echr	intracellular	NR	galacturonate	tagaluronate
tagaluronate reductase	Che, Echr	intracellular	NR	tagaluronate	altronate
altronate reductase	Che, Echr	intracellular	NR	altronate	KDG
deoxyuronate isomerase	Echr	intracellular	NR	DKI	DKII
KDG oxidoreductase	Echr	intracellular	NR	DKII	KDG
KDG kinase	Che, Echr	intracellular	NR	KDG	KDGP
KDGP aldolase	Che, Echr	intracellular	NR	KDGP	pyruvate and G3P

[1]where reported, endo- or exo-enzyme activities are noted; [2]pectin methylesterase and polygalacturonase activities are present in one extracellular protein complex (Van Rijssel *et al.*, 1993); NR, not reported; HG, homogalacturonate; DKI, 5-keto-4-deoxyuronate; DKII, 2,5-diketo-3-deoxyuronate; KDG, 2-keto-3-deoxygluconate; KDGP, 6-phospho-2-keto-3-deoxygluconate; G3P, glyceraldehyde-3-phosphate.

Organisms names abbreviated as follows: *Atum*, *Agrobacterium tumefaciens* (McGuire *et al.*, 1991); *Bcep*, *Burkholderia* (*Pseudomonas*) *cepacia* (Gonzalez *et al.*, 1997); *Bfib*, *Butyrivibrio fibrisolvens* (Wojciechowicz *et al.*, 1982; Heinrichova *et al.*, 1985); *Bfra*, *Bacteroides fragilis* subsp. a (McCarthy and Salyers, 1986); *Bgal*, *Bacteroides galacturonicus* (Jensen and Canale-Parola, 1986); *Bova*, *Bacteroides ovatus* (McCarthy and Salyers, 1986); *Bpec*, *Bacteroides pectinophilus* (Jensen and Canale-Parola, 1986); *Bsp.*, *Bacteroides sp.* strain 3452A (McCarthy and Salyers, 1986); *Bsub*, *Bacillus subtilis* (Nasser *et al.*, 1990; Sakamoto *et al.*, 1997); *Bthe*, *Bacteroides thetaiotaomicron* (McCarthy *et al.*, 1985; Tierny *et al.*, 1994); *Bvul*, *Bacteroides vulgatus* (McCarthy and Salyers, 1986); *Cmul*, *Clostridium multifermentans* (Sheiman *et al.*, 1976); *Cthe*, *Clostridium thermosaccharolyticum* (Van Rijssel *et al.*, 1993; Van Rijssel *et al.*, 1993b); *Ecar*, *Erwinia carotovora* (Tsuyumu and Chatterjee, 1984; Hinton *et al.*, 1989); *Ecry*, *Erwinia chrysanthemi* (Hugouvieux-Cotte-Pattat *et al.*, 1996; Shevchik and Hugouvieux-Cotte-Pattat, 1997); *Koxy*, *Klebsiella oxytoca* (Walker and Pemberton, 1987); *Kpne*, *Klebsiaella pneumoniae* (Chatterjee *et al.*, 1979); *Lmul*, *Lachnospira mutiparus* (Wojciechowicz *et al.*, 1980); *Lpec*, *Lachnospira pectinoschiza* (Cornick *et al.*, 1994); *Pflo*, *Pseudomonas fluorescens* (Liao, 1991; McKie *et al.*, 1997); *Pmar*, *Pseudomonas marginalis* (Sone *et al.*, 1988; Nikaidou *et al.*, 1993); *Psyr*, *Pseudomonas syringae* (Bauer and Collmer, 1997); *Pvir*, *Pseudomonas viridiflava* (Liao *et al.*, 1988); *Rsol*, *Ralstonia* (*Pseudomonas*) *solanacearum* (Huang and Schell, 1990; Schell *et al.*, 1994); *Sbov*, *Streptococcus bovis* (Wojciechowicz and Ziolecki, 1984; *Xcam*, *Xanthomonas campestris* (Liao *et al.*, 1996); *Yent*, *Yersinia enterolitica* (Chatterjee *et al.*, 1979); *Ypes*, *Yersinia pestis* (von Riesen, 1975); *Ypse*, *Yersinia pseudotuberculosis* (Chatterjee *et al.*, 1979).

are then metabolized in *E. chrysanthemi* by two distinctly different enzyme systems to produce 2-keto-3-deoxy-gluconate (KDG) which is further converted to pyruvate and 3-phosphoglyceraldehyde (Hugouvieux-Cotte-Pattat and Robert-Baudouy, 1989). Intracellular di- and tri-galacturonate are hydrolyzed to monomers by oligoglacturonate hydrolase in *Clostridium thermosaccharolyticum*. Galacturonate is further metabolized by a modified KDG pathway similar to that described for *Erwinia* species but involves galacturonate isomerase, tagaturonate reductase and altromate dehydratase (Van Rijssel *et al.*, 1993). While considerable data are available on the uptake and metabolism of pectin hydrolysis products in bacterial systems, no such data are available in fungi.

In most filamentous fungi, pectinases, xylanases, as well as cellulases, are under carbon catabolite repression (Dean and Timberlake, 1989; Dean and Timberlake, 1989; Polizeli *et al.*, 1991; Ho *et al.*, 1995; Reymond-Cotton *et al.*, 1996; Ruijter *et al.*, 1997). Very little is known about the induction phase of these complex and important biochemical activities. We know that induction of cell wall degrading enzymes involves constitutive low level expression of at least some of the structural genes (El-Gogary *et al.*, 1989). In the presence of substrate or partially degraded fragments, these enzymes produce trace amounts of degradation products that are supposedly chemically modified (e.g. via positional isomerization) resulting in a potent inducer. It is assumed that the signal is transduced through several unknown steps and culminates with the massive synthesis of cell wall degrading enzymes (Sternberg and Mandels, 1979; Biely *et al.*, 1980; Polizeli *et al.*, 1991; v. d. Veen *et al.*, 1993). The recognition process remains unclear and no advances have recently been made.

Plant-microbe interactions

SUBSTRATE SPECIFICITY AND PRODUCT DISTRIBUTION

Because of the difficulty in obtaining highly defined pectin substrates and analyzing the exact nature of the products of enzyme digestions, only a few pectin degrading enzymes have been well characterized with respect to their exact substrate specificity and product distribution. Most often, the hydrolytic type of enzymes are assayed by their production of reducing equivalents using a colorimetric assay, and lyases by 236 nm absorbance reflecting the generation of unsaturated GalA residues.

In the last few years, it has become much easier to characterize reaction products because of the high resolving power of high performance liquid chromatography (HPLC) anion exchange columns and the high sensitivity of the pulsed electrochemical, or amperometric, detector commercialized by the Dionex company. It is possible that these will be partially superseded by capillary electrophoresis with fluorescence detection. We have found that fluorescently labelled pectic oligosaccharides are good substrates for most pectin degrading enzymes, and that by capillary electrophoretic analysis the exact nature of the products can be determined quickly, if suitable standards are available (Mort and Chen, 1996; Zhan *et al.*, 1998).

Fungal endopolygalacturonases appear to bind four (or five (Benen *et al.*, 1996)) adjacent GalA residues of the chain and hydrolyze between the last two residues toward the reducing end of the bound cluster (Rexova-Benkova and Markovic, 1976; Thibault, 1983; Mort and Chen, 1996). Some enzymes act randomly, whereas it appears that others exhibit a multiple attack action. Instead of dissociating from the substrate, the enzyme sometimes 'moves' along the chain by one residue towards the non-reducing end and catalyzes repeated glycosidic hydrolysis (Benen *et al.*, 1996).

Exogalacturonases in fungi hydrolyze single GalA residues from the non-reducing end of polymeric, or oligomeric galacturonan (Kester *et al.*, 1996). In contrast, bacteria usually remove disaccharides from the non-reducing end of pectate by an exo-pectate lyase. The substrate specificity of fungal pectin lyases is not well known. Van Houdenhoven (1975) suggests the enzyme binds between 8 and 10 sugar residues. De Vries and collaborators (de Vries *et al.*, 1983) inferred by modelling pectin lyase degradation of apple pectins, that the enzyme needs three, but perhaps four, adjacent esterified residues.

In bacteria, pectate lyases seem to be the most important enzymes in HG degradation. There is good evidence that several enzymes, all capable of degrading pectate, with different specificities are produced at the same time by the same organism. Preston and collaborators (Preston *et al.*, 1992) compared four pectate lyases from *E. chrysanthemi* acting on pectic acid and found that one (*Pla*) produced random cleavage products, two (*Plb*, *Plc*) produced predominantly unsaturated GalA trimers, and another (*Ple*), unsaturated dimers. The three enzymes that cleave in a non-random fashion are likely to function via a multiple attack (scanning) mechanism observed in some fungal EPGs (Benen *et al.*, 1996). Another comparison testing various pectate lyases from *E. carotovora* on differentially methylesterified substrates resulted in synergistic inter-actions, indicating that individual enzymes are differentially affected by the amount of methyl ester groups present on the substrate molecule (Bartling *et al.*, 1996).

BACTERIAL INFECTIONS

Bacteria have variable impacts on agriculture. For example, bacterial pectinases are beneficial in processes like biomass utilization, but the same activities are devastating when acting as virulence factors in plant diseases (He *et al.*, 1993). The insoluble plant cell wall polymers are significant barriers to microorganisms, yet represent an abundant source of carbohydrates. The disease process in general requires pectin degradation and is often dependent on plant sensitivity and environmental conditions such as temperature, nitrogen starvation, osmolarity and oxygen limitations (Hugouvieux-Cotte-Pattat *et al.*, 1992).

To address whether or not pectin degradation was required to establish bacterial pathogenesis, Kelemu and collaborators (Kelemu and Collmer, 1993) created an *E. chrysanthemi* recombinant strain, deficient in all known pectate lyases and polygalacturonases. However, the pectinase deficient mutant was still able to degrade the host plant tissue. This initially disappointing observation led to the discovery of a whole 'new' set of pectate lyases that are induced and secreted *in planta*. The fact that bacteria have evolved alternate sets of functionally redundant cell wall degrading enzymes expressed *in planta* illustrates the importance of these enzymes in the natural life cycle of these apparently simple microorganisms.

FUNGAL INFECTIONS

Most filamentous fungi are outfitted with a specific genetic program that allows them to infect plant cells and colonize specific organs, tissues or entire plants. Although pathogen-host interactions are in many cases determined by very specific biochemical relationships, a more or less morphologically defined developmental programme of events is consistently observed. A typical pathological cycle is initiated when fungal spores land and attach on the surface of a compatible host. During germination, a filamentous cell grows along the plant surface and undergoes several mitotic cell divisions until, in response to physical and chemical signals, a morphologically differentiated cell-type, the *appressorium*, is formed. From the apical vegetative cell, most of the cytoplasm migrates into the forming appressorium (Bourett *et al.*, 1987; Kwon and Hoch, 1991; Kwon *et al.*, 1991; Gross *et al.*, 1993). Penetration of the host is accomplished by the formation of a penetration peg whose production is coordinated by the fully differentiated appressorium. The penetration peg punches and dissolves the thick plant cell wall and seeks the protoplast where the apex penetrates the membrane (Mould *et al.*, 1991; Mould *et al.*, 1991). Once the penetration step has been completed, the infection hypha gains access to host nutrients and returns to the vegetative mode of growth (polar extension).

Compelling physiological evidence for the role of pectin degrading enzymes in pathogenesis has been obtained for many diseases. However, although the physiological evidence is compelling, a satisfactory genetic proof indicating a substantial implication of pectin degradation remains to be produced (Mendgen and Deising, 1993; Walton, 1994; Scott-Craig *et al.*, 1998). There is not much doubt that these enzymes are specifically induced by fungal pathogens during infection and progression of disease (Bateman *et al.*, 1976; Keon and Waksman, 1990; Walton and Cervone, 1990; Van Hoof *et al.*, 1991; Cary *et al.*, 1995; Ho *et al.*, 1995; Wu *et al.*, 1995). However, genetic inactivation (disruption) of individual enzyme-coding loci has shown no detectable phenotypes. The inactivation, in *C. carbonum*, of any one cell wall-degrading enzyme (endo-polygalacturonase, β-$(1\rightarrow4)$-xylanase, exo-$(1\rightarrow3)$-glucanase or a cellulase) does not abolish the infection process as a whole (Scott-Craig *et al.*, 1990; Apel *et al.*, 1993; Schaeffer *et al.*, 1994; Sposato *et al.*, 1995). Moreover, a *C. carbonum* endo-polygalacturonase and exo-polygalacturonase double 'null' mutant is still pathogenic (Scott-Craig *et al.*, 1998). There are many possible interpretations why the disruption of one gene does not inactivate infection: one is that pectin degradation is not needed for infection and another that genetic redundancy and biochemical overlap found in plant cell wall degrading systems compensates the genetic 'null' phenotype (Prade, 1996).

PLANT PROTEINS THAT INHIBIT PGS

Plants produce a class of proteins that inhibit fungal PGs. These polygalacturonase-inhibiting proteins (PGIPs) were reviewed recently (De Lorenzo and Cervone, 1997). PGIPs were discovered in the early '70s, but have only been studied intensively in the '80s. It has been shown that PGIPs are able to inhibit the progress of the invading pathogen in spreading through the infected plant. One possible and likely explanation for the mode of action of PGIPs is that they attenuate the pectin breakdown rate, thus slowing the reduction of oligogalacturonides into trimers, dimers and monomers (De Lorenzo and Cervone, 1997). The reduced efficiency in breaking down HG fragments increases the lifetime of various intermediate sized GalA fragments (12–14mers) and it has been shown that these kinds of fragments are able to activate plant resistance responses (Darvill *et al.*, 1994; Cook *et al.*, 1998). PGIPs have been observed in a wide variety of dicots, and in monocots they have been reported in the onion family. We do not know of any attempts to survey plants for PGIPs to see how widely they are distributed. Constitutive levels of PGIP vary with age and cell type, and in some cases are induced through wounding or the presence of elicitors. In beans, three families of PGIP genes that are differentially expressed have been reported (De Lorenzo and Cervone, 1997). Interestingly, known PGIPs do not inhibit PGs produced by plants or bacteria (De Lorenzo and Cervone, 1997).

Conclusions

Finally, on one hand we are able to provide a fairly detailed account on pectin structure and enzyme systems found in microorganisms. Pectinases, in conjunction with other large sets of degrading activities, are likely to entirely degrade any significant structural components of the invaded host into metabolizable products. This one sided view suggests that microbes are able to assemble an impressive biochemical arsenal. On the other hand, we are beginning to appreciate that a full blown microbial attack is not always successful because hosts have evolved specific biochemical barriers. Even though we know little about PGIPs, their existence alone surely tempt the suggestion that additional, yet unknown biochemical plant (host) fungus (pathogen) communication mechanisms may play determining roles in the outcome of disease. Thus, if biochemical defense reactions between pathogens and hosts are in fact more common and widespread, as currently suspected, plant infections and other related interactions (e.g. rotting) may include biochemical interplay reactions that directly interfere with microbial functions such as nutrition and penetration.

Acknowledgements

We thank Marlee Pierce (Oklahoma State University) for her valuable discussions and precise suggestions. We also thank Wolfgang D. Bauer (Ohio State University) and Greg May (Baylor College of Medicine) for critically reviewing the manuscript. Grants from the United States Department of Agriculture (USDA), National Research Initiative (NRI), and the Department of Energy (DOE) currently support research in the Mort and Prade laboratories.

References

AGUILAR, G. AND HUITRON, C. (1993). Conidial and mycelial-bound exo-pectinase of *Aspergillus sp. FEMS Microbiology Letters* **108**, 127–132.

ALBERSHEIM, P., DARVILL, A.G., O'NEILL, M.A., SCHOLS, H.A. AND VORAGEN, A.G.J. (1996). An hypothesis: the same six polysaccharides are components of the primary cell walls of all higher plants. In *Pectins and Pectinases*. Eds. J. Visser and A.G.J. Voragen, Vol. **14**, pp 47–55. Amsterdam: Elsevier Science.

AN, J., O'NEILL, M.A., ALBERSHEIM, P. AND DARVILL, A.G. (1994). Isolation and structural characterization of β-D-glucosyluronic acid and 4-O-methyl β-D-glucosyluronic acid-containing oligosaccharides from the cell-wall pectic polysaccharide, rhamnogalacturonan I. *Carbohydrate Research* **252**, 235–243.

Apel, P.C., Panaccione, D.G., Holden, F.R. and Walton, J.D. (1993). Cloning and targeted gene disruption of *XYL1*, a beta 1,4 xylanase gene from the maize pathogen *Cochliobolus carbonum*. *Molecular Plant Microbe Interactions* **6**, 467–473.

Bacic, A., Harris, P.J. and Stone, B.A. (1988). Structure and function of plant cell walls. In *The Biochemistry of Plants: A Comprehensive Treatise, vol 14, Carbohydrates*. Ed. J. Preiss, pp. 297–371. New York: Academic Press.

Bartling, S., Derkx, P., Wegener, C. and Olsen, O. (1996). *Erwinia* pectate lyase differences revealed by action pattern analyses. In *Pectins and Pectinases*. Eds. J. Visser and A.G.J. Voragen, Vol. **14**, pp 283–293. Amsterdam: Elsevier Science.

Bateman, D.F., Jones, T.M. and Yoder, O.C. (1976). Degradation of corn cell walls by extracellular enzymes produced by *Helminthosporium maydis* race. *Trends in Phytopathology* **63**, 1523–1529.

Bauer, D.W. and Collmer, A. (1997). Molecular cloning, characterization, and mutagenesis of a *pel* gene from *Pseudomonas syringae* pv. *lachyrmans* encoding a member of the *Erwinia chrysanthemi pelADE* family of pectate lyases. *Molecular Plant Microbe Interactions* **10**, 369–379.

Benen, J.A.E., Kester, H.C.M., Parenicova, L. and Visser, J. (1996). Kinetics and mode of action of *Aspergillus niger* polygalacturonases. In *Pectins and Pectinases*. Eds. J. Visser and A.G. J. Voragen, pp 221–230. Amsterdam: Elsevier.

Biely, P., Vrsanská, M. and Krátky, Z. (1980). Xylan-degrading enzymes of the yeast *Cryptococcus albidus*. Identification and cellular localization. *European Journal of Biochemistry* **108**, 313–321.

Bonnin, E., Lahaye, M., Vigouroux, J. and Thibault, J.F. (1995). Preliminary characterization of a new exo-beta-(1,4)-galactanase with transferase activity. *International Journal of Biological Macromolecules* **17**, 345–351.

Bourett, T., Hoch, H.C. and Staples, R.C. (1987). Association of the microtubule cytoskeleton with the thigmotropic signal for appressorium formation in *Uromyces*. *Mycologia* **79**, 540–545.

Bouveng, H.O. (1965). Polysaccharides in pollen II. The xylogalacturonan from mountain pine (*Pinus mugo* Turra) pollen. *Acta Chem Scandinavia* **19**, 953–963.

Bowen, J.K., Templeton, M.D., Sharrock, K.R., Crowhurst, R.N. and Rikkerink, E.H. (1995). Gene inactivation in the plant pathogen *Glomerella cingulata*: three strategies for the disruption of the pectin lyase gene *pnlA*. *Molecular and General Genetics* **246**, 196–205.

Brown, J.A. and Fry, S.C. (1993). Novel Oβ-galacturonyl esters in the pectic polysaccharides of suspension-cultured plant cells. *Plant Physiology* **103**, 993–999.

Bussink, H.J., Brouwer, K.B., de Graaff, L.H., Kester, H.C. and Visser, J. (1991). Identification and characterization of a second polygalacturonase gene of *Aspergillus niger*. *Current Genetics* **20**, 301–307.

Carpita, N.C. and Gibeaut, D.M. (1993). Structural models of primary cell walls in flowering plants: consistency of molecular structure with the physical properties of the walls during growth. *Plant Journal* **3**, 1–30.

Cary, J.W., Brown, R., Cleveland, T.E., Whitehead, M. and Dean, R.A. (1995). Cloning and characterization of a novel polygalacturonase-encoding gene from *Aspergillus parasiticus*. *Gene* **153**, 129–133.

Centis, S., Dumas, B., Fournier, J., Marolda, M. and Esquerre-Tugaye, M.T. (1996). Isolation and sequence analysis of *Clpg1*, a gene coding for an endopolygalacturonase of the phytopathogenic fungus *Colletotrichum lindemuthianum*. *Gene* **170**, 125–129.

Centis, S., Guillas, I., Sejalon, N. and Esuerre-Tugaye, M.T. (1997). Endopolygalacturonase genes from *Colletotrichum lindemuthianum*: cloning of *CLPG2* and comparison of its expression to that of *CLPG1* during saprophytic and parasitic growth of the fungus. *Molecular Plant Microbe Interactions* **10**, 769–775.

Chambat, G. and Joseleau, J.-P. (1980). Isolation and characterization of a homogalacturonan in the primary cell walls of *Rosa* cells cultures *in vitro*. *Carbohydrate Research* **85**, C10–C12.

Chatterjee, A.K., Buchanan, G.E., Behrens, M.K. and Starr, M.P. (1979). Synthesis and excretion of polygalacturonic acid trans-eliminase in *Erwinia, Yersinia, and Klebsiella* species. *Canadian Journal of Microbiology* **25**, 94–102.

CHEETHAN, N.W.H., CHEUNG, P.C.-K. AND EVANS, A.J. (1993). Structure of the principal non-starch polysaccharide from the cotyledons of *Lupinus angustifolius* (cultivar Gungurru). *Carbohydrate Polymers* **22**, 37–47.

CHEN, E.M.W. AND MORT, A.J. (1996). Nature of sites hydrolyzable by endopolygalacturonase in partially esterified homogalacturonans. *Carbohydrate Polymers* **29**, 129–136.

CHEN, H.J., SMITH, D.L., STARRETT, D.A., ZHOU, D., TUCKER, M.L., SOLOMOS, T. AND GROSS, K.C. (1997). Cloning and characterization of a rhamnogalacturonan hydrolase gene from *Botrytis cinerea. Biochemistry and Molecular Biology International* **43**, 823–838.

CHRISTGAU, S., SANDAL, T., KOFOD, L.V. AND DALBOGE, H. (1995). Expression cloning, purification and characterization of a beta-1,4- galactanase from *Aspergillus aculeatus. Current Genetics* **27**, 135–141.

CLAY, R.P., BERGMANN, C.W. AND FULLER, M.S. (1997). Isolation and characterization of an endopolygalacturonase from *Cochliobolis sativus* and a cytological study of fungal penetration of barley. *Phytopathology* **87**, 1148–1459.

COOK, B.J., CLAY, R.P., BERGMANN, C.W., ALBERSHEIM, P. AND DARVILL, A.G. (Eds.)(1998). Fungal polygalacturonases exhibit different substrate degradation patterns which correlate with susceptibilities to PGIPs. In *Cell Walls '98: 8th International Cell Walls Meeting,* pp 9–13. Norwich, UK: John Innes Centre.

COOK, G.M.W. AND STODDART, R.W. (1973). In *Surface Carbohydrates of the Eukaryotic Cell,* 346 pages. London and New York: Academic Press.

CORNICK, N.A., JENSEN, N.S., STAHL, D.A., HARTMAN, P.A. AND ALLISON, M.J. (1994). *Lachnospira pectinoschiza* sp. *nov.,* an anaerobic pectinophile from the pig intestine. *International Journal of Systematic Bacteriology* **44**, 87–93.

CRAWFORD, M.S. AND KOLATTUKUDY, P.E. (1987). Pectate lyase from *Fusarium solani* f. sp. *pisi,* purification, characterization, *in vitro* translation of the mRNA, and involvement in pathogenicity. *Archives of Biochemistry and Biophysics* **258**, 196–205.

DARVILL, A., BERGMANN, C., CERVONE, F., DE LORENZO, G., HAM, K.S., SPIRO, M.D., YORK, W.S. AND ALBERSHEIM, P. (1994). Oligosaccharins involved in plant growth and host-pathogen interactions. *Biochemical Society Symposium* **60**, 89–94.

DARVILL, A.G., MCNEIL, M. AND ALBERSHEIM, P. (1978). Structure of plant cell walls. VII. A new pectic polysaccharide. *Plant Physiology* **62**, 418–422.

DE LORENZO, G. AND CERVONE, F. (1997). Polygalacturonase-inhibiting proteins (PGIP)s, Their role in specificity and defense against pathogenic fungi. In *Plant Microbe Interactions.* Eds. G. Stacey and N.T. Keen, pp 76–93. New York: Chapman and Hall.

DE VRIES, J.A., HANSEN, M., SØDEBERG, J., GLAHN, P.-E. AND PEDERSEN, J.K. (1986). Distribution of methoxyl groups in pectins. *Carbohydrate Polymers* **6**, 165–176.

DE VRIES, J.A., RAMBOUTS, F.M., VORAGEN, A.G.J. AND PILNIK, W. (1983). Distribution of methoxyl groups in apple pectic substances. *Carbohydrate Polymers* **3**, 245–258.

DEAN, R.A. AND TIMBERLAKE, W.E. (1989). Production of cell wall-degrading enzymes by *Aspergillus nidulans,* a model system for fungal pathogenesis of plants. *Plant Cell* **1**, 265–273.

DEAN, R.A. AND TIMBERLAKE, W.E. (1989). Regulation of the *Aspergillus nidulans* pectate lyase gene *(pelA). Plant Cell* **1**, 275–284.

DI PIETRO, A. AND RONCERO, M.I. (1996). Purification and characterization of an exo-polygalacturonase from the tomato vascular wilt pathogen *Fusarium oxysporum* sp. *lycopersici. FEMS Microbiology Letters* **145**, 295–299.

FEMENIA, A., GAROSI, P., ROBERTS, K., WALDRON, K.W., SELVENDRAN, R.R. AND ROBERTSON, J.A. (1998). Tissue-related changes in methyl-esterification of pectic polysaccharides in cauliflower (*Brassica oleracea* L. var. *botrytis*) stems. *Planta* **205**, 438–444.

FISHMAN, M.L., PFEFFER, P.E., BARFORD, R.A. AND DONER, L.W. (1984). Studies of pectin solution properties by high performance size exclusion chromatography. *Journal of Agricultural and Food Chemistry* **32**, 372–378.

FLIPPHI, M.J., PANNEMAN, H., VAN DER VEEN, P., VISSER, J. AND DE GRAAFF, L.H. (1993). Molecular cloning, expression and structure of the endo-1,5-alpha-L-arabinase gene of *Aspergillus niger. Applied Microbiology and Biotechnology* **40**, 318–326.

FLIPPHI, M.J., VISSER, J., VAN DER VEEN, P. AND DE GRAAFF, L.H. (1994). Arabinase gene expression in *Aspergillus niger* indications for coordinated regulation. *Microbiology* **140**, 2673–2682.

FRY, S.C. (1986). Cross-linking of matrix polymers in the growing cell walls of angiosperms. *Annual Reviews of Plant Physiology* **37**, 165–186.

FU, J. AND MORT, A.J. (1996). Progress toward identifying a covalent crosslink between xyloglucan and rhamnogalacturonan in cotton cell walls. *Plant Physiology* **111 (S)**, 147.

GAO, S., CHOI, G.H., SHAIN, L. AND NUSS, D.L. (1996). Cloning and targeted disruption of *enpg-1*, encoding the major *in vitro* extracellular endopolygalacturonase of the chestnut blight fungus, *Cryphonectria parasitica. Applied Environmental Microbiology* **62**, 1984–1990.

GOLDBERG, R., MORVAN, C. AND ROLAND, J.-C. (1986). Composition, properties and localization of pectins in young and mature cells of the mung bean hypocotyl. *Plant Cell Physiology* **27**, 417–429.

GONZALEZ, C.F., PETTIT, E.A., VALADEZ, V.A. AND PROVIN, E.M. (1997). Mobilization, cloning, and sequence determination of a plasmid-encoded polygalacturonase from a phytopathogenic *Burkholderia (Pseudomonas) cepacia. Molecular Plant Microbe Interactions* **10**, 849–851.

GONZALEZ-CANDELAS, L. AND KOLATTUKUDY, P.E. (1992). Isolation and analysis of a novel inducible pectate lyase gene from the phytopathogenic fungus *Fusarium solani* f. sp. *pisi* (*Nectria haematococca*, mating population VI). *Journal of Bacteriology* **174**, 6343–6349.

GRANT, G.T., MORRIS, E.R., REES, D.A., SMITH, P.J.C. AND THOM, D. (1973). Biological interactions between polysaccharides and divalent cations: the 'egg box' model. *FEBS Letters* **32**, 195–198.

GRASDALEN, H., BAKOY, O.E. AND LARSEN, B. (1988). Determination of the degree of esterification and the distribution of methylated and free carboxyl groups in pectin by 1H-N.M.R. spectroscopy. *Carbohydrate Research* **184**, 183–191.

GROSS, P., JULIUS, C., SCHMELZER, E. AND HAHLBROCK, K. (1993). Translocation of cytoplasm and nucleus to fungal penetration sites is associated with depolymerization of microtubules and defence gene activation in infected, cultured parsley cells. *EMBO Journal* **12**, 1735–1744.

GUO, W., GONZALEZ-CANDELAS, L. AND KOLATTUKUDY, P.E. (1996). Identification of a novel *pelD* gene expressed uniquely *in planta* by *Fusarium solani* f. sp. *pisi* (*Nectria haematococca*, mating type VI) and characterization of its protein product as an endo-pectate lyase. *Archives of Biochemistry and Biophysics* **332**, 305–312.

GYSLER, C., HARMSEN, J.A., KESTER, H.C., VISSER, J. AND HEIM, J. (1990). Isolation and structure of the pectin lyase D-encoding gene from *Aspergillus niger. Gene* **89**, 101–108.

HADFIELD, K.A. AND BENNETT, A.B. (1998). Polygalacturonases: many genes in search of a function. *Plant Physiology* **117**, 337–43.

HAMILTON, A.J., FRAY, R.G. AND GRIERSON, D. (1995). Sense and antisense inactivation of fruit ripening genes in tomato. *Current Topics in Microbiological Immunology* **197**, 77–89.

HART, D.A. AND KINDEL, P.K. (1970a). A novel reaction involved in the degradation of apiogalacturonans from *Lemna minor* and the isolation of apiobiose as a product. *Biochemistry* **9**, 2190–2196.

HART, D.A. AND KINDEL, P.K. (1970b). Isolation and partial characterization of apiogalacturonans from the cell wall of *Lemna minor. Biochemical Journal* **116**, 569–579.

HE, S.Y. AND COLLMER, A. (1990). Molecular cloning, nucleotide sequence, and maker exchange mutagenesis of the exo-poly-alpha-D-polygalacturonosidase-encoding *pheX* gene of *Erwinia chrysanthemi* EC16. *Journal of Bacteriology* **172**, 4988–4995.

HE, S.Y., LINDERBERG, M. AND COLLMER, A. (1993). Protein secretion by plant pathogenic bacteria. In *Biotechnology in Plant Disease Control*. Ed. I. Chet, pp 39–64. New York: Wiley-Liss Incorporated.

HEINRICHOVA, K., WOJCIECHOWICZ, M. AND ZIOLECKI, A. (1985). An exo-D-galacturonanase of *Butyrivibrio fibrisolvens* from the bovine rumen. *Journal of General Microbiology* **131**, 2053–2058.

HINTON, J.C., SIDEBOTHAM, J.M., GILL, D.R. AND SALMOND, G.P. (1989). Extracellular and periplasmic isoenzymes of pectate lyase from *Erwinia carotovora* subspecies *carotovora* belong to different gene families. *Molecular Microbiology* **3**, 1785–1795.

HO, M.C., WHITEHEAD, M.P., CLEVELAND, T.E. AND DEAN, R.A. (1995). Sequence analysis of the *Aspergillus nidulans* pectate lyase *pelA* gene and evidence for binding of promoter regions to CREA, a regulator of carbon catabolite repression. *Current Genetics* **27**, 142–149.

HOTCHKISS, A.T. AND HICKS, K.B. (1990). Analysis of oligogalacturonic acids with 50 or fewer residues by high-performance anion-exchange chromatography and pulsed amperometric detection. *Analytical Biochemistry* **184**, 200–206.

HOU, W.-C. AND CHANG, W.-H. (1996). Pectinesterase-catalyzed firming effects during precooking of vegetables. *Journal of Food Biochemistry* **20**, 397–416.

HUANG, J.H. AND SCHELL, M.A. (1990). DNA sequence analysis of *pglA* and mechanism of export of its polygalacturonase product from *Pseudomonas solanacearum*. *Journal of Bacteriology* **172**, 3879–3887.

HUGOUVIEUX-COTTE-PATTAT, N., CONDEMINE, G., NASSER, W. AND REVERCHON, S. (1996). Regulation of pectinolysis in *Erwinia chrysanthemi*. *Annual Reviews of Microbiology* **50**, 213–57.

HUGOUVIEUX-COTTE-PATTAT, N., DOMINGGUEZ, H. AND ROBERT-BAUDOUY, J. (1992). Environmental conditions affect the transcrption of the pectinase genes of *Erwinia chrysanthemi*. *Journal of Bacteriology* **174**, 7807–7818.

HUGOUVIEUX-COTTE-PATTAT, N. AND ROBERT-BAUDOUY, J. (1989). Isolation of *Erwinia chrysanthemi* mutants altered in pectinolytic enzyme production. *Molecular Microbiology* **3**, 1587–1597.

ISHII, T. (1995). Pectic polysaccharides from bamboo shoot cell-walls. *Mokuzai Gakkaishi* **41**, 669–676.

ISHII, T. (1997). O-acetylated oligosaccharides from pectins of potato tuber cell walls. *Plant Physiology* **113**, 1265–1272.

JARVIS, M.C. (1984). Structure and properties of pectin gels in plant cell walls. *Plant Cell Environment* **7**, 153–164.

JENSEN, N.S. AND CANALE-PAROLA, E. (1986). *Bacteroides pectinophilus* sp. *nov.* and *Bacteroides galacturonicus* sp. *nov.*: two pectinolytic bacteria from the human intestinal tract. *Applied Environmental Microbiology* **52**, 880–887.

KAUPPINEN, S., CHRISTGAU, S., KOFOD, L.V., HALKIER, T., DORREICH, K. AND DALBOGE, H. (1995). Molecular cloning and characterization of a rhamnogalacturonan acetylesterase from *Aspergillus aculeatus*. Synergism between rhamnogalacturonan degrading enzymes. *Journal of Biological Chemistry* **270**, 27172–27178.

KEEGSTRA, K., TALMADGE, K., BAUER, W.D. AND ALBERSHEIM, P. (1973). The structure of plant cell walls. III. A model of the walls of suspension-cultured sycamore cells based on the interconnections of the macromolecular components. *Plant Physiology* **51**, 188–196.

KELEMU, S. AND COLLMER, A. (1993). *Erwinia chysanthemi* EC16 produces a second set of plant-inducible pectate lyase isoenzyme. *Applied Environmental Microbiology* **59**, 1756–1761.

KEON, J.P. AND WAKSMAN, G. (1990). Common amino acid domain among endopolygalacturonases of ascomycete fungi. *Applied Environmental Microbiology* **56**, 2522–2528.

KESTER, C.M., KUSTERS-VAN SOMEREN, M.A., MULLER, Y. AND VISSER, J. (1996). Primary structure and characterization of an exopolgalacturonase from *Aspegillus tubingensis*. *European Journal of Biochemistry* **240**, 738–746.

KESTER, H.C. AND VISSER, J. (1990). Purification and characterization of polygalacturonases produced by the hyphal fungus *Aspergillus niger*. *Biotechnology and Applied Biochemistry* **12**, 150–160.

KHANH, N.Q., RUTTKOWSKI, E., LEIDINGER, K., ALBRECHT, H. AND GOTTSCHALK, M. (1991).

Characterization and expression of a genomic pectin methyl esterase-encoding gene in *Aspergillus niger. Gene* **106**, 71–77.

KIKUCHI, A., EDASHIGE, Y., ISHII, T. AND SATOH, S. (1996). A xylogalacturonan whose level is dependent on the size of cell clusters is present in the pectin from cultured carrot cells. *Planta* **200**, 369–372.

KIM, J.-B. AND CARPITA, N.C. (1992). Changes in esterification of the uronic acid groups of cell wall polysaccharides during elongation of maize coleoptiles. *Plant Physiology* **98**, 646–653.

KOBAYASHI, M., MATOH, T. AND AZUMA, J. (1996). Two chains of rhamnogalacturonan II are crosslinked by borate-diol ester bonds in higher plant cell walls. *Plant Physiology* **110**, 1017–1020.

KOMALAVILAS, P. AND MORT, A.J. (1989). The acetylation at O-3 of galacturonic acid in the rhamnose-rich region of pectins. *Carbohydrate Research* **189**, 261–272.

KOPECNY, J. AND HODROVA, B. (1995). Pectinolytic enzymes of anaerobic fungi. *Letters of Applied Microbiology* **20**, 312–316.

KOTOUJANSKY, A. (1987). Molecular genetics of patheogenesis by soft-rot *Erwinias. Annual Reviews of Phytopathology* **25**, 405–430.

KWON, Y.H. AND HOCH, H.C. (1991). Temporal and spatial dynamics of appressorium formation in *Uromyces appendiculatus. Experimental Mycology* **15**, 116–131.

KWON, Y.H., HOCH, H.C. AND AIST, J.R. (1991). Initiation of appressorium formation in *Uromyces appendiculatus*: organizations of the apex, and the responses involving microtubules and apical vesicles. *Canadian Journal of Botany* **69**, 2560–2573.

LAU, J.M., MCNEIL, M., DARVILL, A.G. AND ALBERSHEIM, P. (1985). Structure of the backbone of rhamnogalacturonan I, a pectic polysaccharide in the primary cell walls of plants. *Carbohydrate Research* **137**, 111–125.

LAU, J. M., MCNEIL, M., DARVILL, A.G. AND ALBERSHEIM, P. (1987). Treatment of rhamnogalacturonan I with lithium in ethylenediamine. *Carbohydrate Research* **168**, 245–274.

LAURENT, F., KOTOUJANSKY, A., LABESSE, G. AND BERTHEAU, Y. (1993). Characterization and overexpression of the *Pme* gene encoding pectin methylesterase of *Erwinia chrysanthemi* strain 3937. *Gene* **131**, 17–25.

LEROUGE, P., O'NEILL, M.A., DARVILL, A.G. AND ALBERSHEIM, P. (1993). Structural characterization of endo-glycanase-generated oligoglycosyl side chains of rhamnogalacturonan I. *Carbohydrate Research* **243**, 359–371.

LIAO, C.H. (1991). Cloning of pectate lyase gene pel from *Pseudomonas fluorescens* and detection of sequences homologous to *pel* in *Pseudomonas viridiflava* and *Pseudomonas putida. Journal of Bacteriology* **173**, 4386–4393.

LIAO, C.H., GAFFNEY, T.D., BRADLEY, S.P. AND WONG, L.C. (1996). Cloning of a pectate lyase gene from *Xanthomonas campestris* pv. *malvacearum* and comparison of its sequence relationship with pel genes of soft-rot *Erwinia* and *Pseudomonas. Molecular Plant Microbe Interactions* **9**, 14–21.

LIAO, C.H., HUNG, H.Y. AND CHATTERJEE, A.K. (1988). An extracellular pectate lyase is the pathogenicity factor of the soft-rotting bacterium *Pseudomonas viridiflava. Molecular Plant Microbe Interactions* **1**, 199–209.

LINERS, F. AND VAN CUTSEN, P. (1992). Distribution of pectic polysaccharides throughout walls of suspension-cultured carrot cells: an immunocytochemical study. *Protoplasma* **170**, 10–21.

LINHARDT, R.J., GALLIHER, P.M. AND COONEY, C.L. (1986). Polysaccharide lyases. *Applied Biochemistry and Biotechnology* **12**, 135–176.

MATOH, T., ISHIGAKI, K., OHNO, K. AND AZUMA, J. (1993). Isolation and characterization of a boron-polysaccharide complex from radish roots. *Plant Cell Physiology* **34**, 639–642.

MAYANS, O., SCOTT, M., CONNERTON, I., GRAVESEN, T., BENEN, J., VISSER, J., PICKERSGILL, R. AND JENKINS, J. (1997). Two crystal structures of pectin lyase A from *Aspergillus* reveal a pH driven conformational change and striking divergence in the substrate-binding clefts of pectin and pectate lyases. *Structure* **5**, 677–689.

MCCANN, M.C., SHI, J., ROBERTS, K. AND CARPITA, N.C. (1994). Changes in pectin structure and localization during the growth of unadapted and NaCl-adapted tobacco cells. *Plant Journal* **5**, 773–785.

MCCARTHY, R.E., KOTARSKI, S.F. AND SALYERS, A.A. (1985). Location and characteristics of enzymes involved in the breakdown of polygalacturonic acid by *Bacteroides thetaiotaomicron*. *Journal of Bacteriology* **161**, 493–499.

MCCARTHY, R.E. AND SALYERS, A.A. (1986). Evidence that polygalacturonic acid may not be a major source of carbon and energy for some colonic *Bacteroides* species. *Applied Environmental Microbiology* **52**, 9–16.

MCGUIRE, R.G., RODRIGUEZ-PALENZUELA, P., COLLMER, A. AND BURR, T.J. (1991). Polygalacturonase production by *Agrobacterium tumefaciens* biovar 3. *Applied Environmental Microbiology* **57**, 660–664.

MCKIE, V.A., BLACK, G.W., MILLWARD-SADLER, S.J., HAZLEWOOD, G.P., LAURIE, J.I. AND GILBERT, H.J. (1997). Arabinanase A from *Pseudomonas fluorescens* subsp. *cellulosa* exhibits both and endo- and an exo- mode of action. *Biochemical Journal* **323**, 547–555.

MCNEIL, M., DARVILL, A.G. AND ALBERSHEIM, P. (1980). Structure of plant cell walls X. Rhamnogalacturonan I, a structurally complex pectic polysaccharide in the cell walls of suspension-cultured sycamore cells. *Plant Physiology* **66**, 1128–1134.

MCNEIL, M., DARVILL, A.G., FRY, S.C. AND ALBERSHEIM, P. (1984). Structure and function of the primary cell walls of plants. *Annual Reviews of Biochemistry* **53**, 625–663.

MENDGEN, K. AND DEISING, H. (1993). Infection structures of fungal plant pathogens – a cytological and physiological evaluation. *New Phytology* **124**, 193–213.

MORT, A.J. AND BAUER, W.D. (1982). Application of two new methods for cleavage of polysaccharides into specific oligosaccharide fragments. Structure of the capsular and extracellular polysaccharides of *Rhizobium japonicum* that bind soybean lectin. *Journal of Biological Chemistry* **257**, 1870–1875.

MORT, A.J. AND CHEN, E.M. (1996). Separation of 8-aminonaphthalene-1,3,6-trisulfonate (ANTS)-labeled oligomers containing galacturonic acid by capillary electrophoresis: application to determining the substrate specificity of endopolygalacturonases. *Electrophoresis* **17**, 379–383.

MORT, A.J., MOERSCHBACHER, B.M., PIERCE, M.L. AND MANESS, N O. (1991). Problems one may encounter during the extraction, purification, and chromatography of pectin fragments and some solutions to them. *Carbohydrate Research* **215**, 219–227.

MORT, A.J., QIU, F. AND MANESS, N.O. (1993). Determination of the pattern of methyl esterification in pectins. Distribution of contiguous non-esterified residues. *Carbohydrate Research* **247**, 21–35.

MOULD, M.J.R., BOLAND, G.J. AND ROBB, J. (1991). Ultrastructure of the *Colletotrichum trifolii-Medicago sativa* pathosystem. I. Pre-penetration events. *Physiology and Molecular Plant Pathology* **38**, 179–194.

MOULD, R.J.M., BOLAND, G.J. AND ROBB, J. (1991). Post penetration events. *Physiology and Molecular Plant Pathology* **38**, 195–210.

MUTTER, M., BELDMAN, G., PITSON, S.M., SCHOLS, H.A. AND VORAGEN, A.G. (1998a). Rhamnogalacturonan alpha-D-galactopyranosyluronohydrolase. An enzyme that specifically removes the terminal nonreducing galacturonosyl residue in rhamnogalacturonan regions of pectin. *Plant Physiology* **117**, 153–163.

MUTTER, M., BELDMAN, G., SCHOLS, H.A. AND VORAGEN, A.G.J. (1994). Rhamnogalaturonan alfa-L-Rhamnopyranohydrolase. *Plant Physiology* **106**, 214–250.

MUTTER, M., COLQUHOUM, I.J., SCHOLS, H.A., BELDAMN, G. AND VORAGEN, A.G.J. (1996). Rhamnogalacturonan alfa-L-rhamnopyranosyl-(1,4)-alfa-D-galactopyranosyluronide lyase: a new enzyme able to cleave RG regions of pectin. *Plant Physiology* **110**, 73–77.

MUTTER, M., COLQUHOUN, I.J., BELDMAN, G., SCHOLS, H.A., BAKX, E.J. AND VORAGEN, A.G. (1998b). Characterization of recombinant rhamnogalacturonan alpha-L-rhamnopyranosyl-(1,4)-alpha-D-galactopyranosyluronide lyase from *Aspergillus aculeatus*. An enzyme that fragments rhamnogalacturonan regions of pectin. *Plant Physiology* **117**, 141–152.

NASSER, W., CHALET, F. AND ROBERT-BAUDOUY, J. (1990). Purification and characterization of extracellular pectate lyase from *Bacillus subtilis*. *Biochimie* **72**, 689–695.

NASUNO, S. AND STARR, M. (1966). Polygalacturonase of *Erwinia carotovora*. *Journal of Biological Chemistry* **241**, 5298–5306.

NEEDS, P.W., RIGBY, N.M., COLQUHOUN, I.J. AND RING, S.G. (1998). Conflicting evidence for non-methyl galacturonoyl esters in *Daucus carota*. *Phytochemistry* **48**, 71–77.

NELSON, M.A., MERINO, S.T. AND METZENBERG, R.L. (1997). A putative rhamnogalacturonase required for sexual development of *Neurospora crassa*. *Genetics* **146**, 531–540.

NIKAIDOU, N., KAMIO, Y. AND IZAKI, K. (1992). Molecular cloning and nucleotide sequence of a pectin lyase gene from *Pseudomonas marginalis* N6301. *Biochemical and Biophysical Research Communications* **182**, 14–19.

NIKAIDOU, N., KAMIO, Y. AND IZAKI, K. (1993). Molecular cloning and nucleotide sequence of the pectate lyase gene from *Pseudomonas marginalis* N6301. *Bioscience, Biotechnology and Biochemistry* **57**, 957–960.

O'NEILL, M.A., WARRENFELTZ, D., KATES, K., PELLERIN, P., DARVILL, A.G. AND ALBERSHEIM, P. (1996). Rhamnogalacturonan-II, a pectic polysaccharide in the cell walls of growing plant cells, forms a dimer that is covalently crosslinked by a borate ester. *In vitro* conditions for the formation and hydrolysis of the dimer. *Journal of Biological Chemistry* **271**, 22923–22930.

PATINO, B., POSADA, M.L., GONZALEZ-JAEN, M.T. AND VAZQUEZ, C. (1997). The course of pectin degradation by polygalacturonases from *Fusarium oxysporum* f. *sp. radicis lycopersici*. *Microbios* **91**, 47–54.

PELLERIN, P. AND BRILLOUET, J.M. (1994). Purification and properties of an exo-(1—>3)-beta-D-galactanase from *Aspergillus niger*. *Carbohydrate Research* **264**, 281–291.

PICKERSGILL, R., JENKINS, J., HARRIS, G., NASSER, W. AND ROBERT-BAUDOUY, J. (1994). The structure of *Bacillus subtilis* pectate lyase in complex with calcium. *Nature Structural Biology* **1**, 717–723.

POLIZELI, M.D.L., JORGE, J.A. AND TERENZI, H.F. (1991). Pectinase production by *Neurospora crassa*: purification and biochemical characterization of extracellular polygalacturonase activity. *Journal of General Microbiology* **137**, 1815–1823.

POWELL, D.A., MORRIS, E.R., GIDLEY, M.J. AND REES, D.A. (1982). Conformations and interactions of pectins. II. Influence of residue sequence on chain association in calcium pectate gels. *Journal of Molecular Biology* **155**, 517–531.

PRADE, R.A. (1996). Xylanases: From Biology to BioTechnology. *Biotechnology and Genetic Engineering Reviews* **13**, 100–131.

PRESTON, J.F., RICE, J.D., INGRAM, L.O. AND KEEN, N.T. (1992). Differential depolymerization mechanisms of pectate lyases secreted by *Erwinia chrysanthemi* EC16. *Journal of Bacteriology* **174**, 2039–2042.

QI, X., BEHRENS, B.X., WEST, P. AND MORT, A.J. (1995). Solubilization and partial characterization of extensin fragments from cell walls of cotton suspension cultures. Evidence for a covalent crosslink between extensin and pectin. *Plant Physiology* **108**, 1691–1701.

RAMON, D., V.D. VEEN, P. AND VISSER, J. (1993). Arabinan degrading enzymes from *Aspergillus nidulans*: induction and purification. *FEMS Microbiology Letters* **113**, 15–22.

RAO, M.N., KEMBHAVI, A.A. AND PANT, A. (1996). Role of lysine, tryptophan and calcium in the beta-elimination activity of a low-molecular-mass pectate lyase from *Fusarium moniliformae*. *Biochemical Journal* **319**, 159–164.

REES, D.A. (1982). Polysaccharide conformation in solutions and gels. Recent results on pectins. *Carbohydrate Polymers* **2**, 254–263.

REES, D.A. AND WRIGHT, A.W. (1971). Polysaccharide conformation. Part VII. Model building computations for α–1,4 galacturonan and the kinking function of L-rhamnose residues in pectic substances. *Journal of the Chemical Society* (B), 1366–1372.

RENARD, C.M.G.C., CREPEAU, M.J. AND THIBAULT, J.-F. (1995). Structure of the repeating units in the rhamnogalacturonic backbone of apple, beet and citrus pectins. *Carbohydrate Research* **275**, 155–165.

REXOVA-BENKOVA, L. AND MARKOVIC, O. (1976). Pectic enzymes. *Advances in Carbohydrate Chemistry and Biochemistry* **33**, 323–385.

REYES, F., ALFONSO, C., MARTINEZ, M.J., PRIETO, A., SANTAMARIA, F. AND LEAL, J.A. (1992). Purification of a new galactanase from *Penicillium oxalicum* catalysing the hydrolysis of beta-(1-5)-galactofuran linkages. *Biochemical Journal* **281**, 657–660.

REYMOND, P., DELEAGE, G., RASCLE, C. AND FEVRE, M. (1994). Cloning and sequence analysis of a polygalacturonase-encoding gene from the phytopathogenic fungus *Sclerotinia sclerotiorum. Gene* **146**, 233–237.

REYMOND-COTTON, P., FRAISSINET-TACHET, L. AND FEVRE, M. (1996). Expression of the *Sclerotinia sclerotiorum* polygalacturonase *pg1* gene: possible involvement of CREA in glucose catabolite repression. *Current Genetics* **30**, 240–245.

RODRIGUES-PALENZUELA, P., BURR, T.J. AND COLLMER, A. (1991). Polygalacturonase is a virulence factor in *Agrobacterium tumefaciens* biovar 3. *Journal of Bacteriology* **173**, 6547–6552.

ROMBOUTS, F.M. AND THIBAULT, J.-F. (1986). Feruloylated pectic substances from sugar-beet pulp. *Carbohydrate Research* **154**, 177–187.

RUIJTER, G.J., VANHANEN, S.A., GIELKENS, M.M., VAN DE VONDERVOORT, P.J. AND VISSER, J. (1997). Isolation of *Aspergillus niger creA* mutants and effects of the mutations on expression of arabinases and L-arabinose catabolic enzymes. *Microbiology* **143**, 2991–2998.

SAKAI, T., SAKAMOTO, T., HALLAERT, J. AND VANDAMME, E.J. (1993). Pectin, pectinase, and protopectinase: Production, properties, and applications. *Advances in Applied Microbiology* **39**, 213–294.

SAKAMOTO, M., SHIRANE, Y., NARIBAYASHI, I., KINURA, K., MORISHITA, N., SAKAMOTO, T. AND SAKAI, T. (1994). Purification and characterization of a rhamnogalacturonase with protopectinase activity from *Tramtes sanguinea. European Journal of Biochemistry* **226**, 285–291.

SAKAMOTO, T., YAMADA, M., KAWASAKI, H. AND SAKAI, T. (1997). Molecular cloning and nucleotide sequence of an endo-1,5-α-L-arabinase gene from *Bacillus subtilis. European Journal of Biochemistry* **245**, 708–714.

SALMOND, G.P.C. (1994). Factors affecting the virulence of sof-rot *Erwinia* species: the molecular biology of an opportunistic phytopathogen. In *Molecular Mechanisms of Bacterial Virulence*. Eds. C. I. Kado and J. H. Crosa, pp 193–206. Dordrecht, The Netherlands: Kluwer Academic.

SCHAEFFER, H.J., LEYKAM, J. AND WALTON, J.D. (1994). Cloning and targeted gene disruption of *EXG1*, encoding exo beta 1, 3 glucanase, in the phytopathogenic fungus *Cochliobolus carbonum. Applied Environmental Microbiology* **60**, 594–598.

SCHELL, M.A., DENNY, T.P. AND HUANG, J. (1994). Extracellular virulence factors of *Pseudomonas solanacearum*: role in disease and their regulation. In *Molecular Mechanisms of Bacterial Virulence*. Eds. C.I. Kado and J.H. Crosa, pp 311–324. Dordrecht, The Netherlands: Kluwer Academic.

SCHOLS, H.A., BAKX, E.J., SCHIPPER, D. AND VORAGEN, A.G.J. (1995). A xylogalacturonan subunit present in the modified hairy regions of apple pectin. *Carbohydrate Research* **279**, 265–279.

SCHOLS, H.A., GERAEDS, C.J.M., SEARLE-VAN LEEUWEN, M.F., KORMELINK, F.J.M. AND VORAGEN, A.G.J. (1990). Rhamnogalturonase: a novel enzyme that degrades the hairy regions of pectin. *Carbohydrate Research* **206**, 105–115.

SCHOLS, H.A., POSTHUMUS, M.A. AND VORAGEN, A.G.J. (1990). Structural features of hairy regions of pectins isolated from apple juice produced by the liquefaction process. *Carbohydrate Research* **206**, 117–129.

SCHOLS, H.A. AND VORAGEN, A.G.J. (1994). Occurrence of pectic hairy regions in various plant cell wall materials and their degradation by rhamnogalacturonase. *Carbohydrate Research* **256**, 83–95.

SCHOLS, H.A. AND VORAGEN, A.G.J. (1996). Complex pectins: Structure elucidation using enzymes. In *Pectins and Pectinases*. Eds. J. Visser and A.G.J. Voragen, Vol. **14**, pp 3–19. Amsterdam: Elsevier Science.

SCHOLS, H.A., VORAGEN, A.G.J. AND COLQUHOUN, I.J. (1994). Isolation and characterization of rhamnogalacturonan oligomers, liberated during degradation of pectic hairy regions by rhamnogalacturonase. *Carbohydrate Research* **256**, 97–111.

SCOTT-CRAIG, J.S., CHENG, Y.Q., CERVONE, F., DE LORENZO, G., PITKIN, J.W. AND WALTON,

J.D. (1998). Targeted mutants of *Cochliobolus carbonum* lacking the two major extracellular polygalacturonases. *Applied Environmental Microbiology* **64**, 1497–503.

SCOTT-CRAIG, J.S., PANACCIONE, D.G., CERVONE, F. AND WALTON, J.D. (1990). Endopolygalacturonase is not required for pathogenicity of *Cochliobolus carbonum* on maize. *Plant Cell* **2**, 1191–200.

SHEIMAN, M.I., MACMILLAN, J.D., MILLER, L. AND CHASE, T., JR. (1976). Coordinated action of pectinesterase and polygalacturonate lyase complex of *Clostridium multifermentans*. *European Journal of Biochemistry* **64**, 565–572.

SHEVCHIK, V.E., CONDEMINE, G., HUGOUVIEUX-COTTE-PATTAT, N. AND ROBERT-BAUDOUY, J. (1996). Characterization of pectin methylesterase B, an outer membrane lipoprotein of *Erwinia chrysanthemi* 3937. *Molecular Microbiology* **19**, 455–466.

SHEVCHIK, V.E. AND HUGOUVIEUX-COTTE-PATTAT, N. (1997). Identification of a bacterial pectin acetyl esterase in *Erwinia chrysanthemi* 3937. *Molecular Microbiology* **24**, 1285–1301.

SONE, H., SUGIURA, J., ITOH, Y., IZAKI, K. AND TAKAHASHI, H. (1988). Production and properties of pectin lyase in *Pseudomonas marginalis* induced by mitomycin C. *Agricultural Biological Chemistry* **52**, 3205–3207.

SPOSATO, P., AHN, J.H. AND WALTON, J.D. (1995). Characterization and disruption of a gene in the maize pathogen *Cochliobolus carbonum* encoding a cellulase lacking a cellulose binding domain and hinge region. *Molecular Plant Microbe Interactions* **8**, 602–609.

STERNBERG, D. AND MANDELS, G.R. (1979). Induction of cellulolytic enzymes in *Trichoderma reesei* by sophorose. *Journal of Bacteriology* **139**, 761–769.

SUTHERLAND, I.W. (1995). Polysaccharide lyases. *FEMS Microbiology Reviews* **16**, 323–347.

SUYKERBUYK, M.E., SCHAAP, P.J., STAM, H., MUSTERS, W. AND VISSER, J. (1995). Cloning, sequence and expression of the gene coding for rhamnogalacturonase of *Aspergillus aculeatus*; a novel pectinolytic enzyme. *Applied Microbiology and Biotechnology* **43**, 861–870.

TAKASAWA, T., SAGISAKA, K., YAGI, K., UCHIYAMA, K., AOKI, A., TAKAOKA, K. AND YAMAMATO, K. (1997). Polygalacturonase isolated from the culture of the psychrophilic fungus *Sclerotinia borealis*. *Canadian Journal of Microbiology* **43**, 417–424.

TAYLOR, A.J. (1982). Intramolecular distribution of carboxyl groups in low methoxyl pectins – a review. *Carbohydrate Polymers* **2**, 9–17.

THAKUR, B.R., SINGH, R.K. AND HANDA, A.K. (1997). Chemistry and uses of pectin – a review. *Critical Reviews in Food Science and Nutrition* **37**, 47–73.

THIBAULT, J.-F., RENARD, C.M.G.C., AXELOS, M.A.V., ROGER, P. AND CRÉPEAU, M.-J. (1993). Studies of the length of homogalacturonic regions in pectins by acid hydrolysis. *Carbohydrate Research* **283**, 271–286.

THIBAULT, J.F. (1983). *Aspergillus niger* endopolygalacturonase: 3. Action pattern on polygalacturonic acid. *Carbohydrate Polymers* **3**, 259–272.

TIERNY, Y., BECHET, M., JONCQUIERT, J.C., DUBOURGUIER, H.C. AND GUILLAUME, J.B. (1994). Molecular cloning and expression in *Escherichia coli* of genes encoding pectate lyase and pectin methylesterase activities from *Bacteroides thetaiotaomicron*. *Journal of Applied Bacteriology* **76**, 592–602.

TSUYUMU, S. AND CHATTERJEE, A.K. (1984). Pectin lyase production of *Erwinia chrysanthemi* and other soft-rot *Erwinia* species. *Physiology and Plant Pathology* **24**, 291–302.

V. D. VEEN, P., FLIPPHI, M.J., VORAGEN, A.G. AND VISSER, J. (1993). Induction of extracellular arabinases on monomeric substrates in *Aspergillus niger*. *Archives of Microbiology* **159**, 66–71.

V.D. VEEN, P., FLIPPHI, M.J., VORAGEN, A.G. AND VISSER, J. (1991). Induction, purification and characterisation of arabinases produced by *Aspergillus niger*. *Archives of Microbiology* **157**, 23–28.

VAN HOOF, A.V., LEYKAM, J., H.J., S. AND WALTON, J.D. (1991). A single β 1,3-glucanase secreted by the maize pathogen *Cochliobolus carbonum* acts by an exolytic mechanism. *Physiology and Molecular Plant Pathology* **39**, 259–267.

VAN HOUDENHOVEN, F.E.A. (1975). Studies on pectin lyase. *PhD Thesis*. Agricultural University, Wageningen.

VAN RIJSSEL, M., GERWIG, G.J. AND HANSEN, T.A. (1993b). Isolation and characterization of an extracellular glycosylated protein complex from *Clostridium thermosaccharolyticum* with pectin methylesterase and polygalacturonate hydrolase activity. *Applied Environmental Microbiology* **59**, 828–836.

VAN RIJSSEL, M., SMIDT, M.P., VAN KOUWEN, G. AND HANSEN, T.A. (1993). Involvement of an intracellular oligogalacturonate hydrolase in metabolism of pectin by *Clostridium thermosacchrolyticum*. *Applied Environmental Microbiology* **59**, 837–842.

VON RIESEN, V.L. (1975). Polypectate digestion by *Yersinia*. *Journal of Clinical Microbiology* **2**, 552–553.

WALKER, M.J. AND PEMBERTON, J.M. (1987). Construction of a transposon containing a gene for polygalacturonate trans-eliminase from *Klebsiella oxytoca*. *Archives of Microbiology* **146**, 390–395.

WALTON, J.D. (1994). Deconstructing the cell wall. *Plant Physiology* **104**, 1113–1118.

WALTON, J.D. AND CERVONE, F. (1990). Endopolygalactorunase from the maize pathogen *Cochliobolus carbonum*. *Physiology and Molecular Plant Pathology* **36**, 351–359.

WHITEHEAD, M.P., SHIEH, M.T., CLEVELAND, T.E., CARY, J.W. AND DEAN, R.A. (1995). Isolation and characterization of polygalacturonase genes (*pecA* and *pecB*) from *Aspergillus flavus*. *Applied Environmental Microbiology* **61**, 3316–3322.

WOJCIECHOWICZ, M., HEINRICHOVA, K. AND ZIOLECKI, A. (1980). A polygalacturonate lyase produced by *Lachnospira multiparus* isolated from the bovine rumen. *Journal of General Microbiology* **117**, 193–199.

WOJCIECHOWICZ, M., HEINRICHOVA, K. AND ZIOLECKI, A. (1982). An exopectate lyase of *Butyrivibrio fibrisolvens* from the bovine rumen. *Journal of General Microbiology* **128**, 2661–2665.

WOJCIECHOWICZ, M. AND ZIOLECKI, A. (1984). A note on the pectinolytic enzyme of *Streptococcus bovis*. *Journal of Applied Bacteriology* **56**, 515–518.

WOJCIECHOWSKI, C.L. AND FALL, R. (1996). A continuous fluorometric assay for pectin methylesterase. *Analytical Biochemistry* **237**, 103–108.

WU, S.C., KAUFFMANN, S., DARVILL, A.G. AND ALBERSHEIM, P. (1995). Purification, cloning and characterization of two xylanases from *Magnaporthe grisea*, the rice blast fungus. *Molecular Plant Microbe Interactions* **8**, 506–514.

YAMAGUCHI, F., INOUE, S. AND HATANAKA, C. (1995). Purification and properties of endo-beta-1,4-D-galactanase from *Aspergillus niger*. *Bioscience, Biotechnology and Biochemistry* **59**, 1742–1744.

YU, L. AND MORT, A.J. (1996). Partial Characterization of Xylogalacturonans from Cell Walls of Ripe Watermelon Fruit: Inhibition of Endopolygalacturonase Activity by Xylosylation. In *Pectins and Pectinases* Eds. J. Visser and A.G.J. Voragen, Vol.14, pp 79–88. Amsterdam: Elsevier Science.

ZHAN, D., JANSSEN, P. AND MORT, A.J. (1998). Scarcity or complete lack of single rhamnose residues interspersed within the homogalacturonan regions of citrus pectin. *Carbohydrate Research* **308**, 373–380.

ZINK, R.T., ENGWALL, J.K., MCEVOY, J.L. AND CHATERJEE, A.K. (1985). *recA* is required in the induction of pectin lyase and carotovoricin in *Erwinia carotovora*. *Journal of Bacteriology* **164**, 390–396.

17
The Unfolding Story of the Chaperonins

ANTHONY R. M. COATES[1]*, BRIAN HENDERSON[2] AND PAOLO
MASCAGNI[3]

[1]*Department of Medical Microbiology, St George's Hospital Medical School,
Cranmer Terrace, London, SW17 0RE, UK, [2]Cellular Microbiology Research
Group, Division of Surgical Sciences, Eastman Dental Institute, University College
London, 256 Gray's Inn Road, London, WC1X 8LD, UK and [3]Italfarmaco
Research Centre, Via Lavoratori, 54, Cinisello Balsamo 20092, Milan, Italy*

Introduction

The chaperonins (cpn) are a family of sequence-related oligomeric proteins found in
all cells which are essential for cell viability, particularly under conditions of environmental stress (Bakau, 1993; Minowanda and Welch, 1995). The genes encoding these
proteins were discovered as part of the unfolding story of how cells respond to stress
and the proteins themselves are now best known for their involvement in the non-
covalent folding of proteins intracellularly (Ellis, 1996). As this article will delineate,
the cpns also have biological actions in addition to those related to protein folding
(Coates, 1996; Henderson *et al.*, 1996; Coates and Henderson, 1998; Lewthwaite *et
al.*, 1998). The cpn family has two subfamilies, GroE and TCP1 (Coates *et al.*, 1993).
The sequence identity within each subfamily is high, about 50%, but there is only
around 20% identity between the subfamilies.

This review focuses on recent scientific advances in the GroE subfamily which
contains two major proteins, cpn 10 and cpn 60 (Gupta, 1996); also known as GroES
and GroEL (in *Escherichia coli*) (Georgopoulos *et al.*, 1972), heat shock protein 10
and 58/60/65 (McMullin and Hallberg, 1988; Gupta, 1995), 10/60/65 kDa antigen
(Shinnick *et al.*, 1988; Baird *et al.*, 1989; Hoffman *et al.*, 1990), early pregnancy
factor (Cavanagh, 1996) and P1 (Jindal *et al.*, 1989). The structural and folding aspects
of the cpns have recently been comprehensively reviewed (Bukau and Horwich,
1998) and this article will concentrate on the non-folding functions of these proteins.

Gene organisation

In prokaryotes, the cpn 10 and cpn 60 genes are often located close to one another
usually, but not always, in an operon (van der Vies and Georgopoulos, 1996). Many

---

*To whom correspondence may be addressed.

Biotechnology and Genetic Engineering Reviews – Vol. 16, April 1999
0264–8725/99/16/393–405 $20.00 + $0.00 © Intercept Ltd, P.O. Box 716, Andover, Hampshire SP10 1YG, UK

organisms contain multiple cpn genes with related sequences, some of which are in operons while others are situated far from one another (Coates *et al.,* 1993). It is not clear what function these non-operon genes might fulfill. In eukaryotes, multiple cpn gene copies are also found (Gupta, 1996), and in the rat the cpn 60 and cpn 10 genes are linked head-to head, in a 14 kb assembly with 14 introns and a shared bidirectional promoter (Ryan *et al.,* 1997). So, whilst the gene arrangement in eukaryotes is different to that in prokaryotes, there are similarities, such as shared promoter regions.

Protein structure and folding properties

THE CHAPERONIN FOLDING MACHINE

Chaperonins belong to a group of proteins called molecular chaperones (Ellis, 1996) which bind non-native proteins and assist them, in an ATP-dependent catalytic process, to fold into the correct three-dimensional form required for a functional protein. The GroE chaperonin folding machine (Bukau and Horwich, 1998) is a large complex of about one million Daltons and assists many different proteins to fold. It consists of a ring of seven cpn 60 molecules which are assembled into a bucket-shaped complex which stands on an upturned bucket of another identical seven cpn 60 monomers. Each bucket has a hole in the bottom. Cpn 10 also forms heptamers, and is a dome-shaped molecule positioned like a lid on the top of one of the cpn 60 buckets. Misfolded protein enters the mouth of one of the buckets, where it binds to the hydrophobic lining of the cpn 60 cavity, the cpn 10 lid is snapped on, assisted folding takes place, the lid is removed and the protein is released. When the cpn10 lid binds to the cpn 60 bucket the volume of the bucket increases 2-fold. This involves a twisting movement of the bucket, which shears the hydrophobic binding surface of the cpn away from the bound polypeptide, releasing it into the cavity. The inner surface of the bucket is now lined by hydrophilic residues which favour the burial of the hydrophobic surfaces in the substrate protein and so help in the formation of correctly-folded functional proteins. The bucket contains two hinges, one of which is between the upper side (termed the apical domain) and the lower side (the intermediate domain); the second hinge is between the intermediate domain and the bottom of the bucket, called the equatorial domain. During the enlargement of the cavity, there is a 25° downward movement of the intermediate domain onto the equatorial domain, which locks ATP into its site in the bottom of the bucket. At the same time, there is a 65° rotation of the upper hinge and a 90° clockwise twist of the apical domain about its long axis that moves the hydrophobic lining of the bucket to a new position in which part of the surface interacts with mobilised neighbouring apical domains and the remainder binds to one edge (with sequence Ile-Val-Leu) of the cpn10 mobile β-hairpin loop, which hangs down and outwards from the lower surface of each of the seven cpn10s.

The folding process is asymmetric. Initially, the cpn10 lid is on one of the buckets and the unfolded or kinetically trapped folding intermediate polypeptide enters the other bucket (*trans*). Then, in the presence of ATP, the cpn 10 lid is released from the bucket and ends up on the other bucket which has the polypeptide and the ATP in it. This is accompanied by a doubling of the cavity size which is described above. Folding of the substrate polypeptide takes place in the enclosed cavity. Then ATP binds to the bottom of the other bucket which results in the ejection of the cpn10 lid, and this allows the native, the committed-to-fold, the uncommitted or the kinetically trapped polypeptide to leave. Once the polypeptide is free of the folding machine, it either regains its function or it binds again to the machine for another refolding cycle.

Not all proteins are folded in this way. Those which are larger than 60–70 kDa may not fit into the bucket with the lid on, although one organism, bacteriophage T4, has solved this problem by evolving a lid with a longer mobile loop and a taller dome which can accommodate its large capsid protein (Hunt, 1997). Alternatively, it is possible that local kinetically trapped regions of large proteins might bind in the trans ring and allow correct folding on release (Gordon, 1994). A few proteins, such as actin and tubulin cannot be folded properly by the GroE chaperonin machine and are folded by the TCP1 chaperonins (Gao *et al.*, 1992; Klumpp *et al.*, 1997).

MONOMERIC AND OLIGOMERIC STATES

In addition to heptamers, cpns can form several different multimeric states and also exist, under certain conditions, as monomers. For example, *E. coli* cpn 10 becomes monomeric below a protein concentration of 0.7 µM (Zondlo *et al.*, 1995). The structure of the subunits of cpn10 heptamers is that of an irregular β-barrel. However, isolated GroES monomers having this structure are energetically unfavourable and only marginally stable at room temperature (Boudker *et al.*, 1997). In contrast, *Mycobacterium tuberculosis* cpn10 forms tetramers which can be detected in bacterial lysates (Fossati *et al.*, 1995), and dimers in dilute buffer (Mascagni *et al.*, 1998). Monomeric *M. tuberculosis* cpn 10 is observed below a concentration of either 4.7 µM or 0.47 µM depending on the concentration of ions such as Mg2+ present in solution. In aqueous buffers these monomers give rise to conformational equilibria. However, in solutions of reduced polarity, these monomers show a preference for a structure which contains both α-helices and β-strands (Mascagni *et al.*, 1998). The biological significance of these findings is unknown, but may have relevance to both the secretion of cpns into the extracellular space, and to the non-folding activities of cpns.

Cellular location

Cpns are located within the cytosol of bacteria and within eukaryotic cells they are found in mitochondria and chloroplasts (Gupta, 1996). For many years it was thought that, because of their crucial intracellular folding role, this was their only location. Recently, however, it has become clear that this is not the case. In bacteria, for example, there is evidence that some cpns, such as the cpn 10 of *M. tuberculosis,* are secreted in large amounts into the extracellular space. About 20% of the total protein content in short-term culture filtrates of logarithmically growing *M. tuberculosis* is cpn 10 (Abou-Zeid *et al.,* 1988). A number of bacteria appear to have cpn 60 associated with their external surfaces. Gentle saline extraction removes surface-associated cpn 60 from the oral pathogen *Actinobacillus actinomycetemcomitans* (Kirby *et al.*, 1995) and this protein has been localised on the outer cell wall by immunogold electron microscopy. The cpn 60 (HspB) of *Helicobacter pylori,* which causes stomach ulceration, also appears on the outside of the cell, in the periplasmic space, in the extracellular space and is believed to act as an adhesin (Eschweiler *et al.*,

1993; Cao *et al.,* 1998). Another pathogen, *Haemophilus ducreyi,* which causes the genital ulcer disease, chancroid, expresses chaperonin 60 on its surface and this protein is involved in the binding of the bacterium to epithelial cells (Frisk *et al.,* 1998). Thus, there appears to be a pattern emerging, with cpn 60 having the ability to act as a bacterial adhesin. This has suggested that the chaperonins may be important bacterial virulence factors (Lewthwaite *et al.,* 1998). A similar story may be emerging with eukaryotic cells. Cpn 60 has been detected by immunogold labelling on the surface of viable chinese hamster ovary cells and the human leukaemic CD4-positive T-cell line CEM-55 (Soltys and Gupta, 1997). In addition to surface expression there is evidence for secretion of cpns from mammalian cells. Thus, human cpn10 is thought to be the same molecule as early pregnancy factor, which is found in the serum of pregnant women (Cavanagh, 1996). In addition, human astrocytes secrete a potent anti-apoptotic cpn 60-like molecule (Brenneman and Gozes, 1996).

How do cpns cross the cell membrane? Generally, transported proteins have an N-terminal signal peptide which is recognised by a cytosolic factor that takes it to a receptor on the membrane (Schatz and Dobberstein, 1996). It is likely that translocation across the membrane is in the monomeric rather than oligomeric form, and so dissociation into monomers is required. Cpns do not have a specific α-helical signal peptide, but those that are secreted, like the *M. tuberculosis* cpn 10 have a signal peptide-like structure at the N-terminus which forms an α-helix in organic solvent/buffer mixtures. It is possible that the N-terminus of some cpns help in the translocation of the molecule across the membrane. It is also possible that there may be direct association of C-terminus of cpn 60 with membranes. Torok and colleagues (Torok *et al.,* 1997) have found that cpn 60 can associate with model lipid membranes, and that proteolytic removal of the C-terminus of the protein prevented association with lipid membranes. Interestingly, the cpn 60 increased the lipid order in the liquid crystalline state, which suggests that cpns can assist the folding of both soluble and membrane-associated proteins whilst, at the same time, stabilising lipid membranes.

Importance in medicine

FOLDING ACTIVITY

Damage to tissues occurs when the blood supply is cut off, and is called ischaemia. Mitochondria in particular are damaged under such conditions. Lau and colleagues (Lau *et al.,* 1997) made cpn 60 and cpn10 adenovirus constructs which they over-expressed in rat neonatal cardiomyocytes and in the myogenic H9 c2 cell line. Simultaneous expression of both of these cpns protected cells from simulated ischaemia, but expression of only one of the cpns led to cellular damage. These data suggest that induction of cpn expression in tissues may protect them from damage due to ischaemia which is seen in organs such as the kidney during the transplantation process.

It has also been suggested (Carrell and Lomas, 1997) that a number of diseases might arise from abnormal folding and subsequent aggregation of specific proteins. For example, accumulation of misfolded proteins occurs in spongiform encephalopathy, α-$_1$ antitrypsin deficiency-associated lung and liver disease, Alzheimer's disease and amyloidosis. It is not known whether cpns are involved in the pathology of these conditions.

NON-FOLDING ACTIVITY

Bone diseases

The skeleton is the largest organ in the body and is prey to many infections and to idiopathic conditions such as osteoporosis. Bone, although composed largely of extracellular inorganic matrix, is not a dead tissue but is permeated by a rich blood and nerve supply and is being continuously destroyed and rebuilt in a dynamic process called bone remodelling. This ongoing remodelling is due to the action of two cell lineages. Bone matrix removal is 'catalysed' by a multinucleate myeloid cell called an osteoclast, while replacement of this resorbed matrix is the responsibility of the mesenchymal cells called osteoblasts. Bacterial infections cause common bone diseases such as the periodontal diseases (with a prevalence of 10–15% worldwide), osteomyelitis, bacterial arthritis and tuberculosis of bone. The first evidence that cpns could cause bone pathology was the finding that the cpn 60 of the oral pathogen, *Actinobacillus actinomycetemcomitans* could cause bone resorption *in vitro* (Kirby *et al.*, 1995). These studies found that groEL was a potent stimulator of bone resorption but that the cpn 60 molecules of *Mycobacterium* spp. (hsp 65) were inactive in this respect. The mechanism of action of groEL appears to be its potent ability to stimulate the growth and development of osteoclasts and to activate bone resorption by the mature cells (Reddi *et al*, 1998).

One of the most striking bone pathologies is vertebral tuberculosis or Pott's disease in which the bacterium causes massive destruction of infected vertebrae and gross spinal deformity. Addition of sonicated *M. tuberculosis* to murine bone in culture causes significant loss of bone matrix. Surprisingly, this bone destruction could be completely blocked by a neutralising antibody to *M. tuberculosis* cpn 10 but not cpn 60, confirming previous findings. Recombinant *M. tuberculosis* cpn 10 was a potent inducer of osteolysis which appeared to act as a stimulator of osteoclast proliferation. GroES is also an active osteolytic agent (Nair *et al*, 1998). Using a series of synthetic peptides to define structure/function relationships, the active domains in cpn 10 appear to be the flexible loop and a small loop structure around the conserved tyrosine at position 71. Interestingly, these loop structures make contact with cpn 60 suggesting that if cell activation by cpn 10 is due to receptor binding, then the receptor may have structural similarity to cpn 60 (Meghji *et al*, 1997).

We have recently demonstrated that human recombinant cpn 60 is also able to stimulate bone resorption (unpublished), raising the possibility that idiopathic diseases such as arthritis and osteoporosis could be caused by release of self-chaperonins.

Arthritis

In animals, inflammatory diseases such as arthritis are associated with imbalances in T-cell populations and T-cells which are specific for conserved epitopes of hsp 60 may regulate inflammatory responses (Gaston, 1998). Passive transfer of a T-cell clone which is specific for the 180–188 amino-acid sequence of mycobacterial hsp 60 induces adjuvant arthritis (Holoshitz *et al.*, 1983; Van Eden *et al.*, 1988). In contrast, a synthetic peptide from mycobacterial hsp 60 suppresses adjuvant arthritis and nonmicrobially induced experimental arthritis (Prakken *et al.*, 1998). Similarly, a synthetic mycobacterial hsp10 peptide delays the onset and severity of adjuvant-induced arthritis (Ragno *et al.*, 1996). These data suggest that hsp10/60 specific T-cells may modulate inflammatory responses in arthritis. Experimental adjuvant arthritis in animals can be induced by the injection of mycobacteria. In humans, a rather similar syndrome occurs after the installation of mycobacteria into the bladder which is part of the treatment for bladder cancer. A rare complication of this treatment is arthritis (Smith *et al.*, 1997; Saporta *et al.*, 1997). The synovial membrane may show cellular infiltrate and cytokine profiles which are similar to those seen in rheumatoid arthritis (Smith *et al.*, 1997). However, the arthritis tends to spontaneously remit, unlike rheumatoid arthritis. In patients with rheumatoid arthritis, hsp 60 is expressed on lymphocytes in peripheral blood and synovial fluid (Sato *et al.*, 1996). These patients also have high levels of antibodies to hsp 60 (Handley *et al.*, 1996) and so, like atherosclerosis (described below), the evidence supports an autoimmune hypothesis. It is possible that antibodies and T-cells arise as a result of an immune response to bacteria such as BCG or *Escherichia coli* (Handley *et al.*, 1996), and that these antibodies recognize cross-reactive epitopes in human hsp 60, so giving rise to microbially-triggered autoimmune damage to the joint. However, complex inter-actions exist between populations of T-cells in rheumatoid arthritis and suppressive T-cell responses may be induced by human but not by bacterial hsp 60 in the synovial fluid (van Roon *et al.*, 1997). In reactive arthritis due to bacteria such as *Yersinia enterocolitica*, hsp 60 is a powerful antigen which induces CD4+ T populations of cells which are specific to bacterial hsp 60 and others which are potentially autoreactive (Mertz *et al.*, 1998). The link between hsp 60 and arthritis is intriguing but its role in human disease remains to be determined.

Arterial disease

In 1992, Xu and colleagues reported the induction of arteriosclerosis in normocholesterolemic rabbits by immunisation with *M. tuberculosis* cpn 60 (hsp 65) (Xu *et al.*, 1992). Immunisation with hsp 65 combined with a cholesterol-rich diet led to more severe atherosclerosis. The hypothesis which has emerged from these findings proposes that an autoimmune response contributes to the inflammation in atheromatous arteries and that a high blood cholesterol is an important risk factor (Wick *et al.*, 1995). The rationale for this hypothesis is the finding that autoantibodies against heat shock protein 60 mediate endothelial cytotoxicity (Schett *et al.*, 1995; Wick *et al.*, 1995). It seems that these autoantibodies cross-react with mycobacterial hsp 65, human hsp 60 and a 60 kDa protein (probably cpn 60) in heat-shocked endothelial cells. Only heat-stressed endothelial cells were lysed by anti-hsp 60/65 antibody in the presence of complement or peripheral blood mononuclear cells. Is this relevant to human atherosclerosis which kills a high proportion of adults in the Western world? Patients with atherosclerosis have elevated levels of anti-hsp 65 antibodies (Mukherjee *et al.*, 1996; Hoppichler *et al.*, 1996). Macrophages in the atherosclerotic lesions of patients express high levels of hsp 60 and human anti-hsp 60 antibodies induce complement-mediated cytotoxicity and antibody-dependent cellular cytotoxicity of stressed peripheral blood derived macrophages and the macrophage-like cell line U937 (Schett *et al.*, 1997). This is not sufficient evidence to prove an association because the raised level of anti-hsp 65 could be a coincidental finding, which is not related to the underlying pathogenesis of atherosclerosis. However, in rabbits, which have been immunized with mycobacterial hsp 65, there is increased expression of hsp 65 in atherosclerotic lesions and these arterial lesions are infiltrated with T-cells which react with hsp 60/65 (Xu *et al.*, 1993). The serum of the immunized rabbits contains elevated levels of antibody to hsp 65. Interestingly, rabbits which are treated with administration of a 0.2% cholesterol diet also generated anti-hsp 65 antibodies and T-cells which are associated with atheromatous arterial lesions. So, in rabbits at least, there is a model of atheroma which is closely associated with an immune response to hsp 65. Another hypothesis (Gupta *et al.*, 1997; Gurfinkel *et al.*, 1997) suggest that bacterial infection may contribute towards the pathogenesis of atherosclerosis. If this were the case, then raised levels of anti-hsp 60 in patients would not be surprising. In rats, treatment with bacterial cell-wall lipopolysaccharide induces coexpression of hsp 60 and intercellular-adhesion molecule-1 and this leads to increased monocyte and T-cell adhesion to aortic endothelium (Seitz *et al.*, 1996). So, microbial infection might induce hsp 60 expression in arteries, and might also induce anti-hsp 60 cross-reacting antibodies and T-cells which could damage the arterial cell walls. Whether human atherosclerosis is an autoimmune or a microbial disease, or a combination of the two, needs further investigation.

Pro-inflammatory effects

Related to 2 and 3 are reports that the chaperonins can stimulate myeloid cells, lymphocytes and endothelial cells to synthesize and secrete the major pro-inflammatory cytokines – interleukin-1 (IL-1), IL-6, IL-8 and tumour necrosis factor (TNF) (Henderson *et al.*, 1996; Verdegaal *et al.*, 1996). This work has been criticised on the basis that activity may be due to residual lipopolysaccharide (a major cytokine-inducing factor from Gram-negative bacterial cell walls) or to the presence of the many peptides and proteins which co-purify with the chaperonins, particularly cpn 60 (Price *et al.*, 1991). However, in a recent study a method has been developed to prepare groEL to homogeneity and in the absence of contaminating peptides and lipopolysaccharide this protein is still a potent stimulator of cytokine synthesis. In addition, proteolysis of groEL resulting in its breakdown into small peptides did not significantly inhibit its ability to stimulate human monocytes to produce pro-inflammatory cytokines (Tabona *et al.*, 1998). This finding suggests that cpn 60, which is a major component in stressed cells such as those causing inflammation in the host, may be an extremely important virulence factor able to stimulate pro-inflammatory signals even after it has been 'destroyed' by proteolysis

Alzheimer's disease

This common condition is characterised by widespread death of neurones in the brain. A potent molecule, called activity-dependent neurotrophic factor (Gozes and Brenneman, 1996) is secreted by vasoactive intestinal peptide-treated astrocytes. This 14 kDa protein protects neurons from death associated with an Alzheimer's disease model. A short 14-amino acid peptide derived from the protein sequence is active at femtomolar concentrations (Brenneman and Gozes, 1996) and protects toxin-treated rats from memory loss (Gozes *et al.,* 1997). Interestingly, the protein is homologous to cpn 60. This observation raises the possibility that extracellular cpns have direct potent actions on human cells and that short peptides derived from the cpn sequence also act directly on cells. It is also possible many other examples of direct cpn action on eukaryotic cells exist but are, so far, undetected. Perhaps whilst intracellular cpns protect cells from stress, extracellular cpns transmit protective messages between cells and even from one organ to another, rather like hormones.

The expression of human cpn 60 is upregulated in various human central nervous system diseases and in experimental animals, particularly in astrocytes and abnormal neurones (Martin *et al.,* 1993; Khanna *et al.,* 1996). It has been suggested that autoimmune inflammation in the central nervous system is associated with increased cpn 60 expression and that this might, in some way, modulate the immune response (Gay *et al.,* 1995). One hypothesis (Birnbaum *et al.,* 1996; Birnbaum and Kotilinek, 1997), is that there is a cross-reacting epitope between cpn 60 and the myelin protein 2', 3' cyclic nucleotide 3' phosphodiesterase, and immunisation with a peptide which contains this epitope modifies the course of acute and chronic experimental autoimmune encephalomyelitis. Cpn 10 can also protect animals against experimental autoimmune encephalomyelitis but, in contrast to the cross-reactive cpn 60, has no shared T-cell epitopes with encephalitogenic proteins (Ben-Nun *et al.,* 1995).

Growth regulation

Pregnancy and cancer are, perhaps, the most dramatic natural examples of mammalian cell growth. Chaperonins are implicated as growth regulators in both. Early pregnancy factor which has been identified as cpn 10, is required for the successful establishment of pregnancy and for the proliferation of both normal and neoplastic cells (Cavanagh, 1996). However, it has been questioned whether early pregnany factor is identical with cpn 10 (Clarke, 1997). A so-far unidentified molecule which is called preimplantation factor is detected in the serum of women shortly after fertilisation (Roussev *et al.,* 1996), but this factor seems to be different to cpn 10.

The cpn 10 of *M. tuberculosis* but not of *E. coli,* increases the proliferation of the rapidly growing mouse P19 teratocarcinoma cells. In contrast, the *M. tuberculosis* cpn 10 increases apoptosis in serum-deprived teratocarcinoma cells. This suggests that this cpn 10 has opposite affects on cell growth depending on the condition of the target cell (Galli *et al.,* 1996). In mice which contain the highly malignant reticulum sarcoma (J774), delivery of the mycobacterial cpn 60 (hsp 65) gene DNA in liposomes results in regression of the tumour (Lukacs *et al.,* 1997). The mechanism of action of the cpn 60 is not clear, but the injected animals also produce antibodies against the tumour cells which suggest the tumour becomes more antigenic after gene transfer, and so, perhaps is rejected by the immune system.

Vaccines

Some small molecules such as the Vi Ag of *Salmonella typhi* are poorly immunogenic T-independent antigens. Cpn 60 peptides serve as immunogenic carriers for such antigens and can greatly increase their immunogenicity (Konen-Waisman *et al.,* 1995).

The future

FOLDING

Co-expression of cpns with recombinant proteins may help to increase the yield of correctly folded protein and so make industrial production of proteins more efficient. A rather different line of thought lies behind the idea of pre-stressing transplantation organs, so that they increase their content of cpns and so become more resistant to the rigours of existing without a blood supply.

NON-FOLDING

The potential use of cpns to treat disease is now a growth area. The potency of cpns to directly affect the function of cells such as neurones and bone cells suggests that new pathways of action will be discovered and may prove amenable to therapeutic intervention. The indirect affects of cpns, operating through the immune system, either by suppressing immunity or by enhancing it, will continue to be an active area. New vaccine formulations may well emerge to boost the immune response to specific peptides by combination with cpns.

References

ABOU-ZEID, C., SMITH, I., GRANGE, J.M., RATLIFF, T.L., STEELE, J. AND ROOK, G.A. (1988). The secreted antigens of *Mycobacterium tuberculosis* and their relationship to those recognized by the available antibodies. *Journal of General Microbiology* **134**, 531–538.

BAIRD, P.N., HALL, L.M. AND COATES, A.R.M. (1989). Cloning and sequence analysis of the 10K antigen gene of *Mycobacterium tuberculosis*. *Journal of General Microbiology* **135**, 931–939.

BEN-NUN, A., MENDEL, I., SAPPLER, G., KERLERO, D.E. AND ROSBO, N. (1995). A 12-kDa protein of *Mycobacterium tuberculosis* protects mice against experimental autoimmune encephalomyelitis. Protection in the absence of shared T cell epitopes with encephalitogenic proteins. *Journal of Immunology* **154**, 2939–2948.

BIRNBAUM, G. AND KOTILINEK, L. (1997). Heat shock or stress proteins and their role as autoantigens in multiple sclerosis. *Annals of the New York Academy of Sciences* **835**, 157–167.

BIRNBAUM, G., KOTILINEK, L., SCHLIEVERT, P., CLARK, H.B., TROTTER, J., HORVATH, E., GAO, E., COX, M. AND BRAUN, P.E. (1996). Heat shock proteins and experimental autoimmune encephalomyelitis (EAE): 1. Immunization with a peptide of the myelin protein 2',3' cyclic nucleotide 3' phosphodiesterase that is cross-reactive with a heat shock protein alters the course of EAE. *Journal of Neuroscience Research* **44**, 381–396.

BOUDKER, O., TODD, M.J. AND FREIRE, E.(1997). The structural stability of the co-chaperonin GroES. *Journal of Molecular Biology* **272**, 770–779.

BRENNEMAN, D.E. AND GOZES, I. (1996). A femtomolar-acting neuroprotective peptide. *Journal of Clinical Investigation* **97**, 2299–2307.

BUKAU, B. (1993). Regulation of the Escherichia coli heat-shock response. *Molecular Microbiology* **9**, 671–80.

BUKAU, B. AND HORWICH, A.L. (1998). The Hsp70 and Hsp60 chaperone machines. *Cell* **92**, 351–366.

CARRELL, R.W. AND LOMAS, D.A. (1997). Conformational disease. *Lancet* **350**, 134–138

CAVANAGH, A.C. (1996). Identification of early pregnancy factor as chaperonin 10: implications for understanding its role. *Reviews in Reproduction* **1**, 28–32.

CAO, P., MCCLAIN, M.S., FORSYTH, M.H. AND COVER, T.L. (1998). Extracellular release of antigenic proteins by *Helicobacter pylori*. *Infection and Immunity* **66**, 2984–2986.

CLARKE, F.M. (1997). Controversies in assisted reproduction and genetics. Does 'EPF' have an identity? *Journal of Assist Reproductive Genetics* **14**, 489–491.

COATES, A.R.M. (1996). Immunological Aspects of Chaperonins. In *The Chaperonins*. Ed. R.J. Ellis, pp 267–296. London: Academic Press.

COATES, A.R.M. AND HENDERSON, B. (1998). Chaperones in Health and Disease. In *Stress of Life from Molecules to Man*. Ed. P. Csermely. *Annals of the New York Academy of Sciences* **851**, 48–53.

COATES, A.R.M., SHINNICK, T.M. AND ELLIS, R.J. (1993). Chaperonin nomenclature. *Molecular Microbiology* **8**, 787.

ESCHWEILER, B., BOHRMANN, B., GERSTENECKER, B., SCHILTZ, E. AND KIST, M. (1993). *In situ* localization of the 60 k protein of *Helicobacter pylori*, which belongs to the family of heat shock proteins, by immuno-electron microscopy. *International Journal of Medical Microbiology, Virology, Parasitology and Infectious Diseases* **280**, 73–85.

ELLIS, R.J. (1996). Chaperonins: Introductory perspective. In *The Chaperonins*. Ed. R.J. Ellis, pp 1–25. London: Academic Press.

FOSSATI, G., LUCIETTO, P., GIULANI, P., COATES, A.R., HARDING, S., COLFEN, H., LEGNAME, G., CHAN, E., ZALIANI, A. AND MASCAGNI, P. (1995). *Mycobacterium tuberculosis* chaperonin 10 forms stable tetrameric and heptameric structures. Implications for its diverse biological activities. *Journal of Biological Chemistry* **270**, 26159–26167.

FRISK, A., ISON, C.A. AND LAGERGARD, T. (1998). GroEL heat shock protein of *Haemophilus ducreyi*: association with cell surface and capacity to bind to eukaryotic cells. *Infection and Immunity* **66**, 1252–1257.

GALLI, G., GHEZZI, P., MASCAGNI, P., MARCUCCI, F. AND FRATELLI, M. (1996). *Mycobacterium tuberculosis* heat shock protein 10 increases both proliferation and death in mouse P19 teratocarcinoma cells. *In Vitro Cell Developmental Biology of Animals* **32**, 446–450.

GAO, Y., THOMAS, J.O., CHOW, R.L., LEE, G.H. AND COWAN, N.J. (1992). A cytoplasmic chaperonin that catalyzes beta-actin folding. *Cell* **69**, 1043–1050.

GASTON, J.S. (1998). Role of T-cells in the development of arthritis. *Clinical Science* **95**, 19–31.

GAY, Y.L., BROSNAN, C.F., RAISE, C.S. (1995). Experimental autoimmune encephalomyelitis. Qualitative and semiquantitative differences in heat shock protein 60 expression in the central nervous system. *Journal of Immunology* **154**, 3548–3556.

GEORGOPOULOS, C.P., HENDRIX, R.W., KAISER, A.D. AND WOOD, W.B. (1972). Role of the host cell in bacteriophage morphogenesis: effects of a bacterial mutation on T4 head assembly. *Nature New Biology* **239**, 38–41.

GORDON, C.L., SATHER, S.K., CASJENS, S. AND KING, J. (1994). Selective *in vivo* rescue by GroEL/ES of thermolabile folding intermediates to phage P22 structural proteins. *Journal of Biological Chemistry* **269**, 27941–27951.

GOZES, I., BARDEA, A. BECHAR, M., PEARL, O., RESHEF, A., ZAMOSTIANO, R., DAVIDSON, A., RUBINRAUT, S., GILADI, E., FRIDKIN, M. AND BRENNEMAN, D.E. (1997). Neuropeptides and neuronal survival: neuroprotective strategy for Alzheimer's disease. *Annals of the New York Academy of Sciences* **814**, 161–166.

GOZES, I. AND BRENNEMAN, D.E. (1996). Activity-dependent neurotrophic factor (ADNF). An extracellular neuroprotective chaperonin? *Journal of Molecular Neuroscience* **7**, 235–244.

GUPTA, R.S. (1995). Review: Evolution of the chaperonin families (Hsp60, Hsp10 and Tcp-1) of proteins and the origin of eukaryotic cells. *Molecular Microbiology* **15**, 1–11.

GUPTA, R.S. (1996). Evolutionary Relationships of Chaperonins. In *The Chaperonins*. Ed. R.J. Ellis, pp 27–64. London: Academic Press.

GUPTA, S, LEATHAM, E.W., CARRINGTON, D., MENDALL, M.A., KASKI, J.C. AND CAMM, A.J. (1997). Elevated *Chlamydia pneumoniae* antibodies, cardiovascular events, and azithromycin in male survivors of myocardial infarction. *Circulation* **96**, 404–407.

GURFINKEL, E., BOZOVICH, G., DAROCA, A., BECK, E. AND MAUTNER, B. (1997). Randomised trial of roxithromycin in non-Q-wave coronary syndromes: ROXIS Pilot Study. ROXIS Study Group. *Lancet* **350**, 404–407.

HANDLEY, H.H., YU, J., YU, D.T., SINGH, B., GUPTA, R.S. AND VAUGHAN, J.H. (1996). Autoantibodies to human heat shock protein (hsp)60 may be induced by *Escherichia coli* groEL. *Clinical Experimental Immunology* **103**, 429–435.

HENDERSON, B., NAIR, S.P. AND COATES, A.R.M. (1996). Review. Molecular chaperones and disease. *Inflammation Research* **45**, 155–158.

HOFFMAN, P.S., HOUSTON, L. AND BUTLER, C.A. (1990). Legionella pneumophila htpAB heat shock operon: nucleotide sequence and expression of the 60-kilodalton antigen in L. pneumophila-infected HeLa cells. *Infection and Immunity* **58** (10), 3380–3387

HOLOSHITZ, J., NAPARSTEK, Y., BEN-NUN, A. AND COHEN, I.R. (1983). Lines of T lymphocytes induce or vaccinate against autoimmune arthritis. *Science* **219**, 56–58.

HOPPICHLER, F., LECHLEITNER, M., TRAWEGER, C., SCHETT, G., DZIEN, A., STURM, W. AND XU, Q. (1996). Changes of serum antibodies to heat-shock protein 65 in coronary heart disease and acute myocardial infarction. *Atherosclerosis* **126**, 333–338.

HUNT, J.F., VAN DER VIES, S.M., HENRY, L. AND DEISENHOFER, J. (1997). Structural adaptations in the specialized bacteriophage T4 co-chaperonin Gp31 expand the size of the Anfinsen cage. *Cell* **90**, 361–371.

JINDAL, S., DUDANI, A.K., SINGH, B., HARLEY, C.B. AND GUPTA, R.S. (1989). Primary structure of a human mitochondrial protein homologous to the bacterial and plant chaperonins and to the 65-kilodalton mycobacterial antigen. *Molecular Cell Biology*, **9**, 2279–2283.

KHANNA, N., SHANKAR, S.K, CHANDRAMUKI, A. AND JAGANNATH, C. (1996). Immunohistochemical study of the expression of human groEL-stress protein in human nervous tissue. *Indian J Med Res* **103**, 103–111.

KIRBY, A.C., MEGHJI, S., NAIR, S.P, WHITE, P., REDDI, K., NISHIHARA, T., NAKASHIMA, K., WILLIS, A.C., SIM, R. AND WILSON, M. (1995). The potent bone-resorbing mediator of *Actinobacillus actinomycetemcomitans* is homologous to the molecular chaperone GroEL. *Journal of Clinical Investigation* **96**, 1185–1194.

KLUMPP, M., BAUMEISTER, W. AND ESSEN, L.O. (1997). Structure of the substrate binding domain of the thermosome, an archaeal group II chaperonin. *Cell* **91**, 263–270.

KONEN-WAISMAN, S., FRIDKIN, M. AND COHEN, I.R. (1995). Self and foreign 60-kilodalton heat shock protein T cell epitope peptides serve as immunogenic carriers for a T cell-independent sugar antigen. *Journal of Immunology* **154**, 5977–5985.

LAU, S., PATNAIK, N., SAYEN, M.R. AND MESTRIL, R. (1997). Simultaneous overexpression of two stress proteins in rat cardiomyocytes and myogenic cells confers protection against ischemia-induced injury. *Circulation* **96**, 2287–2294.

LEWTHWAITE, J., SKINNER, A. AND HENDERSON, B. (1998). Are molecular chaperones microbial virulence factors? *Trends in Microbiology* **6**, 426–428.

LUKACS, K.V., NAKAKES, A., ATKINS, C.J., LOWRIE, D.B. AND COLSTON, M.J. (1997). *In vivo* gene therapy of malignant tumours with heat shock protein-65 gene. *Gene Therapy* **4**, 346–350.

MARTIN, J.E., SWASH, M., MATHER, K. AND LEIGH, P.N. (1993). Expression of the human groEL stress-protein homologue in the brain and spinal cord. *Journal of Neurological Sciences* **118**, 202–206.

MASCAGNI, P., FOSSATI, G., LUCIETTO, P., COATES, A.R.M., JUMEL, K., ERRINGTON, N.E., HARDING, S.E., ZALIANI, A. AND RIZZI, E.(1998). Self-association equilibria of the *Mycobacterium tuberculosis* chaperonin 10. (submitted).

MASCAGNI, P., FOSSATI, G., LUCIETTO, P., COATES, A.R.M., MODENA, D., GANCIA, E., ZALIANI, A. AND RIZZI, E. (1998). The solution conformational equilibria of the MT chaperonin 10 monomer. (submitted).

MCMULLIN, T.W. AND HALLBERG, R.L. (1988). A highly evolutionarily conserved mitochondrial protein is structurally related to the protein encoded by the *Escherichia coli* groEL gene. *Molecular Cell Biology* **8**, 371–380.

MEGHJI, S., WHITE, P.A, NAIR, S.P., REDDI, K., HERON, K., HENDERSON, B., ZALIANI, A., FOSSATI, G., MASCAGNI, P., HUNT, J.F., ROBERTS, M.M. AND COATES, A.R.M. (1997). *Mycobacterium tuberculosis* chaperonin 10 stimulates bone resorption: A potential contributory factor in Pott's disease. *Journal of Experimental Medicine* **186**, 1241–1246.

MERTZ, A.K., UGRINOVIC, S., LAUSTER, R., WU, P., GROLMS, M., BOTTCHER, U., APPEL, H., YIN, Z., SCHILTZ, E., BATSFORD, S., SCHAUER-PETROWSKI, C., BRAUN, J., DISTLER, A. AND SIEPER, J. (1998). Characterization of the synovial T cell response to various recombinant Yersinia antigens in Yersinia enterocolitica-triggered reactive arthritis. Heat-shock protein 60 drives a major immune response. *Arthritis and Rheumatism* **41**, 315–326.

MINOWANDA, G. AND WELCH, W.J. (1995). Clinical implications of the stress response. *Journal of Clinical Investigation* **95**, 3–12.

MUKHERJEE, M., DE BENEDICTIS, C., JEWITT, D. AND KAKKAR, V.V. (1996). Association of antibodies to heat-shock protein-65 with percutaneous transluminal coronary angioplasty and subsequent restenosis. *Thrombosis and Haemostasis* **75**, 258–60.

NAIR, S.P., MEGHJI, S., REDDI, K., POOLE, S., MILLER, A.D. AND HENDERSON, B. (1999). Molecular Chaperones Stimulate Bone Resorption. *Calcified Tissue International* **64**, 214–218.

PRAKKEN, B., WAUBEN, M., VAN KOOTEN, P., ANDERTON, S., VAN DER ZEE, R., KUIS, W. AND VAN EDEN, W. (1998). Nasal administration of arthritis-related T cell epitopes of heat shock protein 60 as a promising way for immunotherapy in chronic arthritis. *Biotherapy* **10**, 205–211.

PRICE, N., KELLY, S.M., WOOD, S. AND AUF DER MAUER, A. (1991). The aromatic amino acid content of the bacterial chaperone protein groEL (cpn 60): evidence for the presence of a single tryptophan. *FEBS Letters* **320**, 83–84.

RAGNO, S., WINROW, V.R., MASCAGNI, P., LUCIETTO, P, DI PIERRO, F., MORRIS, C.J. AND BLAKE, D.R. (1996). A synthetic 10-kD heat shock protein (hsp10) from *Mycobacterium tuberculosis* modulates adjuvant arthritis. *Clinical Experimental Immunology* **103**, 384–390.

REDDI, K., MEGHJI, S., NAIR, S.P., ARNETT, T.R., MILLER, A.D., PREUSS, M., WILSON, M., HENDERSON, B. AND HILL, P. (1998). The *Escherichia coli* chaperonin 60 (groEL) is a potent stimulator of osteoclast formation. *Journal of Bone Mineral Research* **13**, 1260–1266.

ROUSSEV, R.G., COULAM, C.B. AND BARNEA, E.R. (1996). Development and validation of an assay for measuring preimplantation factor (PIF) of embryonal origin. *American Journal of Reproductive Immunology* **35**, 281–287.

RYAN, M.T., HERD, S.M., SBERNA, G., SAMUEL, M.M., HOOGENRAAD, N.J. AND HOJ, P.B. (1997). The genes encoding mammalian chaperonin 60 and chaperonin 10 are linked head-to-head and share a bidirectional promoter. *Gene* **196**, 9–17.

SAPORTA, L., GUMUS, E., KARADAG, H., KURAN, B. AND MIROGLU, C. (1997). Reiter syndrome following intracavitary BCG administration. *Scandinavian Journal of Urology and Nephrology* **31**, 211–212.

SATO, H., MIYATA, M. AND KASUKAWA, R. (1996). Expression of heat shock protein on lymphocytes in peripheral blood and synovial fluid from patients from rheumatoid arthritis. *Journal of Rheumatology* **23**, 2027–2032.

SCHATZ, G. AND DOBBERSTEIN, B. (1996). Common principles of protein translocation across membranes. *Science* **271**, 1519–1526.

SCHETT, G., XU, Q., AMBERGER, A., VAN DER ZEE, R., RECHEIS, H., WILLEIT, J. AND WICK, G. (1995). Autoantibodies against heat shock protein 60 mediate endothelial cytotoxicity. *Journal of Clinical Investigation* **96**, 2569–2577.

SCHETT, G., METZLER, B., MAYR, M., AMBERGER, A., NIEDERWIESER, D., GUPTA, R.S., MIZZEN, L., XU, Q. AND WICK, G. (1997). Macrophage-lysis mediated by autoantibodies to heat shock protein 65/60. *Atherosclerosis* **128**, 27–38.

SEITZ, C.S., KLEINDIENST, R., XU, Q. AND WICK, G. (1996). Co-expression of heat-shock protein 60 and intercellular-adhesion molecule-1 is related to increased adhesion of monocytes and T cells to aortic endothelium of rats in response to endotoxin. *Laboratory Investigation* **74**, 241–52.

SHINNICK, T.M. (1987). The 65-kilodalton antigen of Mycobacterium tuberculosis. *Journal of Bacteriology* **169**, 1080–1088.

SMITH, M.D., CHANDRAN, G., PARKER, A., YOUSSEF, P.P., AHERN, M., COLEMAN, M., MACARDLE, P. AND ROBERTS-THOMSON, P. (1997). Synovial membrane cytokine profiles in reactive arthritis secondary to intravesical bacillus Calmette-Guerin therapy. *Journal of Rheumatology* **24**, 752–758.

SOLTYS, B.J. AND GUPTA, R.S. (1997). Cell surface localization of the 60 kDa heat shock chaperonin protein (hsp60) in mammalian cells. *Cell Biology International* **21**, 315–320.

TABONA, P., REDDI, K., KHAN, S., NAIR, S.P., CREAN, ST. J.V., MEGHJI, S., WILSON, M., PREUSS, M., MILLER, A.D., POOLE, S., CARNE, S. AND HENDERSON, B. (1998). Homogeneous *Escherichia coli* chaperonin 60 induces IL-1 beta and IL-6 gene expression in human monocytes by a mechanism independent of protein conformation. *Journal of Immunology* **161**, 1414–1421.

TOROK, Z., HORVATH, I., GOLOUBINOFF, P., KOVACS, E., GLATZ, A., BALOGH, G. AND VIGH, L. (1997). Evidence for a lipochaperonin: association of active protein-folding GroESL oligomers with lipids can stabilize membranes under heat shock conditions. *Proceedings of the National Academy of Sciences, USA* **94**, 2192–2197.

VAN DER VIES, S. AND GEORGOPOULOS, C. (1996). Regulation of Chaperonin Gene Expression. In *The Chaperonins*. Ed. R.J. Ellis, pp 1–25. London: Academic Press.

VAN EDEN, W., THOLE, J.E., VAN DER ZEE, R., NOORDZIJ, A., VAN EMBDEN, J.D., HENSEN, E.J. AND COHEN, I.R. (1988). Cloning of the mycobacterial epitope recognized by T lymphocytes in adjuvant arthritis. *Nature* **33**, 171–173.

VAN ROON, J.A., VAN EDEN, W., VAN ROY, J.L., LAFEBER, F.J. AND BIJLSMA, J.W. (1997). Stimulation of suppressive T cell responses by human but not bacterial 60-kD heat-shock protein in synovial fluid of patients with rheumatoid arthritis. *Journal of Clinical Investigation* **100**, 459–463.

VERDEGAAL, M.E., ZEGVELD, S.T. AND VAN FURTH, R. (1996). Heat shock protein 65 induces CD62e, CD106, and CD54 on cultured human endothelial cells and increases their adhesiveness for monocytes and granulocytes. *Journal of Immunology* **157**, 369–376.

WICK, G., SCHETT, G., AMBERGER, A., KLEINDIENST, R. AND XU, Q. (1995). Is atherosclerosis an immunologically mediated disease? *Immunology Today* **16**, 27–33.

XU, Q., DIETRICH, H., STEINER, H.J., GOWN, A.M., SCHOEL, B., MIKUZ, G., KAUFMANN, S.H. AND WICK, G. (1992). Induction of arteriosclerosis in normocholesterolemic rabbits by immunization with heat shock protein 65. *Arteriosclerosis & Thrombosis* **12**, 789–799.

XU, Q., KLEINDIENST, R., WAITZ, W., DIETRICH, H. AND WICK, G. (1993). Increased expression of heat shock protein 65 coincides with a population of infiltrating T lymphocytes in atherosclerotic lesions of rabbits specifically responding to heat shock protein 65. *Journal of Clinical Investigation* **91**, 2693–2702.

ZONDLO, J., FISHER, K.E., LIN, Z., DUCOTE, K.R. AND EISENSTEIN, E. (1995). Monomerheptamer equilibrium of the *Escherichia coli* chaperonin GroES. *Biochemistry* **34**, 10334–10339.

Index